Natural Killer Cells

Jacques Zimmer
Editor

Natural Killer Cells

At the Forefront of Modern Immunology

Springer

Editor
Dr. Jacques Zimmer
Centre de Recherche Public de la Santé
Lab. Immuno-Allergologie
84 Val Fleuri
1526 Luxembourg
Luxembourg
jacques.zimmer@crp-sante.lu

ISBN 978-3-642-02308-8   e-ISBN 978-3-642-02309-5
DOI 10.1007/978-3-642-02309-5
Springer Heidelberg Dordrecht London New York

Library of Congress Control Number: 2009930465

© Springer-Verlag Berlin Heidelberg 2010
This work is subject to copyright. All rights are reserved, whether the whole or part of the material is concerned, specifically the rights of translation, reprinting, reuse of illustrations, recitation, broadcasting, reproduction on microfilm or in any other way, and storage in data banks. Duplication of this publication or parts thereof is permitted only under the provisions of the German Copyright Law of September 9, 1965, in its current version, and permission for use must always be obtained from Springer. Violations are liable to prosecution under the German Copyright Law.
The use of general descriptive names, registered names, trademarks, etc. in this publication does not imply, even in the absence of a specific statement, that such names are exempt from the relevant protective laws and regulations and therefore free for general use.

Cover design: WMXDesign GmbH, Heidelberg, Germany

Printed on acid-free paper

Springer is part of Springer Science+Business Media (www.springer.com)

Preface

It is an interesting and even exciting experience to be the editor of a scientific book. After the decision of producing the book has been made together with the publisher, the first step is to establish the list of topics to be covered and the list of potential authors. Regarding the latter, the editor is in the same situation as the director of a movie in that he wants to have as much stars in the field as possible to contribute a chapter. For that, the potential authors have to be contacted and invited. They can be subdivided into three groups: (1) those who reply and accept to write a chapter, (2) those who reply but do not accept to write a chapter, and finally (3) those who do not even reply. In my case, I was lucky that the vast majority of authors I contacted with my request were in group 1, and so I did not face too many difficulties in filling the table of contents. However, there was a small subgroup containing one author who first submitted his chapter but subsequently decided to publish it in another book. In group 2, people claimed that they were too busy, which is a valuable argument.

In any case, I am very grateful to all the authors who contributed for the time and energy they spent in doing so.

Why publish a book about NK cells in 2009? Simply because they are still, nearly two decades after becoming fashionable, at the forefront of modern immunology. Nearly every month, new and exciting findings about NK cells are published, and they have not yet revealed all their secrets. They are by far not only of academic interest, as it becomes increasingly clear that they can be exploited in the therapy of human diseases, in particular in the fight against cancer and infections.

This book covers the state of the art of most aspects of NK cell biology, including most recent topics like NK cell development and education, NK cell trogocytosis, NK cells and allergy, regulatory NK cells and interactions between NK cells and regulatory T cells, another type of immunological forefront actors.

I am confident that the book will be useful to everyone who is interested in those fascinating NK cells.

April 2009   Jacques Zimmer

Contents

Natural Killer Cells: Deciphering Their Role, Diversity and Functions 1
Vicente P.C. Peixoto deToledo, Renato Sathler-Avelar,
Danielle Marquete Vitelli-Avelar, Vanessa Peruhype-Magalhães,
Denise Silveira-Lemos, Ana Carolina Campi-Azevedo,
Mariléia Chaves Andrade, Andréa Teixeira-Carvalho,
and Olindo Assis Martins-Filho

Dissecting Human NK Cell Development and Differentiation 39
Nicholas D. Huntington, Jean-Jacques Mention, Christian Vosshenrich,
Naoko Satoh-Takayama, and James P. Di Santo

Diversity of KIR Genes, Alleles and Haplotypes 63
D. Middleton, F. Gonzalez-Galarza, A. Meenagh, and P.A. Gourraud

NK Cell Education and *CIS* Interaction Between Inhibitory NK Cell Receptors and Their Ligands 93
Jacques Zimmer, François Hentges, Emmanuel Andrès,
and Anick Chalifour

Trogocytosis and NK Cells in Mouse and Man 109
Kiave-Yune HoWangYin, Edgardo D. Carosella, and Joel LeMaoult

Virus Interactions with NK Cell Receptors 125
Vanda Juranić Lisnić, Iva Gasparović, Astrid Krmpotić,
and Stipan Jonjić

The Role of NK Cells in Bacterial Infections 153
Brian P. McSharry, and Clair M. Gardiner

NK Cells and Autoimmunity 177
Hanna Brauner, and Petter Höglund

NK Cells and Allergy 191
Tatiana Michel, Maud Thérésine, Aurélie Poli, François Hentges,
and Jacques Zimmer

**Natural Killer Cells and Their Role in Hematopoietic Stem
Cell Transplantation** 199
Deborah L.S. Goetz, and William J. Murphy

NK Cells, NKT Cells, and KIR in Solid Organ Transplantation 221
Cam-Tien Le and Katja Kotsch

NK Cells in Autoimmune and Inflammatory Diseases 241
Nicolas Schleinitz, Nassim Dali-Youcef, Jean-Robert Harle,
Jacques Zimmer, and Emmanuel Andres

Natural Killer Cells and the Skin 255
Dagmar von Bubnoff

NK Cells in Oncology 267
Sigrid De Wilde, and Guy Berchem

The Role of KIR in Disease 275
Salim I Khakoo

Interactions Between NK Cells and Dendritic Cells 299
Guido Ferlazzo

**Modulation of T Cell-Mediated Immune Responses
by Natural Killer Cells** 315
Alessandra Zingoni, Cristina Cerboni, Michele Ardolino,
and Angela Santoni

Interactions Between NK Cells and Regulatory T Cells 329
Magali Terme, Nathalie Chaput, and Laurence Zitvogel

**Interactions Between B Lymphocytes and NK Cells:
An Update** 345
Dorothy Yuan, Ning Gao, and Paula Jennings

The Regulatory Natural Killer Cells 369
Zhigang Tian and Cai Zhang

NK Cells and Microarrays 391
Esther Wilk and Roland Jacobs

Natural Killer Cells in the Treatment of Human Cancer 405
Karl-Johan Malmberg and Hans-Gustaf Ljunggren

Index 423

Natural Killer Cells: Deciphering Their Role, Diversity and Functions

Vicente P.C. Peixoto deToledo, Renato Sathler-Avelar, Danielle Marquete Vitelli-Avelar, Vanessa Peruhype-Magalhães, Denise Silveira-Lemos, Ana Carolina Campi-Azevedo, Mariléia Chaves Andrade, Andréa Teixeira-Carvalho, and Olindo Assis Martins-Filho

Abstract Natural killer (NK) cells represent the third largest lymphoid cell population in mammals and are critical in innate immune responses. These cells express a large repertoire of receptors, named inhibitors and activators that mediate their function. NK cells occur naturally, do not require previous sensitization to engage their activity and are distributed to blood, as circulating cells, and to other organs of the body as resident cells. No longer considered simple "killing machines," NK cells have gained recognition for their abilities to secrete cytokines/chemokines that influence the differentiation of the adaptive immune responses, control viral/parasitic infections and participate in pathological and physiological mechanisms such as transplant rejection and the vascularization of implanting embryos during pregnancy. Here, we describe in detail the ontogeny of NK cells, their role in innate immunity from the point of view of their phenotypic features and functional activities as well as their function in health and disease. We also discuss the role of NK cells in immunological events in murine models. This review aims to highlight what is currently known and what remains to be understood about these essential innate immune cells.

V.P.C. Peixoto de Toledo (✉)
Departamento de Análises Clínicas e Toxicológicas, Faculdade de Farmácia, Universidade Federal de Minas Gerais, Avenida Antônio Carlos, 6627 Belo Horizonte, MG, Brasil, 31270-901
e-mail: vtoledo@ufmg.br

R. Sathler-Avelar, D. Marquete Vitelli-Avelar, V. Peruhype-Magalhães, D. Silveira-Lemos, A. Carolina Campi-Azevedo, M. Chavs Andrade, A. Teixeira-Carvalho and O. Assis Martins-Filho
Laboratório de Biomarcadores de Diagnóstico e Monitoração, Centro de Pesquisas René Rachou, Fundação Oswaldo Cruz, Av. Augusto de Lima, 1715 Barro Preto, Belo Horizonte, MG, Brasil, 30190-002

V. Peruhype-Magalhães
Laboratório de Pesquisas Clínicas, Centro de Pesquisas René Rachou, Fundação Oswaldo Cruz, Av. Augusto de Lima, 1715 Barro Preto, Belo Horizonte, MG, Brasil, 30190-002

D. Silveira-Lemos
Laboratório de Imunopatologia, NUPEB, Departamento de Análises Clínicas, Escola de Farmácia, Universidade Federal de Ouro Preto, Rua Costa Sena s/n, Ouro Preto, MG, Brasil, 35400-000

1 Introduction

Natural killer (NK) cells represent the third largest lymphoid cell population in mammals and are critical in innate immune responses [1]. They are characterized by the expression of a varied repertoire of receptors, named inhibitors and activators, that mediate their function [2]. These cells are large, granular, bone-marrow- as well as lymph node-derived lymphocytes. However, NK cells are distinct from T cells or B cells and have distinct morphologic, phenotypic and functional properties. These cells occur naturally and unlike T cells or B cells, do *not* require sensitization for the expression of their activity [3]. NK cells are distributed to blood as circulating cells and to other organs of the body as resident cells. In peripheral blood, they are characteristically described as having the morphology of large granular lymphocytes (LGL) [4], whereas in tissues, the microenvironment of the organ has influence on phenotype and activity of NK cells as demonstrated in lung and spleen [5].

NK cell functions can be classified in three categories: (A) *Cytotoxicity* – NK cells can kill certain virally infected cells and tumor target cells regardless of their MHC expression [6]. NK cells possess relatively large numbers of cytolytic granules, which are secretory lysosomes containing perforin and various granzymes. Upon contact between an NK cell and its target, these granules travel to the contact zone with the susceptible target cell (the so-called immunological synapse), and the contents are extruded to effect lysis. Perforin-dependent cytotoxicity is the major mechanism of NK cell lysis, although NK cells can also kill in a perforin-independent manner utilizing FAS ligand, TNF or TNF-related apoptosis-inducing ligand (TRAIL), albeit less efficiently and in a slower time kinetic; (B) *Cytokine and chemokine secretion* – NK cells are best noted for their ability to produce IFN-γ but also produce a number of other cytokines and chemokines including TNF-α, GM-CSF, interleukin(IL)-5, IL-13, CCL3/MIP-1α, CCL4/MIP-1β and CCL5/RANTES [7–9]. Killing and cytokine secretion are probably mediated by two different subsets of human NK cells characterized by the intensity of expression of CD56 on their surface; and (C) *Contact-dependent cell costimulation*: NK cells express several costimulatory ligands including CD40L (CD154) and OX40L, which allow them to provide a costimulatory signal to T cells or B cells [10, 11]. Thus, NK cells may serve as a bridge in an interactive loop between innate and adaptive immunity. Dendritic cells (DC) stimulate NK cells, which then deliver a costimulatory signal to T or B cells allowing for an optimal immune response.

The current model for NK cell activation and inhibition is one based upon the balance of function between specific activating and inhibitory receptors. If the balance favors inhibitory signaling, then intracellular events leading to cell function will not progress. If the balance favors activation signals, NK cells can then progress through a series of intracellular stages and checkpoints to exert their function. The balance of inhibitory and stimulatory signals received by a NK cell determines the outcome of interactions with target cells. Normal target cells are protected from killing by NK cells when signals delivered by stimulatory ligands are balanced by inhibitory signals delivered by self MHC class I molecules.

If, however, a target cell loses expression of self MHC class I molecules (as a result of transformation or infection), then the stimulatory signals delivered by the target cell are unopposed, resulting in NK-cell activation and target-cell lysis (known as missing-self recognition). Transformation or infection might also induce expression of stimulatory ligands such that constitutive inhibition delivered by inhibitory receptors is overcome (known as induced-self recognition). In many contexts, it is probable that both missing-self and induced-self recognition operate simultaneously to provide NK cells with the maximal ability to discriminate between normal cells and transformed or infected target cells [12].

In the past, the main function of NK cells was associated with their capacity to kill susceptible virus-infected or tumor target cells. Currently, these cells have been also shown to play key roles in host defense against intracellular pathogens as well as parasitic diseases, rejection of bone marrow transplants, reproduction and maternal tolerance of the fetus [3].

2 Human Natural Killer Cell Development

2.1 Development of Human NK Cells

Despite the fact that our knowledge of human NK cell development lags far behind that of other hematopoietic cells, as T and B cells, research focused on this area in the last decades has brought new light for this incomplete picture. NK cells are believed to be relatively short-lived, and at any one time there are likely more than 2 billions circulating in an adult [13]. Thus, one would think that these cells must be continually replenished to maintain homeostasis [14]. While it is clear that NK cells are derived from the same $CD34^+$ hematopoietic progenitor cells (HPC) as B and T cells, the sites of their development/maturation have not yet been totally elucidated [15, 16], although recent observations do indicate that both the bone marrow (BM) and lymph nodes (LN) are important [17, 18].

2.1.1 NK Cell Development in the Bone Marrow

For the past three decades, it has been a general consensus that NK cell development primarily occurs within the BM and that its milieu (other hematopoietic cells and stroma) is crucial to promote their survival, apoptosis, proliferation, and/or differentiation into cellular intermediates [14, 19]. In this context, human BM analyses have demonstrated an enriched microenvironment with $CD34^+$ HPC, including a fraction of NK cell precursors (pre-NK) [20]. Furthermore, it has been noted that long-term BM cultures with BM-derived $CD34^+$ HPC and BM stroma were able to support the differentiation of NK cells [21, 22], although the critical factors necessary for development are unknown.

In parallel to the identification of several phenotypic and functional stages of NK cell development, recent studies have shed new light on the role of stromal cells in driving functional maturation of NK cells [23]. It is important to note that the cytokine milieu is crucial for cell development in BM. IL-2 has been used extensively to study NK cell development from CD34$^+$ HPC in vitro [20, 24, 25]; however, this cytokine is produced exclusively by antigen-activated T cells and is not found within the BM stroma [24, 26, 27], suggesting that other factors that bind to the IL-2R are critical for NK cell development. Studies focusing on IL-15 function have demonstrated that this cytokine, produced by human BM stromal cells, facilitated the differentiation of cytolytic NK cells from CD34$^+$ HPC derived from fetal liver, BM, thymus, cord blood (CB), adult blood or LN [27–29]. These findings were further clarified by experiments that used IL-2 to generate NK cells from HPC on the basis that IL-15 shares common signaling receptor subunits with IL-2, including the IL-2/IL-15Rβ and common γ chain (CD122), which together form an intermediate-affinity heterodimeric receptor complex, IL-2/IL-15R$\beta\gamma$ [30]. In this manner, Fehniger and Caligiuri [31] also confirmed that IL-15 is a critical physiological cytokine present in BM, which mediates its effects through the IL-15R complex. Consistent with the essential role of IL-15 and its receptor in NK cell development, mice deficient in components of the IL-15R signaling pathway, such as Jak3 and STAT5a/b, exhibit defects in pre-NK [32–34]. Although the expression of IL-2/IL-15R$\beta\gamma$ complex is confirmed as a pre-NK marker in mice [29], CD122 expression on human CD34$^+$ cells is *not* easily detectable, necessitating the use of surrogate antigens (such as CD7, CD10, and/or CD45RA) to distinguish CD34$^+$ pre-NK from other CD34$^+$ populations [24, 35, 36]. It has been demonstrated that only the CD34$^+$CD45RA$^+$ phenotype is all-inclusive for human IL-2/IL-15-responsive pre-NK cells [35, 37].

Other BM stromal cell factors, c-kit ligand (KL) and flt3 ligand (FL), the ligands for members of the class III receptor tyrosine kinase family that includes c-kit and flt3, have been shown to enhance significantly the expansion of NK cells from CD34$^+$ HPC in combination with IL-15, but alone have no effect on NK cell differentiation [27, 31, 38]. Yu et al. [24] have explored the mechanism by which FL induces IL-15-mediated NK cell development. This group found that FL induces CD122 expression on CD34$^+$ HPC. The CD34$^+$CD122$^+$ cell coexpressed CD38, but lacked expression of CD7, CD56, NK cell receptors (NKR) or the ability to mediate cytotoxic activity in the absence of IL-15. Altogether this suggests that human NK cell development may be divided into an early phase in which FL acts synergistically with IL-15 to generate a unique CD34$^+$CD122$^+$CD38$^+$ NK cell intermediate from CD34$^+$ HPC, and that IL-15 is also required to stimulate the mature NK cell features such as CD56 and NKR expression, cytotoxic activity and the ability to produce abundant cytokines and chemokines [24].

2.1.2 NK-Cell Development in Lymph Nodes

Despite the observation that BM is a well-defined NK-cell development site, recent research also suggests that LN may be another important compartment for NK cell differentiation. Recently, Freud et al. [35] have observed that LN are also naturally and selectively enriched with $CD34^{dim}CD45RA^+$ HPC and are capable to differentiate into $CD56^{bright}$ NK cells in the presence of either IL-2 or IL-15 [35]. In addition, $CD34^{dim}CD45RA^+$ HPC express a high density of adhesion markers (L-selectin and alpha4/beta7-integrin). These investigators also demonstrated that $CD34^{dim}CD45RA^+$ HPC constitute approximately 1%, 6% and 95% of $CD34^+$ HPC pool from BM, blood and LN, respectively [35]. Furthermore, within LN $CD34^{dim}CD45RA^+$ HPC are colocalized with $CD56^{bright}$ NK cells, in the parafollicular T cell-rich area [31, 35]. Based on this data [35], coculture of LN activated T cells with autologous LN $CD34^{dim}$ HPC induced the emergence of $CD56^{bright}$ NK cells after 7 days. Moreover, Fehniger et al. [39] showed that endogenous T cell-derived IL-2 may trigger, through the NK high-affinity IL-2 receptor, $CD56^{bright}$ NK cells to secrete IFN-γ. Thus, this selective enrichment of both $CD34^{dim}CD45RA^+$ HPC and $CD56^{bright}$ NK cells within LN compared with the BM or blood suggested that LN may be a site for NK cell development in vivo.

There are two distinct blood subsets of human NK cells identified by cell surface density of CD56 [40]. The majority of human NK cells are $CD56^{dim}$ and express high levels of CD16 and killer cell immunoglobulin-like receptor (KIR), whereas a minority are $CD56^{bright}CD16^{dim/neg}$, with low cytotoxic activity and higher cytokine secretion, such as IFN-γ, in response to monokine costimulation [40]. This NK cell subset also expresses c-kit, a molecular marker typically present on $CD34^+$ HPC cell surface that enhances IL-2-induced proliferation and stimulates the upregulation of the bcl-2 antiapoptotic protein [41, 42]. It is important to note that $CD56^{bright}$ NK cells appear to be the only lymphocytes with constitutive expression of the high-affinity IL-2R, while $CD56^{dim}$ NK cells express only the intermediate affinity IL-2R and do not express c-kit [40]. These data could explain, in part, why $CD34^+$ HPC in LN differ preferentially in $CD56^{bright}$ NK cells, but not $CD56^{dim}$ NK cells that proliferate weakly in response to high doses of IL-2 [43, 44]. Since the discovery of $CD56^{bright/dim}$ NK cell subsets, it has been postulated that $CD56^{dim}$ NK cells may represent a subsequent stage of $CD56^{bright}$ NK cell maturation. This hypothesis is supported by some data, as mentioned above that only $CD56^{bright}$ NK cells express the c-kit immature cell marker. Furthermore, Romagnani et al. [45] have demonstrated that $CD56^{dim}$ NK cells from peripheral blood display shorter telomeres than peripheral and LN-derived $CD56^{bright}$ NK cells. Along this line, Ferlazzo et al. [46] have observed distinct NK cell subset distribution, with more than 95% blood $CD3^-CD56^+$ cells being $CD56^{dim}CD16^+$ NK cells, while in LN approximately 90% of those cells have mainly a $CD56^{bright}CD16^-$ phenotype. Together, these data allowed to presume a mechanism by which the NK cells may develop. However, this does not reject a contribution from the BM to generation of the later mature $CD56^+$ NK cells, mainly in the resting state. Thus, the initial stage of NK cell development occurs in the BM under the influence of stromal cell factors that drive the early $CD34^+$ HPC differentiation, generating $CD34^+CD45RA^+alpha4/beta7^+$ NK cell precursors. These cells, expressing cellular adhesion molecules promptly travel to LN where under the influence of IL-2 secreted by T cells they differentiate into $CD56^{bright}$ NK cells that may further differentiate

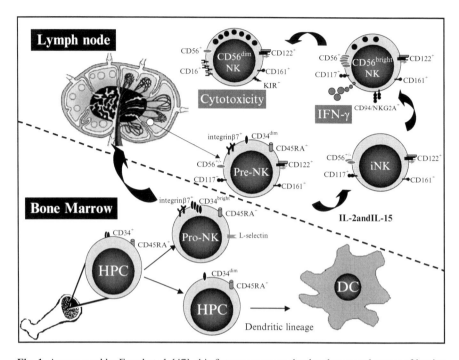

Fig. 1 As proposed by Freud et al. [47], this figure represents the developmental stages of in vivo human NK cell ontogeny. The cells are believed to mature from BM-derived HSC to last stages NK cells within LN. As NK cells progress from stage 1 to stage 3, they become committed to the NK cell lineage and lose the capacity for T cell or DC development. From stages 3–5, NK cells undergo functional maturation and transitioning, such that in vivo stage 3 iNK cells may produce growth factors, stage 4. CD56^{bright} NK cells may preferentially contribute with IFN-γ secretion, and stage 5 CD56^{dim} NK cells are mainly cytotoxic.

into CD56^{dim} NK cells. This journey from BM to LN seems to be important for NK cell differentiation, since the development of higher numbers of mature CD56^{bright} NK cells from peripheral blood CD34^{+} HPC culture than from BM was observed in the presence of IL-2 or IL-15, suggesting that peripheral blood contains a much higher percentage of pre-NK cells than BM [35] (Fig. 1).

The NK-cell development stages in human are not yet well established. Since human NK cells can give rise from CD34^{+} HPC in vitro, such stages of differentiation could be elucidated. In this context, some investigators have hypothesized that NK cells developed from CD34^{+} HPC and like T- and B-lymphocytes present a different expression kinetics of surface markers that could define the distinct stages of their development [14, 17, 18]. Freud et al. [47] based on the premise that (1) more than 99% of NK cells within LN express at least CD34, CD117, and/or CD94; (2) CD34 and CD94 are independent antigens, indicating that NK cells intermediate stages would first lose CD34 and then express CD94; and (3) NK cell functional maturity (cytotoxic and IFN-γ secretion) as well as acquisition of surface CD56 in humans are acquired at a later stage of development [48, 49], have proposed a marker panel set using the combination of CD34, CD117, and CD94 to differentiate the functionally distinct stages of human $CD56^{bright}$ NK cell development within LN as following: stage 1 – NK cell progenitor ($CD34^+CD117^-CD94^-$), stage 2 – Pre-NK cell ($CD34^+CD117^+CD94^-$), stage 3 – Immature NK cell ($CD34^-CD117^+CD94^-$), stage 4 – $CD56^{bright}$ NK cell ($CD34^-CD117^{+/-}CD94^+$), and stage 5 – $CD56^{dim}$ NK cell ($CD34^-CD117^-CD94^+$).

The differentiation of the first stages of NK cells is dependent of concomitant IL-15 responsiveness [19]. As mentioned above, the BM stromal cell factors KL and FL may stimulate $CD34^+CD122^-$ progenitor NK cells (pro-NK) to upregulate CD122 expression, a prerequisite for those cells to give rise to $CD34^+CD117^+$ pre-NK cells, which can differentiate into the following stages after stimulation with IL-15 [19, 47]. Furthermore, it is important to note that not all $CD34^+ CD117^+CD94^-$ cells are compromised with NK cell lineage. It has been proposed that $CD34^{dim}$ human HSC are committed progenitors to the type 1 IFN-γ-producing plasmacytoid DC, while $CD34^{bright}$ cells are more involved to lymphoid lineage, as confirmed by CD2, CD7, CD10, HLA-DR and integrin β7 expression [15, 35]. In addition, Freud et al.[47] have observed that $CD34^+CD117^{+/-}CD94^-$ cells in appropriate cultures also may differentiate into immature T cells. On the other hand, $CD34^-CD117^{+/-}CD94^{+/-}$ cells (stage 3 and stage 4) did not give rise to them. These data suggest that stromal milieu and membrane antigen expression are crucial to drive the lineage plasticity cell (stage 2) toward subsequent NK cell differentiation stages within LN (Fig. 2).

Immature NK cells (iNK cells), different from the first developmental NK cell stages, are completely incapable to generate T cells and DC and have been committed with NK cell lineage [18, 47]. Those cells express NK cell antigens including CD2, CD7, CD56, CD161 and NKp44, besides lack of CD10, integrin β7 and HLA-DR which further distinguishes the iNK cell phenotype from pre-NK cells.

Stage 1 (Pro-NK)	Stage 2 (Pre-NK)	Stage 3 (iNK)	Stage 4 ($CD56^{brigh}$)	Stage 5 ($CD56^{dim}$)
CD34 (+)	CD34(+)	CD34 (−)	CD34(−)	CD34 (−)
CD117 (−)	CD117 (+)	CD117 (+)	CD117 (+/−)	CD117 (−)
CD94 (−)	CD94 (−)	CD94 (−)	CD94 (+)	CD94 (+/−)
CD16 (−)	CD16 (−)	CD16 (−)	CD16(−)	CD16 (+)

Enrollment to NK-cells lineage →

NK-cell maturation →

Cytotoxicity acquisition →

Fig. 2 Phenotypic features used to discriminate the main NK cell subsets during the ontogenic process. This figure brings the more important markers acquired during the five ontogenic stages

Despite that iNK cells fall into an exclusively NK cell branch they are not yet capable to produce IFN-γ or mediate perforin-dependent cellular cytotoxicity against MHC-I negative target cells [47]. In addition, Grzywacz et al. [50] described the importance of CD117 and CD94 in the segregation of distinct intermediate NK cell stages. They demonstrated that $CD56^+$ cells can express different levels of CD117 and are subdivided in $CD56^+CD117^{high}$ and $CD56^+CD117^{low/-}$ subsets. These populations are considered different since the latter expresses NKp30, NKp46, NKG2D, NKG2A and CD94, while $CD56^+CD117^{high}$ cell do not. It is important to note that the fact of CD94 to be expressed exclusively on $CD56^+CD117^{low/-}$ cells allowed its combination with CD117 to distinguish two discrete subsets of differentiating $CD56^+$ NK cells. These authors showed that $CD56^+CD117^{high}$ $CD94^-$ cells are not cytotoxic, whereas $CD56^+CD117^{low/-}$ $CD94^+$ effectively kill target cells and express high levels of FasL and IFN-γ, representing differentiation at stage 4, as proposed by Freud et al. [47]. It is important to emphasize that, despite Grzywacz et al. [50] having demonstrated an effective killer profile for $CD56^+CD117^{low/-}$ $CD94^+$ cells, Freud et al. [47] did not observe the same pattern for all $CD56^+CD117^{low/-}CD94^+$ mature NK cells in LN, suggesting different cell subsets within stage 4 or their presence at a subsequent stage.

It has been hypothesized for the past two decades that $CD56^{bright}$ and $CD56^{dim}$ NK cells represent different stages of NK cell differentiation, and that $CD56^{dim}$ NK cells could derive directly from $CD56^{bright}$ [51]. However, until recently, evidence for this development has been lacking because there was no approach to analyze a complex set of markers. The advance of new tools that allow us to evaluate a more detailed cellular antigen panel, brings some light on the developmental differences between $CD56^{bright}$ and $CD56^{dim}$ human NK cell subsets. Freud et al. [47] assessing the expression of CD16 and CD94 within the total $CD3^-CD56^+$ populations cells from LN and peripheral blood, could support the theory that the $CD56^{dim}$ cell subset represents the last developmental NK cell stage (stage 5), since blood $CD56^{bright}$ NK cells are $CD94^+CD16^{+/-}$ and blood $CD56^{dim}$ NK cells are mainly $CD94^{+/-}CD16^+$. In addition, KIR^+ NK cells are primarily within the $CD56^{dim}CD94^{+/-}CD16^+$ fraction of cells in both LN and peripheral blood, whereas the $CD56^{bright}CD94^+CD16^{+/-}$ do not express KIR, which is consistent with evidence indicating that KIR acquisition is rather a late event during NK cell maturation both in vitro and in vivo [52–54].

Interestingly, the $CD56^{bright}$ NK cells are dominant in LN (75% median value), whereas in peripheral blood and spleen, the majority of NK cells are $CD56^{dim}$ (95% and 85%, respectively) [46]. Furthermore, while the $CD56^{dim}$ subsets in the spleen and in the peripheral blood express CD16, $CD56^{bright}$ NK cells in LN are negative for CD16 and express low levels of activation markers (HLA-DR and CD69) [46]. Cooper et al. [40] have also shown that $CD56^{bright}CD16^{+/-}$ NK cells represent the minority within human peripheral blood $CD56^+$ NK cells, whereas $CD56^{dim}CD16^+$ NK cells are the majority. These data together suggest that the $CD56^{bright}$ cells are more immature and may give rise to $CD56^{dim}CD16^+$ NK cells (the last stage of NK cell life) inside the LN and then go to peripheral blood. It is important to mention that despite these stages representing a possible developmental NK cell pathway, it is still possible that some cells from each stage may still be terminally differentiated with essential functions for body homeostasis.

3 NK Cells in Innate Immunity

3.1 Phenotypic Features of NK Cells

3.1.1 Human NK Receptors: Recognition and Killing of Target Cells

NK cells, morphologically classified as large granular lymphocytes (LGL), in contrast to cells of the adaptive immune system, can quickly lyse target cells without prior sensitization. Although first identified by their cytotoxic activity against tumor and virally infected cells [55], there is now increasing evidence that NK cells are important mediators of the innate resistance to a variety of pathogenic microorganisms, including intracellular bacteria [56]. Functionally, they exhibit cytolytic activity against a variety of allogeneic targets in a nonspecific, contact-dependent, nonphagocytic process, which does not require prior sensitization to an antigen. NK cells express a variety of receptors that serve to either activate or suppress their cytolytic activity. These receptors bind to various ligands on target cells, both endogenous and exogenous, and have an important role in regulating the NK cell response [57]. These cells also have a regulatory role in the immune system through the release of cytokines, which in turn stimulate other immune functions [58].

NK cells and T cells similarly express CD2 and utilize LFA-1 surface antigen to enhance effector cell adhesion to target cells. However, human NK cells can be distinguished from T lymphocytes by the expression of distinct phenotypic markers such as CD56 and CD16 and the lack of rearranged T cell receptor gene products and CD3. Once considered relatively homogeneous, it is now known that NK cells are highly diverse. Within an individual, expression of different combinations of receptors creates a diverse NK cell repertoire, which exhibits some specificity in the immune response [59].

It has been shown that almost 80% of human NK cells express the CD2 antigen, and a CD2-negative population resides within peripheral blood NK cells [60]. In addition to its adhesive properties, CD2 also acts as a costimulatory molecule on NK cells [61]. LFA-1 is an integrin and is expressed as a heterodimer of α_L (CD11a) and β_2 (CD18) polypeptides. LFA-1 binds intercellular adhesion molecule (ICAM)-1 through ICAM-5 [62, 63]. Although NK cells express other integrins, LFA-1 is believed to play a dominant role in target cell lysis [64]. Antibody blocking of LFA-1–ICAM interactions impairs Antibody Dependent Cellular Cytotoxicity (ADCC) and natural cytotoxicity by human NK cells [65]. Studies suggest a signaling role of LFA-1 in perforin release by a subset of IL-2-pulsed primary NK cells [66], and in cytolytic granule polarization in long term IL-2-cultured NK cells [67].

CD16 is expressed on a large subset of peripheral blood NK cells, which also express intermediate levels of CD56 (CD56^{dim}). A phenotypically and functionally distinct CD56^{bright}CD16$^-$ subset of NK cells constitutes <10% of peripheral blood NK cells [40]. CD56^{dim} NK cells mediate ADCC through engagement of CD16.

3.1.2 Receptors on Resting NK Cells

Studies of Warren and Skipsey [68] demonstrated that resting NK cells are small cells unresponsive to high concentrations of IL-2. The resting subpopulation of NK cells contained two distinct populations based on the intensity of CD16 expression, namely CD16^{dim} and CD16^{bright}. Both populations expressed similar levels of CD56. These authors showed that the differentiation of resting NK cells does not involve the CD16 molecule. Thus CD16 interaction using solid-phase bound-linked anti-CD16 mAb did not induce proliferation of resting NK cells in the presence of IL-2, and neither enhanced nor inhibited the differentiation process triggered by the stimulator cells and IL-2. However, CD16 interaction did reduce the cytotoxic activity of the culture-generated cells, and did stimulate them to produce IFN-γ. These results are consistent with the studies of others showing that the effector functions of NK cells are influenced by CD16 interaction [69, 70], but that such interaction does not induce NK cell proliferation [71]. Thus, it can be concluded that CD16 functions only as an effector stage molecule for NK cells. Like resting T cells resting NK cells express the p70 IL-2R but do not respond to high concentrations of IL-2 [68]. Freshly isolated, resting NK cells are generally less lytic against target cells than in vitro IL-2-activated NK cells [72]. These authors showed that target-cell lysis by IL-2-activated NK cells in a redirected, ADCC assay was triggered by a number of receptors. In contrast, cytotoxicity by resting NK cells was induced only by CD16, and not by NKp46 or NKG2D.

Bryceson et al. [73] demonstrated that resting NK cells were induced to secrete TNF-α and IFN-γ, and to kill target cells by engagement of specific, pair-wise combinations of receptors. Therefore, natural cytotoxicity by resting NK cells is induced only by mutual costimulation of activating receptors, revealing distinct and specific patterns of synergy among receptors on resting NK cells. NK cells use specific receptors to mediate killing through the recognition of distinct ligands expressed on target cells. These receptors fall into two functional types, inhibitory and stimulatory [74].

3.1.3 Activating NK Cell Receptors

A number of structurally distinct receptors have been implicated in activation of NK cell effector functions. It is not yet clear if any one receptor is necessary or sufficient to activate NK cells and if activation receptors may be redundant. Activation receptors are grouped in three categories: receptors that signal through immunoreceptor tyrosine-based activation motif (ITAM)–containing subunits (e.g., CD16, NKp46, NKp44), the DAP10- associated receptor NKG2D, and several other receptors (e.g., 2B4, CD2, DNAM-1) that signal by different pathways [73, 75].

ITAM-Bearing NK Receptor Complexes

NK cells are unique among hematopoietic cells in that all mature NK cells constitutively express FcεRI-γ, CD3-ζ and DAP12 type I transmembrane-anchored proteins that exist as either disulfide-bonded homodimers or, in the case of FcεRI-γ and CD3-ζ, as disulfide-bonded heterodimers. All have minimal extracellular regions comprising only a few amino acids, principally the cysteine residues through which they dimerize. Most importantly, these proteins contain ITAM motifs, defined by the sequence (D/E)XXYXX(L/I)X_{6-8}YXX(L/I) (where X_{6-8} denotes any 6–8 amino acids between the two YXX(L/I) elements) in their cytoplasmic domains. DAP12 and FcεRI-γ have a single ITAM, and CD3-ζ has three ITAM per chain [75]. ITAM-receptor activation induces actin cytoskeleton reorganization, which is required for cell polarization and release of the cytolytic granules containing perforin and granzymes, and results in the transcription of many cytokine and chemokine genes.

CD16 (FcγRIII), a low-affinity receptor for IgG, is associated with the ITAM-containing FcεRI-γ chain and T cell receptor (TCR) ζ chain. NKp46 and NKp30 are associated with the TCR ζ chain [55]. NKp44, KIR2DS, and CD94/NKG2C are associated with the ITAM-containing DAP12. Natural cytotoxicity receptors (NCR), which include NKp46, NKp44, and NKp30, play a major role in NK-cell cytotoxicity against transformed cells [58]. Although ligands of NCR have not been identified, antibodies against NCR have been used to block lysis of tumor cells by IL-2-activated and resting NK cells [76, 77]. However, in the mouse, Syk/ZAP70-independent natural cytotoxicity by NK cells was observed, implying that natural cytotoxicity can occur independently of ITAM-based activation signals [78].

NKG2D–DAP10 Receptor Complexes

A single gene with little polymorphism encodes the C-type lectin–like superfamily member NKG2D [79], which is a type II transmembrane-anchored glycoprotein expressed as a disulfide-bonded homodimer on the surface of NK cells, γδ T cells and $CD8^+$ T cells [80]. An arginine residue centrally located within the transmembrane region of NKG2D associates with the aspartate residue within the transmembrane domain of the DAP10 signaling subunit [81]. NKG2D can signal through both DAP10 and DAP12 in mice [82, 83] whereas human NKG2D associates only with DAP10 [81, 84, 85]. DAP10 is a signaling subunit that carries a phosphatidylinositol-3 kinase-binding motif [81]. Ligands for NKG2D, such as MICA and ULBP, are expressed on some tumor cells and on infected or stressed cells [86]. Experiments have suggested that NKG2D signals are sufficient to activate NK-cell functions [87, 88]. Lysis of certain tumor cells by resting NK cells and by IL-2-activated NK cells can be blocked by antibodies to NKG2D [76, 89]. The importance of ligands for NKG2D in immune defense is underscored by strategies developed by viruses to interfere with their expression [86, 90, 91]. The cytoplasmic domain of DAP10 is very small, with only 21 amino acids, and contains only one known signaling motif: the sequence YINM, which when phosphorylated is able to bind either the p85 subunit of phosphatidylinositol-3-OH kinase (PI (3) K, through YXXM) or the adaptor Grb2 (through YXNX). Because these two binding sites overlap, a single DAP10 chain will bind either p85 or Grb2, but not both. Unlike the ITAM-containing receptors in NK cells, NKG2D–DAP10 receptor complexes do not require Syk family kinases or LAT (linker for activation of T cells) for signaling, as demonstrated by biochemical studies [88] and by the ability of NK cells from mice lacking both Syk and ZAP-70, or LAT and NTAL (non–T cell activation linker), to mediate NKG2D-dependent cytolysis [92]. Another candidate proposed for the phosphorylation of DAP10 is the kinase Jak3 [93]. Recent data suggest that IL-15 stimulates Jak3-mediated phosphorylation of DAP10 and that this process is necessary to enable the NKG2D receptor to initiate killing of NKG2D ligand–bearing targets [93]. Bryceson et al. [94] had previously demonstrated the necessity for human NK cells to be "primed" by stimulation with cytokines before certain NK cell receptors, including NKG2D, are competent to trigger cytotoxicity. Similarly, the ability of human $CD8^+$ T cells to kill when activated through the NKG2D receptor also requires previous stimulation by high concentrations of IL-2 or IL-15 [95].

Although there is a consensus that engaging the NKG2D–DAP10 receptor complex on NK cells efficiently initiates cell-mediated cytotoxicity, the influence of NKG2D–DAP10 complexes on other effector functions, such as cytokine production, is less well understood [75]. In human NK cells, cross-linking of NKG2D triggers cytotoxicity but not cytokine secretion [88]. In contrast, human NK cells stimulated with soluble recombinant NKG2D ligands (MICA, ULBP-1, or ULBP-2) secrete cytokines, including IFN-γ, GM-CSF and CCL4/MIP-1β [96].

CD48: A Non-MHC Class I Ligand for 2B4 (CD244)

Receptor 2B4 generally is considered as a coactivator of NK cell cytotoxicity, because it enhances NK cell responses under limiting ITAM-mediated activation [97, 98]. 2B4 is a member of the CD2 family of immunoglobulin-related proteins and its ligand, CD48, is a glycosylphosphatidylinositol-linked also CD2-related molecule expressed widely on haematopoietic cells [98–100].

Although 2B4–CD48 interactions influence activation of leukocyte effector function, the outcomes are not clearly understood. NK cell activation through 2B4 is accompanied by phosphorylation of tyrosine-based motifs in the cytoplasmic tail and recruitment of signaling by activation molecule associated protein and the Src-family kinase Fyn [101, 102]. However, it is still unclear how 2B4 provides co-activation signals in the context of other receptor and ligand interactions and whether 2B4 is capable of triggering cytotoxicity independently of other receptor signals. In human NK cells, 2B4 can activate cytotoxicity and IFN-γ production through SLAM associated protein (SAP), an SH2 domain-containing adaptor molecule, which results in severe immunodeficiency when defective [103]. In contrast, in the absence of SAP, 2B4 delivers inhibitory signals in murine and human NK cells [103].

NCR Receptors: Receptors that do not Recognize HLA class I Ligands

The activatory NCR and NKG2D receptors have received most recent attention [104]. Their expression is predominantly restricted to NK cells, and they have the ability to activate NK cells in the absence of additional stimuli. NKp46 [105] and NKp30 [77] are expressed by all NK cells, while NKp44 [106] expression is restricted to activated NK cells. Although not an absolute phenomenon, NK cell clones can be identified as NCR^{bright} or NCR^{dull} based on the surface receptor densities and high levels of NCR correlated with high natural cytotoxicity against many target cells [77, 105]. Indeed, a complementary relationship between NCR and NKG2D appears to exist, and NCR^{dull} clones kill tumor cells in an NKG2D-dependent manner [58]. Together, NCR and NKG2D accounted for virtually all cytotoxicity mediated by NK cells against a wide variety of tumor target cell types. While the ligands for NKG2D are known to be the stress-induced antigens MICA, MICB [80] and ULBP, the endogenous ligands for the NCR remain to be identified [58]. It appears, however, that NCR may have been coopted by NK cells to recognize pathogen-specific moieties, as both NKp46 and NKp44 recognize virus-specific haemagglutinins and facilitate NK cell lysis of virally infected cells [107, 108]. In contrast, NKp30 appears to have been targeted by the pp65 protein of human cytomegalovirus (HCMV), which binds NKp30 and inhibits NK cell cytotoxicity [109]. The selection of NK cell receptors by the immune system for specific recognition of pathogen and their targeting by specific viral proteins highlight their importance in the immune response to these infections [110].

TLR: Direct and Indirect Pathogen Recognition

TLR are a recently described family of innate immune receptors which recognize conserved pathogen-associated molecular patterns (PAMP) [111]. TLR3, TLR7, TLR8, and TLR9 have been identified on NK cells [112, 113]. TLR3 recognizes double-stranded RNA (dsRNA) produced during viral replication [114]. TLR7 and TLR8 recognize single-stranded RNA (ssRNA), whereas TLR9 recognizes unmethylated CpG motifs [115]. Indeed, a role for TLR agonists in NK cell activation has been demonstrated recently with reports that TLR2 [116], TLR3 [113, 117, 118] and TLR9 [113] agonists stimulate NK cell functions.

It is interesting that despite the obvious importance of TLR in viral infection, the first evidence for TLR function in NK cells came from models of parasitic and bacterial infections. Lipophosphoglycan from *Leishmania major* stimulated NK cells to produce the proinflammatory cytokines IFN-γ and TNF-α through cell surface ligation of TLR2 [116]. Subsequently, NK cells were shown to respond to the outer membrane protein A from *Klebsiella pneumoniae* and flagellin from *Escherichia coli* through TLR2 and TLR5, respectively.

The discovery that TLR3 is a receptor for polyinosinic-polycytidylic acid (poly I:C – a synthetic analog of dsRNA) [114] and the observation that poly (I:C) activates NK cells in vivo [119, 120] clearly implicate it as a possible receptor through which NK cells are activated. While the original study reported direct activation of human NK cells by poly (I:C) [118], a second study reported that IL-12 was required for this to occur, i.e., there was a requirement for accessory cell-derived cytokine [113]. O'Connor et al. [110] showed both a direct and an indirect role of poly (I:C) activation through TLR3.

NK cells respond to the TLR7/8 agonist R848 and demonstrate increased cytotoxicity and cytokine production. While NK cells required priming with cytokine to directly transduce signal in response to R848, they were potently activated by accessory cell-derived cytokines. In fact, IFN-γ production by NK cells in response to R848 was entirely IL-12, and therefore accessory cell dependent. The increased cytotoxicity observed was also partly attributable to the accessory cell involvement [112].

TLR9 is activated by double-stranded unmethylated CpG motifs which are present in bacterial and, occasionally, viral genomes [111] . CpG has been reported to activate human NK cells in vitro, although this is a dependent accessory cell as IL-12 is required for this effect [114]. When the immune response to MCMV was examined in vivo, it was surprising to find that infections were more severe in TLR9$^{-/-}$ than TLR3$^{-/-}$ mice [121, 122]. In fact, although TLR3$^{-/-}$ mice did have an impaired immune response to MCMV, IFN-γ production by NK cells was relatively normal while it was almost completely absent in TLR9$^{-/-}$ animals [122].

3.1.4 Inhibitory NK Cell Receptors

NK cells express a variety of inhibitory receptors that recognize MHC class I molecules and block NK cell-mediated cytotoxicity [123]. In humans, these receptors include the KIR family, the leukocyte immunoglobulin-like receptor (LILR) family, and the family of CD94/NKG2 lectin-like receptors [124]. All known ligands for these inhibitory receptors are MHC class I or molecules of host or pathogen origin that are homologous to MHC class I. Ligation of such molecules by MHC I on target cells results in inhibition of the NK cytotoxic activity through the immunoreceptor tyrosine-based inhibitory motifs (ITIM) [58]. These ITIM are defined by the sequence (I/L/V/S)XYXX(L/V), where X represents any amino acid, and slashes separate alternative amino acids that may occupy a given position. Phosphorylated ITIM in the cytoplasmic tails of such inhibitory receptors recruit the tyrosine phosphatases SHP-1 and SHP-2 [124, 125]. These tyrosine phosphatases suppress NK cell responses by dephosphorylating the protein substrates of the tyrosine kinases linked to activating NK receptors [126].

Killer Immunoglobulin-Like Receptors (KIR)

KIR belong to a multigene family of more recently-evolved immunoglobulin-like extracellular domain receptors in human and are the main receptors for classical MHC I. KIR are specific for certain HLA subtypes (HLA-A, HLA-B, HLA-C) [127].

KIR may have long or short cytoplasmic tails (L or S forms, respectively). The L form KIR has two or more tyrosine-based inhibitory motifs within its cytoplasmic domain and sends a transient inhibitory signal to the NK cell upon binding its cognate ligand. The S forms of KIR (KIR2DS and KIR3DS) lack these domains and send an activating signal following ligand binding [58, 127]. In general, KIR2D receptors recognize HLA-C alleles, whereas KIR3D receptors recognize HLA-A and HLA-B alleles.

Inhibition by KIR blocks NK cell activation at a very proximal step, which precedes actin-dependent processes [124]. For instance, binding of inhibitory KIR to MHC class I on target cells prevents the tyrosine phosphorylation of activation receptors 2B4 and NKG2D, as well as their recruitment to detergent-resistant membrane microdomains [128]. Engagement of ITIM-containing inhibitory receptors blocks the accumulation of filament of actin at NK cell immune synapses [129]. Reorganization of the actin cytoskeleton is essential for the cytotoxic activity of NK cells. Inhibitors of actin polymerization prevent cytolytic activity, hinder accumulation of receptors at activating immune synapses and block phosphorylation of NK cell activation receptors [128, 129]. Given that actin cytoskeleton rearrangement is inhibited by ITIM-containing receptors, it is generally assumed that KIR engagement at an inhibitory event prevents the delivery of activation signals by blocking the cytoskeleton-dependent movement of activating receptors.

Heterodimeric C-Type Lectin Receptors (CD94/NKG2)

CD94 is expressed as a disulphide-linked heterodimer with members of the NKG2 family, including NKG2A, which transduces inhibitory signals, and NKG2C and NKG2E, which transduce activating signals [130]. CD94/NKG2 is a C-type lectin family receptor, conserved in primates and identifies nonclassical (also nonpolymorphic) MHC I molecules like HLA-E. As with KIR molecules, the inhibitory receptor has a higher binding affinity and inhibition dominates mediated via the cytoplasmic ITIM [131, 132].

As CD94/NKG2 complements KIR functionally, this is an indirect way to survey the levels of classical (polymorphic) HLA class I molecules, because expression of HLA-E at the cell surface is dependent upon the presence of classical MHC class I leader peptides [133, 134]. Consequently, downregulation of MHC I expression leads to a reduction in HLA-E expression at the cell surface. Accordingly, the interaction between CD94/NKG2 and HLA-E represents a central innate mechanism by which NK cells indirectly monitor the expression of other MHC I molecules within a cell [133].

NK cells can kill activated T cells, unless the T cells express sufficient amounts of classical or nonclassical MHC class I molecules. As a consequence, blockade of CD94/NKG2A inhibitory receptors leads to NK cell cytotoxicity against activated $CD4^+$ T cells, suggesting the use of blocking antibodies to NKG2A to prevent $CD4^+$ T cell–dependent autoimmunity [135].

3.2 Functional Characteristics of NK Cells

3.2.1 Cytokine Production and Killing

The cytokines play a crucial role in NK cell activation. As these are stress molecules, released by cells upon viral infection, they serve to signal to the NK cell the presence of viral pathogens. NK cell cytokine production may be governed in part by the monocyte-derived cytokines induced during the early proinflammatory response to infection and by the subset of NK cells present at the site of inflammation.

NK cells produce a range of cytokines, including haematopoietic factors such as IL-3 and granulocyte–macrophage colony-stimulating factor (GM-CSF), TNF-α and regulatory cytokines such as transforming growth factor (TGF-β) and IFN-γ [56]. In both viral and bacterial models of infection, IFN-γ production by NK cells has been shown to be a key event in successful resolution of infection [136]. As a general rule, IL-12, produced very early in infection, is responsible for driving NK cells to produce IFN-γ [137]. The $CD56^{bright}$ NK cell subset produced significantly more IFN-γ following IL-18 and IL-12 stimulation compared with $CD56^{dim}$ NK cells [137].

Isolated NK cells from human lymphoid tissues that express CD4 molecule on their surface efficiently mediate NK cell cytotoxicity and CD4 expression does not appear to alter lytic function. $CD4^+$ NK cells are more likely to produce the cytokines IFN-γ and TNF-α than are $CD4^-$ NK cells [138].

3.2.2 Steps in NK Cell Activation, Adhesion and Degranulation

Signals are required to activate normal, resting NK cells to migrate to sites of infection and to acquire effector functions, such as production of IFN-γ and cellular cytotoxicity [119]. The rapid response of resting NK cells to cytokines is well documented. Type I IFN, produced during virus infections, stimulates the proliferation of NK cells and augments their cytotoxic activity. IL-15, secreted by a number of different cell types, activates NK cell proliferation, cytotoxicity, and cytokine production [31]. NK cells secrete large amounts of IFN-γ in response to IL-12 and IL-18, which are produced during infections by various intracellular pathogens [40]. In addition to signals received from soluble mediators, NK cells are also activated through cell contact by receptors that recognize ligands on other cells.

T and B cells possess a single antigen receptor that dominates their development and activation. Signals initiated through these antigen receptors are augmented by costimulatory molecules. In contrast, NK cells do not possess one dominant receptor, but instead rely on a vast combinatorial array of receptors to initiate effector functions. Several structurally distinct activation receptors have been implicated in NK cell cytotoxicity, a complex process that involves adhesion, synapse formation, and granule polarization and exocytosis. The first contact between an NK cell and its target is a cellular association that may be similar to tethering. The early interactions could then result in a longer-lasting association that leads to initial adhesion. These events probably contribute to NK cell activation, as receptors that are potentially engaged at this point, such as CD2, may participate in activation signaling [138]. The next step in the formation of the lytic synapse is firm adhesion, which is facilitated by receptor–ligand interactions of higher affinity. The integrin family of adhesion molecules is important in firm adhesion. Some of the integrins that are expressed by NK cells are well studied and include lymphocyte function-associated antigen 1 (LFA-1; CD11a–CD18) and macrophage receptor 1 (MAC-1; CD11b–CD18). Although these integrins rapidly cluster at the NK cell synapse following initiation [139–141] ,they probably function in adhesion and participate in signaling (even in resting NK cells) before their rearrangement to the synapse [142]. Integrin signaling can fully activate some NK cells [142] and partially activate others [72]. Several in vitro assays have indicated an essential role for the αLβ2 integrin LFA-1 and β1 integrins in the conjugation of NK cells with target cells. LFA-1 transduces signals that induce tyrosine phosphorylation and activation of the guanosine triphosphate (GTP), guanosine diphosphate (GDP) and exchange factor (GEF) Vav1 [142], leading to actin polymerization, cytoskeletal rearrangements, and clustering of lipid rafts. These LFA-1-mediated signals not only promote adhesion but also can be sufficient to trigger NK cell-mediated cytotoxicity [67]. Engagement of β1 and β2 integrins also induces phosphorylation and activation of the proline-rich tyrosine kinase-2 (PYK-2) [143]. Moreover, PYK-2 activates extracellular signal-regulated kinases 1 and 2 (ERK1/2), which trigger cytotoxicity through an as yet undefined pathway [144]. NK cells derived from patients with leukocyte adhesion deficiency I carry a genetic mutation of the β2 integrin (CD18) and exhibit defects in cytotoxicity [145]. Recent studies have shown that human NK cells express additional adhesion molecules, called CD226 (DNAX accessory molecule 1), CD96 (Tactile), and CRTAM, which recognize nectins and nectin-like (Necl) molecules on target cells [146]. Therefore, genetic deletion of individual adhesion molecules may not ultimately lead to defects in cytotoxicity, if other adhesion molecules remain intact. The main pathway of NK-cell mediated cytolysis is dependent on perforin and granzymes. However, other mechanisms triggered by FasL (CD178) and tumor-necrosis factor-related apoptosis-inducing ligand (TRAIL)-dependent receptors may occur [147].

3.2.3 Antibody Dependent Cellular Cytotoxicity

Cytotoxicity is probably the best characterized effector function of NK cells, and targets include tumor cells, virally infected cells, cells infected with intracellular bacterial pathogens and as more recently reported, immature DC (iDCs) [56]. NK cells, along with macrophages and several other cell types, express the FcR molecule, an activating biochemical receptor that binds the Fc portion of antibodies. This allows NK cells to target cells against which a humoral response has been mobilized and to lyse cells through ADCC [148].

Several activating receptors, including CD16, NKG2C, NKG2D, LFA-1, 2B4 and the NCR (NKp30, NKp44, and NKp46) exist on NK cells, and ligation of these receptors can also lead to activation and increased cytotoxicity. ITAM-receptor activation induces actin cytoskeleton reorganization, which is required for cell polarization and release of the cytolytic granules containing perforin and granzymes, and results in the transcription of many cytokine and chemokine genes [58, 104]. Recent studies suggest that CD16 signals for degranulation, and engagement of CD16 and LFA-1 leads to efficient target cell lysis. Coengagement of 2B4 and CD16 enhances the number of degranulating NK cells, and can induce cytotoxicity in the absence of LFA-1–ICAM-1 interaction [72] (Fig. 3).

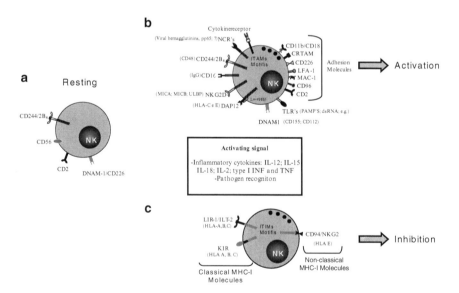

Fig. 3 Activating and inhibitory receptors in human NK cells and their ligands. Resting NK cells (**a**) interact with target cells expressing ligands for both (**b**) activating receptors with an immunoreceptor tyrosine-based activation motif (ITAM) and (**c**) inhibitory receptors with an immunoreceptor tyrosine-based inhibition motif (ITIM)

4 Role of Human Natural Killer Cells in Health and Disease

4.1 The Role of NK Cells in Health

Any discussion of the normal physiology of NK cells is incomplete if mention is not made about their 'putative role' in pregnancy. It has long been recognized that large granular lymphocytes of NK cell origin are the major lymphocyte population lining the pregnant uterus. The potential role of NK cells in pregnancy is suggested by the fact that approximately 70% of lymphocytes residing in the pregnant decidua are NK cells. These are phenotypically distinct from peripheral blood NK cells and may result from a distinct NK cell lineage or unique microenvironment which is $CD56^{+bright}/CD16^-/CD3^-$, has low direct cytotoxicity [149, 150], and proliferates through production of IL-15 by placental macrophages and the IL-15 receptor on $CD56^{+bright}$ uterine NK cells [151].

Proposed functions of NK cells in the gravid uterus are: (A) *Production of cytokines to facilitate decidualization and control of the invading trophoblast* (13) – IFN-γ modulates the NK cell pool resident in the uterus. The uterine NK cell-derived IFN-γ is not necessary for initiation of decidualization but appears essential for decidual maintenance in the second trimester; (B) *Protection of fetal tissues from maternal immune attack* [152] – Since NK cells have receptors for nonclassical MHC class I molecules, some authors have hypothesized that HLA-G protects fetal cells from lysis by maternal NK cells through this mechanism. In this regard, trophoblasts express the nonclassical MHC class I molecule HLA-G [153], which can ligate the KIR2DL4 receptor expressed on almost all NK cells [154]. KIR2DL4 is particularly interesting in that it can inhibit NK cell cytotoxicity but promote NK cell cytokine production upon ligation [155, 156]. Recent reports suggest that unfavorable combinations between maternal KIR and fetal HLA expression may be associated with preeclampsia [157] and recurrent miscarriage [158]; and (C) *Protection of the fetus from infectious diseases* – Acute or chronic infection of the uterus during pregnancy needs active innate immunity to rapidly prevent spreading of pathogens to the fetus and possible, subsequent spontaneous abortion. Virus-infected uterine cells might be recognized by some activating uterine NK receptors through as yet undefined protein ligands, leading to the killing of infected target cells. Other uterine NK activating receptors when triggered by specific ligands may also play a role in local antiviral immunity by inducing the secretion of proinflammatory cytokines. During viral infections, NK cells were shown to produce IFN-γ, CCL3/MIP-1α and CCL4/MIP-1β important to generate an inflammatory environment by recruiting eosinophils, macrophages, and activated DC [159]. Dendritic cells are likely to interact with uterine NK cells, priming innate immunity by triggering NK cell effector functions and thus providing a link between innate and adaptive immune responses [160]. Furthermore, NK cell inhibitory receptors that trigger dominant negative signals in normal pregnancy can be targeted by viral immune evasion strategies [94, 161]. Moreover, some NK cell activating receptor ligands can be upregulated in cells infected with CMV [161].

In addition, abnormal functions of NK cells in women with recurrent spontaneous abortions have been postulated. There seems to be a correlation with recurrent spontaneous abortion and increased NK cell cytotoxicity and an abnormally high decidual $CD56^+/CD16^+$ to $CD56^+/CD16^-$ ratio compared to normal pregnancies [162]. A better understanding of NK cells in pregnancy may lead to important new therapies to prevent fetal loss.

4.2 The Role of NK Cells in Infections

4.2.1 NK Cells and Viruses

Although NK cells were initially identified in the 1970s on the basis of their ability to kill tumor cells, soon thereafter, activated NK cells were also detected in virus-infected hosts. Studies from several laboratories showed that NK cells isolated after infection with any of several different viruses had increased cytolytic activity in vitro against tumor-cell targets, and that viral infection can result in NK cell proliferation and recruitment to the infected tissues and organs [86, 163]. In many cases, the increased NK–cell-mediated killing was attributed to activation of the NK cells by the production of type I IFN induced in the host by viral infection, because it was known that NK cells could be directly stimulated by exposure to IFN-α or IFN-β. Although type I IFN are not mitogenic for NK cells, they do induce the production of IL–15 [164], in addition to augmenting the cytotoxicity and cytokine production of NK cells. Hence, the induction of type I IFN and IL–15 by viral infection could well account for the presence of activated NK cells in virus-infected hosts.

The first clear evidence indicating cognate recognition of a viral pathogen by NK cells that is important to host protection emerged from studies with HCMV. A patient selectively lacking NK cells, but with normal B and T cells, was found to suffer life-threatening illness after infection with HCMV [165]. There are intriguing hints that NK cells expressing DAP12-associated CD94/NKG2C receptor complex preferentially proliferate after coculture with HCMV-infected cells [166]. Both the activating CD94/NKG2C receptor complex and the highly related inhibitory CD94/NKG2A receptor complex recognize HLA-E as their ligand [133], although these functionally diverse receptors might discriminate between different peptides presented by HLA-E [167], potentially allowing the selective activation of CD94/NKG2C-bearing NK cells to mediate immunity against HCMV. However, the leader peptide derived from the HCMV protein UL40 can assemble into the peptide binding groove of HLA–E and when this UL40 peptide is over-expressed with HLA–E it can engage the inhibitory CD94/NKG2A receptor complex and suppress NK cell activation [168–170]. UL142 has been shown to block the expression of proteins encoded by some, but not all, alleles of MICA [171]. The extraordinary polymorphism of the genes encoding MICA (61 alleles) and MICB (30 alleles) might reflect the host's attempt to counter the viral immune-evasion strategies that thwart NKG2D-mediated immune responses. UL18 engages the inhibitory ILT2 receptor and suppresses immune responses against HCMV-infected cells that have downregulated their expression of MHC class I proteins. The UL141 protein encoded by HCMV specifically targets and prevents the expression of CD155, a cell-surface glycoprotein that functions as a ligand for the activating NK cell receptors DNAM1 (CD226) and CD96 [172]. The HCMV structural protein pp65 has been reported to interact with the activating NKp30 receptor on NK cells and dissociate the receptor from its CD3ζ signaling adaptor protein, thereby preventing NK cells from recognizing and attacking HCMV-infected cells [109]. NK cells also participate in immune responses against other herpesviruses. It has been observed that a patient lacking NK cells suffered from severe infections not only with HCMV, but also with varicella zoster virus (VZV) and herpes simplex virus (HSV) [165]. Although there is extensive older literature documenting the activation of NK cells during HSV–1 infection [163], it remains unclear whether there is cognate recognition of the HSV-1-infected cells by NK cells or whether the effects of HSV-1 on NK cells are an indirect result of the high-level production of type I IFN and other cytokines that are induced by HSV infection. Epstein–Barr virus (EBV), another human herpesvirus, transforms B cells, causing B cell lymphomas in immunosuppressed patients. EBV-transformed B cell lines are relatively resistant to NK cell-mediated attack, mainly owing to their high level of expression of MHC class I molecules. However, a recent study showed that when latent EBV was reactivated in transformed B cells, they became susceptible to NK cell-mediated lysis, which was partially inhibited by blocking the activating NKG2D and DNAM-1 receptors on NK cells [173]. In addition, a patient with a selective immunodeficiency of NK cells, caused by a mutant gene mapping to chromosome 8, has been reported to have an EBV-associated lymphoproliferative disorder [174]. Epidemiological studies have indicated a protective effect for the activating KIR2DS1 and KIR3DS1 genes in patients with EBV-associated Hodgkin's lymphoma [175], although there is as yet no direct evidence for the involvement of these receptors in the recognition of EBV-transformed cells.

In the immunity to poxviruses, human NK cells lyse vaccinia-virus-infected cells and this can be partially blocked by antibodies specific for NKp30, NKp44 and NKp46 [176]. On infected cells, vaccinia virus seems to cause downregulation of HLA-E, but not of other MHC class I proteins, thereby causing the preferential lysis of these target cells by NK cells that express the inhibitory CD94/NKG2A receptor [177].

In influenza virus infection, the NK cells are activated and produce IFN-γ when cocultured with autologous DC infected with influenza A virus, and this NK cell activation is blocked in the presence of neutralizing antibodies specific for NKG2D and NKp46 [178]. The activation of NK cells by influenza A-virus-infected DC also depends on the secretion of IFN-α and IL-12 by the DC, which promotes NK cell-mediated cytotoxicity and IFN-γ production, respectively.

Epidemiological studies have shown an intriguing relationship between the highly polymorphic HLA and KIR genes and the resolution of hepatitis C virus (HCV) infection [179]. Previous studies reported that many of the inhibitory KIR proteins expressed by NK cells bind to HLA-C proteins on the surface of target cells and thereby prevent killing of these targets. HLA-C proteins have been divided into two categories, group 1 and group 2, on the basis of a dimorphic polymorphism in the α1 domain of the HLA-C heavy chain. These group 1 and group 2 HLA-C proteins are specifically recognized by different members of the KIR family [180]. Khakoo and colleagues observed that the ability to clear HCV infections is associated with individuals possessing the KIR2DL3 and HLA-C group 1 alleles [179]. Although functional studies to precisely detail the mechanism have not yet been reported, these investigators speculate that an individual expressing an inhibitory KIR with low affinity for its ligands (as is the case with KIR2DL3 and HLA-C) might mount a stronger attack against viral infection, thereby resolving the disease. That HCV might be trying to evade NK cells is also indicated by the finding that E2, the main HCV envelope protein, binds to CD81 on the surface of human NK cells and suppresses NK cell cytokine production, proliferation and cytotoxicity [181, 182].

As for HCV, KIR and HLA polymorphisms have been reported to significantly influence the progression to AIDS in HIV-infected individuals [183]. Previous studies had shown that HIV-infected individuals expressing HLA-B alleles with the Bw4 epitope have delayed onset of AIDS [184]. However, protection is significantly increased when the individual concerned also possesses certain alleles of the KIR3DL1 locus [183]. Different alleles of the KIR3DL1 gene either encode inhibitory KIR3DL1 receptors with ITIM that suppress NK cell activation when encountering target cells expressing an HLA-Bw4 ligand [185], or encode an activating receptor, designated KIR3DS1, that lacks ITIM and transmits activating signals through its association with the ITAM-bearing DAP12 adaptor [186, 187]. Possession of either an inhibitory KIR3DL1 receptor or the activating KIR3DS1 receptor, together with an HLA-Bw4 allele encoding an isoleucine residue at position 80 (HLA-Bw4-I80) significantly slows progression to AIDS [183]. HIV-infected individuals with this compound KIR3DS1 plus HLA-Bw4-I80 haplotype are protected against opportunistic infection, but not AIDS-associated malignancies [188]. The inhibitory KIR3DL1 molecules clearly recognize HLA-Bw4 proteins [185, 189], but attempts to show direct interactions between KIR3DS1 and HLA-Bw4 proteins have failed [186, 190]. Although evidence for the direct binding of KIR3DS1 to HLA-Bw4 was lacking, Alter and colleagues have reported that KIR3DS1$^+$ NK cells, but not KIR3DS1$^-$ NK cells, suppress HIV replication in HLA-Bw4-infected cells in vitro [191]. However, how the inhibitory KIR3DL1 confers protection is counterintuitive. Remarkably, the KIR allele shown to be most protective, KIR3DL1*004 [183], encodes a protein that cannot be expressed on the cell surface of NK cells and is probably degraded intracellularly [192]. Interestingly, the Nef (negative factor) protein encoded by HIV has been shown to preferentially downregulate the expression of HLA-A and HLA-B, but not HLA-C [193]. Potentially, this provides a mechanism for HIV-infected cells to avoid recognition by HLA-A and HLA-B restricted HIV-specific CTL, but to engage the inhibitory KIR expressed by NK cells in the infected individual. A recent study has shown that Nef might also downregulate expression of NKG2D ligands [194], which indicates another potential evasion mechanism.

4.2.2 NK Cells and Intracellular Mycobacteria

NK cells represent an important cellular population that has been implicated in early immune responses to a variety of intracellular pathogens and is capable of rapidly producing IFN-γ and other immunoregulatory cytokines, as well as lysing specific target cells in the absence of prior activation. In vitro studies showed that NK cells from peripheral blood contribute for protective immunity through IFN-γ or through cytotoxic mechanisms [195], these being highly bactericidal against *Mycobacterium tuberculosis*.

The role of NK cell receptors contributing to cytotoxic-mediated killing of mononuclear phagocytes infected with an intracellular bacterium has been reported [196]. Human NK cells are known to directly lyse *M. tuberculosis*-infected monocytes and macrophages in vitro [196, 197]. The recognition is thought to be mediated by NKG2D and the NCR NKp46 that bind to the stress-induced ligands ULPB1 and vimentin, respectively [196, 198]. Another recent study showed direct binding of NKp44 to the mycobacterial surface [199], suggesting that ligands for other NK cell receptors may play a role in the specific NK-mediated recognition of *M. tuberculosis*. Human NK cells not only lyse *M. tuberculosis*-infected cells but also actively restrict mycobacterial growth in an apoptosis-dependent but Fas/FasL independent manner [200, 201] and killing can be further enhanced by addition of IL-2, IL-12 and glutathione [200].

Consistent with the protective role of NK cells in tuberculosis (TB), reduced activity of NK cells has been reported in pulmonary TB patients who are ill [195]. In addition, higher levels of pre-NK cells were observed in both positive tuberculin skin test (TST$^+$) and TB patients; TST$^+$ subjects presented particularly higher levels of these cells in comparison to TB [202]. Moreover, mainly putative activated NK cells selectively increase in TST$^+$ individuals. This observation could be important in the immunopathogenic context, since these cells would contribute for protective mechanisms considering their great ability to proliferate and their potential to differentiate into CD3$^-$CD16$^+$CD56$^+$ cells with higher cytotoxic activity. It has been proposed that NK cells can be activated by cytokines (IL-12, IL-15, and IL-18) or by exposure to *M. tuberculosis*-stimulated monocytes [203]. Indeed, interplay between *M. tuberculosis*, antigen-presenting cells and NK cells triggering IFN-γ are thought to play a beneficial role in TB by helping to maintain a type 1 immune profile [203]. Barcelos and colleagues [202] demonstrated for TST$^+$ and TB patients an outstanding distinct correlation profile between NK cells and macrophage-like monocytes, suggesting that high levels of activated NK cells aside macrophage-like monocytes may be involved in protective mechanisms in putative TB-resistant individuals.

4.2.3 NK Cells and Intracellular Protozoan Infections

Destruction of protozoa by NK cells could occur either by the lysis of extracellular organisms or by the destruction of infected cells. Both types of NK cell activity have been described. Functional studies have demonstrated that $CD16^-CD56^+$ cells showed higher proliferate capacity and synthesized increased cytokine levels, whereas $CD16^+CD56^+$ cells were mainly cytotoxic [204].

In the context of *Trypanosoma cruzi* infection, the activation of the innate immune response during early stages of the disease in human, through recruitment of pre-NK cells ($CD16^+CD56^-$), could be an important bridge that might control the transient immunological events during acute and chronic infection. It has been observed an equivalent number of IFN-γ^+, TNF-α^+ and IL-4^+ NK cells were detected in infected children when compared with uninfected. These results revealed that the ex vivo innate immune compartment at the peripheral blood of the early-indeterminate chagasic patients resembled one of healthy, uninfected children. On the other hand, the infected children's cytokine pattern displays a shift towards a mixed immune profile with high levels of IFN-γ^+, TNF-α^+ and IL-4^+ $CD16^+$ cells upon in vitro stimulation with *T. cruzi* antigens. These findings may suggest that $CD16^+$ cells (NK and NKT cells) could provide protection against *T. cruzi* not only in the acute phase but also in the early-indeterminate [205, 206]. Recently, it was demonstrated, in a pioneering study, that an increased frequency of circulating NK cells ($CD3^-CD16^+CD56^{-/+}$) can be found also in the peripheral blood of all patients chronically infected with *T. cruzi* bearing distinct clinical forms of the disease. However, only indeterminate patients (IND) showed a higher percentage of $CD3^-CD16^+CD56^{+dim}$ NK cells [207]. The authors hypothesized that the higher cytotoxic activity of these NK cell subsets in IND patients could be important in helping to suppress parasitism to very low levels, resulting in avoidance of the development of a strong acquired immune response against parasite-specific antigens and the outcome of severe chagasic disease.

NK cell effector functions in leishmaniasis are cytokine-mediated rather than cytotoxicity-mediated. It has been demonstrated that the NK cells of healthier individuals, after *Leishmania donovani* and *L. aethiopica* stimulation produce high levels of IFN-γ, nevertheless, in the active visceral leishmaniasis (ACT), the antigen-specific activation occurs in the absence of IFN-γ-producing NK cells [208]. This finding might contribute to the inefficient immune response, favoring the parasite growth. Other authors observed that ACT individuals had lower frequency of circulating NK cells and that the IFN-γ expression by these cells after soluble *L. chagasi* antigen (SLA) stimulation was increased in asymptomatic individuals (AS) and cured individuals (CUR). Furthermore, AS and ACT groups had significant numbers of IL-4^+ cells. The data demonstrated a type 2, type 1/type 2 and type 1 immune response in ACT, AS and CUR individuals, respectively [209]. Individuals who carry the ACT presented a functional deficiency in NK cells [210]. It is known that the NK cells are regulated by IL-4, IL-10 and TGF-β. These cytokines inhibit the IL-12 and TNF-α synthesis by macrophages and inhibit indirectly, the IFN-γ and TNF-α synthesis by NK cells [211]. Manna and colleagues [212] showed that the plasmatic constituents of ACT carrying individuals suppress the lytic activity of NK cells, and this effect is owing to the presence of high levels of circulating immunoglobulins, rheumatoid factor and triglycerides in patients with ACT. Chakrabarti and colleagues [213] also described that the monocytes from ACT carrying individuals suppress the NK cell function, reinforcing the hypothesis that IL-10 and TGF-β participate as suppression mediators of this mechanism.

In nonimmune donors, NK cells are among the first cells in peripheral blood to produce IFN-γ in response to *Plasmodium falciparum*-infected red blood cells [214]. In this in vitro setting it was observed that NK cells are activated during the first 18 h of exposure to infected erythrocytes. This activation was dependent on IL-12, and to a lesser extent IL-18, released from accessory cells [214]. The results of a study in children with acute *P. falciparum* infection suggested a positive correlation between the lytic activity of NK cells towards the human leukemia line K562 ex vivo and the extent of parasitaemia [215], and Orago and Facer [216] have provided evidence that purified NK cells from both healthy and *P. falciparum*-infected individuals directly lyse parasitized erythrocytes in vitro. Parasitized red blood cells undergo massive structural alterations during trophozoite and schizont development [217], characterized by abnormal exposure on the cell surface of erythrocyte membrane components such as spectrin and Band 3 [218] and surface expression of parasite-encoded neo-antigens [219]. Thus, it is of interest that lysis by NK cells of erythrocytes harboring schizont-stage *P. falciparum* was found to be significantly more efficient than killing of uninfected erythrocytes, suggesting a specific recognition of the altered erythrocyte surface [216]. In support of this idea, it was demonstrated that optimal activation of NK-derived IFN-γ production in vitro requires contact between NK cells and *P. falciparum*-infected erythrocytes [214, 220]. Thus, the activation of human NK cells by blood stages of *P. falciparum* appears to depend on at least two signals, i.e., cytokines released by bystander cells such as monocyte–macrophages or DC and direct recognition of the infected red blood cell by NK cell receptors. The ability for specific recognition of malaria-infected erythrocytes could be explained by the abnormal expression of ligands for stimulatory NK cell receptors, e.g., activating KIR, NCR or TLR, or alternatively by the downregulation or complete loss of a ligand for inhibitory receptors such as Siglec-7.

4.3 The Role of NK Cells in Cancer

The major function of NK cells in fighting cancer is likely to be in surveillance and elimination of cells that become malignant before they can cause a tumor. A large body of evidence argues that enhancement of NK cell numbers and function in human cancer patients is associated with increases in tumor clearance and duration of clinical remission. How can NK cells distinguish between tumor cells and untransformed autologous cells in cancer patients?

The "missing self" hypothesis proposed by Ljunggren and Karre [221] suggests that NK cells recognize self MHC class I molecules and are thus inhibited from killing host cells. The recognition of self MHC class I molecules by KIR inhibits NK cell killing of autologous cells, even in the presence of activating signals.

On the other hand, downregulation of MHC class I molecules in transformed cells removes the inhibitory signal and killing occurs, provided that activating signals are present. In support of this concept, human NK cells kill Epstein–Barr virus-transformed B-lymphoblastoid cell lines lacking class I, whereas re-expression of class I in these lines inhibits cytolysis [222].

Alternatively, in the setting of HLA haplotype-mismatched hematopoietic stem cell transplantation (HSCT), incompatibility between KIR on donor NK cells and MHC class I on tumor cells of the recipient results in killing in the presence of an activating signal [223].

NK cells can eradicate tumors through multiple killing pathways, including direct tumor lysis and apoptosis. In addition, NK cells coordinate the innate and adaptive immune responses to tumor cells via the production of cytokines.

ADCC may enhance the activity of NK cells to specifically recognize and kill their targets. A large proportion of human NK cells express CD16, the low-affinity Fcγ receptor IIIa (FCGR3A), which binds to the constant (Fc) region of immunoglobulin. Thus, CD16 enables NK cells to recognize antibody-coated tumor cells, resulting in NK cell degranulation and perforin-dependent killing [224]. This ADCC is one proposed mechanism for the efficacy of tumor-directed antibody therapies, such as rituximab (α-CD20) and trastuzumab (α-Her2/Neu) [225].

NK cell killing of tumor cells is mediated by exocytosis of cytotoxic granules containing perforin and granzymes and also by NK surface expression of ligands that engage death receptors of the TNF superfamily on tumor cells, thereby triggering apoptosis. The role of perforin in NKG2D-dependent cytotoxicity was established. The tumoricidal effects of IL-2 and IL-12 have been shown to require NK cell NKG2D and perforin [226]. In contrast, the antimetastatic effect of IL-18 relies principally on the expression of FasL (CD95L) by NK cells and is independent of NKG2D [226]. Predictably, the combination of IL-12 and IL-18 synergistically activates perforin-dependent and FasL-dependent killing and produces potent metastasis suppression [226]. NK expression of a related TNF family member, TNF-α–related apoptosis-inducing ligand (TRAIL), is implicated in the suppression of liver metastasis by resident NK cells [227].

Considering the observations described above, the eradication of tumor cells by NK cells is achieved directly by cytolytic activity or indirectly through the production of cytokines such as IFN-γ and TNF-α. TNF-α has been implicated in NK eradication of peritoneal tumors [228]. IFN-γ is the principal NK-derived cytokine implicated in tumor surveillance and the suppression of growth and metastasis by established tumors. Administration of IL-12 and low-dose IL-2 has been proved to increase IFN-γ production in patients with melanoma and renal cell carcinoma [229]. Finally, the combination of IL-12 and trastuzumab results in increased IFN-γ levels in patients with HER2$^+$ breast cancer [230].

On the other hand, TGF-β has been associated with failure of NK cells in killing tumor cells. Regulatory T (Treg) cells, defined in humans as CD4$^+$CD25^{high}-FOXP3$^+$, are enriched in tumors [231], and have been shown to exert a profound inhibitory effect on NK cell function, in part via their membrane-bound expression of TGF-β.

5 NK Cells in Mice

Currently, it is known that there is a vast heterogeneity of NK cells in mice determined by morphology, antigenic phenotype and functional activity [232]. In a brief history, the murine diversity of NK cells was first inferred by Roeder and colleagues in 1978 in studies of the mechanism of killing in vitro comparing cold-target inhibition of lysis of tumor cells and a variety of other susceptible cells [233]. Afterward, Stutman and colleagues demonstrated a capacity of cytolysis inhibition of susceptible target cells mediated by NK cells from mice using simple sugars [234] and a few years later Gorczynski and his research group described a heterogeneity of NK effector functions by panel analysis with a variety of simple sugars in spleen fragment cultures which was a characteristic of the strain of mice under investigation [235]. At this time, the mechanism of inhibition of NK lysis by sugars was not elucidated. In another work, these authors also demonstrated that the repertoire of NK cells is influenced by mouse development and that there is considerably less diversity in neonates than in adults [232].

The first nomenclature of murine NK cells was based in the identification of cell surface antigens selectively expressed in the plasmatic membrane using alloantisera. Based on the chronology of their discovery, these markers were designated NK1.1, NK2.1, NK3.1 and NK 4.1 [236], and are expressed in different strains of mice, with no distinct function when analyzed in the same organ. Two major inhibitory receptors, CD94/NKG2A and Ly49, are expressed in mouse NK cells. The acquisition of the CD94/NKG2A receptor seems to be an early event, whereas Ly49 receptor expression is considered a relatively late event during NK cell ontogeny [237].

Other membrane molecules have an important role on NK cell functions by mediating adhesion in susceptible tumor targets. During cell-to-cell contact, supramolecular activating clusters (SMAC) or supramolecular inhibitory clusters (SMIC) accumulate at the contacting plasma membrane between NK cells and target cells [238]. The Ly5 expression intensity, a surface glycoprotein expressed by NK cells, has its production induced by IFN-α and -β and is highly related with cytolytic mechanisms against certain targets [239].

5.1 Phenotypic Properties of Mouse NK Cell Subsets

In humans, CD16 and CD56 are surface markers to identify subpopulations of NK cells. However, CD56 is not expressed in rodents and subsets have proven difficult to identify [240]. Recently, Hayakawa and Smyth [241] identified CD27 as a molecule that distinguishes subsets of mature mouse NK cells. These authors showed that the mature Mac-1^{high} NK cell pool can be divided into two functionally distinct CD27^{high} and CD27^{low} subsets. These subsets have distinct effector functions, proliferative capacity, tissue organization and response to chemokines related with different intensity of inhibitory receptors expressed on the cell surface. Comparatively, the Mac-1^{high}CD27^{low} NK cell population contained a greater proportion of Ly49 when compared with Mac-1^{high}CD27^{high} subset. The latter demonstrated greater effective cytotoxicity against tumor target cells even in the presence of MCH class I expression, besides greater responsiveness to DC and produced higher amounts of IFN-γ upon in vitro stimulation. This research group has also verified that CD27^{high} and CD27^{low} subsets with distinct cell surface phenotypes also exist in human peripheral blood. The majority of peripheral blood human NK cells were CD27^{low}CD56^{dim} NK cells, whereas the minor CD27^{high} NK cell population correspondingly displayed a CD56^{bright} phenotype. Distinctions between CD27^{low} and CD27^{high} NK cells in their receptor expression and typical NK cell functions such as cytotoxicity and cytokine production can be easily delineated [240].

5.2 NK Cell Receptors in Mouse

In contrast with B and T lymphocytes, NK cells do not possess one dominant receptor but vast combinatory receptors to initiate effector functions. A balance between negative and positive signals mediated by a repertoire of inhibitory receptors specific for MHC class I and stimulatory receptors, respectively, regulate the activation of NK cells [75].

Inhibitory receptors bind classical and nonclassical MHC class I molecules and may inhibit NK cell killing of MHC class I-bearing target cells. These receptors are crucial for maintaining tolerance to self. Cells that reduce the expression of MHC class I in plasma membranes are targets for NK cells. In mouse, these inhibitory receptors are represented by Ly49 and CD94/NKG2 molecules, which use a common strategy to inhibit NK cell activation [242]. The inhibitory receptors mediate their effects through one or more ITIM motifs, which are located in the cytoplasmic domain of the receptor. When ITIM-bearing receptors engage their ligands, the tyrosine residues are phosphorylated resulting in the recruitment of phosphatases to the interface between the NK cell and its potential target cell suppressing NK cell responses by dephosphorylating the protein substrates of the tyrosine kinases linked to activating NK receptors [75, 242].

The discovery of MHC class I-specific inhibitory receptors clarified the molecular basis of this important NK cell function. However, the triggering receptors responsible for positive NK cell stimulation remained elusive until recently. CD16 is able to elicit cytolytic activity and cytokine secretion. However, association of CD16 with other receptors induces additive effects on NK cell activation. These coactivating receptors are constitutively expressed in all mature NK cells and are represented by FcεRI-γ, CD-3ζ, and DAP12 type I-transmembrane anchored proteins, which contain ITAM motifs. Engagement of these receptors causes the phosphorylation of the ITAM tyrosines, which bind the Syk tyrosine kinases. The activation induces cytoskeleton reorganization, which is required for cell polarization and release of cytolytic granules containing perforin and granzymes, and results in the transcription of many cytokine and chemokine genes [75].

5.3 Development of Mouse NK Cell Subsets

NK cells are generated by the lymphoid branch during hematopoiesis, a tightly regulated process involved in the production of blood cells. There appear to be two layers of regulation: the first layer of control has been referred to as basal hematopoiesis, regulated by cytokines produced within the microenvironment of bone marrow. The second regulatory layer is called amplified hematopoiesis, regulated by endocrine system caused by physiological stress [243]. The multiple stages of NK cell development have been proposed on the basis of their function, phenotype and proliferative capacity [241]. Progenitor cells in the fetal blood, with a phenotype of $NK1.1^+CD117^+$ (c-kit) can give rise to NK cells. The expression of the β subunit of IL-2 and IL-15 receptor occurs early in their development. IL-15 is required for NK cell differentiation and, in addition, stimulates expression of at least one of the inhibitory receptor families expressed by NK cells [244]. This phase is highly dependent of lymphotoxin α (LTα), a cytokine belonging to the TNF ligand superfamily. LTα can be secreted or can form a membrane-bound $LT\alpha_1\beta_2$ heterotrimer that signals through the LTβ receptor (LTβR) [245] Mice LTα knock-out (LTα–/–) and LTβR–/– mice present reduced percentages of NK cells in spleen, bone marrow and blood. It is possible that a contact of membrane-bound $LT\alpha_1\beta_2$-expressing hematopoietic progenitor cells with bone marrow stromal cells activates the latter, which in turn induce IL-15 receptor expression on NK precursor cells. This cytokine would be sufficient for the IL-15-responsive precursors to differentiate to immature NK cells [246]. The acquisition of NK1.1 and CD94/NKG2 receptors are sequential events, and following this step, NK cells then express Ly49 molecules [241]. It is possible that NK cells in mice can mature in other organs as shown Veinotte and colleagues [247] who identified two populations of NK progenitors, one in the thymus and the other in the LN.

5.4 Tissue Distribution and Functional Diversity of Mouse NK Cell Subsets

Mature NK cells which express Mac-1^+ are predominantly found in peripheral organs such as the spleen and continually increase in number from birth until they reach a constant level at adulthood due to homeostatic expansion [1]. When compared with Mac-1^{high}CD27^{low}, the subset Mac-1^{high}CD27^{high} are preferential distributed in lymphoid organs and have a predominant role in cross talk with other immune cells, particularly DC [241].

In recent years, a number of studies have highlighted the novel concept that the actual role of NK cells is not only confined to the destruction of virus-infected cells or tumors. Although NK cells reside in naive LN at a very low frequency, they can be recruited into LN draining sites of infection, inflammation, or immunization where they potentially influence adaptive immunity [248]. Interaction of NK cells with myeloid DC during the early phases of inflammation, appear to play a crucial role in shaping both innate immune reactions (within inflamed peripheral tissues) and adaptive immune responses (in secondary lymphoid compartments) [249]. NK cells have also been shown to affect adaptive immune responses by their production of both pro- and antiinflammatory cytokines. It has been proposed that NK cells are the main source of IFN-γ that supports either Th1 or CTL priming [248].

5.5 *Different Functions of the Murine and Human Receptors on NK Cells*

2B4 (CD244) was initially discovered on murine NK cells and T cells displaying non-MHC dependent cytotoxicity. Human 2B4 was cloned based on sequence homology with mouse 2B4. Recent evidence suggests that the function of this receptor might be different in the two species [250]. In mature mouse NK cells, 2B4 reportedly has the capacity to either stimulate or inhibit NK cell activation [247]. In vitro and in vivo studies using 2B4-deficient mice suggest that the major function of mouse 2B4 is to inhibit murine NK cell functions when triggered by CD48, a high affinity ligand, on target cells, although there are reports of activating function of murine 2B4 [250]. Engagement of 2B4 on NK cell surfaces with specific antibodies or CD48 can trigger cell-mediated cytotoxicity, IFN-γ secretion, phosphoinositol turnover and NK cell invasiveness [251]. The inhibitory function of murine 2B4 is mediated by EAT-2, ERT and possibly other phosphatases like SHP-1 and SHIP. 2B4-SAP interaction in mouse NK cells might be a low affinity one and might not be physiologically relevant considering the inhibitory function of 2B4 [250]. As an inhibitory receptor, 2B4 is unconventional as it is not regulated by MHC class I molecules.

6 Future Challenges

NK cells represent a lymphoid cell population in mammals critical in innate immunity and crucial for surveillance against pathogens and tumors. Although the basic mechanisms through which NK cells work have been well established, many questions are still unresolved. Are all NK cells equal or can we identify subsets with distinct developmental origin and function? How do NK cells interact with other components of the immune system, such as DC, regulatory/effector T cells to elicit effective immune responses? How do NK cells become tolerant to self and preclude autoimmunity? Can we design novel therapeutic avenues for cancer treatment and transplant procedure based on NK cells? Answers for these issues represent future perspectives in research involving this cell-type. Although we have gained valuable insights about the processes orchestrating NK cell diversity, we are still faced with many serious challenges in understanding how NK cell functional diversity integrates into innate and adaptive immune responses as well as noninfectious inflammation and tissue remodeling and as tissue environments affect the development, homeostasis and biological roles of distinct NK cell subsets [252–259].

Acknowledgments The authors VPCPT, ATC and OAMF are grateful for Conselho Nacional de Pesquisa e Desenvolvimento Tecnológico-CNPq for financial supporting (PQ) fellowships. The author RSA is grateful for FAPEMIG for financial supporting postdoctoral fellowship.

References

1. Huntington ND, Vosshenrich CA, Di Santo JP (2007) Nat Rev Immunol 7:703
2. Biassoni R (2008) Adv Exp Med Biol 640:35
3. Orange SJ, Ballas KZ (2006) Clin Immunol 118:1
4. Timonen T, Ortaldo JR, Herberman RB (1981) J Exp Med 153:569
5. Stein-Streilein J (1983) J Immunol 131:1748
6. Hokland M, Kuppen PJ (2005) Mol Immunol 42:381
7. Loza MJ, Zamai L, Azzoni L, Rosati E, Perussia B (2002) Blood 99:1273
8. Dorner BG, Smith HR, French AR, Kim S, Poursine-Laurent J, Beckman DL, Pingel JT, Kroczek RA, Yokoyama WM (2004) J Immunol 172:3119
9. Robertson MJ (2002) J Leukoc Biol 7:173
10. Zingoni A, Sornasse T, Cocks BG, Tanaka Y, Santoni A, Lanier LL (2004) J Immunol 173:3716
11. Blanca IR, Bere EW, Young HA, Ortaldo JR (2001) J Immunol 167:6132
12. Raulet DH, Vance RE (2006) Nat Rev Immunol 6:520
13. Blum KS, Pabst R (2007) Immunol Lett 108:45
14. Yokoyama WM, Kim S, French AR (2004) Annu Rev Immunol 22:405
15. Galy A, Travis M, Cen D, Chen B (1995) Immunity 3:459
16. Miller JS, Alley KA, McGlave P (1994) Blood 1:2594
17. Caligiuri MA (2008) Blood 112:461
18. Freud AG, Caligiuri MA (2006) Immunol Rev 214:56
19. Colucci F, Caligiuri MA, Di Santo JP (2003) Nat Rev Immunol 3:413
20. Srour EF, Brandt JE, Briddell RA, Leemhuis T, van Besien K, Hoffman R (1991) Blood Cells 17:287
21. Miller JS, Verfaillie C, McGlave P (1992) Blood 80:2182
22. Seaman WE, Gindhart TD, Greenspan JS, Blackman MA, Talal N (1979) J Immunol 122:2541
23. Roth C, Rothlin C, Riou S, Raulet DH, Lemke G (2007) J Mol Med 85:1047
24. Yu H, Fehniger TA, Fuchshuber P, Thiel KS, Vivier E, Carson WE, Caligiuri MA (1998) Blood 92:3647
25. Shibuya A, Nagayoshi K, Nakamura K, Nakauchi H (1995) Blood 85:3538
26. Smith KA (1988) Science 240:1169
27. Mrózek E, Anderson P, Caligiuri MA (1996) Blood 87:2632
28. Jaleco AC, Blom B, Res P, Weijer K, Lanier LL, Phillips JH, Spits H (1997) J Immunol 159:694

29. Sánchez MJ, Muench MO, Roncarolo MG, Lanier LL, Phillips JH (1994) J Exp Med 180:569
30. Waldmann T, Tagaya Y, Bamford R (1998) Int Rev Immunol 16:205
31. Fehniger TA, Caligiuri MA (2001) Blood 97:14
32. Park SY, Saijo K, Takahashi T, Osawa M, Arase H, Hirayama N, Miyake K, Nakauchi H, Shirasawa T, Saito T (1995) Immunity 3:771
33. Moriggl R, Topham DJ, Teglund S, Sexl V, McKay C, Wang D, Hoffmeyer A, van Deursen J, Sangster MY, Bunting KD, Grosveld GC, Ihle JN (1999) Immunity 10:249
34. Imada K, Bloom ET, Nakajima H, Horvath-Arcidiacono JA, Udy GB, Davey HW, Leonard WJ (1998) J Exp Med 188:2067
35. Freud AG, Becknell B, Roychowdhury S, Mao HC, Ferketich AK, Nuovo GJ, Hughes TL, Marburger TB, Sung J, Baiocchi RA, Guimond M, Caligiuri MA (2005) Immunity 22:295
36. Shibuya A, Kojima H, Shibuya K, Nagayoshi K, Nagasawa T, Nakauchi H (1993) Blood 81:1819
37. Canque B, Camus S, Dalloul A, Kahn E, Yagello M, Dezutter-Dambuyant C, Schmitt D, Schmitt C, Gluckman JC (2000) Blood 96:3748
38. Lyman SD, Jacobsen SE (1998) Blood 91:1101
39. Fehniger TA, Cooper MA, Nuovo GJ, Cella M, Facchetti F, Colonna M, Caligiuri MA (2003) Blood 101:3052
40. Cooper M, Fehniger TA, Caligiuri MA (2001) Trends Immunol. 22:633
41. Carson WE, Haldar S, Baiocchi RA, Croce CM, Caligiuri MA (1994) Proc Natl Acad Sci USA 91:7553
42. Matos ME, Schnier GS, Beecher MS, Ashman LK, William DE, Caligiuri MA (1993) J Exp Med 178:1079
43. André P, Spertini O, Guia S, Rihet P, Dignat-George F, Brailly H, Sampol J, Anderson PJ, Vivier E (2000) Proc Natl Acad Sci USA 97:3400
44. Voss SD, Daley J, Ritz J, Robertson MJ (1998) J Immunol 160:1618
45. Romagnani C, Juelke K, Falco M, Morandi B, D'Agostino A, Costa R, Ratto G, Forte G, Carrega P, Lui G, Conte R, Strowig T, Moretta A, Münz C, Thiel A, Moretta L, Ferlazzo G (2007) J Immunol 178:4947
46. Ferlazzo G, Thomas D, Lin SL, Goodman K, Morandi B, Muller WA, Moretta A, Münz C (2004) J Immunol 172:1455
47. Freud AG, Yokohama A, Becknell B, Lee MT, Mao HC, Ferketich AK, Caligiuri MA (2006) J Exp Med 203:1033
48. Loza MJ, Perussia B (2001) Nat Immunol 2:917
49. Zamai L, Ahmad M, Bennett IM, Azzoni L, Alnemri ES, Perussia B (1998) J Exp Med 188:2375
50. Grzywacz B, Kataria N, Sikora M, Oostendorp RA, Dzierzak EA, Blazar BR, Miller JS, Verneris MR (2006) Blood 108:3824
51. Lanier LL, Le AM, Civin CI, Loken MR, Phillips JH (1986) J Immunol 136:4480
52. Vitale C, Chiossone L, Morreale G, Lanino E, Cottalasso F, Moretti S, Dini G, Moretta L, Mingari MC (2004) Eur J Immunol 34:455
53. Sivori S, Cantoni C, Parolini S, Marcenaro E, Conte R, Moretta L, Moretta A (2003) Eur J Immunol 33:3439
54. Miller JS, McCullar V (2001) Blood 98:705
55. Bottino C, Moretta L, Pende D, Vitale M, Moretta A (2004) Mol Immunol 41:569
56. Newman KC, Riley EM (2007) Nat Rev Immunol 7:279
57. Ritz J, Schmidt RE, Michon J, Hercend T, Schlossman SF (1988) Adv Immunol 42:181
58. Moretta A, Bottino C, Vitale M, Pende D, Cantoni C, Mingari MC, Biassoni R, Moretta L (2001) Annu Rev Immunol 19:197
59. McQueen KL, Parham P (2002) Curr Opin Immunol 14:615
60. Robertson MJ, Caligiuri MA, Manley TJ, Levine H, Ritz J (1990) J Immunol 145:3194
61. Yang J, Ye Y, Carroll A, Yang W, Lee H (2001) Curr Protein Pept Sci 2:1
62. Hogg N, Laschinger M, Giles K, McDowall A (2003) J Cell Sci 116:4695

63. van Kooyk Y, Figdor CG (2000) Curr Opin Cell Biol 12:542
64. Carman CV, Springer TA (2003) Curr Opin Cell Biol 15:547
65. Bunting M, Harris ES, McIntyre TM, Prescott SM, Zimmerman GA (2002) Curr Opin Hematol 9.30
66. Perez OD, Mitchell D, Jager GC, Nolan GP (2004) Blood 104:1083
67. Barber DF, Faure M, Long EO (2004) J Immunol 173:3653
68. Warren HS, Skipsey LJ (1991) Immunology 72:150
69. Macintyre EA, Wallace DW, O'flynn K, Abdul-Gaffar R, Tetteroo PAT, Morgan G, Linch DC (1989) Immunology 66:459
70. Anégon I, Cuturi MC, Trincheri G, Perussia B (1988) J Exp Med 167:452
71. Lanier L, Ruitenberg JJ, Phillips JH (1988) J Immunol 141:3478
72. Bryceson YT, March ME, Barber DF, Ljunggren HG, Long EO (2005) J Exp Med 202:1001
73. Bryceson YT, March ME, Ljunggren HG, Long EO (2006) Blood 107:159
74. Raulet DH, Vance RE, McMahon CW (2001) Annu Rev Immunol 19:291
75. Lanier L (2008) Nat Immunol 9:495
76. Pende D, Cantoni C, Rivera P, Vitale M, Castriconi R, Marcenaro S, Nanni M, Biassoni R, Bottino C, Moretta A, Moretta L (2001) Eur J Immunol 31:1076
77. Pende D, Parolini S, Pessino A, Sivori S, Augugliaro R, Morelli L, Marcenaro E, Accame L, Malaspina A, Biassoni R, Bottino C, Moretta L, Moretta A (1999) J Exp Med 190:1505
78. Colucci F, Schweighoffer E, Tomasello E, Turner M, Ortaldo JR, Vivier E, Tybulewicz VL, Di Santo JP (2002) Nat Immunol 3:288
79. Houchins JP, Yabe T, McSherry C, Bach FH (1991) J Exp Med 1:1017
80. Bauer S, Groh V, Wu J, Steinle A, Phillips JH, Lanier LL, Spies T (1999) Science 285:727
81. Wu J, Song Y, Bakker AB, Bauer S, Spies T, Lanier LL, Phillips JH (1999) Science 285:730
82. Diefenbach A, Tomasello E, Lucas M, Jamieson AM, Hsia JK, Vivier E, Raulet DH (2002) Nat Immunol 3:1142
83. Gilfillan S, Ho EL, Cella M, Yokoyama WM, Colonna M (2002) Nat Immunol 3:1150
84. Rosen DB, Araki M, Hamerman JA, Chen T, Yamamura T, Lanier LL (2004) J Immunol 173:2470
85. Coudert JD, Zimmer J, Tomasello E, Cebecauer M, Colonna M, Vivier E, Held W (2005) Blood 106:1711
86. Lodoen MB, Lanier LL (2005) Nat Rev Microbiol 3:59
87. Sutherland CL, Chalupny NJ, Schooley K, VandenBos T, Kubin M, Cosman D (2002) J Immunol 168:671
88. Billadeau DD, Upshaw JL, Schoon RA, Dick CJ, Leibson PJ (2003) Nat Immunol 4:557
89. Pende D, Rivera P, Marcenaro S, Chang CC, Biassoni R, Conte R, Kubin M, Cosman D, Ferrone S, Moretta L, Moretta A (2002) Cancer Res 62:6178
90. Dunn C, Chalupny NJ, Sutherland CL, Dosch S, Sivakumar PV, Johnson DC, Cosman D (2003) J Exp Med 197:1427
91. Cosman D, Müllberg J, Sutherland CL, Chin W, Armitage R, Fanslow W, Kubin M, Chalupny NJ (2001) Immunity 14:123
92. Chiesa S, Mingueneau M, Fuseri N, Malissen B, Raulet DH, Malissen M, Vivier E, Tomasello E (2006) Blood 15:2364
93. Horng T, Bezbradica JS, Medzhitov R (2007) Nat Immunol 8:1289
94. Bryceson YT, March ME, Ljunggren HG, Long EO (2006) Immunol Rev 214:73
95. Meresse B, Chen Z, Ciszewski C, Tretiakova M, Bhagat G, Krausz TN, Raulet DH, Lanier LL, Groh V, Spies T, Ebert EC, Green PH, Jabri B (2004) Immunity 21:357
96. André P, Castriconi R, Espéli M, Anfossi N, Juarez T, Hue S, Conway H, Romagné F, Dondero A, Nanni M, Caillat-Zucman S, Raulet DH, Bottino C, Vivier E, Moretta A, Paul P (2004) Eur J Immunol 34:961
97. Nakajima H, Cella M, Langen H, Friedlein A, Colonna M (1999) Eur J Immunol 29:1676
98. Korínek V, Stefanová I, Angelisová P, Hilgert I, Horejsí V (1991) Immunogenetics 33:108
99. Latchman Y, McKay PF, Reiser H (1998) J Immunol 161:5809

100. Sivori S, Parolini S, Falco M, Marcenaro E, Biassoni R, Bottino C, Moretta L, Moretta A (2000) Eur J Immunol 30:787
101. Chen R, Relouzat F, Roncagalli R, Aoukaty A, Tan R, Latour S, Veillette A (2004) Mol Cell Biol 24:5144
102. Eissmann P, Beauchamp L, Wooters J, Tilton JC, Long EO, Watzl C (2005) Blood 105:4722
103. Tangye SG, Phillips JH, Lanier LL, Nichols KE (2000) J Immunol 165:2932
104. Bottino C, Castriconi R, Moretta L, Moretta A (2005) Immunology 26:221
105. Sivori S, Pende D, Bottino C, Marcenaro E, Pessino A, Biassoniv R, Moretta L, Moretta A (1999) Eur J Immunol 29:1656
106. Cantoni C, Bottino C, Vitale M, Pessino A, Augugliaro R, Malaspina A, Parolini S, Moretta L, Moretta A, Biassoni R (1999) J Exp Med 189:787
107. Arnon TI, Lev M, Katz G, Chernobrov Y, Porgador A, Mandelboim O (2001) Eur J Immunol 31:2680
108. Mandelboim O, Lieberman N, Lev M, Paul L, Arnon TI, Bushkin Y, Davis DM, Strominger JL, Yewdell JW, Porgador A (2001) Nature 409:1055
109. Arnon TI, Achdout H, Levi O, Markel G, Saleh N, Katz G, Gazit R, Gonen-Gross T, Hanna J, Nahari E, Porgador A, Honigman A, Plachter B, Mevorach D, Wolf DG, Mandelboim O (2005) Nat Immunol 6:515
110. O'Connor GM, Hart OM, Gardiner CM (2005) Immunology 117:1
111. Takeda K, Akira S (2005) Int Immunol 17:1
112. Hart OM, Athie-Morales V, O'Connor GM, Gardiner CM (2005) J Immunol 175:1636
113. Sivori S, Falco M, Della Chiesa M, Carlomagno S, Vitale M, Moretta L, Moretta A (2004) Proc Natl Acad Sci USA 101:10116
114. Alexopoulou L, Holt AC, Medzhitov R, Flavell RA (2001) Nature 413:732
115. Bauer S, Kirschning CJ, Häcker H, Redecke V, Hausmann S, Akira S, Wagner H, Lipford GB (2001) Proc Natl Acad Sci USA 98:9237
116. Becker I, Salaiza N, Aguirre M, Delgado J, Carrillo-Carrasco N, Kobeh LG, Ruiz A, Cervantes R, Torres AP, Cabrera N, González A, Maldonado C, Isibasi A (2003) Mol Biochem Parasitol 130:65
117. Pisegna S, Pirozzi G, Piccoli M, Frati L, Santoni A, Palmieri G (2004) Blood 104:4157
118. Schmidt KN, Leung B, Kwong M, Zarember KA, Satyal S, Navas TA, Wang F, Godowski PJ (2004) J Immunol 172:138
119. Biron CA (1997) Curr Opin Immunol 9:24
120. Gidlund M, Orn A, Wigzell H, Senik A, Gresser I (1978) Nature 273:759
121. Krug A, French AR, Barchet W, Fischer JA, Dzionek A, Pingel JT, Orihuela MM, Akira S, Yokoyama WM, Colonna M (2004) Immunity 21:107
122. Tabeta K, Georgel P, Janssen E, Du X, Hoebe K, Crozat K, Mudd S, Shamel L, Sovath S, Goode J, Alexopoulou L, Flavell RA, Beutler B (2004) Proc Natl Acad Sci USA 101:3516
123. Bryceson YT, Long EO (2008) Curr Opin Immunol 20:344
124. Long EO (2008) Immunol Rev 224:70
125. Burshtyn DN, Scharenberg AM, Wagtmann N, Rajagopalan S, Berrada K, Yi T, Kinet JP, Long EO (1996) Immunity 4:77
126. Burshtyn DN, Lam AS, Weston M, Gupta N, Warmerdam PA, Long EO (1999) J Immunol 162:897
127. Stebbins CC, Watzl C, Billadeau DD, Leibson PJ, Burshtyn DN, Long EO (2003) Mol Cell Biol 23:6291
128. Natarajan K, Dimasi N, Wang J, Mariuzza RA, Margulies DH (2002) Annu Rev Immunol 20:853
129. Endt J, McCann FE, Almeida CR, Urlaub D, Leung R, Pende D, Davis DM, Watzl C (2007) J Immunol 178:5606
130. Watzl C, Long EO (2003) J Exp Med 197:77
131. Vales-Gomez M, Reyburn HT, Erskine RA, Lopez-Botet M, Strominger JL (1999) EMBO J 18:4250

132. Brostjan C, Bellon T, Sobanov Y, Lopez-Botet M, Hofer E (2002) J Immunol Methods 264:109
133. Kaiser BK, Barahmand-Pour F, Paulsene W, Medley S, Geraghty DE, Strong RK (2005) J Immunol 174:2878
134. Braud VM, Allan DS, O'Callaghan CA, Soderstrom K, D'Andrea A, Ogg GS, Lazetic S, Young NT, Bell JI, Phillips JH, Lanier LL, McMichael AJ (1998) Nature 391:795
135. Petrie EJ, Clements CS, Lin J, Sullivan LC, Johnson D, Huyton T, Heroux A, Hoare HL, Beddoe T, Reid HH, Wilce MC, Brooks AG, Rossjohn J (2008) J Exp Med 205:725
136. Lu L, Ikizawa K, Hu D, Werneck MB, Wucherpfennig KW, Cantor H (2007) Immunity 26:593
137. Byrne P, McGuirk P, Todryk S, Mills KH (2004) Eur J Immunol 34:2579
138. Fehniger TA, Shah MH, Turner MJ, VanDeusen JB, Whitman SP, Cooper MA, Suzuki K, Wechser M, Goodsaid F, Caligiuri MA (1999) J Immunol 15(162):4511
139. Inoue H, Miyaji M, Kosugi A, Nagafuku M, Okazaki T, Mimori T, Amakawa R, Fukuhara S, Domae N, Bloom ET (2002) Eur J Immunol 32:2188
140. Orange JS, Harris KE, Andzelm MM, Valter MM, Geha RS, Strominger JL (2003) Proc Natl Acad Sci USA 100:14151
141. Davis DM, Chiu I, Fassett M, Cohen GB, Mandelboim O, Strominger JL (1999) Proc Natl Acad Sci USA 96:15062
142. Vyas YM, Mehta KM, Morgan M, Maniar H, Butros L, Jung S, Burkhardt JK, Dupont B (2001) J Immunol 15:4358
143. Riteau B, Barber DF, Long EOJ (2003) Exp Med 198:469
144. Gismondi A, Bisogno L, Mainiero F, Palmieri G, Piccoli M, Frati L, Santoni A (1997) J Immunol 159:4729
145. Tassi I, Klesney-Tait J, Colonna M (2006) Immunol Rev 214:92
146. Fischer A, Lisowska-Grospierre B, Anderson DC, Springer TA (1988) Immunodefic Rev 1:39
147. Fuchs A, Colonna M (2006) Semin Cancer Biol 16:359
148. Smyth MJ, Hayakawa Y, Takeda K, Yagita H (2002) Nat Rev Cancer 2:850
149. Miedema F, Tetteroo PA, Hesselink WG, Werner G, Spits H, Melief CJ (1984) Eur J Immunol 14:518
150. Koopman LA, Kopcow HD, Rybalov B, Boyson JE, Orange JS, Schatz F, Masch R, Lockwood CJ, Schachter AD, Park PJ, Strominger JL (2003) J Exp Med 98:1201
151. Faust Z, Laskarin G, Rukavina D, Szekeres-Bartho J (1999) Am J Reprod Immunol 42:71
152. Parham P (2004) J Exp Med 200:951
153. Agrawal S, Pandey MK (2003) J Hematother Stem Cell Res 12:749
154. Loke YW, Hiby S, King A (1999) J Reprod Immunol 43:235
155. Rajagopalan S, Long EO (1999) J Exp Med 189:1093
156. Rajagopalan S, Fu J, Long EO (2001) J Immunol 167:1877
157. Kikuchi-Maki A, Catina TL, Campbell KS (2005) J Immunol 174:3859
158. Hiby SE, Walker JJ, O'Shaughnessy M, Redman CW, Carrington M, Trowsdale J, Moffett A (2004) J Exp Med 200:957
159. Varla-Leftherioti M, Spyropoulou-Vlachou M, Niokou D, Keramitsoglou T, Darlamitsou A, Tsekoura C, Papadimitropoulos M, Lepage V, Balafoutas C, Stavropoulos-Giokas C (2003) Am J Reprod Immunol 49:183
160. Salazar-Mather TP, Hamilton TA, Biron CA (2000) J Clin Invest 105:985
161. Moretta A, Marcenaro E, Parolini S, Ferlazzo G, Moretta L (2008) Cell Death Differ 15:226
162. Guma M, Angulo A, Lopez-Botet M (2006) Curr Top Microbiol Immunol 298:207
163. Yamamoto T, Takahashi Y, Kase N, Mori H (1999) Am J Reprod Immunol 41:337
164. Welsh RM, Vargas-Cortes M (1992) The natural killer cell: the natural immune system. C.E. Lewis and J.O. McGee, Oxford, p 107
165. Zhang X, Sun S, Hwang I, Tough DF, Sprent J (1998) Immunity 8:591
166. Biron CA, Byron KS, Sullivan JL (1998) N Eng J Med 320:1731

167. Guma M, Budt M, Sáez A, Brckalo T, Hengel H, Angulo A, López-Botet M (2005) Blood 107:3624
168. Llano M, Lee N, Navarro F, García P, Albar JP, Geraghty DE, López-Botet M (1998) Eur J Immunol 28:2854
169. Tomasec P, Braud VM, Rickards C, Powell MB, McSharry BP, Gadola S, Cerundolo V, Borysiewicz LK, McMichael AJ, Wilkinson GW (2000) Science 287:1031
170. Ulbrecht M, Martinozzi S, Grzeschik M, Hengel H, Ellwart JW, Pla M, Weiss EH (2000) J Immunol 164:5019
171. Wang EC, McSharry B, Retiere C, Tomasec P, Williams S, Borysiewicz LK, Braud VM, Wilkinson GW (2002) Proc Natl Acad Sci USA 99:7570
172. Chalupny NJ, Rein-Weston A, Dosch S, Cosman D (2006) Biochem Biophys Res Commun 346:175
173. Tomasec P, Wang ECY, Davison AJ, Vojetesek B, Armstrong M, Griffin C, McSharry BP, Morris RJ, Llewellyn-Lacey S, Rickards C, Nomoto A, Sinzger C, Wilkinson GWG (2005) Nat Immunol 6:181
174. Pappworth IY, Wang EC, Rowe M (2007) J Virol 81:474
175. Eidenschenk C, Dunne J, Jouanguy E, Fourlinnie C, Gineau L, Bacq D, McMahon C, Smith O, Casanova JL, Abel L, Feighery C (2006) Am J Hum Genet 78:721
176. Besson C, Roetynck S, Williams F, Orsi L, Amiel C, Lependeven C, Antoni G, Hermine O, Brice P, Ferme C, Carde P, Canioni D, Brière J, Raphael M, Nicolas JC, Clavel J, Middleton D, Vivier E, Abel L (2007) PLoS ONE 2:406
177. Chisholm SE, Reyburn HT (2006) J Virol 80:2225
178. Brooks CR, Elliott T, Parham P, Khakoo SI (2006) J Immunol 176:1141
179. Draghi M, Pashine A, Sanjanwala B, Gendzekhadze K, Cantoni C, Cosman D, Moretta A, Valiante NM, Parham P (2007) J Immunol 178:2688
180. Khakoo SI, Thio CL, Martin MP, Brooks CR, Gao X, Astemborski J, Cheng J, Goedert JJ, Vlahov D, Hilgartner M, Cox S, Little AM, Alexander GJ, Cramp ME, O'Brien SJ, Rosenberg WM, Thomas DL, Carrington M (2004) Science 305:872
181. Vilches C, Parham P (2002) Annu Rev Immunol 20:217
182. Crotta S, Stilla A, Wack A, D'Andrea A, Nuti S, D'Oro U, Mosca M, Filliponi F, Brunetto RM, Bonino F, Abrignani S, Valiante NM (2002) J Exp Med 195:35
183. Tseng CT, Klimpel GR (2002) J Exp Med 195:43
184. Martin MP, Qi Y, Gao X, Yamada E, Martin JN, Pereyra F, Colombo S, Brown EE, Shupert WL, Phair J, Goedert JJ, Buchbinder S, Kirk GD, Telenti A, Connors M, O'Brien SJ, Walker BD, Parham P, Deeks SG, McVicar DW, Carrington M (2007) Nat Genet 39:733
185. Flores-Villanueva PO, Yunis EJ, Delgado JC, Vittinghoff E, Buchbinder S, Leung JY, Uglialoro AM, Clavijo OP, Rosenberg ES, Kalams SA, Braun JD, Boswell SL, Walker BD, Goldfeld AE (2001) Proc Natl Acad Sci USA 98:5140
186. Litwin V, Gumperz J, Parham P, Phillips JH, Lanier LL (1994) J Exp Med 180:537
187. Carr WH, Rosen DB, Arase H, Nixon DF, Michaelsson J, Lanier LL (2007) J Immunol 178:647
188. Trundley A, Frebel H, Jones D, Chang C, Trowsdale J (2007) Eur J Immunol 37:780
189. Qi Y, Martin MP, Gao X, Jacobson L, Goedert JJ, Buchbinder S, Kirk GD, O'Brien SJ, Trowsdale J, Carrington M (2006) PLoS Pathog 2:79
190. Gumperz JE, Litwin V, Phillips JH, Lanier LL, Parham P (1995) J Exp Med 181:1133
191. Gillespie GM, Ashirova A, Dong T, McVicar DW, Rowland-Jones SL, Carrington M (2007) AIDS Res Hum Retrovir 23:451
192. Alter G, Martin MP, Teigen N, Carr WH, Suscovich TJ, Schneidewind A, Streeck H, Waring M, Meier A, Brander C, Lifson JD, Allen TM, Carrington M, Altfeld M (2007) J Exp Med 204:3027
193. Pando MJ, Gardiner CM, Gleimer M, McQueen KL, Parham P (2003) J Immunol 171:6640
194. Cohen GB, Gandhi RT, Davis DM, Mandelboim O, Chen BK, Strominger JL, Baltimore D (1999) Immunity 10:661

195. Cerboni C, Neri F, Casartelli N, Zingoni A, Cosman D, Rossi P, Santoni A, Doria M (2007) J Gen Virol 88:242
196. Nirmala R, Narayanan PR, Mathew R, Maran M, Deivanayagam CN (2001) Tuberculosis 81:343
197. Vankayalapati R, Wizel B, Weis SE, Safi H, Lakey DL, Mandelboim O, Samten B, Porgador A, Barnes PF (2002) J Immunol 168:3451
198. Denis M (1994) Cell Immunol 156:529
199. Vankayalapati R, Garg A, Porgador A, Griffith DE, Klucar P, Safi H, Girard WM, Cosman D, Spies T, Barnes PF (2005) J Immunol 175:4611
200. Esin S, Batoni G, Counoupas C, Stringaro A, Brancatisano FL, Colone M, Maisetta G, Florio W, Arancia G, Campa M (2008) Infect Immun 76:1719
201. Millman AC, Salman M, Dayaram YK, Connell ND, Venketaraman V (2008) J Interferon Cytokine Res 28:153
202. Brill KJ, Li Q, Larkin R, Canaday DH, Kaplan DR, Boom WH, Silver RF (2001) Infect Immun 69:1755
203. Barcelos W, Sathler-Avelar R, Martins-Filho OA, Carvalho BN, Guimarães TMPD, Miranda SS, Andrade HM, Oliveira MHP, Toledo VPCP (2008) Scand J Immunol 68:92
204. Roy S, Barnes PF, Garg A, Wu S, Cosman D, Vankayalapati R (2008) J Immunol 180:1729
205. Sondergaard SR, Ullum H, Pedersen BK (2000) APMIS 108:831
206. Sathler-Avelar R, Lemos EM, Reis DD, Medrano-Mercado N, Araújo-Jorge TC, Antas PR, Corrêa-Oliveira R, Teixeira-Carvalho A, Elói-Santos SM, Favato D, Martins-Filho OA (2003) Scand J Immunol 8:655
207. Sathler-Avelar R, Vitelli-Avelar DM, Massara RL, Borges JD, Lana M, Teixeira-Carvalho A, Dias JC, Elói-Santos SM, Martins-Filho OA (2006) Scand J Immunol 64:554
208. Vitelli-Avelar DM, Sathler-Avelar R, Dias JC, Pascoal VP, Teixeira-Carvalho A, Lage PS, Elói-Santos SM, Corrêa-Oliveira R, Martins-Filho OA (2005) Scand J Immunol 62:297
209. Nylen S, Maasho K, Soderstrom K, Ilg T, Akuffo H (2003) Clin Exp Immunol 131:457
210. Peruhype-Magalhães V, Martins-Filho OA, Prata A, Silva LA, Rabello A, Teixeira-Carvalho A, Figueiredo RM, Guimarães-Carvalho SF, Ferrari TC, Correa-Oliveira R (2005) Scand J Immunol 62:487
211. Manna PP, Bharadwaj D, Bhattacharya S, Chakrabarti G, Basu D, Mallik KK, Bandyopadhyay S (1993) Infect Immun 61:3565
212. Spagnoli GC, Juretic A, Schultz-Thater E, Dellabona P, Filgueira L, Horig H, Zuber M, Garotta G, Heberer M (1993) Cell Immunol 146:391
213. Manna PP, Chakrabarti G, Bhattacharya S, Mallik KK, Basu D, Bandyopadhyay S (1994) Trans R Soc Trop Med Hyg 88:247
214. Chakrabarti G, Basu A, Manna PP, Bhattacharya S, Sen S, Bandyopadhyay S (1996) Trans R Soc Trop Med Hyg 90:582
215. Artavanis-Tsakonas K, Riley E (2002) J Immunol 169:2956
216. Ojo-Amaize E, Salimonu L, Williams A, Akinwolere O, Shabo R, Alm G, Wigzell H (1981) J Immunol 127:2296
217. Orago A, Facer C (1991) Clin Exp Immunol 86:22
218. Cooke BM, Mohandas N, Coppel RL (2001) Adv Parasitol 50:1
219. Crandall I, Sherman IW (1994) Parasitology 108:257
220. Craig A, Scherf A (2001) Mol Biochem Parasitol 115:129
221. Artavanis-Tsakonas K, Eleme K, McQueen K, Cheng N, Parham P, Davis D, Riley E (2003) J Immunol 171:5396
222. Ljunggren HG, Karre K (1990) Immunol Today 11:237
223. Storkus WJ, Alexander J, Payne JA, Dawson JR, Cresswell P (1989) Proc Natl Acad Sci USA 86:2361
224. Ruggeri L, Capanni M, Urbani E, Perruccio K, Shlomchik WD, Tosti A, Posati S, Rogaia D, Frassoni F, Aversa F, Martelli MF, Velardi A (2002) Science 295:2097
225. Sconocchia G, Titus JA, Segal DM (1997) Blood 90:716

226. Iannello A, Ahmad A (2005) Cancer Metastasis Rev 24:487
227. Smyth MJ, Swann J, Kelly JM, Cretney E, Yokoyama WM, Diefenbach A, Sayers TJ, Hayakawa Y (2004) J Exp Med 200:1325
228. Takeda K, Hayakawa Y, Smyth MJ, Kayagaki N, Yamaguchi N, Kakuta S, Iwakura Y, Yagita H, Okumura K (2001) Nat Med 7:94
229. Smyth MJ, Kelly JM, Baxter AG, Körner H, Sedgwick JD (1998) J Exp Med 188:1611
230. Gollob JA, Veenstra KG, Parker RA, Mier JW, McDermott DF, Clancy D, Tutin L, Koon H, Atkins MB (2003) J Clin Oncol 21:2564
231. Parihar R, Nadella P, Lewis A, Jensen R, De Hoff C, Dierksheide JE, VanBuskirk AM, Magro CM, Young DC, Shapiro CL, Carson WE (2004) Clin Cancer Res 10:5027
232. Wang HY, Wang RF (2007) Curr Opin Immunol 19:217
233. Gorczynski MR, Harris JF, Kennedy M, Macrae S, Chang M-P (1984) Immunology 53:731
234. Roder JC, Kiessling R, Biberfeld P, Andersson B (1978) J Immunol 121:2509
235. Stutman O, Dien P, Wisun RE, Lattime EC (1980) Proc Natl Acad Sci USA 77:2895
236. Gorczynski RM, Kennedy M, Chang MP, MacRae S (1983) Cell Immunol 80:335
237. Burton RC, Smart YC, Koo GC, Winn HJ (1991) Cell Immunol 135:445
238. Hayakawa Y, Watt SV, Takeda K, Smyth MJ (2008) J Leukoc Biol 83:106
239. Brilot F, Strowig T, Roberts SM, Arrey F, Münz C (2007) J Clin Invest 117:3316
240. Nogusa S, Ritz BW, Kassim SH, Jennings SR, Gardner EM (2008) Mech. Ageing Dev 129:223
241. Silva A, Andrews DM, Brooks AG, Smyth MJ, Hayakawa Y (2008) Int Immunol 20:625
242. Hayakawa Y, Smyth MJ (2006) J Immunol 176:1517
243. Backström E, Kristensson K, Ljunggren HG (2004) J Immunol 60:14
244. Richards JO, Chang X, Blaser BW, Caligiuri MA, Zheng P, Liu Y (2006) Blood 108:246
245. Raulet DH (1999) Curr Opin Immunol 11:129
246. Ware CF, VanArsdale TL, Crowe PD, Browning JL (1995) Curr Top Microbiol Immunol 198:175
247. Stevenaert F, Van Beneden K, De Colvenaer V, Franki AS, Debacker V, Boterberg T, Deforce D, Pfeffer K, Plum J, Elewaut D, Leclercq G (2005) Blood 106:956
248. Veinotte LL, Halim TY, Takei F (2008) Blood 111:4201
249. Watt SV, Andrews DM, Takeda K, Smyth MJ, Hayakawa Y (2008) J Immunol 181:5323
250. Marcenaro E, Ferranti B, Moretta A (2005) Autoimmun Rev 4:520
251. Vaidya SV, Mathew PA (2006) Immunol Lett 105:180
252. Rosmaraki EE, Douagi I, Roth C, Colucci F, Cumano A, Di Santo JP (2001) Eur J Immunol 31:1900
253. Hesslein DG, Takaki R, Hermiston ML, Weiss A, Lanier LL (2006) Proc Natl Acad Sci USA 103:7012
254. Mason LH, Willette-Brown J, Taylor LS, McVicar DW (2006) J Immunol 176:6615
255. Huntington ND, Xu Y, Nutt SL, Tarlinton DM (2005) J Exp Med 201:1421
256. Orange JS, Fassett MS, Koopman LA, Boyson JE, Strominger JL (2002) Nat Immunol 3:1006
257. Bernstein HB, Plasterer MC, Schiff SE, Kitchen CM, Kitchen S, Zack JA (2006) J Immunol 177:3669
258. Verma S, Hiby SE, Loke YW, King A (2000) Biol Reprod 62:959
259. Boles KS, Stepp SE, Bennett M, Kumar V, Mathew PA (2001) Immunol Rev 181:234

Dissecting Human NK Cell Development and Differentiation

Nicholas D. Huntington, Jean-Jacques Mention, Christian Vosshenrich, Naoko Satoh-Takayama, and James P. Di Santo

Abstract Our understanding of human NK cell biology lags behind that of the mouse NK cell biology; this is in a large part because of the ethical and logistical restrictions to the access of healthy human lymphoid tissue and the experimental manipulation in vivo. Nevertheless, in-depth analyses in genetically modified mice have provided us with models for NK cell development, differentiation, and function that guide our thinking about the role of NK cells in immune defense. Collectively, mouse and human studies have unveiled a number of conserved transcription factors, cytokines, cell surface receptors, and associated signaling proteins that are essential for normal NK cell development. Still, human and mouse NK cells differ with regard to expression of several key cell surface receptors, kinetics of development, and frequency in adult lymphoid organs (Huntington ND, Vosshenrich CA, Di Santo JP. Nat Rev Immunol 7:703–714, 2007). Accordingly, the specific biological roles for NK cells in human immune responses remain poorly described. New preclinical animal models that allow the analysis of human immune system development in function may provide a means to further our understanding of the biology of human NK cells in vivo.

1 General Introduction

1.1 Transcription Factors and NK Cell Lineage Commitment

Like all haematopoietic cells, human NK cells are derived from $CD34^+$ haematopoietic stem cells (HSC). During fetal life, HSC are predominately found in the liver but they begin to seed the bone marrow towards the end of the first trimester.

N.D. Huntington, J.-J. Mention, C. Vosshenrich, N. Satoh-Takayama and J.P. Di Santo (✉)
Cytokines and Lymphoid Development Unit Institut Pasteur, Paris, France
e-mail: disanto@pasteur.fr

Inserm U668, Paris, France

HSC from both these locations can generate NK cells in vitro and in vivo provided adequate growth factors are available [1–4]. Nevertheless, the molecular mechanisms that guide the development of NK cells remain poorly understood. Transcription factors (TFs) play a critical role in specifying lymphoid cell fates, and can reinforce identities of differentiated lymphoid cells subsets [5–8]. The concept that "master" TFs can play a dominant role in determining lymphoid fate finds support in the T cell lineage (Notch1, Gata3), B cell lineage (EBF, Pax5), and NK-T cell lineage (PLZF; [5–7, 9]. Does a specific transcriptional program determine NK cell development or do NK cells result as a default pathway when other lymphoid fates fail? The only TF that appears to strongly affect NK cells but not other lymphocytes is Id2 [10–13], but as Id2 is a transcriptional repressor, the only conclusion that can be taken is that E-box proteins act as strong negative regulators of NK cell development in the mouse [10–13]. Still, the "toxicity" of E-box proteins for NK cells appears conserved in man as over-expression of Id3 can convert bi-potent fetal thymus T/NK precursors towards the NK cell lineage at the expense of T lymphopoiesis [14]. Other TFs are known to promote early lymphoid precursors in general (PU.1, Ikaros), while a distinct set of TFs appears important for NK cell maturation in the BM (IRF-2, Gata3, Tbx21, Dlx) [15–18].

As IL-15 is essential [19] and CD122 marks NK precursors in mouse [20] and is highly expressed by immature NK cells in man [21], knowledge on the TFs that regulate CD122 or its signaling components may provide clues to NK cell commitment and lineage specification. Several different stromal cell-derived signals have been suggested as major inducers of commitment to the NK cell lineage, including cytokines (Flk2L/Flt3L), growth factor (osteopontin), and inflammatory mediators (LTα). Nevertheless, it has not been formally demonstrated that lineage commitment (for NK cells or other lymphocytes) is altered in the absence of these factors. As such, the reduction in overall NK cell homeostasis that has been reported in mice deficient in Flk2L/Flt3L [22], LTα [23], or OPN [24] may reflect a maintenance role for these soluble factors during different stages of NK development. Alternatively, these factors may promote homeostasis of multiple hematopoietic precursors, with a stronger relative effect in the NK lineage. Along these lines, it should be noted that hematopoietic lineages other than NK cells are affected by these mutations [22–24].

1.1.1 Cytokines Driving NK Cell Development

In adults, the bone marrow is the predominant site for NK generation, although it cannot be ruled out that NK cell development can occur to some degree outside of the bone marrow (lymph node, thymus, liver, gut) [25]. In order to appreciate the when and where of human NK cell development, it is first important to reflect on the soluble factors responsible for driving this development.

Starting at an immature stage, NK cells require the expression of the common gamma chain (γ_c) and IL-2Rβ chains of the IL-2/IL-15 receptor as well as their associated signaling proteins (Jak1/Jak3 and STAT5a/b) in order to survive, expand, and differentiate [26]. This homeostatic requirement of IL-15 continues for mature NK cells in the peripheral tissues [27]. An absence of either IL-15 or the alpha chain of its receptor (IL-15Rα) prevents NK cell development from HSC, with little effect on the generation of most conventional B and T cells. [28–32]. Curiously, transfer of IL-15- or IL-15Rα-deficient HSC allowed for a large degree of NK cell development, suggesting a role for both IL-15 and IL-15Rα "in trans" [28–32].

Initial in vitro experiments suggested that IL-15Rα might act as a molecular chaperone that brings IL-15 to the cell surface [30]. This initial model of "IL-15 transpresentation" has been subsequently confirmed by an elegant series of murine studies using gene targeting and bone marrow chimeras [28–32]. IL-15 expressing stromal cells or myeloid cells use IL-15Rα to bring IL-15 to the cell surface where it is bioactive and significantly more potent in inducing activation and proliferation of IL-15-dependent cells [33].

In light of the obvious absence of gene targeting technology in humans, naturally occurring mutations do arise within the human genome and have provided insight into factors involved in human NK cell development. In man, mutations in IL-15 or IL-15Rα have not been reported; however, NK cells are dramatically reduced in severe combined immunodeficiency patients carrying mutations in CD132 (γ_c, used in IL-15 as well as IL-2, -4, -7, -9, and -21 receptors), Jak3 or CD122 (IL-2Rβ). As NK cells develop in IL-7Rα deficient patients who are largely devoid of mature T cells, it is likely that IL-15 is a key regulator of human NK cell development [34, 35] as has been documented in mice.

Several hematopoietic and nonhematopoietic cells express IL-15, including stromal cells, monocytes, macrophages, and dendritic cells (DC) [19, 31, 36]. While these various cell types are found in human bone marrow, they also reside in many other tissues. Thus, extra-marrow NK cell development could be driven by these cells that "transpresent" IL-15/IL-15Rα. To this end, it has been previously shown in vitro that CD34$^+$ HSC can respond to IL-15 with differentiation into immature NK cells [37, 38]. Nevertheless, while IL-15 is likely the major factor for human NK cell development, the critical demonstration of this is still lacking.

Other growth factors, such as stem cell factor (SCF or c-kit ligand) and IL-7, could synergize with IL-15 to promote NK cell survival and/or differentiation. The ability of SCF to synergize with IL-15 and IL-7 in enhancing NK cell development from CD34$^+$ cells may involve distinct mechanisms (for example, through proliferation of stem cells, survival of developing NK cells, or other mechanisms). Indeed, it has been shown that while SCF alone does not promote proliferation of a subset of human NK cells with the CD56^{hi}c-kit$^+$ phenotype, it does increase expression of the antiapoptotic factor Bcl-2 in the case of growth factor withdrawal [39]. Thus, SCF may indirectly enhance the IL-15-induced expansion of NK cells derived from CD34$^+$ precursors in man [38]. Notch signaling in human progenitor cells is thought to be critical for T-lineage commitment and if constitutively expressed, suppresses NK cell development [40]. Other studies have shown that the presence of TEC-derived soluble factors could inhibit NK cell development and differentiation from CD34$^+$ thymocytes in the presence of IL-15 [41].

2 NK Cell Precursors and Immature NK Cells in Man

Committed NK cell precursors (NKP) are defined as cells that have the potential to become NK cells but no other hematopoietic cell lineage. Unlike the situation in mice, where $CD122^+$ and IL-15-responsive NKP have been identified and characterized [20], lineage-$CD122^+$ hematopoietic precursors that are enriched for NK cell potential have not been found in human fetal liver or adult bone marrow (Huntington et al., unpublished). Moreover, IL-15-responsive hematopoietic precursors in man appear to lack CD122 expression when examined by flow cytometry (Huntington et al., unpublished). Thus, the precise phenotype of NKP in humans remains undefined. Early work reported that CD7 expression within the adult $CD34^+Lin^-$ bone marrow cells identified a population with enhanced NK cell cloning efficiency with the level of CD7 expression being proportional to their NK precursor potential [42]. In contrast, $CD34^+Lin^-$ bone marrow cells expressing CD10 appear to be preferentially restricted to the B cell lineage [21, 43–45]. Still, neither CD10, nor CD7 identifies a homogeneous precursor population in the bone marrow with NK cell restricted potential.

A population of apparently "immature" human NK cells with the $CD3^-CD161^+CD56^-$ phenotype can be induced from $Lin^- CD34^+$ cord blood cells in vitro [3, 46–48]. These cells lack expression of mature NK cell markers, and show poor cytolytic activity compared with the more mature $CD161^+CD56^+$ cells. When further cultured with IL-2 and IL-12, these immature NK cells acquire CD56 expression and high levels of natural cytotoxicity [46, 48]. A similar population of $CD161^+CD56^-$ immature NK cells could also be identified in umbilical cord blood [49], although their ability to differentiate into other cell types has not been thoroughly tested. We also observe $Lin^-CD161^+CD56^-$ cells in human fetal bone marrow ex vivo and IL-15 stimulated fetal bone marrow in vitro, with these cells often expressing other NK cell surface antigens such as CD122 and NKp46 (Huntington, unpublished). Collectively, these results are consistent with a model of human NK cell development that involves lineage-restricted lymphoid precursors that acquire functional capacities through a stepwise differentiation process (Fig. 1).

3 NK Cell Education

Both murine and human NK cells express polymorphic inhibitory receptors that recognize MHC class I ligands or peptides derived from MHC class I molecules. These inhibitory receptors include the C-lectin receptors of the Ly49 family in mice, the KIR (killer cell immunoglobulin-like receptors) in humans, and the CD94/NKG2A complex in both species [50–52]. The identification and functional characterization of these receptors provided a molecular explanation for Kärre's "missing self hypothesis" [53] that stated that NK cells are poised to scrutinize target cell MHC expression and only target cells lacking the expression of MHC I

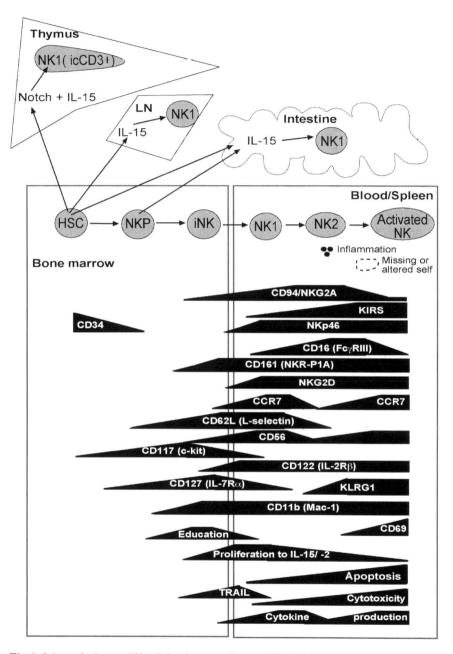

Fig. 1 Scheme for human NK cell development. Human NK cell development can be considered as a linear process. For each stage of development, the expression of surface markers used to distinguish different stages is shown. HSC and NK progenitors can also be found outside the bone marrow, such as the thymus, lymph node (LN), and intestine and in the presence of IL-15, these cells have the potential to develop into NK cells

will evoke a response from host NK cells [53]. In contrast, MHC-competent target cells would be spared from NK cell-mediated destruction as a result of engagement of the relevant inhibition receptor. KIR expression in human NK cells (and Ly49 expression in the mouse) generates a "repertoire" with NK cells expressing varying numbers of different inhibitory receptor alleles. KIR are predominantly expressed on the $CD56^{lo}CD16^+$ subset of NK cells that harbor abundant intracellular perforin and granzymes and display potent spontaneous cytotoxicity [25, 54]. Given the importance of KIR expression in regulating NK cell function, knowledge of elements influencing KIR acquisition would improve our understanding and clinical approaches to diseases where KIR and HLA haplotypes influence susceptibility, progression, or outcome such as autoimmune/inflammatory disease, cancer, infections (HIV, HCV), and bone marrow transplants/graft versus leukemia effects [55, 56].

While much effort has been directed to understanding the molecular mechanisms that control KIR expression [57], we still lack essential information in this area. While some TF have been identified that control Ly49 expression (Tcf1) [58], the transcriptional control of KIR expression remains enigmatic. Moreover, the ligands (cell surface expressed or soluble) that are implicated in the induction or regulation of KIR/Ly49 expression are basically unknown. While IL-15 regulates inhibitory receptor expression in both man and mouse [59–61], definitive proof that IL-15 directly induces KIR/Ly49 expression, rather than acting indirectly through maintenance of cell proliferation, has yet to be provided. As such, our knowledge of NK cell "education" is still in preschool.

The mechanisms involved in NK cell receptor repertoire formation have recently received increasing attention, as engagement of inhibitory KIR by self-MHC class I ligands has been shown to influence the functional maturation of developing NK cells [62–64]. Evidence from MHC I-deficient mice and humans first suggested that MHC ligands were important for the functional differentiation of NK cells. NK cells from a TAP-1 deficient patient showed no cytotoxicity against target cells lacking MHC I molecules despite the presence of class I-recognizing NK receptors [65]. In a similar fashion, NK cells from MHC I deficient mice fail to be activated when triggered through activating receptors [63, 64]. Studies from the Yokoyama lab showed that engagement of Ly49 receptors in developing mouse NK cells was associated with increased functionality (cytokine production and cytotoxicity). Curiously, the ITIM motif present in the cytoplasmic domain of the Ly49 inhibitory receptor was required for acquisition of higher functional activity, although the mechanism through which the SHP1 phosphatase that binds the ITIM motif confers this activity is not known. Around the same time, a population of MHC I inhibitory receptor negative NK cells was reported in both mice [63, 64] and man [62]. Human KIR^-NKG2A^- NK cells are distributed evenly between the $CD56^{hi}CD16^-$ and $CD56^{dim}CD16^+$ NK subsets, and they are less sensitive to activation by cytokines or activating receptors compared to the KIR^+NKG2A^+ NK cells as determined by induction of CD107a and IFN-γ expression [62]. Thus, NK cells that had not "seen" self-MHC class I were clearly less functional compared to NK cells in the same individuals that expressed inhibitory receptors for self-MHC.

These observations form the foundations of "licensing" or "education" models for achieving NK cell tolerance, where an inhibitory receptor/MHC I interaction will either activate directly or modify other signals to developing NK cells, allowing them to gain full functionality [52, 62, 64–66]. Accordingly, NK cells expressing at least one inhibitory receptor recognizing self MHC-I would have a lower threshold of activation and appear more functionally active than NK cells expressing no self-MHC receptors or those only expressing inhibitory receptors recognizing non-self MHC-I ligands [62, 67].

4 Functional Diverse NK Cell Subsets in Man

Once their functional maturation is finished, NK cells take up residence in the peripheral lymphoid tissues and circulate through the blood and lymph. Mature NK cells represent a substantial fraction of the circulating lymphocytes in humans (up to 20%) [68–70]. Lanier and colleagues first proposed the notion of subsets of human peripheral blood NK cells (on the basis of differential CD16 expression) [71], and subsequently an extensive functional and phenotypic analysis has been compiled (reviewed in [72, 73]). Two human NK cell subsets can be defined by the differential cell surface expression of CD16 (FcγRIII) and CD56 (N-CAM) on circulating CD3$^-$ lymphocytes : CD56^{lo}CD16$^+$ NK cells represent ~90% of blood NK cells, while the remainder are CD56^{hi}CD16$^-$ NK cells [54]. Phenotypically, CD56^{hi}CD16$^-$ NK cells express CD94/NKG2A, CCR7, CD62L, CD25, and CD117 but mostly lack KIR expression, whereas CD56^{lo}CD16$^+$ NK cells are CD94/NKG2A$^{+/-}$ and KIR$^+$ but fail to express CCR7, CD62L, CD25, or CD117 [74–81]. These two NK cell subsets also show differential functions as CD56^{hi}CD16$^-$ NK cells produce greater amounts and a wider range of cytokines (IFN-γ, TNF-α, granulocyte/macrophage colony-stimulating factor (GM-CSF), IL-13, and IL-10) than CD56^{lo}CD16$^+$ NK cells [75, 82, 83]. In contrast, CD56^{lo}CD16$^+$ NK cells possess abundant intracellular perforin and granzyme and display enhanced cytotoxicity [54, 78, 83]. While these two subsets are often contrasted as "cytokine producers" (CD56^{hi}CD16$^-$) versus "killers" (CD56^{lo}CD16$^+$), it should be emphasized that both subsets have clearly demonstrable capacities to make IFN-γ and to lyse NK-sensitive targets [54]. The functional distinction between these two subsets is therefore only relative.

Whether these two well-defined human NK cell subsets share a precursor–product relationship or are the result of independent pathways of NK cell development is still being debated. Several lines of evidence suggest that CD56^{hi}CD16$^-$ NK cells are precursors of CD56^{lo}CD16$^+$ NK cells. First, the proportion of CD56^{hi}CD16$^-$ is elevated in the blood of patients shortly after bone marrow transplantation and declines over time in these patients, when the CD56^{lo} population becomes more prominent [84–86]. Second, telomere length, which progressively diminishes with cell division and "age", is significantly decreased in CD56^{lo}CD16$^+$ NK cells compared to that in CD56^{hi}CD16$^-$ NK cells from the same donor [87]. Third, transfer of

CD56^{hi}CD16$^-$ NK cells into NOD/SCID mice generates mostly CD56^{lo}CD16$^+$ NK cells [88]. This is also observed (albeit to a lesser extent) in vitro where CD56^{hi}CD16$^-$ NK cells adopt the CD56^{lo}CD16$^+$ phenotype and demonstrate increased perforin expression with CD62L downregulation upon coculture with fibroblasts [88]. Taken together, these findings strongly suggest that CD56^{hi}CD16$^-$ NK cells are precursors to CD56^{lo}CD16$^+$ NK cells; however, a common precursor that gives rise to both of these subsets cannot be excluded.

Understanding the factors that regulate the differentiation of CD56^{hi}CD16$^-$ NK cells to CD56^{lo}CD16$^+$ NK cells in vivo could help tailor human immune therapies to certain diseases, for example by modulating the relative ratios of "cytotoxic" to "cytokine-producing NK cells." In a similar fashion, NK cells expressing at least one KIR to self MHC-I ligand appear more functional than NK cells expressing no KIR or NK cells expressing KIR to non-self MHC-I ligands [62, 63, 67] indicating that KIR-self MHC-I interactions regulate NK cell function. With further knowledge of molecular mechanisms influencing NK cell differentiation, one could envisage future therapeutic modalities that would allow modulation of NK cell function in vivo that might impact on our clinical approaches to human diseases.

5 NK Cell Development and Differentiation Outside of the Bone Marrow

While the bone marrow is acknowledged as a primary site for NK cell development [26], recent evidences from mouse and human systems indicate that NK cells may develop via a local process of differentiation within various tissues, including the thymus, lymph node (LN), and intestine. In some cases, this tissue-specific NK cell differentiation results in the generation of functionally distinct NK cell subsets that could play unique roles during immune responses [89].

5.1 NK Cells in Tissues: Lymph Nodes

In humans, recent evidence suggests that signals present in the LN may provide an appropriate environment for development of CD56^{hi}CD16$^-$ NK cells [90, 91]. LN-resident hematopoietic precursors were identified with the Lin$^-$CD34^{lo} $\alpha_4\beta_7^+$CD117^{hi}CD56$^-$CD94$^-$CD16$^-$ phenotype that could develop in a stepwise manner into CD56^{hi} NK cells with classical NK-cell markers (CD122, CD11b, NKG2D and NKp46) [91]. One intermediate in the identified developmental process included cells with the CD161$^+$CD117^{hi}CD94$^-$ phenotype [91]; a similar population had been identified following culture of cord blood CD34$^+$Lin$^-$ cells with IL-3, IL-7, IL-15, SCF, and Flt3L [92]. Importantly, both in vitro-derived and ex vivo-sorted CD117^{hi}CD94$^-$ cells could differentiate into CD56$^+$CD117^{lo} CD94$^+$ cells in vitro [91, 92], indicating the presence of NKP in these populations.

However, whether this LN population was uniquely restricted to the NK cell lineage is unknown and the clonal frequency of NKP was not determined. Although it has been suggested that these LN NKP derive from a circulating bone-marrow precursor [21, 42, 43], it remains possible that these cells have a thymic origin. Recently, a LN-resident lymphocyte precursor was identified in mice (with striking phenotypic similarities to early thymocyte precursors) that harbored NKP [93]. On the basis of these reports, a dual origin (BM- and/or thymus-derived) lymphoid precursor with NK cell potential may exist in the LN in both mice and man.

When comparing LN NK cell development with the more conventional BM-derived pathway, it is likely that the LN microenvironment provides some (but potentially not all) of the signals required for NK cell maturation. As LNs are not a known repository for HSC, the earliest steps of lymphoid lineage and NK cell commitment from HSC likely take place elsewhere with LN seeding of these NKP. As IL-15/IL-15Rα "transpresentation" is critical for NK cell development and homeostasis, a source of IL-15 must be present in the LN and could be provided by activated DC and/or macrophages [19]. Finally, as most LN NK cells are KIR, it is possible that the specific BM-derived signals that drive KIR expression are lacking in the LN. Alternatively, the presence of LN NK cells could predominately rely on specific migration of $CD56^{hi}CD16^-$ NK cells via CCR7 and CD62L; these cells may be short lived in this location and experience no further differentiation explaining the virtual lack of KIR+ NK cells in LN. Taken together, these observations suggest that LN NK cell development may represent a "window" of the developmental process that normally occurs in the BM.

In addition to this more conventional pathway of NK cell development, alternative pathways may exist. Mebius et al. first described a population of fetal LN cells with the $CD45^{lo}CD4^+CD3^-CD127^+$ phenotype that harbored NK cell as well as DC potential [94]. These cells were later identified as lymphoid tissue inducer (LTi) cells that are now recognized to play an important role in the organization of secondary lymphoid structures (LN), as well as intestinal cryptopatches, Peyers' patches, and isolated lymphoid follicles [95]. More recently, Cupedo and colleagues further characterized fetal and adult LTi cells and showed that they give rise to a peculiar subset of $CD127^+$ NK cells in vitro [96]. While the physiological relevance of this alternate pathway is unclear, the NK cells developing from LTi cells in vitro express intracellular granzyme B and perforin, and thus could participate in local immune responses.

NK cells found in the human spleen resemble that of the peripheral blood in terms of phenotype and subset frequency (90% $CD56^{lo}CD16^+$, 10% $CD56^{hi}CD16^-$). This is not surprising given that most splenic NK cells appear to be excluded from B and T cell follicles of the white pulp, and instead reside in the splenic red pulp. In contrast, LN appears to preferentially recruit $CD56^{hi}CD16^-$ NK cells and as a result the ratio of $CD56^{lo}CD16^+$ to $CD56^{hi}CD16^-$ NK cells in the LN is low (1:10) [97]. The predominant $CD56^{hi}CD16^-$ NK cells in the LN express higher levels of CD62L, CCR7, and CXCR3, which may mediate their preferential recruitment to the LN under steady-state and inflammatory conditions [98–100]. IL-18 can induce NK cell expression of CCR7 and also reduces NK cell cytotoxicity towards activated DC, thereby promoting local increases in IL-12 and IFN-γ production [101]. Enhanced cytokine production by CD56^{hi}CD16$^-$ NK cells has been documented in the draining LN following interactions with activated macrophages or DC [82, 102] and may play a role in promoting T helper 1 (T$_H$1)-polarized immune responses [100, 103]. Thus, the preferential development and/or retention of CD56^{hi}CD16$^-$ NK cells within the LN may potently condition the local immune responses and allow for more robust Th1 immunity in the face of infection or inflammation.

5.2 NK Cells in Tissues: Thymus

About 20 years ago, thymus-resident NK cells were identified in the human and mouse fetal and adult thymus by the Phillips, Ritz, Kumar, and Moretta labs [104–107]. Thymic NK cells are an infrequent population representing about 0.5% of total thymocytes and enriched in the early immature thymocyte population (CD3$^-$CD4$^-$CD8$^-$ cells). In the human thymus, there is an increased frequency of CD56^{hi}CD16$^-$ NK cells and during fetal development, this subset can represent up to 70% of total thymic NK cells [107]. A proportion of thymic NK cells in humans show a phenotype consistent with peripheral blood NK cells, and it is likely that some thymic NK cells represent circulating mature NK cells that have developed elsewhere and can reenter the thymus. Along these lines, previous studies have shown that in vitro cultured thymic NK cells in mice and humans have a similar capacity as peripheral blood NK cells to lyse NK-sensitive target cells [104, 106, 107].

While some thymic NK cells may represent recirculating BM-generated NK cells, a local process of thymic NK cell development in situ is also supported by experimental evidence. Early studies by Spits and colleagues and Zuniga-Pflucker and colleagues showed that early thymocyte precursors in humans and mice have bipotent T and NK cell potential that could be revealed at the clonal level [4, 108–110]. More recently, the molecular mechanisms that drive NK development in the thymus in mice were characterized, and were found to include the TF GATA-3 and the cytokine IL-7 [111]. Moreover, CD127 (IL-7Rα chain) expression was found on murine NK cells generated in situ in the thymus compared with those deriving from recirculating peripheral BM-derived NK cells. Phenotypically, CD127$^+$ thymic NK cells bore lower frequencies of Ly49 receptors and were CD16$^-$. Functionally, thymic NK cells showed reduced cytotoxicity but heightened cytokine production compared to their splenic counterparts [111]. Collectively, these results identified essential pathways and signals that condition development of NK cells within the mouse thymus.

Previous studies had demonstrated selective CD127 expression on human CD56^{hi}CD16$^-$ NK cells [112, 113]. Considering the similarities in phenotype (CD16$^-$, CD11b^{lo}, KIR/Ly49$^-$) and function (enhanced cytokine capacity and reduced cytotoxicity) between mouse CD127$^+$ NK cells and human CD56^{hi}CD16$^-$NK cells, we previously proposed that a functional diversification of NK cells existed in humans and mice that was evolutionarily conserved [114]. The observations in mice also suggested that a fraction of human $CD56^{hi}CD16^-$ NK cells might have a thymic origin.

One phenotypic difference between human thymic NK cells and their peripheral counterparts is the expression of intracellular CD3ε (iCD3ε) [106, 112]. A large fraction of thymic NK cells including both $CD16^+$ and $CD16^-$ subsets express iCD3ε, whereas peripheral blood NK cells do not [106, 112]. The significance of iCD3ε expression by thymic NK cells is unclear, but several explanations are possible. First, iCD3ε expression may result from signals present in the thymic microenvironment. One obvious signal could involve Notch receptors and their ligands. Alternatively, iCD3ε expression could represent a remnant of a previous T cell lineage differentiation event. The putative relationship between T cells and thymic NK cells will be addressed below. While iCD3ε expression by thymic but not peripheral NK cells could indicate that thymic NK cells, once generated, remain in the thymus, previous studies have shown that thymic NK cells can be exported and populate secondary lymphoid organs including the spleen and LN [111]. As such, iCD3ε expression might be extinguished in thymic NK cells once they exit the thymus.

Recent in vitro studies have identified that signaling via Notch proteins (Notch 1–4) on hematopoietic progenitors can influence NK cell development. In the murine system, several reports demonstrated that transient exposure of fetal or adult HSC to Notch ligands strongly enhanced the generation of NK cells [115–117]; in contrast, longer (or constant) exposure reduced overall output of NK cells but promoted T cell development. Similar studies using human $CD34^+$ HSC showed that while limited Notch receptor triggering was compatible with NK cell development (but only in the presence of high concentrations of IL-15), sustained Notch stimulation (especially via the ligand delta-like-4) generally favored T cell development. Interestingly, human NK cell generation from cord blood $CD34^+$ progenitors in the presence of Delta-like 1 and IL-15 resulted in NK cells with almost uniform expression of iCD3ε. In contrast, NK cells generated on stroma lacking Delta-like 1 were mostly negative for iCD3ε expression [112]. These results would suggest that Notch signals within the thymus are critical for generating phenotypic (and functional?) characteristics of thymic NK cells.

Are Notch signals critical for NK cell development in the thymus (or elsewhere)? Previous studies using a conditional Notch-1 allele demonstrated that absence of Notch-1 in murine HSC had no obvious effect on BM or splenic NK cell development or homeostasis [118]. As deletion of Notch-1 (or deletion of the common Notch receptor signaling intermediate RBPJ) completely eliminates T cell progenitors in the thymus (Anne Wilson personal communication), the analysis of thymic NK cell development in the absence of Notch-1 or RBPJ may shed light on the putative developmental relationship between T cells and thymic NK cells. Interestingly, thymic NK cell phenotype and homeostasis are completely intact in the absence of signaling through all Notch receptors in the hematopoietic system in mice (Di Santo, unpublished). Thus, while Notch signals within the thymic microenvironment may condition some phenotypic properties of human (or mouse) thymic NK cells, they are clearly not essential for thymic NK cell development.

Moreover, development of thymic NK cells and T cells are clearly distinct and likely involve nonoverlapping hematopoietic precursors.

Whether thymic NK cells participate in immune responses or play a role in thymopoiesis is currently unknown. In any case, it appears that there is significant NKP frequency within the immature $CD3^-CD4^-CD56^-CD1^-CD7^+$ thymocyte population [106, 119] and subsequent exposure to IL-15 via thymic epithelial cells, myeloid cells, or IL-2 via T cells may be adequate to drive NK cell differentiation in this environment. That said, the low frequency of thymic NK cells indicates this process is not favored and/or inefficient. Further work is required to identify the unique roles for thymic NK cells in mice and humans.

5.3 NK Cells in Tissues: Intestine

The intestinal immune system is an organized collection of diverse innate and adaptive immune cells that reside and circulate through distinct lymphoid structures including Peyers' patches, cryptopatches, isolated lymphoid follicles, and mesenteric LN, as well as less-defined structures such as the intestinal epithelium and lamina propria. While most adaptive immune responses in the gut are tuned towards IgA secretion, a potential role for NK cells in this tissue could be envisaged at the level of epithelial cell elimination under situations of stress and/or infection.

Several reports have documented the presence of NK cells in the intestine of humans and mice [120–122]. These NK cells were found among both lamina propria lymphocytes (LPL) and intraepithelial lymphocytes (IEL) with NK cells from these two sources appearing largely identical in phenotype except for more frequent expression of integrin β_7 and α_E by intraepithelial NK cells. In terms of human NK cell subset distribution, the intestine resembles LN, with very few (~10%) $CD56^{lo}CD16^+$ NK cells present and subsequently a marked depletion of KIR^+ NK cells. The marked differences in chemokine receptor and integrin expression between the IEL and LPL NK cells and $CD56^{hi}CD16^-$ NK cells from peripheral blood may help to explain the preferential recruitment and retention of NK cell subsets within the intestine. IEL and LPL NK cells express CXCR3 and heterogeneously express the integrins β_7 and α_E, whereas their expression of integrin α_M (CD11b) and integrin α_X (CD11c) is reduced compared to that of $CD56^{hi}CD16^-$ NK cells from peripheral blood. In terms of function, IEL and LPL NK cells spontaneously express intracellular granzyme B and perforin, although less than $CD56^{lo}CD16^+$ NK cells, and efficiently lyse K562 cells ex vivo. In addition, IEL and LPL NK cells secrete similar levels of IFN-γ and TNF-α to peripheral blood $CD56^{hi}CD16^-$ NK cells in response to IL-12 and IL-18 [123].

Recently, an unusual population of innate lymphocytes bearing some NK cell-related cell surface markers (including the natural cytotoxicity receptor NKp46) have been identified in the intestinal tract of mice [124–126]. While these $NKp46^+$ cells clearly lacked many "classical" NK cell markers (including NK1.1 and Ly49 receptors) and functions (noncytotoxic, no IFN-γ secretion), they did express the

TF RORγt and constitutively produced IL-22 [124–128]. Moreover, these NKp46$^+$ cells differed from NK1.1$^+$ cells in the gut by their normal development in the absence of IL-15. Curiously, microbial flora were required to maintain these IL-22$^+$ NKp46$^+$ cells that played an important role in intestinal defense against the pathogen *Citrobacter rodentium* [125–127]. While a similar population of IL-22$^+$ NK cells has not been identified in the gut in humans, IL-22$^+$ NKp44$^+$ NK cells are found in adult tonsils juxtaposed to the epithelium [127]. These observations suggest that NK cell subsets present at mucosal surfaces may have unique functions and open the possibility that NK cell diversification may be conditioned by local environmental signals.

Further studies have clearly shown that hematopoietic precursors are contained within the intestinal compartments. CD3$^-$CD7$^+$CD34$^+$ cells are detected in the fetal intestine from 7 weeks of gestation and a similar population can be found in adult lamina propria (Lin$^-$CD7$^+$CD117$^+$CD33$^+$). The presence of these hematopoietic precursors poses the question of whether LPL and IEL NK cells (and their precursors) are generated in situ or are seeded from the peripheral circulation. Whatever the case, their presence indicates that intraintestinal NK lymphopoiesis may be possible. Considering the soluble factors (IL-7, IL-15) that drive NK cell development at other sites (see above), transcripts for IL-7 and IL-15 have been detected in mouse fetal intestine at levels equal to or greater than that in thymus [129]. Another interesting observation concerning CD3$^-$CD7$^+$ cells in the human fetal intestine is that over half of these cells express iCD3ε. As many of these cells bear NK cell markers, this indicates that iCD3ε NK cells are found (and may be generated) outside the thymus throughout the life of the organism [107, 112]. Notch receptor triggering may drive iCD3ε expression (see above), and abundant transcripts for Notch ligands (Jagged 1 and 2 and Delta-like 1) are present in intestine throughout development [130, 131]. In adult LPL, a clear population of lymphoid lineage negative, small side light scatter, c-kit expressing cells were observed and extensively characterized for TF and differentiation potential. This population was largely CD34$^-$CD33$^+$CD38$^+$CD45RA$^+$, expressed high levels of Id2, PU.1, Spi-B1, lymphotoxins α and β, and possessed NKP potential. The same study went on further to show that NK cells were increased among LPL and IEL in Crohn's diseases and that NK cell differentiation from c-kit$^+$ precursors from Crohn's diseases LPL was accelerated [123]. Given their production of IFN-γ and TNF-α, which are involved in the pathogenesis of Crohn's diseases, the authors concluded that NK cells may be involved in the chronic inflammation presented in Crohn's diseases.

6 Human Immune System Mice to Study Human NK Cell Biology

In vivo studies of NK cells have been largely restricted to mice and while this line of experimentation is valuable, some of this knowledge will have limited impact on our understanding of human NK cell biology. A clear example of this is NK cell development, where the kinetics, frequency, and phenotype are clearly different between these two species [25]. A hybrid intermediate between murine and human in vivo studies exists in the form of human immune system (HIS) mice. HIS mouse model can be generated by engraftment of newborn Balb/c $Rag2^{-/-}\gamma_c^{-/-}$ mice with purified human $CD34^+$ hematopoietic stem cells (HSC) from fetal liver or cord blood [2, 132]. HIS mice represent a practical model with high human cell chimerism, most lymphoid and myeloid lineages generated (although to varying degrees), the capacity to elicit adaptive immune responses (with humoral responses being more robust than T cell responses), and unlike earlier models, a notable lack of spontaneous thymoma development [2, 132–134].

HIS mice provide a novel system to dissect human lymphopoiesis and immune responses; however, little is known about the mechanisms that drive development of human hematopoietic lineages in HIS mice. Human HSC, immature, mature, and fully differentiated myeloid and lymphoid cells are observed in Balb/c $Rag2^{-/-}\gamma_c^{-/-}$ HIS mice, suggesting that sufficient quantities of the different growth factors required to initiate human hematopoiesis are available in this context. While initially most if not all cytokines encountered by engrafted HSC are murine in origin, human cytokines, produced by developing human hematopoietic cells could eventually influence human lymphoid differentiation in HIS mice. Whether there is adequate cross-species reactivity or sufficient concentrations of murine cytokines for human hematopoiesis in HIS mice is unknown.

NK cell development is generally poor in previous and current models of HIS mice. Various approaches have been attempted to improve human NK cell homeostasis in human $CD34^+$ engrafted mice, including engraftment of fetal liver and autologous thymic tissue [135] and repeated high dose (10 μg) injections of hIL-15/Flt3L/SCF [136] in NOD/SCID mice, although human chimerism and development of other lymphoid lineages were diminished in the latter study. Furthermore, human NK cells generated following treatment with hIL-15/Flt3L/SCF appeared strongly activated with uniform CD69 expression while only 20% expressed NKp46 [136]. This result suggested that the hIL-15/Flt3L/SCF cocktail, while boosting NK cell numbers, was potentially altering NK cell homeostasis. Clearly, other methods to improve NK cell numbers, in HIS mice, that would be simpler and less perturbing should be identified.

We recently demonstrated that a problem with human NK cell homeostasis in HIS mice is the availability of human IL-15 [60]. We considered that murine IL-15, while abundantly available in the recipient mice, might not have adequate biological activity on human IL-15-responsive cells. Using mixtures of murine and human IL-15 and IL-15Rα, we were able to show that human IL-15 is much more potent than mouse IL-15 in driving human NK cell survival and proliferation. When human IL-15/IL-15Rα agonists were injected in HIS mice, improved NK cell homeostasis and NK cell differentiation were observed [60]. Our findings that hIL-15R agonists result in an accumulation of $CD56^{lo}CD16^+$ NK cells in vivo is consistent with the model that these cells represent a terminal stage in NK cell development (reviewed in [73]). While not ruled out, it is highly unlikely that this accumulation of $CD56^{lo}CD16^+$ NK cells represents a specific expansion of this subset in response to IL-15 stimulation, as this subset appears refractory to IL-15 stimulation in vitro [59, 137]. In contrast, CD56^{hi}CD16$^-$ NK cells readily proliferate to IL-15 and the failure to observe an accumulation of this subset 7 days after the final injection of hIL-15/IL-15Rα agonists indicates they have already differentiated to CD56^{lo}CD16$^+$ NK cells by this time.

In addition to increasing overall human NK cell homeostasis in HIS mice, IL-15R agonists also markedly increased the frequency of KIR$^+$ NK cells [60]. The failure to observe accumulation of CD16$^+$KIR$^+$ NK cells in HIS mice transduced with human CD34$^+$ HSC ectopically expressing Bcl–X$_L$ suggests that NK cell survival alone is not sufficient for differentiation and IL-15 mediated NK cell proliferation (which is enhanced with hIL-15R agonists treatment) may be required for KIR expression [60]. KIR can be acquired by KIR$^-$ NK cells in vitro in the presence of stromal cells and IL-15 [59, 138]. Our findings demonstrate that this maturation process can occur in vivo in HIS mice in the absence of nonhematopoietic self MHC-I expression. One practical application of this knowledge could involve the treatment of patients recovering from hematopoietic cell transfer (who have reduced KIR expression compared to healthy adults) [139, 140] with hIL-15R agonists to accelerate NK cell reconstitution and maturation, thereby potentially improving clinical outcomes. Use of hIL-15R agonists could promote graft recovery in patients with deficiencies in the transporter associated with antigen processing (TAP) that show a reduced population of CD56^{lo}CD16$^+$ NK cells in the early period after graft [141]. Lastly, hIL-15R agonists may find clinical applications during various types of cancer immunotherapy.

The efficiency of IL-15R agonists in augmenting human NK cell development in HIS mice will enable us to more readily dissect the developmental pathways and signals involved in the generation of IL-15 dependent lymphocytes (NK cells, memory CD8 T cells, NK T cell, and TCR$\gamma\delta$ T cells). Given obvious crosstalk between innate and adaptive immune cells, having robust reconstitution of IL-15-dependent cells improves the accuracy and application of HIS mice for studying human immune responses to infectious pathogens and cellular transformation events in vivo.

7 Concluding Remarks

Knowledge of the rules governing innate lymphocyte development (including NK cells) has lagged behind that of T and B cells of the adaptive immune system. Even in the late 1990s, schematic models for NK cell development in the mouse were extremely simplistic with an NKP in the bone marrow and a mature NK cell in the periphery (blood, spleen, etc.). Models for human NK cell development were based mostly on in vitro culture systems and clearly underestimated the complexity of the process. We still have much to learn about the signals and mechanisms that control human NK cell development in vivo. Nevertheless, lessons from mice and novel xenograft models that recapitulate some aspects of human NK cell development in vivo should provide a means to further dissect this process. A better understanding of human NK cell development may provide avenues for developing novel therapeutic approaches that could impact in the clinic.

Acknowledgements This work is supported by grants from Institut Pasteur, INSERM, Ligue Nationale contre le Cancer, Human Frontiers Science Program, National Health and Medical Research Council of Australia, and a Grand Challenges in Global Health grant from the Bill and Melinda Gates Foundation.

References

1. Blom B, Spits H (2006) Development of human lymphoid cells. Annu Rev Immunol 24: 287–320
2. Gimeno R, Weijer K, Voordouw A, Uittenbogaart CH, Legrand N, Alves NL, Wijnands E, Blom B, Spits H (2004) Monitoring the effect of gene silencing by RNA interference in human CD34+ cells injected into newborn RAG2-/- gammac-/- mice: functional inactivation of p53 in developing T cells. Blood 104:3886–3893
3. Jaleco AC, Blom B, Res P, Weijer K, Lanier LL, Phillips JH, Spits H (1997) Fetal liver contains committed NK progenitors, but is not a site for development of CD34+ cells into T cells. J Immunol 159:694–702
4. Spits H, Lanier LL, Phillips JH (1995) Development of human T and natural killer cells. Blood 85:2654–2670
5. Cobaleda C, Schebesta A, Delogu A, Busslinger M (2007) Pax5: the guardian of B cell identity and function. Nat Immunol 8:463–470
6. Laiosa CV, Stadtfeld M, Graf T (2006) Determinants of lymphoid-myeloid lineage diversification. Annu Rev Immunol 24:705–738
7. Maillard I, Fang T, Pear WS (2005) Regulation of lymphoid development, differentiation, and function by the Notch pathway. Annu Rev Immunol 23:945–974
8. Pear WS, Radtke F (2003) Notch signaling in lymphopoiesis. Semin Immunol 15:69–79
9. Bendelac A, Savage PB, Teyton L (2007) The biology of NKT cells. Annu Rev Immunol 25:297–336
10. Boos MD, Yokota Y, Eberl G, Kee BL (2007) Mature natural killer cell and lymphoid tissue-inducing cell development requires Id2-mediated suppression of E protein activity. J Exp Med 204:1119–1130
11. Ikawa T, Fujimoto S, Kawamoto H, Katsura Y, Yokota Y (2001) Commitment to natural killer cells requires the helix-loop-helix inhibitor Id2. Proc Natl Acad Sci USA 98:5164–5169
12. Spits H, Couwenberg F, Bakker AQ, Weijer K, Uittenbogaart CH (2000) Id2 and Id3 inhibit development of CD34(+) stem cells into predendritic cell (pre-DC)2 but not into pre-DC1. Evidence for a lymphoid origin of pre-DC2. J Exp Med 192:1775–1784
13. Yokota Y, Mansouri A, Mori S, Sugawara S, Adachi S, Nishikawa S, Gruss P (1999) Development of peripheral lymphoid organs and natural killer cells depends on the helix-loop-helix inhibitor Id2. Nature 397:702–706
14. Heemskerk MH, Blom B, Nolan G, Stegmann AP, Bakker AQ, Weijer K, Res PC, Spits H (1997) Inhibition of T cell and promotion of natural killer cell development by the dominant negative helix loop helix factor Id3. J Exp Med 186:1597–1602
15. Lohoff M, Duncan GS, Ferrick D, Mittrucker HW, Bischof S, Prechtl S, Rollinghoff M, Schmitt E, Pahl A, Mak TW (2000) Deficiency in the transcription factor interferon regulatory factor (IRF)-2 leads to severely compromised development of natural killer and T helper type 1 cells. J Exp Med 192:325–336

16. Samson SI, Richard O, Tavian M, Ranson T, Vosshenrich CA, Colucci F, Buer J, Grosveld F, Godin I, Di Santo JP (2003) GATA-3 promotes maturation, IFN-gamma production, and liver-specific homing of NK cells. Immunity 19:701–711
17. Sunwoo JB, Kim S, Yang L, Naik T, Higuchi DA, Rubenstein JL, Yokoyama WM (2008) Distal-less homeobox transcription factors regulate development and maturation of natural killer cells. Proc Natl Acad Sci USA 105:10877–10882
18. Townsend MJ, Weinmann AS, Matsuda JL, Salomon R, Farnham PJ, Biron CA, Gapin L, Glimcher LH (2004) T-bet regulates the terminal maturation and homeostasis of NK and Valpha14i NKT cells. Immunity 20:477–494
19. Ma A, Koka R, Burkett P (2006) Diverse functions of IL-2, IL-15, and IL-7 in lymphoid homeostasis. Annu Rev Immunol 24:657–679
20. Rosmaraki EE, Douagi I, Roth C, Colucci F, Cumano A, Di Santo JP (2001) Identification of committed NK cell progenitors in adult murine bone marrow. Eur J Immunol 31:1900–1909
21. Rossi MI, Yokota T, Medina KL, Garrett KP, Comp PC, Schipul AH Jr, Kincade PW (2003) B lymphopoiesis is active throughout human life, but there are developmental age-related changes. Blood 101:576–584
22. McKenna HJ, Stocking KL, Miller RE, Brasel K, De Smedt T, Maraskovsky E, Maliszewski CR, Lynch DH, Smith J, Pulendran B et al (2000) Mice lacking flt3 ligand have deficient hematopoiesis affecting hematopoietic progenitor cells, dendritic cells, and natural killer cells. Blood 95:3489–3497
23. Iizuka K, Chaplin DD, Wang Y, Wu Q, Pegg LE, Yokoyama WM, Fu YX (1999) Requirement for membrane lymphotoxin in natural killer cell development. Proc Natl Acad Sci USA 96:6336–6340
24. Chung JW, Kim MS, Piao ZH, Jeong M, Yoon SR, Shin N, Kim SY, Hwang ES, Yang Y, Lee YH et al (2008) Osteopontin promotes the development of natural killer cells from hematopoietic stem cells. Stem Cells 26:2114–2123
25. Huntington ND, Vosshenrich CA, Di Santo JP (2007) Developmental pathways that generate natural-killer-cell diversity in mice and humans. Nat Rev Immunol 7:703–714
26. Di Santo JP (2006) Natural killer cell developmental pathways: a question of balance. Annu Rev Immunol 24:257–286
27. Ranson T, Vosshenrich CA, Corcuff E, Richard O, Muller W, Di Santo JP (2003) IL-15 is an essential mediator of peripheral NK-cell homeostasis. Blood 101:4887–4893
28. Burkett PR, Koka R, Chien M, Chai S, Boone DL, Ma A (2004) Coordinate expression and trans presentation of interleukin (IL)-15Ralpha and IL-15 supports natural killer cell and memory CD8+ T cell homeostasis. J Exp Med 200:825–834
29. Cooper MA, Bush JE, Fehniger TA, VanDeusen JB, Waite RE, Liu Y, Aguila HL, Caligiuri MA (2002) In vivo evidence for a dependence on interleukin 15 for survival of natural killer cells. Blood 100:3633–3638
30. Dubois S, Mariner J, Waldmann TA, Tagaya Y (2002) IL-15Ralpha recycles and presents IL-15 In trans to neighboring cells. Immunity 17:537–547
31. Koka R, Burkett PR, Chien M, Chai S, Chan F, Lodolce JP, Boone DL, Ma A (2003) Interleukin (IL)-15R[alpha]-deficient natural killer cells survive in normal but not IL-15R[alpha]-deficient mice. J Exp Med 197:977–984
32. Sandau MM, Schluns KS, Lefrancois L, Jameson SC (2004) Cutting edge: transpresentation of IL-15 by bone marrow-derived cells necessitates expression of IL-15 and IL-15R alpha by the same cells. J Immunol 173:6537–6541
33. Mortier E, Woo T, Advincula R, Gozalo S, Ma A (2008) IL-15Ralpha chaperones IL-15 to stable dendritic cell membrane complexes that activate NK cells via trans presentation. J Exp Med 205:1213–1225
34. Buckley RH (2004) Molecular defects in human severe combined immunodeficiency and approaches to immune reconstitution. Annu Rev Immunol 22:625–655
35. Gilmour KC, Fujii H, Cranston T, Davies EG, Kinnon C, Gaspar HB (2001) Defective expression of the interleukin-2/interleukin-15 receptor beta subunit leads to a natural killer cell-deficient form of severe combined immunodeficiency. Blood 98:877–879

36. Schluns KS, Nowak EC, Cabrera-Hernandez A, Puddington L, Lefrancois L, Aguila HL (2004) Distinct cell types control lymphoid subset development by means of IL-15 and IL-15 receptor alpha expression. Proc Natl Acad Sci USA 101:5616–5621
37. Giuliani M, Giron-Michel J, Negrini S, Vacca P, Durali D, Caignard A, Le Bousse-Kerdiles C, Chouaib S, Devocelle A, Bahri R et al (2008) Generation of a novel regulatory NK cell subset from peripheral blood CD34+ progenitors promoted by membrane-bound IL-15. PLoS ONE 3:e2241
38. Mrozek E, Anderson P, Caligiuri MA (1996) Role of interleukin-15 in the development of human CD56+ natural killer cells from CD34+ hematopoietic progenitor cells. Blood 87:2632–2640
39. Carson WE, Haldar S, Baiocchi RA, Croce CM, Caligiuri MA (1994) The c-kit ligand suppresses apoptosis of human natural killer cells through the upregulation of bcl-2. Proc Natl Acad Sci USA 91:7553–7557
40. De Smedt M, Hoebeke I, Reynvoet K, Leclercq G, Plum J (2005) Different thresholds of Notch signaling bias human precursor cells toward B-, NK-, monocytic/dendritic-, or T-cell lineage in thymus microenvironment. Blood 106:3498–3506
41. Le PT, Adams KL, Zaya N, Mathews HL, Storkus WJ, Ellis TM (2001) Human thymic epithelial cells inhibit IL-15- and IL-2-driven differentiation of NK cells from the early human thymic progenitors. J Immunol 166:2194–2201
42. Miller JS, Alley KA, McGlave P (1994) Differentiation of natural killer (NK) cells from human primitive marrow progenitors in a stroma-based long-term culture system: identification of a CD34+7+ NK progenitor. Blood 83:2594–2601
43. Galy A, Travis M, Cen D, Chen B (1995) Human T, B, natural killer, and dendritic cells arise from a common bone marrow progenitor cell subset. Immunity 3:459–473
44. Haddad R, Guardiola P, Izac B, Thibault C, Radich J, Delezoide AL, Baillou C, Lemoine FM, Gluckman JC, Pflumio F, Canque B (2004) Molecular characterization of early human T/NK and B-lymphoid progenitor cells in umbilical cord blood. Blood 104:3918–3926
45. Ryan DH, Nuccie BL, Ritterman I, Liesveld JL, Abboud CN, Insel RA (1997) Expression of interleukin-7 receptor by lineage-negative human bone marrow progenitors with enhanced lymphoid proliferative potential and B-lineage differentiation capacity. Blood 89:929–940
46. Bennett IM, Zatsepina O, Zamai L, Azzoni L, Mikheeva T, Perussia B (1996) Definition of a natural killer NKR-P1A+/CD56-/CD16- functionally immature human NK cell subset that differentiates in vitro in the presence of interleukin 12. J Exp Med 184:1845–1856
47. Lanier LL, Chang C, Phillips JH (1994) Human NKR-P1A. A disulfide-linked homodimer of the C-type lectin superfamily expressed by a subset of NK and T lymphocytes. J Immunol 153:2417–2428
48. Zamai L, Ahmad M, Bennett IM, Azzoni L, Alnemri ES, Perussia B (1998) Natural killer (NK) cell-mediated cytotoxicity: differential use of TRAIL and Fas ligand by immature and mature primary human NK cells. J Exp Med 188:2375–2380
49. Loza MJ, Zamai L, Azzoni L, Rosati E, Perussia B (2002) Expression of type 1 (interferon gamma) and type 2 (interleukin-13, interleukin-5) cytokines at distinct stages of natural killer cell differentiation from progenitor cells. Blood 99:1273–1281
50. Borrego F, Masilamani M, Kabat J, Sanni TB, Coligan JE (2005) The cell biology of the human natural killer cell CD94/NKG2A inhibitory receptor. Mol Immunol 42:485–488
51. Parham P (2006) Taking license with natural killer cell maturation and repertoire development. Immunol Rev 214:155–160
52. Yokoyama WM, Kim S (2006) Licensing of natural killer cells by self-major histocompatibility complex class I. Immunol Rev 214:143–154
53. Karre K, Ljunggren HG, Piontek G, Kiessling R (1986) Selective rejection of H-2-deficient lymphoma variants suggests alternative immune defence strategy. Nature 319:675–678
54. Lanier LL, Le AM, Civin CI, Loken MR, Phillips JH (1986) The relationship of CD16 (Leu-11) and Leu-19 (NKH-1) antigen expression on human peripheral blood NK cells and cytotoxic T lymphocytes. J Immunol 136:4480–4486

55. Carrington M, Martin MP (2006) The impact of variation at the KIR gene cluster on human disease. Curr Top Microbiol Immunol 298:225–257
56. Ruggeri L, Capanni M, Urbani E, Perruccio K, Shlomchik WD, Tosti A, Posati S, Rogaia D, Frassoni F, Aversa F et al (2002) Effectiveness of donor natural killer cell alloreactivity in mismatched hematopoietic transplants. Science 295:2097–2100
57. Anderson SK (2006) Transcriptional regulation of NK cell receptors. Curr Top Microbiol Immunol 298:59–75
58. Held W, Clevers H, Grosschedl R (2003) Redundant functions of TCF-1 and LEF-1 during T and NK cell development, but unique role of TCF-1 for Ly49 NK cell receptor acquisition. Eur J Immunol 33:1393–1398
59. Cooley S, Xiao F, Pitt M, Gleason M, McCullar V, Bergemann TL, McQueen KL, Guethlein LA, Parham P, Miller JS (2007) A subpopulation of human peripheral blood NK cells that lacks inhibitory receptors for self-MHC is developmentally immature. Blood 110:578–586
60. Huntington ND, Legrand N, Alves NL, Jaron B, Weijer K, Plet A, Corcuff E, Mortier E, Jacques Y, Spits H, Di Santo JP (2009) IL-15 trans-presentation promotes human NK cell development and differentiation in vivo. J Exp Med 206:25–34
61. Williams NS, Kubota A, Bennett M, Kumar V, Takei F (2000) Clonal analysis of NK cell development from bone marrow progenitors in vitro: orderly acquisition of receptor gene expression. Eur J Immunol 30:2074–2082
62. Anfossi N, Andre P, Guia S, Falk CS, Roetynck S, Stewart CA, Breso V, Frassati C, Reviron D, Middleton D et al (2006) Human NK cell education by inhibitory receptors for MHC class I. Immunity 25:331–342
63. Fernandez NC, Treiner E, Vance RE, Jamieson AM, Lemieux S, Raulet DH (2005) A subset of natural killer cells achieves self-tolerance without expressing inhibitory receptors specific for self-MHC molecules. Blood 105:4416–4423
64. Kim S, Poursine-Laurent J, Truscott SM, Lybarger L, Song YJ, Yang L, French AR, Sunwoo JB, Lemieux S, Hansen TH, Yokoyama WM (2005) Licensing of natural killer cells by host major histocompatibility complex class I molecules. Nature 436:709–713
65. Furukawa H, Yabe T, Watanabe K, Miyamoto R, Miki A, Akaza T, Tadokoro K, Tohma S, Inoue T, Yamamoto K, Juji T (1999) Tolerance of NK and LAK activity for HLA class I-deficient targets in a TAP1-deficient patient (bare lymphocyte syndrome type I). Hum Immunol 60:32–40
66. Bix M, Liao NS, Zijlstra M, Loring J, Jaenisch R, Raulet D (1991) Rejection of class I MHC-deficient haemopoietic cells by irradiated MHC-matched mice. Nature 349:329–331
67. Yu J, Heller G, Chewning J, Kim S, Yokoyama WM, Hsu KC (2007) Hierarchy of the human natural killer cell response is determined by class and quantity of inhibitory receptors for self-HLA-B and HLA-C ligands. J Immunol 179:5977–5989
68. Ferlazzo G, Thomas D, Lin SL, Goodman K, Morandi B, Muller WA, Moretta A, Munz C (2004) The abundant NK cells in human secondary lymphoid tissues require activation to express killer cell Ig-like receptors and become cytolytic. J Immunol 172:1455–1462
69. Trinchieri G (1989) Biology of natural killer cells. Adv Immunol 47:187–376
70. Whiteside TL, Herberman RB (1994) Role of human natural killer cells in health and disease. Clin Diagn Lab Immunol 1:125–133
71. Lanier LL, Le AM, Phillips JH, Warner NL, Babcock GF (1983) Subpopulations of human natural killer cells defined by expression of the Leu-7 (HNK-1) and Leu-11 (NK-15) antigens. J Immunol 131:1789–1796
72. Farag SS, Caligiuri MA (2006) Human natural killer cell development and biology. Blood Rev 20:123–137
73. Freud AG, Caligiuri MA (2006) Human natural killer cell development. Immunol Rev 214:56–72
74. Andre P, Spertini O, Guia S, Rihet P, Dignat-George F, Brailly H, Sampol J, Anderson PJ, Vivier E (2000) Modification of P-selectin glycoprotein ligand-1 with a natural killer

cell-restricted sulfated lactosamine creates an alternate ligand for L-selectin. Proc Natl Acad Sci USA 97:3400–3405
75. Caligiuri MA, Zmuidzinas A, Manley TJ, Levine H, Smith KA, Ritz J (1990) Functional consequences of interleukin 2 receptor expression on resting human lymphocytes. Identification of a novel natural killer cell subset with high affinity receptors. J Exp Med 171: 1509–1526
76. Campbell JJ, Qin S, Unutmaz D, Soler D, Murphy KE, Hodge MR, Wu L, Butcher EC (2001) Unique subpopulations of CD56+ NK and NK-T peripheral blood lymphocytes identified by chemokine receptor expression repertoire. J Immunol 166:6477–6482
77. Frey M, Packianathan NB, Fehniger TA, Ross ME, Wang WC, Stewart CC, Caligiuri MA, Evans SS (1998) Differential expression and function of L-selectin on CD56bright and CD56dim natural killer cell subsets. J Immunol 161:400–408
78. Jacobs R, Hintzen G, Kemper A, Beul K, Kempf S, Behrens G, Sykora KW, Schmidt RE (2001) CD56bright cells differ in their KIR repertoire and cytotoxic features from CD56dim NK cells. Eur J Immunol 31:3121–3127
79. Matos ME, Schnier GS, Beecher MS, Ashman LK, William DE, Caligiuri MA (1993) Expression of a functional c-kit receptor on a subset of natural killer cells. J Exp Med 178:1079–1084
80. Nagler A, Lanier LL, Phillips JH (1990) Constitutive expression of high affinity interleukin 2 receptors on human CD16-natural killer cells in vivo. J Exp Med 171:1527–1533
81. Voss SD, Daley J, Ritz J, Robertson MJ (1998) Participation of the CD94 receptor complex in costimulation of human natural killer cells. J Immunol 160:1618–1626
82. Cooper MA, Fehniger TA, Turner SC, Chen KS, Ghaheri BA, Ghayur T, Carson WE, Caligiuri MA (2001) Human natural killer cells: a unique innate immunoregulatory role for the CD56(bright) subset. Blood 97:3146–3151
83. Nagler A, Lanier LL, Cwirla S, Phillips JH (1989) Comparative studies of human FcRIII-positive and negative natural killer cells. J Immunol 143:3183–3191
84. Chklovskaia E, Nowbakht P, Nissen C, Gratwohl A, Bargetzi M, Wodnar-Filipowicz A (2004) Reconstitution of dendritic and natural killer-cell subsets after allogeneic stem cell transplantation: effects of endogenous flt3 ligand. Blood 103:3860–3868
85. Gottschalk LR, Bray RA, Kaizer H, Gebel HM (1990) Two populations of CD56 (Leu-19)+/CD16+ cells in bone marrow transplant recipients. Bone Marrow Transplant 5: 259–264
86. Jacobs R, Stoll M, Stratmann G, Leo R, Link H, Schmidt RE (1992) CD16- CD56+ natural killer cells after bone marrow transplantation. Blood 79:3239–3244
87. Romagnani C, Juelke K, Falco M, Morandi B, D'Agostino A, Costa R, Ratto G, Forte G, Carrega P, Lui G et al (2007) CD56brightCD16- killer Ig-like receptor- NK cells display longer telomeres and acquire features of CD56dim NK cells upon activation. J Immunol 178:4947–4955
88. Chan A, Hong DL, Atzberger A, Kollnberger S, Filer AD, Buckley CD, McMichael A, Enver T, Bowness P (2007) CD56bright human NK cells differentiate into CD56dim cells: role of contact with peripheral fibroblasts. J Immunol 179:89–94
89. Di Santo JP (2008) Natural killer cells: diversity in search of a niche. Nat Immunol 9: 473–475
90. Freud AG, Becknell B, Roychowdhury S, Mao HC, Ferketich AK, Nuovo GJ, Hughes TL, Marburger TB, Sung J, Baiocchi RA et al (2005) A human CD34(+) subset resides in lymph nodes and differentiates into CD56bright natural killer cells. Immunity 22:295–304
91. Freud AG, Yokohama A, Becknell B, Lee MT, Mao HC, Ferketich AK, Caligiuri MA (2006) Evidence for discrete stages of human natural killer cell differentiation in vivo. J Exp Med 203:1033–1043
92. Grzywacz B, Kataria N, Sikora M, Oostendorp RA, Dzierzak EA, Blazar BR, Miller JS, Verneris MR (2006) Coordinated acquisition of inhibitory and activating receptors and functional properties by developing human natural killer cells. Blood 108:3824–3833

93. Veinotte LL, Halim TY, Takei F (2008) Unique subset of natural killer cells develops from progenitors in lymph node. Blood 111:4201–4208
94. Mebius RE, Rennert P, Weissman IL (1997) Developing lymph nodes collect CD4+CD3-LTbeta+ cells that can differentiate to APC, NK cells, and follicular cells but not T or B cells. Immunity 7:493–504
95. Eberl G, Littman DR (2003) The role of the nuclear hormone receptor RORgammat in the development of lymph nodes and Peyer's patches. Immunol Rev 195:81–90
96. Cupedo T, Crellin NK, Papazian N, Rombouts EJ, Weijer K, Grogan JL, Fibbe WE, Cornelissen JJ, Spits H (2009) Human fetal lymphoid tissue-inducer cells are interleukin 17-producing precursors to RORC+ CD127+ natural killer-like cells. Nat Immunol 10:66–74
97. Fehniger TA, Cooper MA, Nuovo GJ, Cella M, Facchetti F, Colonna M, Caligiuri MA (2003) CD56bright natural killer cells are present in human lymph nodes and are activated by T cell-derived IL-2: a potential new link between adaptive and innate immunity. Blood 101:3052–3057
98. Gallatin WM, Weissman IL, Butcher EC (1983) A cell-surface molecule involved in organ-specific homing of lymphocytes. Nature 304:30–34
99. Kim CH, Pelus LM, Appelbaum E, Johanson K, Anzai N, Broxmeyer HE (1999) CCR7 ligands, SLC/6Ckine/Exodus2/TCA4 and CKbeta-11/MIP-3beta/ELC, are chemoattractants for CD56(+)CD16(-) NK cells and late stage lymphoid progenitors. Cell Immunol 193:226–235
100. Martin-Fontecha A, Thomsen LL, Brett S, Gerard C, Lipp M, Lanzavecchia A, Sallusto F (2004) Induced recruitment of NK cells to lymph nodes provides IFN-gamma for T(H)1 priming. Nat Immunol 5:1260–1265
101. Mailliard RB, Alber SM, Shen H, Watkins SC, Kirkwood JM, Herberman RB, Kalinski P (2005) IL-18-induced CD83+CCR7+ NK helper cells. J Exp Med 202:941–953
102. Cooper MA, Fehniger TA, Caligiuri MA (2001) The biology of human natural killer-cell subsets. Trends Immunol 22:633–640
103. Morandi B, Bougras G, Muller WA, Ferlazzo G, Munz C (2006) NK cells of human secondary lymphoid tissues enhance T cell polarization via IFN-gamma secretion. Eur J Immunol 36:2394–2400
104. Garni-Wagner BA, Witte PL, Tutt MM, Kuziel WA, Tucker PW, Bennett M, Kumar V (1990) Natural killer cells in the thymus. Studies in mice with severe combined immune deficiency. J Immunol 144:796–803
105. Michon JM, Caligiuri MA, Hazanow SM, Levine H, Schlossman SF, Ritz J (1988) Induction of natural killer effectors from human thymus with recombinant IL-2. J Immunol 140:3660–3667
106. Mingari MC, Poggi A, Bellomo R, Pella N, Moretta L (1991) Thymic origin of some natural killer cells: clonal proliferation of human CD3-16+ cells from CD3-4-8- thymocyte precursors requires the presence of H9 leukemic cells. Int J Clin Lab Res 21:176–178
107. Sanchez MJ, Spits H, Lanier LL, Phillips JH (1993) Human natural killer cell committed thymocytes and their relation to the T cell lineage. J Exp Med 178:1857–1866
108. Barcena A, Muench MO, Galy AH, Cupp J, Roncarolo MG, Phillips JH, Spits H (1993) Phenotypic and functional analysis of T-cell precursors in the human fetal liver and thymus: CD7 expression in the early stages of T- and myeloid-cell development. Blood 82:3401–3414
109. Carlyle JR, Michie AM, Furlonger C, Nakano T, Lenardo MJ, Paige CJ, Zuniga-Pflucker JC (1997) Identification of a novel developmental stage marking lineage commitment of progenitor thymocytes. J Exp Med 186:173–182
110. Michie AM, Carlyle JR, Schmitt TM, Ljutic B, Cho SK, Fong Q, Zuniga-Pflucker JC (2000) Clonal characterization of a bipotent T cell and NK cell progenitor in the mouse fetal thymus. J Immunol 164:1730–1733
111. Vosshenrich CA, Garcia-Ojeda ME, Samson-Villeger SI, Pasqualetto V, Enault L, Richard-Le Goff O, Corcuff E, Guy-Grand D, Rocha B, Cumano A et al (2006) A thymic pathway of

mouse natural killer cell development characterized by expression of GATA-3 and CD127. Nat Immunol 7:1217–1224
112. De Smedt M, Taghon T, Van de Walle I, De Smet G, Leclercq G, Plum J (2007) Notch signaling induces cytoplasmic CD3 epsilon expression in human differentiating NK cells. Blood 110:2696–2703
113. Hanna J, Bechtel P, Zhai Y, Youssef F, McLachlan K, Mandelboim O (2004) Novel insights on human NK cells' immunological modalities revealed by gene expression profiling. J Immunol 173:6547–6563
114. Di Santo JP, Vosshenrich CA (2006) Bone marrow versus thymic pathways of natural killer cell development. Immunol Rev 214:35–46
115. Carotta S, Brady J, Wu L, Nutt SL (2006) Transient Notch signaling induces NK cell potential in Pax5-deficient pro-B cells. Eur J Immunol 36:3294–3304
116. Huang J, Garrett KP, Pelayo R, Zuniga-Pflucker JC, Petrie HT, Kincade PW (2005) Propensity of adult lymphoid progenitors to progress to DN2/3 stage thymocytes with Notch receptor ligation. J Immunol 175:4858–4865
117. Rolink AG, Balciunaite G, Demoliere C, Ceredig R (2006) The potential involvement of Notch signaling in NK cell development. Immunol Lett 107:50–57
118. Radtke F, Wilson A, Ernst B, MacDonald HR (2002) The role of Notch signaling during hematopoietic lineage commitment. Immunol Rev 187:65–74
119. Poggi A, Biassoni R, Pella N, Paolieri F, Bellomo R, Bertolini A, Moretta L, Mingari MC (1990) In vitro expansion of CD3/TCR- human thymocyte populations that selectively lack CD3 delta gene expression: a phenotypic and functional analysis. J Exp Med 172:1409–1418
120. Leon F, Roldan E, Sanchez L, Camarero C, Bootello A, Roy G (2003) Human small-intestinal epithelium contains functional natural killer lymphocytes. Gastroenterology 125:345–356
121. Lundqvist C, Baranov V, Hammarstrom S, Athlin L, Hammarstrom ML (1995) Intraepithelial lymphocytes: evidence for regional specialization and extrathymic T cell maturation in the human gut epithelium. Int Immunol 7:1473–1487
122. Tagliabue A, Befus AD, Clark DA, Bienenstock J (1982) Characteristics of natural killer cells in the murine intestinal epithelium and lamina propria. J Exp Med 155:1785–1796
123. Chinen H, Matsuoka K, Sato T, Kamada N, Okamoto S, Hisamatsu T, Kobayashi T, Hasegawa H, Sugita A, Kinjo F et al (2007) Lamina propria c-kit+ immune precursors reside in human adult intestine and differentiate into natural killer cells. Gastroenterology 133:559–573
124. Luci C, Reynders A, Ivanov II, Cognet C, Chiche L, Chasson L, Hardwigsen J, Anguiano E, Banchereau J, Chaussabel D et al (2009) Influence of the transcription factor RORgammat on the development of NKp46+ cell populations in gut and skin. Nat Immunol 10:75–82
125. Sanos SL, Bui VL, Mortha A, Oberle K, Heners C, Johner C, Diefenbach A (2009) RORgammat and commensal microflora are required for the differentiation of mucosal interleukin 22-producing NKp46+ cells. Nat Immunol 10:83–91
126. Satoh-Takayama N, Vosshenrich CA, Lesjean-Pottier S, Sawa S, Lochner M, Rattis F, Mention JJ, Thiam K, Cerf-Bensussan N, Mandelboim O et al (2008) Microbial flora drives interleukin 22 production in intestinal NKp46+ cells that provide innate mucosal immune defense. Immunity 29:958–970
127. Cella M, Fuchs A, Vermi W, Facchetti F, Otero K, Lennerz JK, Doherty JM, Mills JC, Colonna M (2009) A human natural killer cell subset provides an innate source of IL-22 for mucosal immunity. Nature 457:722–725
128. Zenewicz LA, Yancopoulos GD, Valenzuela DM, Murphy AJ, Stevens S, Flavell RA (2008) Innate and adaptive interleukin-22 protects mice from inflammatory bowel disease. Immunity 29:947–957
129. Murray AM, Simm B, Beagley KW (1998) Cytokine gene expression in murine fetal intestine: potential for extrathymic T cell development. Cytokine 10:337–345
130. Sander GR, Powell BC (2004) Expression of notch receptors and ligands in the adult gut. J Histochem Cytochem 52:509–516

131. Schroder N, Gossler A (2002) Expression of Notch pathway components in fetal and adult mouse small intestine. Gene Expr Patterns 2:247–250
132. Traggiai E, Chicha L, Mazzucchelli L, Bronz L, Piffaretti JC, Lanzavecchia A, Manz MG (2004) Development of a human adaptive immune system in cord blood cell-transplanted mice. Science 304:104–107
133. Legrand N, Cupedo T, van Lent AU, Ebeli MJ, Weijer K, Hanke T, Spits H (2006) Transient accumulation of human mature thymocytes and regulatory T cells with CD28 superagonist in "human immune system" Rag2(-/-)gammac(-/-) mice. Blood 108:238–245
134. Shultz LD, Ishikawa F, Greiner DL (2007) Humanized mice in translational biomedical research. Nat Rev Immunol 7:118–130
135. Melkus MW, Estes JD, Padgett-Thomas A, Gatlin J, Denton PW, Othieno FA, Wege AK, Haase AT, Garcia JV (2006) Humanized mice mount specific adaptive and innate immune responses to EBV and TSST-1. Nat Med 12:1316–1322
136. Kalberer CP, Siegler U, Wodnar-Filipowicz A (2003) Human NK cell development in NOD/SCID mice receiving grafts of cord blood CD34+ cells. Blood 102:127–135
137. Carson WE, Giri JG, Lindemann MJ, Linett ML, Ahdieh M, Paxton R, Anderson D, Eisenmann J, Grabstein K, Caligiuri MA (1994) Interleukin (IL) 15 is a novel cytokine that activates human natural killer cells via components of the IL-2 receptor. J Exp Med 180:1395–1403
138. Miller JS, McCullar V (2001) Human natural killer cells with polyclonal lectin and immunoglobulinlike receptors develop from single hematopoietic stem cells with preferential expression of NKG2A and KIR2DL2/L3/S2. Blood 98:705–713
139. Cooley S, McCullar V, Wangen R, Bergemann TL, Spellman S, Weisdorf DJ, Miller JS (2005) KIR reconstitution is altered by T cells in the graft and correlates with clinical outcomes after unrelated donor transplantation. Blood 106:4370–4376
140. Shilling HG, McQueen KL, Cheng NW, Shizuru JA, Negrin RS, Parham P (2003) Reconstitution of NK cell receptor repertoire following HLA-matched hematopoietic cell transplantation. Blood 101:3730–3740
141. Zimmer J, Bausinger H, Andres E, Donato L, Hanau D, Hentges F, Moretta A, de la Salle H (2007) Phenotypic studies of natural killer cell subsets in human transporter associated with antigen processing deficiency. PLoS ONE 2:e1033

Diversity of KIR Genes, Alleles and Haplotypes

D. Middleton, F. Gonzalez-Galarza, A. Meenagh, and P.A. Gourraud

Abstract In this chapter, we discuss the vast polymorphism of the killer cell immunoglobulin-like receptors (KIR), of the natural killer (NK) cell, which rivals that of the HLA complex. There are several aspects of this polymorphism. Initially, there is the presence/absence of individual KIR genes, with four of these genes termed framework genes, being omnipresent, with very few exceptions, in all individuals tested to date. Within each gene, alleles are present at different frequencies. We show how these frequencies vary in different world-wide populations.

Another concept of the KIR complex is the division of an individual genotype into A and/or B haplotypes, the former having a more inhibitory role and the latter, a more activating role on the function of the NK cell. Family studies have been used to ascertain the make-up of these haplotypes, inclusion of allele typing enabling determination of whether one or two copies of a particular gene is present.

The KIR gene complex is rapidly evolving not only at the genetic level but additionally at the functional level with different alleles having different protein expression levels and different avidity with their HLA ligand.

We provide details of a new website, which enables convenient searching for data on KIR gene, allele and genotype frequencies in different populations. Finally, we expand on our original algorithms, using additional family data, to give details on mathematical methods for estimation of haplotypes from genotype data.

D. Middleton (✉)
Transplant Immunology, Royal Liverpool and Broadgreen University Hospital and School of Infection and Host Defence Liverpool University, Prescott Street, Liverpool, UK
e-mail: Derek.Middleton@rlbuht.nhs.uk

F. Gonzalez-Galarza and A. Meenagh
Northern Ireland Region Histocompatibility and Immunogenetics Laboratory, Belfast City Hospital, Belfast, Northern Ireland, UK

P.A. Gourraud
INSERM, Unit 558, University of Toulouse, Toulouse, France

Department of Neurology, University of California at San Francisco, 513 Parnassus Avenue, San Francisco, CA 94143-043, USA

1 Introduction

Natural killer (NK) cells are an intrinsic component within the immune response, specifically modulating the innate targeting of infected and transformed malignant cells [1].

NK cells were discovered because of their ability to kill certain tumor cell lines, which expressed little or no major histocompatibility complex (MHC) class I molecules [2]. This led to the "missing-self" hypothesis which formulated that NK cells recognize and, thereafter, eliminate cells that fail to express self-MHC molecules.

The cytolytic activity of human NK cells is modulated by the interaction of inhibitory and activating membrane receptors, expressed on their surface, with MHC class I antigens expressed by host cells. The receptors belong to two distinct families, C-type lectins-like group (CD94/NKG2) mapping to chromosome 12q1.3–13.4 and the immunoglobulin-like superfamily consisting of the killer cell immunoglobulin-like receptors (KIR), leucocyte immunoglobulin-like receptors (LIR) and the leukocyte-associated immunoglobulin-like receptors mapping to chromosome 19q13.4 [3–5]. KIR are the most variable of these receptors, interacting with the very polymorphic human leukocyte antigen (HLA) system and thus have been the subject of intense investigation.

Through KIR and other receptors NK cells also maintain wide-ranging interactions with other immune cells such as macrophages and dendritic cells, producing more extensive effects on the immune system as a whole, through stimulation of cytokine production and induction of cytotoxicity [6]. This intracellular crosstalk not only provides a beneficial role but can also produce a detrimental effect, which may play a role in the development of autoimmune disorders [7–9] and less common conditions such as preeclampsia [10]. It has also been shown that haematopoietic stem cell engraftment may be improved by careful selection of KIR mismatches in relation to the HLA types of the donor and recipient [11], or that presence of certain KIR receptors leads to improved survival [12].

KIR have highly variable genetic make-up and are one of, or may actually be, the most rapidly evolving human families. Variation is due to gene and allele content, giving rise to haplotype diversity and leading to a staggering number of different genotypes. This diversity is compounded by functional diversity (variegated expression, ligand- binding specificity and inhibitory strength).

2 KIR Genes

2.1 Nomenclature

Interactions between KIR and their appropriate ligands on target cells result in the production of positive or negative signals, which, in effect, regulate NK cell function [13, 14]. Intense research interest over the last few years has seen a clearer picture emerge on the genomic organization of the KIR [15, 16] and the extent of KIR diversity within the human population [17, 18]. To date, 15 distinct KIR gene loci have been identified which vary with respect to their presence or absence on different KIR haplotypes, creating considerable diversity in the number of KIR genotypes observed in the population. This extensive diversity caused by genomic recombination events means that the distinction between separate genes or alleles of the same KIR is *not always* clear. Although sometimes in this review we refer to them separately, KIR3DL1 and KIR3DS1 are designated as KIR3DL1/S1 and KIR2DL2 and KIR2DL3 as KIR2DL2/3 respectively. Each KIR gene encodes either an inhibitory or an activating KIR, except KIR3DL1/S1 which encodes either, depending on which allele is present. KIR have either two or three extra cellular Ig domains, called 2D or 3D and either a long (L) or a short (S) intracellular domain. A subcommittee of the WHO Nomenclature Committee for Factors of the HLA System has reported on the naming of the genes and the alleles of the genes encoding the KIR [19]. The names given to the KIR genes are based on the structures of the molecules they encode (Fig. 1). The first digit following the KIR acronym corresponds to the number of Ig-like domains in the molecule and the "D" denotes "domain." The D is followed by either an "L," indicating a "long" cytoplasmic tail, (these proteins have inhibitory function), or "S" indicating a "short" cytoplasmic tail, (these proteins have activating function), or a "P" for "pseudogene." The final digit indicates the number of the gene encoding a protein with this structure. Where two or more genes have very similar structures and have very similar sequences, they may be given the same number but distinguished by a *final letter*, for example, the KIR2DL5*A* and KIR2DL5*B* genes [20]. KIR2DL4 shares structure features with *both* inhibitory and activating KIR and although initial

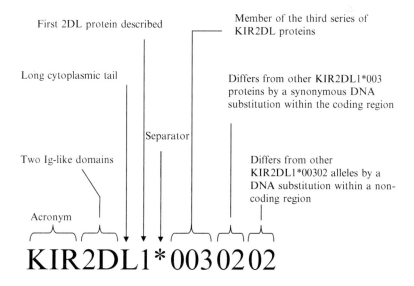

Fig. 1 Nomenclature of KIR genes and alleles

observations suggested only an inhibitory role, evidence has also been produced later for an activating function [21]. KIR alleles are named in a similar fashion to alleles of the HLA system (Fig. 1). Thus, the first three digits distinguish alleles differing in exon sequences that lead to nonsynonymous changes; the next two digits indicate alleles that differ in exon sequences leading to synonymous changes and the last two digits are used for those alleles that only differ in an intron, promoter or other noncoding region.

2.2 Ligands

The HLA class I molecules are the ligands for several of the KIR genes. Two groups of HLA-Cw alleles differ by the amino acid present at position 80 of the molecule. Approximately, 50% of known HLA-Cw alleles are in each group. HLA-C group 1 with asparagine at position 80 provides the ligand for KIR2DL2 and KIR2DL3, whilst HLA-C group 2 with lysine at position 80 provides the ligand for KIR2DL1. However, a very recent report has shown that whereas KIR2DL1 has only interaction with HLA-C2 group, KIR2DL2 and to a weaker extent KIR2DL3, bind to HLA-C2 group [22]. Although the binding of KIR2DL2 to HLA-C2 group is usually weaker than the corresponding binding of KIR2DL1 to HLA-C2 group, some alleles, notably HLA-Cw*0501, a HLA-C2 group allele, bind similarly to HLA-C1 alleles. Two positions, one in each of the domains D1 and D2, determine binding differences between KIR2DL2 and KIR2DL3. The authors believed that the binding of KIR2DL2 and KIR2DL3 to HLA-C2 group should be considered in disease analysis.

KIR3DL1 has specificity for the HLA-B molecules with an HLA-Bw4 epitope at residues 77–83. The HLA-Bw4 epitope is also present on some HLA-A molecules. KIR3DL2 has specificity for the HLA–A3 and –A11 allele families but only when certain virally derived peptides are loaded. Finally, HLA-G is the receptor for KIR2DL4. HLA-C would appear to be the most important in the regulation of NK cells as all individuals will carry an HLA-Cw allele while approximately 25% of individuals will not have the HLA-A or HLA-B ligands [23].

2.3 A and B Haplotypes

Two haplotype groups, A and B, have been identified based on KIR gene content. These haplotypes were initially described using HindIII digestion and Southern blot analyses resulting in a 24-kb band being present in group B and absent in group A [24]. The basis of each A or B haplotype consists of four framework genes: KIR2DL4, KIR3DL2, KIR3DL3 and KIR3DP1. Duplication and deletion of genes have led to many different haplotypes (Fig. 2). The A haplotype is generally less variable in its gene organization, utilizing up to eight genes: those of the

Diversity of KIR Genes, Alleles and Haplotypes

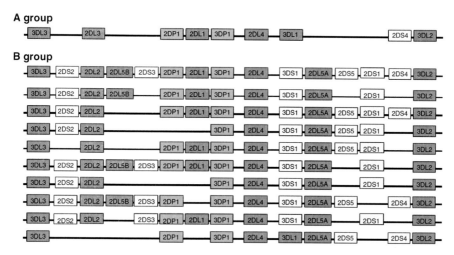

Fig. 2 Two groups of KIR haplotypes (A and B)

Table 1 Number of alleles for each KIR gene

2DL1	15	2DS1	12	2DP1	5
2DL2	5	2DS2	10	3DP1	7
2DL3	7	2DS3	7		
2DL4	26	2DS4	11		
2DL5A	8	2DS5	10		
2DL5B	13	3DS1	14		
3DL1	49				
2DL2	38				
3DL3	55				

framework and KIR2DL1, KIR2DL3, KIR2DS4 and KIR3DL1. Variability within the A haplotype is produced by the tendency towards allelic diversity of the genes. Of the 292 alleles reported to date, 216 are from the nine inhibitory genes, whereas 64 are from the six activating genes (Table 1) [19, 25]. The remaining 12 alleles are contributed by the pseudogenes KIR2DP1 and KIR3DP1. The only activating gene, bar KIR2DL4, on the A haplotype is KIR2DS4.

The B haplotype is defined by the presence of one or more of the genes encoding activating KIR, KIR2DS1/2/3/5, KIR3DS1 and the genes encoding inhibitory KIR, KIR2DL5A/B and KIR2DL2. Those genes (KIR2DL1, KIR2DL3, KIR2DS4 and KIR3DL1) normally associated with the A haplotype can also be found on the B haplotype. Variability on the B haplotype is created mainly by the presence or absence of the genes and, to a lesser extent, by allele variability. The genes encoding inhibitory KIR are nearly always present in populations at frequencies greater than 90%. The exceptions are those on the B haplotypes; KIR2DL2 (which is believed to be allelic with KIR2DL3) and the KIR2DL5 genes, KIR2DL5A and KIR2DL5B.

Full haplotype-length sequencing has been performed for KIR haplotypes [15, 16, 26]. The order of the genes on each haplotype has been mapped with KIR3DL3 at centromeric end, KIR3DL2 at telomeric end and KIR2DL4 in the middle.

2.4 Methods to Detect KIR Genes

The most popular methods used to date for determination of presence or absence of KIR genes have been sequence specific primers (SSP) [24, 27–29] and sequence specific oligonucleotide probes (SSOP) [30]. These methods do not distinguish between homozygotes and heterozygotes at a given locus. During the development of the techniques variation in results were found between laboratories [31, 32]. Virtually all these difficulties have been overcome and many laboratories now take part with 100% success in the typing of samples from the UCLA cell exchange [33]. Several groups have provided KIR typing information on lymphoblastoid cell-lines to act as reference material for these typing methodologies. Results of typing of cell lines, including allele typing, are included on the website www.allelefrequencies.net [34]. Data are available for 84 International Histocompatibility Workshop (IHW) cell lines and 12 CEPH families from the 13th IHW.

The growth in the number of alleles, discovered to date, and the similarities in sequences of alleles, even those of different KIR genes, indicate that SSOP and SSP methods, although initially applied, will no longer be sufficient for allele determination. A method to separate the two haplotypes present in an individual would prove very beneficial, not only for allele typing but to determine the order of genes on a haplotype. Such a method has been reported [35] but unfortunately has not yet enjoyed much success in other laboratories. Some laboratories have resorted to sequencing for allele determination whilst others have used mass spectrometry or pyrosequencing [12, 23, 36]. We are attempting in our laboratory to use real-time RT-PCR, which not only could prove useful for allele determination but also for determining copy number of either gene or allele.

2.5 Gene Frequencies in Different Populations

There is a wide variation in the frequencies of KIR genes in world-wide populations. We have examined the website www.allelefrequencies.net which contains data for 136 populations [34]. Most of the population data is taken from publications and information on the publication and demographic details of the populations are given on the website. The populations have been named according to country name, region and ethnicity. Analysis shows that the framework genes are present with very few exceptions in all individuals. The inhibitory KIR are present in the majority of individuals, whereas activating KIR show great variation in presence/absence in the populations studied.

To date, publications have reported only four individuals negative for KIR2DL4; One CEPH family member [29], one from the Bubi population on Bioko Island Equatorial Guinea [37] and two from South Asia [38]. However, in a study in 77 Caucasian families, we found two haplotypes on different individuals which were missing KIR2DL4 [39]. Individuals have been reported to the website as being negative for KIR2DL4 and the other framework genes KIR3DL2 and KIR3DL3. Unfortunately, it is not possible to verify if some of these reports are due to inaccurate typing. KIR2DL4 has been reported as negative, in addition to the occasions mentioned previously, in one Brazilian individual. Six populations have a total of 13 individuals without KIR3DL2, five populations have a total of 10 individuals without KIR3DL3 and 8 populations have a total of 15 individuals without KIR3DP1. The other pseudogene, KIR2DP1 is present in all populations at greater than 90% with two exceptions, Amerindian populations from Argentina: the Salta Wichis have 84.2% and the Chiraguanos 84.0%.

The inhibitory KIR, KIR2DL1 and KIR2DL3 have very high frequencies (90%) in most populations but there are exceptions. Reports of 136 populations show that 15 populations have less than 90% frequency for KIR2DL1 notably Taiwan (65%), Australian Aborigines (72%), Venezuela Yucpa (70.5%). Frequencies of KIR2DL3 are usually between 80% and 100%, although six populations have frequencies less than 70%. Three of these populations are from Africa – South Africa San, South Africa Xhosa and Congo Mbut. KIR2DL2, which is believed to be allelic to KIR2DL3 tends to have frequencies in the range 40–60%. Again, there are marked exceptions with much lower frequencies being found in all six Japanese populations reported (8.5–15.4%) and much higher in the San, Xhosa and Australian Aborigine populations (72–79%). One very high frequency (95.5%) has been reported in Papua New Guinea Nasioi, which correspondingly has the lowest frequency (59%) reported to date for KIR2DL3. The Taiwan Taroko Atayal population does not have KIR2DL2 or its activating counterpart KIR2DS2. A very small percentage of individuals (0.21%) on the website are negative for both KIR2DL2 and KIR2DL3.

KIR3DL1 has generally very high frequencies (>90%) with 10 populations having a frequency of 100%, and 16 populations having frequencies of 80–90%. Five populations have lower frequencies. Again, these populations tend to be indigenous – Australia Aborigines (55%), Papua New Guinea Nasioi (59.1%), Brazil Amazon (65%), Venezuela Yucpa (70.5%), Mexico Pima (76.8%).

Very little data is available for the other genes encoding inhibitory KIR, KIR2-DL5A and KIR2DL5B, mainly because until recently these genes were not differentiated and reports gave a combined frequency for the two genes. In the 11 populations with data for these two genes the tendency is for frequencies for both genes to be in the 20–40% region, exceptions being China Zhejiang Han (5.8% for KIR2DL5B) and Comoros (13% for KIR2DL5A).

There is a much wider range of frequencies for the activating KIR, indicating a much wider frequency of B haplotypes. There is also within individual populations marked variation in the frequency of each of the activating KIR. B haplotypes have previously been shown to be more prevalent in non-Caucasian populations such as

Australia Aborigines and Asian Indians [38, 40–42], whereas in Caucasian populations approximately 55% of the population will have A haplotypes and 30% have two A haplotypes [43]. It is believed that populations with higher frequencies of B haplotypes will be those under strong pressure from infectious diseases. KIR2DS1 has four populations with greater than 80% frequency :Australia Aborigines, Brazil Amazon, Brazil Rodonia Province Karitiana and Papua New Guinea Nasioi but three African populations with less than 10%: Central Africa Republic Bagandu Biaka, Ghana and Nigeria Enugu Ibo. Similarly, KIR2DS2 has high frequencies (>70%) in nine populations (e.g., Australia Aborigines, South Africa San and Xhosa and populations from India) but very low frequencies in Japan (8.5–16%), S. Korea (16.9%) and China (17.3%). The low frequencies of KIR2DL2 and KIR2DS2 in Asian populations would correspond to an earlier report of high frequency (75%) of the A haplotypes in Japanese [44]. In all populations, the frequencies of KIR2DL2 and KIR2DS2 are very similar owing to the high linkage disequilibrium (LD) between these genes.

KIR2DS3 tends to have lower frequencies – only eight populations have >40%; these are in India, and Australian Aborigines which has the highest at 81%. Many populations have very low frequencies (<10%) and in some of the South American Amerindian populations KIR2DS3 is absent – Argentina Salta Wichis, Mexico Tarahumaras, Venezuela Bari and Venezuela Yucpa [45, 46]. The frequency of this gene is also low in Japan and China.

The KIR2DS4 gene is present in seven populations at 100% – either from Africa or African Americans in USA. However, it has also low frequencies – Costa Rica (31%), Australian Aborigines (52%) Taiwan (59.4%). KIR2DS5 is nearly always present in frequencies from 20% to 60%, with a tendency to be the highest in India, Amerindian populations from Argentina, Mexico Venezuela and San and Xhosa from S. Africa. Notably the population Brazil Amazon has a frequency of 90%.

KIR3DS1 has much lower frequencies than KIR3DL1. However, 13 populations have been reported to have frequencies > 60%. These populations are from Mexico, India, Brazil, Australia Aborigines, and Venezuela. In Indian populations differences range from 37% to 69% [47] (it is interesting that the Indian populations have similar frequencies for many of the genes but not for others; in addition to KIR3DS1, KIR2DS3 varies between 20% and 63% and KIR2DS5 between 39% and 66%). Conversely, some African populations have very low frequencies (<10%) of KIR3DS1: San (2.2%), Xhosa (4%) Nigeria (3.4% and 6.3%) Senegal (4%) Kenya (0.7%) Ghana (4.9%) Central Africa Republic Bagandu Biaka (2.9%). Selection against having KIR3DS1 has been reported in African populations [23]. Obviously there is a close inverted correspondence between the frequencies of KIR3DL1 and KIR3DS1 in an individual population. A very small percentage of individuals (0.34%) are negative for both KIR3DL1 and KIR3DS1. Thus the estimation made by Gardiner and colleagues in 2001 that haplotypes without KIR3DL1 and KIR3DS1 would be present at 7% giving a frequency of individuals not having either KIR3DL1 or KIR3DS1 at 0.5% has proven very accurate [48].

More detailed analysis can be performed on the website (see Sect. 4) but in general it can be seen that it is the indigenous populations especially Aborigines and Amerindians who have outlying frequencies. Interestingly, despite each population having a few haplotypes, diversity of the Venezuelan Amerindian populations is high [46]. All KIR genes are present at >70% with the exception of KIR2DS3, which is actually absent in two of the three populations investigated.

Such extensive diversity between modern populations may indicate that geographically distinct diseases have exerted recent or perhaps on going selection on KIR repertoires. The differences in frequencies thus make the choice of controls for disease studies very important for all populations.

We attempted to link the published data by analyzing all populations submitted to the website www.allelefrequencies.net which had data for 13 KIR genes (excluding KIR2DP1 and KIR3DP1) [49]. The 56 populations examined grouped with a few exceptions according to a geographical gradient, using neighbor-joining dendrograms and correspondence analysis. Subsequent to this publication we recently selected 38 of the 56 populations which we considered to be well defined in the anthropological sense. We found that based on KIR haplotype B genes (i.e., mainly genes encoding activating KIR) the populations were related to geography like a good anthropological marker such as HLA or Y chromosome. However, the results based on the KIR haplotype A (i.e., mainly genes encoding inhibitory KIR) did not show such a correlation [50].

2.6 Rapid Diversification

The different population distribution from these studies indicates that KIR genes have been through rapid diversification and KIR alleles may have been under selection due to functional significance. Several studies have found that balancing selection, which maintains diversity, is the principle selective force acting on the KIR locus [23, 46]. Different mechanisms created and maintained diversity within the KIR genes including point mutations and meiotic recombination events such as gene conversions and crossovers [51]. Domain shuffling, due to meiotic recombination, has also been advanced as the mechanism for formation of new KIR receptors [52]. There is little conservation of KIR genes between species and indeed only two KIR genes (KIR2DL4 and 2DL5) have been preserved through hominoid evolution [52]. The diversification is thought to be more rapid for KIR genes than HLA, as HLA genes in humans and chimpanzees are more similar in sequence than their KIR counterparts [14, 53]. Even the CD94–NKG2 receptors are much more similar in chimpanzees and humans than KIR. Interestingly, the HLA ligands for KIR genes are highly polymorphic whereas those for CD94–NKG2 are not.

The KIR genes are very similar in their sequence. Misalignment of homologous chromosomes preceding meiotic recombination may lead to unequal crossing-over. This, in turn, can lead to duplication or deletion of genes resulting in shorter or longer haplotypes and facilitating rapid diversification of the KIR gene complex.

Initially it was believed that receptors KIR3DL1 and KIR3DS1 were products of different loci but it is now established that they segregate as alleles of the same KIR3DL1/S1 locus [54, 55]. This has been noted by the nomenclature committee who, although they still name alleles as either KIR3DL1 or KIR3DS1, use a noncoinciding numbering system for these alleles. However, two studies identified two copies of both KIR2DL4 and KIR3DL1/S1 on one haplotype [56, 57]. Further work on this topic showed that 4.5% of Caucasian individuals had a recombinant allele of the pseudogene KIR3DP1 which associated strongly with gene duplications of KIR2DL4 and KIR3DL1/S1 and was possibly formed by recombination of KIR3DP1 and KIR2DL5A [58]. The reciprocal haplotype lacking the KIR3DL1/S1 and KIR2DS4 has also been found [58]. A practical issue from this recombination event is that it would not be possible (without family studies) to ascertain if individuals have two copies of the one allele or when using SSOP as the detection system if the probe pattern for one allele was contained within the probe pattern of another allele. Several individuals have been reported to have two KIR2DL5 alleles on the same haplotypes, one being KIR2DL5A, the other KIR2DL5B [20]. Again, emphasizing possible unequal recombination we have reported a haplotype which has two alleles of KIR2DL5A [39].

3 KIR Alleles

Few studies have looked at alleles of the KIR genes. One exception and probably the first study examined the alleles of KIR2DL1, -2DL3, -3DL1 and -3DL2 in 34 families [59]. The authors reported LD between the alleles in different haplotypes. Strong LD existed between KIR2DL1 and 2DL3 alleles in the centromeric half and KIR 3DL1 and 3DL2 alleles in the telomeric half, but these two sets of pairs had little LD between them and appeared to define the two halves of the KIR gene complex. This study was the first to show that in addition to gene content, diversity of KIR was due to allele polymorphism and the combination of gene content and allele differences resulted in the vast majority of individuals having different KIR genotypes. A further study on individuals from North India determining alleles of KIR2DL1, 2DL3, 2DL5, 3DL1 and 3DL2 showed that all individuals had different KIR genotypes [42].

Emphasizing the growth in the discovery of KIR alleles is that 87 alleles were reported in the first KIR nomenclature report in 2002 but 292 alleles reported by September 2008 on the IPD-KIR database (Table 1) [19, 25]. This fully curated database is the location for the sequence of all the KIR alleles. Notably, growth in the number of recognized alleles can be seen for some genes more than others due to the fact that some laboratories have concentrated on investigations on one gene. For example, the number of alleles at KIR3DL1/S1 has increased from 15 to 63 mainly due to the work performed by the group of Parham who identified 34 new alleles in one study on KIR3DL1/S1 in many populations [23].

3.1 Allele Level Haplotypes

We previously examined KIR genes and alleles in 77 Northern Ireland families [39] and include herein a further 27 families, in addition to further allele systems as described below. Gene content was first ascertained [30] and those genes present were allele typed [60–66]. Since our initial report we have added SSOP typing systems for alleles of 2DL1, 2DL2, [65, 66] and 2DS3 (unpublished results). We have also expanded the typing system for KIR2DL4 in order to differentiate the run of nine Adenines (2DL4-9A) and ten Adenines (2DL4-10A) at the end of the transmembrane domain sequence of this gene. By virtue of a frame shift the 9A alleles have a premature stop colon and are not transcribed.

By using families whose inheritance had been confirmed by HLA typing and with knowledge of the alleles of a gene it was possible to ascertain if an individual had one or two copies of the gene, although it was necessary to make some assumptions [39]. Of the 209 genotypes of the parents of the 104 families (haplotype information was derived from three parents in one family), there were 26 different genotypes using gene content alone, but there were 188 (90%) different genotypes allowing for allele information, not surprising considering the previous reports of Rajalingam et al. [42] and Shilling et al. [59]. It is worth emphasizing that the Northern Ireland population is very homogeneous and drawn from a Caucasian population of 1.5 million, with very little immigration. These genotypes were devolved from 122 different haplotypes from the total 418 haplotypes (Table 2). Of these, 48 were A and 74 were B. Sixty-six haplotypes only occurred on one occasion. In total, 230 (55%) of haplotypes were A and 188 (45%) were B. The percentage of individuals who were homozygous for the A haplotype was 32.3%, the percentage homozygous for the B haplotype was 12.1% and 55.6% of individuals had both A and B haplotypes.

Interestingly those genes normally thought of being on A haplotypes, KIR3DL1, KIR2DS4, KIR2DL1, KIR2DL3, were also present on 45, 42, 47 and 20 different B haplotypes and 102, 99, 113 and 52 of the total B haplotypes, respectively. Ninety-six B haplotypes had both KIR3DL1 and KIR2DS4. However, in A haplotypes, KIR2DL1, KIR2DL3, KIR3DL1, KIR2DS4 were always present. The exception was one haplotype which we later showed to have these genes deleted, similar to a previous report [38] and this was reassigned as a B haplotype. This led us to define a genotype as AA, AB, or BB according to the following. If none of the following genes (2DL2, 2DL5A, 2DL5B, 2DS1, 2DS2, 2DS3, 2DS5, 3DS1) are present we term the genotype AA. If any of these genes were present we term the genotype AB, as long as KIR2DL1, KIR2DL3, KIR2DS4 and KIR3DL1 are all present. However, if any of these four genes are missing, we term the genotype BB. A very small minority of genotypes show exception to this definition – see Sect. 4.

Some alleles of the two framework genes that we allele typed (KIR2DL4, KIR3DL2), occurred more frequently on the different B haplotypes ($n = 74$) than the different A haplotypes ($n = 48$). Most notable of these was the occurrence of KIR2DL4*00501, which is expressed at the cell surface, on 43.6% of B but absent

Table 2 Haplotypes identified in Northern Ireland families

KIR2 DL4*	KIR 3DL2*	KIR 2DL3*	KIR2 DL1*	KIR 3DL1*/S1*	KIR2 DS4*	KIR 2DS2*	KIR 2DL2*	KIR 2DL5B*	KIR2 DS3*	KIR 2DL5A*	KIR 2DS5*	KIR 2DS1*	Hap ID	Haplo type	Total	Freq. %
011	001	001	00302	005	003								1	A	34	8.13
00801	001	001	00302	00101	003								3	A	22	5.26
00102	002		002	002	00101	+	003						5	B	22	5.26
00802	005	002	00302	004	006								2	A	19	4.55
00102	002	001	00302	01502	00101								6	A	17	4.07
00501	007	001	00302	013						001	002	002	4	B	17	4.07
00802	003	001	00302	004	006								7	A	16	3.82
00103	009	001	00302	008	003								13	A	13	3.11
00102	002	002	002	01502	00101								12	A	10	2.39
00801	001			00101	003	+	003						30	B	8	1.91
00802	003	002	002	004	006								8	A	8	1.91
006	008	002	002	007	004								17	A	8	1.91
00501	007		00401	013		+	001	002	00103	001	002	002	24	B	8	1.91
00501	007		00401	013		+	001	002	00103+002	005	002	002	15	B	8	1.91
00801	011	002	002	00101	003								14	A	7	1.68
011	001	002	002	005	003								11	A	7	1.68
00501	007	002	002	013						001	002	002	25	B	7	1.68
00801	001	005	001	00101	003								23	A	6	1.44
00501	007	003	00302	013				002	00103	001	002	002	16	B	6	1.44
00801	001	002	002	00101	003								10	A	5	1.20
00501	007			013		+	003	002	00103	001	002	002	9	B	5	1.20
00801	001		00401	00101	003	+	001						22	B	4	0.96
00801	011			00101	003	+	001						40	B	4	0.96
00802	005			004	006	+	001						36	B	4	0.96
00102	002		00401	01502	00101	+	001						26	B	4	0.96
00102	002	002	002	002	00101								19	A	4	0.96
00102	002	001	00302	002	00101								18	A	4	0.96
00103	009	002	002	020	00101								20	A	4	0.96
00501	007			013		+	001			001	002	002	83	B	4	0.96
00501	007			013		+	003			001	002	002	99	B	4	0.96
00801	010			00101	003								39	B	3	0.72
00801	011	001	00302	00101	003								31	A	3	0.72

Diversity of KIR Genes, Alleles and Haplotypes

00802	003		00401	004	006	+	001				21	B	3	0.72
00802	005			004	006	+	003				122	B	3	0.72
00802	005	001	00302	004	006		001				29	A	3	0.72
00102	002		00401	002	00101						34	B	3	0.72
00103	007	001	00302	008	003	+		002			27	A	3	0.72
00103	009	002	002	008	003		001	002			28	A	3	0.72
00501	007	001	00302+00401	013							43	B	3	0.72
00501	007	002	002	049N					001+005	002	108	B	3	0.72
011	010		00402	005	003		001	002	001	002	32	B	3	0.72
00102 +00501	002		00401	002+013	00101						60	B	2	0.48
00801	011		00401	009	003	+	001	002	00103		41	B	2	0.48
00801	011	002	002	009	003						52	A	2	0.48
00802	003	003	00302	004	006						35	A	2	0.48
00802	005	002	008	004	006			002	00103		97	A	2	0.48
00802	005	001	00302	019	006						37	A	2	0.48
00103	009		00401	008	003	+	001	002	00103		56	B	2	0.48
006	008			007	004	+	001				46	B	2	0.48
006	008		007	007	004	+	001	002	00103		115	B	2	0.48
006	008		00401	007	004	+	001	002	00103		47	B	2	0.48
011	001			005	003	+	001				42	B	2	0.48
00501	007	005	001	013					002	002	44	B	2	0.48
011	010	002	002	005	003						111	A	2	0.48
011	010	001	00302	005	003			008			33	A	2	0.48
00501	007			00101+013	003+00101	+	001		005		95	B	2	0.48
00501 +00801	001			004+013	006	+	001			002	78	B	2	0.48
00501 +00802	003		00401				003		00103		61	B	1	0.24
	007		004v				003			002	68	B	1	0.24
00801	007	001	004v	00101		+	001			002	96	B	1	0.24
00801	001			00101	003	+	003				69	B	1	0.24
00801	001			00101						002	101	B	1	0.24
00801	001	001	004v	00101	003		001				70	B	1	0.24
00801	001	001	00302v	00101	003	+					73	A	1	0.24
00801	003				003		003				71	B	1	0.24

(continued)

Table 2 (continued)

KIR2 DL4*	KIR 3DL2*	KIR 2DL3*	KIR2 DL1*	KIR 3DL1*/ S1*	KIR2 DS4*	KIR 2DS2*	KIR 2DL2*	KIR 2DL5B*	KIR2 DS3*	KIR 2DL5A*	KIR 2DS5*	KIR 2DS1*	Hap ID	Haplo type	Total	Freq. %
00801	005	002	002	004	006								84	A	1	0.24
00801	005	002	002	00101	003	+							85	A	1	0.24
00801	009	005		00101	003		003						72	B	1	0.24
00801	010		001	00101	003								74	A	1	0.24
00801	011			009	003	+	003						76	B	1	0.24
00801	011			00101		+	001						75	B	1	0.24
00801	011	002	002	009	003							002	91	B	1	0.24
00801	011	001	00302	009	003								77	A	1	0.24
00802	003			004	006	+	001						63	B	1	0.24
00802	003		00401	004	006	+	001	002					64	B	1	0.24
00802	005			019	006	+	003		00103				103	B	1	0.24
00802	005	005	001	004	006					001	002	002	65	A	1	0.24
00802	009			004		+	003						66	B	1	0.24
00802	009	002	002	004	006								121	A	1	0.24
00802	011		00402	004	006	+	001	002	00103				67	B	1	0.24
00802	012	001	00302	004	006								38	A	1	0.24
00802	002v	002	002	020	00101								62	A	1	0.24
00103	001		00401	01502	00101	+	001	002	00103				48	B	1	0.24
00102	002			002	00101	+	003						120	B	1	0.24
00102	002		00401	01502	00101	+	003	002					98	B	1	0.24
00102	002			002	00101	+	001	002	00103				119	B	1	0.24
00102	002		00401	01502	00101	+	001		00103				49	B	1	0.24
00102	002	005	001	002	00101								118	A	1	0.24
00102	002	003	00302	002	00101								117	A	1	0.24
00102	002	002	008	01502	00101								116	A	1	0.24
00102	002	002	002	007	00101								51	A	1	0.24
00102	002	002	002	005	00101								50	A	1	0.24
00103	009		00401	020	00101	+	001	002					53	B	1	0.24
00103	009		00402	020	00101	+	001	002	00103				54	B	1	0.24
00103	009		00401	008					00103				55	B	1	0.24
00103	009	001	00302	020	00101								57	A	1	0.24
00103	009v	005	001	008	003								59	A	1	0.24

Diversity of KIR Genes, Alleles and Haplotypes

00103	009v	001	00302	008	003							58	A	1	0.24
006	008			007	004	+	001					114	B	1	0.24
006	008	001	00302	007	004							94	A	1	0.24
011	001			005	003							79	B	1	0.24
011	002	002	002	005	003							80	A	1	0.24
011	005	001	00302	005	003							104	A	1	0.24
011	006	002	008	005	003							82	A	1	0.24
00501	006	002	002	013						001	002	105	B	1	0.24
00501	006	001	00302	013					003N	005	002	81	B	1	0.24
00501	007		00401	013		+	003	002	00103	001	002	110	B	1	0.24
00501	007			013		+	001	002	00103	001	002	109	B	1	0.24
00501	007		007	013	003		001	002	00103	001	002	86	B	1	0.24
00501	007	002	008	013						001	002	100	B	1	0.24
00501	007	002	002	013v						001	002	92	B	1	0.24
00501	007	001		013						001	002	106	B	1	0.24
00501	007	001	00302	049N	003				002	005	002	90	B	1	0.24
00501	007	001	00302	013					00103	001	002	88	B	1	0.24
00501	007	001	00302+004v	013	00101				00103+002	005		87	B	1	0.24
00501	007	001	00302	013				002	00103	001	002	89	B	1	0.24
00501	007	001	00302	013							002	107	B	1	0.24
011	010		002	005	003	+	003					93	B	1	0.24
011	010		00401	005	003	+	001	002	00103			102	B	1	0.24
011	010	002	008	005	003							112	A	1	0.24
00501	010	002	002	013						001	002	45	B	1	0.24
00501	006+009		00401	013		+	001	002	00103+002	005	002	113	B	1	0.24

Blank equals gene not present

from A and KIR3DL2*007 on 43.6% of B but only 1.3% of A. In those genes that have been thought to be on A haplotypes (KIR2DL1, 2DL3, 3DL1, 2DS4) but which we found at a high occurrence on B haplotypes, there was little difference in the frequency of specific alleles on an A compared to a B haplotype, except the absence of KIR2DL1*00401 on A haplotypes, this allele being the most common allele of KIR2DL1 on B haplotypes at 27.7%.

Another interesting variation in the frequency of the KIR alleles is the occurrence of the two versions of KIR2DS4, one with the full sequence and one with a short deletion. The deleted version has a 22 bp deletion in exon 5 which causes a frame shift leading to a stop codon in exon 7 [69] and it is believed that this version is not expressed at the cell surface. The deleted version (either KIR2DS4*003, 004, 006) is quite common, at 80% in the Northern Ireland population and nearly 60% of the population will only have the deleted KIR2DS4. There is a trend for decreased frequency of the deleted version in those populations which are homozygous for the A haplotypes [70]. Interestingly, we found that 30 (62.5%) of the different A haplotypes and 155 (67.4%) of total A haplotypes contained both a deleted version of KIR2DS4 and one of the 2DL4 -9A alleles. Indeed all A haplotypes with 2DL4-9A also had a deleted version of 2DS4, although not all A haplotypes with a deleted version of 2DS4 had 2DL4-9A. In those individuals who have the genotype AA, 43.1% do not have an activating KIR, leading to 13.9% in the overall population not having an activating receptor.

In general, the typing of additional genes to allele level or increasing allele resolution after the initial publication of original data from 77 families [39] did not dramatically increase the number of different haplotypes, although it did change our understanding of their content. Addition of KIR2DL1, KIR2DL2 and KIR2DS3 allele data produced four, three and zero new different haplotypes, respectively. However, the addition of 27 new families to the haplotype study resulted in the definition of 19 new individual haplotypes, some of which occurred more than once. This would indicate that even in a small homogeneous population, the number of families (77 in the original report) needs to be greatly increased to cover all potential haplotype variation.

A report of allele frequency data in a Japanese population showed that for the KIR genes KIR2DL1, KIR2DL2/2DL3, KIR2DL4, KIR3DL1/S1, KIR3DL2 and KIR2DS4, one allele at each gene was at a very high-frequency (44–89%) compared to the next frequent allele [67]. This is not the case in the Northern Ireland population, emphasizing the conclusion reached by Yawata and Parham of the skewed distribution of KIR variants in the Japanese population, which reflected a distinct history of directional and balancing selection [67]. What is also worth noting in our study is the difference in homozygosity of alleles of some of the high frequency genes [39]. This is not due to one allele being represented at a high frequency relative to other alleles. It may be that in addition to being useful enough to be present in nearly all individuals, there is an advantage in being heterozygous at the allele level for KIR2DL4, KIR3DL1 and KIR3DL2, whereas this is not the case for KIR2DL3 or KIR2DS4.

3.2 Other Populations

We reported the allele frequencies in five populations selected from various regions of the world [49]. More than 90% of individuals from four populations (Brazil, Cuba, Oman and Xhosa) had unique genotypes. However, in the Hong Kong population only 64% of individuals carried a unique genotype. Indeed, there were three genotypes that were present in 11%, 9% and 6% of this population. This suggests that this population may have been drawn from a relatively small pool. However, allele typing at other KIR loci or the discovery of new alleles may prove that even in this population, all genotypes are unique – we did not examine alleles of KIR3DL3 ($n = 37$) or KIR2DL1 ($n = 14$).

The framework genes (KIR2DL4 and KIR3DL2) and genes normally found on the A haplotypes (KIR2DL3, KIR2DS4 and KIR3DL1) were represented by several alleles [49]. Whereas there is a more equal distribution in frequency of alleles of these genes in the Brazilian, Omani and Cuban populations, the Xhosa and Hong Kong populations have usually one allele at a much higher frequency than the others, although this does not happen to the same extent at the KIR2DL4 gene. Certain alleles had high frequencies in one population but not in the other populations. Although many new alleles were found, several known alleles were not found in any of the populations and it may be that these alleles are very rare or were originally incorrectly sequenced. Sequencing of alleles of the KIR genes is indeed problematic – the size of the introns makes it difficult to sequence from genomic DNA and similarities in the sequences of alleles from different genes also cause problems.

There was a marked difference in the number of alleles found according to whether the gene is normally found on A or B haplotype. The limited polymorphism in genes found only on B haplotypes has been previously reported [55, 67]. It is not clear if the B haplotype, with its many gene arrangements, does not require allele polymorphism or if natural selection has acted against variability at the allele level of these genes because of possible autoimmune destruction. More likely is that the genes encoding activating KIR evolved from genes encoding inhibitory KIR and are short-lived in comparison to the genes encoding the inhibitory KIR [68]. Thus, there may not have been enough time for polymorphism to develop, or perhaps more polymorphic genes encoding the activating KIR no longer exist.

Apart from the studies previously mentioned, there is little data on allele frequencies in different populations, with one exception, the large population study in 28 populations on 50 KIR3DL1/S1 alleles by Norman and colleagues [23]. Whereas KIR3DS1*013 accounted for 97% of occasion when KIR3DS1 was found, and was found in all 28 populations examined, other KIR3DS1 alleles were very rare and restricted to one population. This, in contrast to KIR3DL1, where KIR3DL1*1502 was the most frequent allele at 34% [23]. Allele data from a further six to nine populations are available on www.allelefrequencies.net for KIR2DL1, 2DL3, 2DL4, 3DL2, 2DS4.

Determination of alleles is also useful for positioning of KIR genes on a haplotype. Recently, KIR2DS3*00103 has been shown to map to the centromeric side, and KIR2DS3*002 and KIR2DS3*003N to the telomeric sides of the haplotype [71]. KIR2DS5*002 was also shown to map to the same telomeric position as KIR2D3*002/003N thereby implying that these alleles belong to a single locus. We have extrapolated this work to our family data by determining the KIR2DS3 alleles. KIR2DS3 was present on 67 (16%) of the 418 haplotypes. None of the four haplotypes positive for KIR2DS3*002 or KIR2DS3*003N had KIR2DS5. Ten haplotypes which had two copies of KIR2DS3 (*00103 and *002), demonstrating that these variants are not true alleles of each other, were negative for KIR2DS5. In 53 haplotypes positive for KIR2DS3*00103, KIR2DS5*002 was present in 17. KIR2DS5*002 is the only KIR2DS5 allele found in the Northern Ireland population [66]. Thus, it would appear that KIR2DS3 alleles *002 and *003N are allelic to KIR2DS5*002 and KIR2DS3*001 forms a separate gene, emphasizing that we have still much to learn of the generic make-up of KIR.

4 Website and KIR Genotypes

At present, on the website there is data for KIR gene and allele frequencies in 159 populations (18,152 individuals) with 81 populations having genotype information. The data are available in two formats; KIR gene or allele frequencies (Fig. 3) and KIR genotypes (i.e., presence or absence of KIR genes) (Fig. 4). Phenotypic frequencies (individuals in a population having that gene or allele) are stored as percentages and allele frequencies are stored in three decimal format.

KIR genotype database consists of 280 different genotypes found in 8,246 individuals from 81 populations distributed around the world. Western Europe,

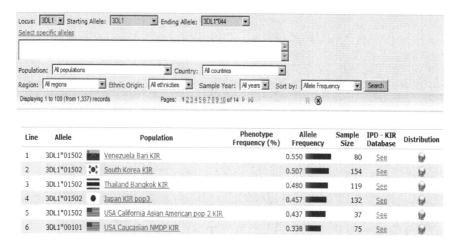

Fig. 3 KIR allele frequency search

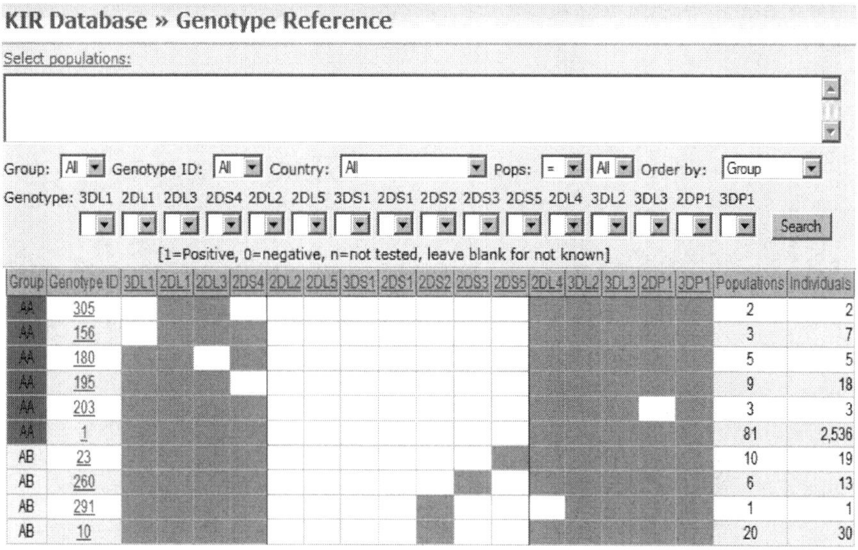

Diversity of KIR Genes, Alleles and Haplotypes 81

Fig. 4 KIR genotype search

Asia and South and Central Americas cover 69% of total individuals tested (Table 3). It is notable that Eastern Europe presents 111 distinct genotypes compared with the 136 genotypes reported in Western Europe despite the difference in the number of populations tested (5 and 18, respectively). Only two genotypes occur in all ten geographic regions, whereas 18 of the genotypes appear in only a single (different) region. According to the analysis, many genotypes only occur in one individual in one population (Table 4). Only seven genotypes are very common being reported in more than 50 of the 81 populations and representing 5,659 (68.7%) of the total of individuals tested (Table 4).

Six distinct AA genotypes are present in 2,573 individuals representing 31.2% of individuals. AB has 90 genotypes in 4,375 (53.1%) individuals whereas BB has 184 genotypes but only in 1,298 (15.7%) individuals. This again emphasizes the variation in gene content in B haplotypes. One AA genotype is present in all 81 populations, a total of 2536 (30.8%) individuals. Other common genotypes are shown in Fig. 5. The KIR genotypes were grouped into AA, AB, and BB, as mentioned previously, according to the definition we derived from our family data [39]. However, there are exceptions to this definition. Four of the genotypes labeled AA had none of the genes associated with the B haplotype present but also had missing one or two of the four genes (KIR 2DL1,2DL3,2DS4, and 3DL1) used to define an A haplotype (Fig. 4). This is probably due to deletions of these genes but the possibility of inaccurate typing cannot be ruled out. Neither can inaccurate typing be ruled out in those genotypes which only occur in one individual. In order to bring this to the attention of individuals submitting data, if the genotype they input is not listed in the database, the genotype closest in content will be shown. It is hoped that the list of genotypes will lead to a common nomenclature system for the genotypes.

Table 3 Distribution of KIR genotypes by Geographic Region

Region	Pops	Genotypes	Individuals
Asia	18	117	1,711
Australia	1	6	42
East Europe	5	111	853
Middle East	3	48	324
North Africa	1	22	67
North America	5	66	599
Pacific	4	55	194
South and Central America	20	132	1,754
Sub-Saharan Africa	6	55	406
Western Europe	18	136	2,296
Total	81		8,246

Table 4 Distribution of KIR genotypes by population

Genotypes	Populations	Individuals
129	1	142
30	2	81
31	3	149
11	4	76
8	5	68
39	6–10	608
26	11–50	1,463
7	>50	5,659
Total 281		8,246

Group	Genotype ID	3DL1	2DL1	2DL3	2DS4	2DL2	2DL5	3DS1	2DS1	2DS2	2DS3	2DS5	2DL4	3DL2	3DL3	2DP1	3DP1	Pops	Indiv
1	AA	1																81	2,536
2	AB	2																71	847
3	AB	4																75	818
4	AB	5																64	516
5	AB	3																65	447
6	AB	6																62	332
7	AB	7																49	225
8	AB	8																51	163
9	BB	71																43	155
10	AB	9																43	109

Fig. 5 Most common KIR genotypes reported in www.allelefrequencies.net

5 KIR Coevolution with HLA

There is evidence that HLA class I is influencing the frequency of KIR expression by NK cells. In a comparison of 85 sibling pairs, Shilling and colleagues showed that KIR-identical and HLA-different siblings had similar KIR repertoire expression but HLA-identical, KIR-different siblings were almost as disparate as HLA-different, KIR-different siblings [72]. However, although KIR genotype had the major effect, comparison of HLA-identical, KIR-identical siblings with HLA-different, KIR-identical siblings showed the former to have more similar repertoires than the latter. Expression of KIR receptors is also influenced by the presence of HLA ligand. Individuals with KIR2DL1 or KIR3DL1 had greater numbers of NK cells expressing these genes, if the HLA-C2 group or HLA-Bw4 ligands were respectively present in the individual [67]. Furthermore, the effect of the ligand on its specific KIR diminished with the number of additional KIR that also had their ligand present, suggesting cooperation between receptor and ligand pairs.

Evidence of coevolution is also suggested by disease studies [55, 73] and population genetics [23, 74]. In a study in preeclampsia, Hiby and colleagues showed that the increased prevalence of the AA genotype in women with preeclampsia, compared to AB or BB genotypes, only happened when the fetus carried the HLA-C2 group [10]. An inverse correlation exists in populations between the frequencies of the KIR A haplotype and the HLA-C2 group [10], thus reducing the frequencies of potential preeclampsia pregnancies and indicating that the bias in frequencies of KIR and HLA-C is due to natural selection. The extreme variability of both HLA and KIR genes giving rise to many HLA and KIR combinations to provide protection against different pathogens is a good example of genetic epistasis.

A study of world KIR3DL1/S1 diversity showed that positive selection was focused to the residues that interact with HLA and also identified reciprocal evolution of the HLA-Bw4 motif [23]. A further study into 30 geographical distinct populations found strong negative correlations between KIR3DS1 and its presumed HLA-Bw4 ligand [74]. Although this was the only association with statistical significance the tendency was for inhibitory KIR to have positive correlations with their ligands and activating KIR to have negative correlation with theirs. The one exception was the negative correlation between KIR2DL2 and HLA-C1 group. In a recent project of the 15th International Histocompatibility Workshop (Brazil, September 2008) this same negative correlation was found in data from 23 world-wide populations (Hollenbach and Middleton, unpublished studies). However, on the reverse side, this is a positive correlation between KIR2DL2 and HLA-C2 group homozygosity. Further studies are on-going on the correlation of the alleles of KIR3DL1 and KIR2DL2 in these populations.

6 KIR Expression

A further level of diversity is provided by variation in expression of KIR genes on the NK cell and the avidity of the reaction between the KIR gene and its ligand. Knowing the KIR and HLA genes present does not help guage the size of the NK cell repertoire [75]. Individual NK cells express only some of the KIR genes of that individual and this expression varies from NK cell to NK cell. The frequency of expression of combinations of two KIR genes is the product of their individual frequencies of expression. However, these patterns of recognition are stable [76]. Whereas KIR2DL4 is expressed on all NK cells, other KIR are only expressed on some NK cells, due to patterns of KIR gene methylation [76, 77]. Promoter regions are unmethylated in genes expressed but methylated in those not expressed. The KIR gene promoters have been shown to be polymorphic and display significant structural and functional differences [78]. KIR2DL4 and KIR3DL3 have promoters different from each other and from the other KIR genes. A bidirectional promoter has been found in KIR genes predicting that this would determine the number of NK cells expressing a specific KIR protein [79].

The proteins of some alleles, (3DS1*049N) are not expressed while those of other alleles have high or low expression [80]. For example, 3DL1*004 is poorly expressed at the cell surface. Normal levels of protein are produced but retained within the cell due to polymorphisms at two positions, one in each of the immunoglobulin domain (D0 and D1) [81]. The frequency of KIR3DL1*004 would indicate the possibility that not having a functional allele of KIR3DL1 provides an advantage in certain situations. Indeed, KIR3DL1*004 has been shown to be the most protective allele against HIV disease progression when present with the HLA-Bw4 ligand [82].

Much of the work on expression of individual KIR alleles has been performed on KIR3DL1/S1, mainly by the group of Parham. KIR3DL1 alleles have been divided into three groups according to the level of binding of their protein with the KIR3DL1 specific mAb DX9 [48]. No binding is found with KIR3DL1*004 whereas 3DL1*001, 002, 008, 1502, bind at high levels and 3DL1*005,006,007, at low levels. (The allele name KIR3DL1*003, originally mentioned in Gardiner's report, was subsequently changed to KIR3DL1*1502). Later studies showed that the differences in binding were due to the abundance on the cell surface of KIR3DL1 and that alleles with high binding were expressed by a greater percentage of NK cells [67]. Individuals with two alleles had nearly a doubling of expression. Different alleles also had different KIR3DL2 expression, with KIR3DL2*008 being high-binding and KIR3DL2*007 and KIR3DL2*00902 being low-binding [67]. These differences appear to be important in human disease. KIR3DL1 alleles, with high expression gave stronger protection than those with low expression in HIV, except the previously mentioned KIR3DL1*004 [82].

7 KIR and Ligand Interaction

The receptor KIR3DL1/S1 recognizes the HLA-Bw4 epitope, although residues outside the HLA-Bw4 epitope which determine peptide binding, contribute to KIR3DL1 binding to HLA-B [83]. Difference in which amino acid is present at position 80 of the HLA-Bw4 epitope (threonine or isoleucine) affects strength of inhibition by KIR3DL1 resulting in better protection against NK cell-mediated cytolysis, isoleucine being the strongest [84]. The HLA-Bw4 epitope is also present on HLA-A alleles. Some of these HLA-A alleles deliver strong inhibitory signals (HLA-A*2402,-A*3201) but others are weak ligands (HLA-A*2501,-A*2301), although the HLA-Bw4 epitope on HLA-B is more potent than if carried by

HLA-A [85]. The same study also found that HLA-B*1301 and HLA-B*1302, which carry the HLA-Bw4 epitope, provide poor protection against lysis. KIR2DL1 binding with HLA-C2 group is stronger and more specific than KIR2DL2 with HLA-C1 group, which is stronger than KIR2DL3 with HLA-C1 group. Despite very similar structures KIR2DS1 binds HLA-C group 2 very weakly and there is no evidence that KIR2DS2 binds HLA-C group 1. These differences would appear to be due to one different amino acid in each of these genes encoding activating KIR, compared to their corresponding genes encoding inhibitory KIR [55].

In addition to having higher levels of protein expression on NK cell surface, KIR3DL1*002 is a stronger inhibitory receptor for HLA-Bw4 ligands than KIR3DL1*007, even when transduced human cell lines expressing equal levels of each allele are used [86]. Interestingly, the two residues which contribute to this difference are not predicted to interact directly with the HLA-Bw4 ligand.

8 Algorithm Models

8.1 *Mathematical Method for Study of Haplotypes*

Because families are seldom available for study, several groups have tried to use or extend mathematical methods, which use unrelated individuals as input data, in order to compute haplotype frequencies. To overcome the lack of phase information provided by the typing techniques, likelihood-based calculations have been formalized by Dempster and further developed in the general framework of Expectation Maximization algorithm (EM) [87]. The EM algorithm is primarily set to handle only missing phase information. However, it was shown that it could be adapted to handle at the same time, complete and incomplete genotypes (missing phase information and missing values within a genotype) [87, 88]. Due to typical KIR ambiguities, further efforts were made to extend the likelihood models and the EM algorithms to be able to estimate KIR haplotype frequencies using unrelated data.

In our studies the validation of haplotype estimation was achieved by simulation of phase known data, samples being drawn from fixed haplotype frequency distribution. Hardy Weinberg proportions were assumed. Frequencies were derived from the 104 families, 418 founder haplotypes mentioned previously, with simulations on phase-known samples. In order to assess the influence of the sample size, different sample sizes were simulated. A priori list of possible haplotypes was not used. Details on the method can be found in the preliminary study [89].

The results show that KIR unrelated genotypes can be successfully handled. It provides good estimation of the expected haplotype frequencies. For example, for a 250-individual sample, the average difference between the haplotype estimation and its expected value is 0.52% (S.D. 0.11%). Unfortunately, the algorithm cannot resolve the full ambiguity of the genotype as it adds many unexpected haplotypes with very low frequency estimation. However, the vast majority of these are so rare that their frequency is below the 1 in 2N threshold (i.e., occurring at least once in the sample). A sample size of 500–1,000 individuals allows the haplotype diversity to be covered in the simulation.

The algorithms were used to combine phase known KIR genotype data (the combined 209 founder individuals of the families) and ambiguous phase unknown KIR genotype data. Results are shown in Table 5 for the most frequent haplotypes and Table 6 for the rarest haplotypes.

8.2 KIR Genotype Diversity Resulting from Combination of KIR Haplotypes

A diplotype is a pair of haplotypes with phase known, whereas in a genotype the phase is not known. KIR genetics is so complex that several pairs of haplotypes (diplotypes) can result in the observation of the same KIR genotype. As observed with other genes like HLA, this is due to the fact that in a given genotype it is not possible to know which allele of one locus segregates with which allele of the other locus (this is sometimes referred to as *cis* and *trans* positions). In KIR, the absence of gene and the typing ambiguities make even more numerous the possible diplotypes that result in the same genotype. In this section we only used the haplotype distribution previously published to assess the diplotype-genotype relationship [39].

Briefly, we inferred the distribution of the diplotypes from the haplotype frequency distribution. In Table 7, in which 60 genotypes were studied, the number of possible diplotypes for a given genotype ranges from 1 to 21. Eighteen genotypes corresponded to a single diplotype. Taken together these genotypes represent 3.70% of the genotypes. On the contrary however, there is one genotype which could correspond to 21 different pairs of haplotypes. This genotype represents 3.1% of the genotypes.

When multiple diplotypes can be found for a single KIR genotype, we studied the distribution of the most frequent diplotype for a given genotype, in order to assess how important the confusion bias can be when taking a genotype as a single genetic configuration (Table 7). All analyses were also weighted by the diplotype-expected frequencies. In conclusion, we believe that to fully understand the affect of KIR polymorphism it will be necessary to be able to derive the haplotypes. Depending on the number of KIR genes typed and the typing resolution of KIR genes at allele level, algorithms may require an a priori list of possible haplotypes. A very recent publication discusses the methodological issues involved in the analysis of KIR gene content [90].

9 Summary

We have shown the vast diversity in KIR haplotypes when allele typing is included and that the allele variation appears to affect expression levels of a given KIR. With the two haplotypes of each individual combined, leading in all probability to a

Table 5 Thirty most frequent haplotype frequencies estimated from combination of 300 unrelated and 209 phase-known unrelated founders of families

2DS2–2DL3–2DL2–2DL5b–2DS3–2DL1–2DL4–3DL1S1–2DL5a–2DS5–2DS1–2DS4–3DL2 KIR Haplotype	Haplotype frequency estimation (%)	Counting frequency in founders (%)
NEG–001–NEG–NEG–NEG–NEG–00302–005–005–NEG–NEG–NEG–003–001	6.93	8.13
POS–NEG–003–NEG–NEG–NEG–00102–002–NEG–NEG–NEG–00101–002	5.05	5.26
NEG–001–NEG–NEG–NEG–NEG–00302–00202–00101–NEG–NEG–NEG–003–001	4.90	5.26
NEG–001–NEG–NEG–NEG–NEG–00302–005–013–001–002–002–NEG–007	4.61	4.07
NEG–002–NEG–NEG–NEG–NEG–002–00201–004–NEG–NEG–NEG–006–005	3.99	4.55
NEG–001–NEG–NEG–NEG–NEG–00302–00102–01502–NEG–NEG–NEG–00101–002	3.81	4.07
NEG–001–NEG–NEG–NEG–NEG–00302–00201–004–NEG–NEG–NEG–006–003	3.21	3.83
NEG–001–NEG–NEG–NEG–NEG–00302–00102–008–NEG–NEG–NEG–003–009	2.46	3.11
NEG–002–NEG–NEG–NEG–NEG–002–005–013–001–002–002–NEG–007	2.27	2.39
POS–NEG–001–002–00103–00401–005–013–001–002–002–NEG–007	2.15	1.91
NEG–002–NEG–NEG–NEG–NEG–002–00202–00101–NEG–NEG–NEG–003–001	1.85	1.20
POS–NEG–003–NEG–NEG–NEG–00202–00101–NEG–NEG–NEG–003–001	1.70	1.91
NEG–002–NEG–NEG–NEG–NEG–00102–01502–NEG–NEG–NEG–00101–002	1.66	2.39
NEG–001–NEG–NEG–NEG–NEG–00302–00102–002–NEG–NEG–NEG–00101–002	1.63	0.96
NEG–002–NEG–NEG–NEG–NEG–002–00201–004–NEG–NEG–NEG–006–003	1.62	1.91
POS–NEG–001–002–00103–00401–00102–002–NEG–NEG–NEG–00101–002	1.51	0.72
NEG–005–NEG–NEG–NEG–NEG–001–00202–00101–NEG–NEG–NEG–003–001	1.40	1.44
NEG–002–NEG–NEG–NEG–NEG–002–006–007–NEG–NEG–NEG–004–008	1.26	1.91
NEG–002–NEG–NEG–NEG–NEG–002–005–005–NEG–NEG–NEG–003–001	1.19	1.67
POS–NEG–003–NEG–NEG–NEG–005–013–001–002–002–NEG–007	1.07	1.20
POS–001–NEG–NEG–NEG–NEG–00302–00202–00101–NEG–NEG–NEG–003–001	1.07	0
NEG–002–NEG–NEG–NEG–NEG–002–00102–002–NEG–NEG–NEG–00101–002	1.04	0.96
POS–NEG–001–002–00103+002–00401–005–013–005–NEG–002–NEG–007	1.02	1.91
NEG–002–NEG–NEG–NEG–NEG–002–00202–00101–NEG–NEG–NEG–003–011	1.01	1.67
NEG–002–NEG–NEG–NEG–NEG–002–00102–008–NEG–NEG–NEG–003–009	0.98	0.72
POS–NEG–001–NEG–NEG–NEG–00202–00101–NEG–NEG–NEG–003–011	0.88	0.96
NEG–003–NEG–NEG–NEG–NEG–00302–005–013–001–002–002–NEG–007	0.87	1.44
POS–NEG–001–002–00103–00401–00201–004–NEG–NEG–NEG–006–003	0.86	0.72
POS–NEG–001–002–00103–00401–00202–00101–NEG–NEG–NEG–003–001	0.83	0.96
NEG–002–NEG–NEG–NEG–NEG–002–00102–007–NEG–NEG–NEG–00101–009	0.82	0.96

Table 6 Ten rarest over 1 in 2N and 10 rarest below 1 in 2N

2DS2–2DL3–2DL2–2DL5B–2DS3–2DL1–2DL4–3DL1S1–2DL5A–2DS5–2DS1–2DS4–3DL2 KIR Haplotype	Haplotype Frequency estimation (%)	Counting Frequency in founders (%)	Number of haplotypes observed in founders
POS–NEG–001–002–00103–NEG–00102–01502–NEG–NEG–NEG–00101–002	0.13	0.24	1
NEG–002–NEG–NEG–NEG–002–005–013v–001–002–002–NEG–007	0.13	0.24	1
POS–NEG–003–NEG–NEG–NEG–00202–00101–NEG–NEG–NEG–003–003	0.13	0.24	1
POS–NEG–003–NEG–NEG–NEG–00201–004–NEG–NEG–NEG–NEG–009	0.13	0.24	1
NEG–002–NEG–NEG–NEG–002–00102–005–NEG–NEG–NEG–00101–002	0.13	0.24	1
NEG–002–NEG–NEG–NEG–008–005–005–NEG–NEG–NEG–003–006	0.13	0.24	1
POS–NEG–001–002–00103–00401–00102–01502–NEG–NEG–NEG–NEG–002	0.13	0.24	1
NEG–002–NEG–NEG–NEG–002–005–013–001–002–002–NEG–006	0.13	0.24	1
POS–NEG–001–008–NEG–NEG–00102–NEG–NEG–002–002–NEG–007	0.13	0.24	1
POS–NEG–001–002–00103–00401–00201–004–001–002–002–006–003	0.13	0.24	1
POS–NEG–NEG–NEG–NEG–NEG–00202–005–NEG–NEG–NEG–003–001	0.13	0	0
POS–002–NEG–002–002–002–005–013–005–NEG–002–NEG–007	0.11	0	0
NEG–002–NEG–NEG–NEG–002–005–013–001–002–002–00102–007	0.11	0	0
NEG–002–NEG–NEG–NEG–002–005–004–NEG–NEG–NEG–006–003	0.11	0	0
POS–002–NEG–NEG–002–002–005–013–005–NEG–NEG–NEG–007	0.11	0	0
NEG–001–NEG–NEG–NEG–00302–NEG–007–NEG–NEG–NEG–NEG–009	0.10	0	0
NEG–002–NEG–NEG–NEG–00302–00201–004–NEG–NEG–NEG–006–009	0.10	0	0
NEG–002–NEG–NEG–NEG–002–00102–004–NEG–NEG–NEG–006–005	0.08	0	0
POS–NEG–003–NEG–NEG–004v–00102–01502–001–002–NEG–00101–002	0.08	0	0
NEG–003–NEG–NEG–NEG–NEG–00102–007–NEG–NEG–NEG–00101–009	0.07	0	0

Table 7 Distribution of the number of diplotypes corresponding to a single genotype, and the probability of the most frequent diplotype corresponding to a single genotype

Number of diplotypes corresponding to a single genotype	Number of genotypes	Sum of genotype frequencies in population (%)	Median of the prediction	Interquartile range [p25–p50]	Min–Max [Min–Max]
1	18	3.70	1	[1–1]	[1–1]
2	15	47.01	0.857	[0.81–0.95]	[0.5–0.98]
3	6	2.84	0.709	[0.71–0.96]	[0.5–0.96]
4	2	1.65	0.816	[0.48–0.82]	[0.48–0.82]
5	4	2.46	0.429	[0.33–0.48]	[0.33–0.48]
6	2	5.37	0.543	[0.43–0.54]	[0.43–0.54]
7	4	8.62	0.664	[0.66–0.66]	[0.29–0.81]
8	1	3.97	0.355	[0.36–0.36]	[0.36–0.36]
9	1	7.42	0.456	[0.46–0.46]	[0.46–0.46]
10	3	6.49	0.692	[0.69–0.73]	[0.19–0.73]
11	1	0.91	0.238	[0.24–0.24]	[0.24–0.24]
20	2	6.46	0.294	[0.29–0.29]	[0.12–0.29]
21	1	3.10	0.182	[0.18–0.18]	[0.18–0.18]
Total	*60*	*100%*	*0.81*	*[0.46–0.86]*	*[0.11–1]*

The cumulative frequencies were computed in a weighted way, according to each diplotype expected frequency

different set of alleles, the level of expression and functional capability for each individual could vary remarkably. Whether this will have any quantitative differences on predisposition to disease remains to be determined, but it is possible that the diversity is the result of natural selection by pathogens. Indeed, the diversity of KIR has led to a search for their role in human disease, especially considering that many of the KIR ligands are MHC class I molecules, which have already been implicated in many clinical conditions. However, functional studies, which have lagged behind genetic studies, are urgently needed to show the immunological relevance of the statistical association found between KIR and some diseases.

Acknowledgements We are very grateful to Dr. P. Norman, Stanford and Dr. R. Rajalingam, Los Angeles for their very diligent, constructive and critical reading of the manuscript.

References

1. Caligiuri MA (2008) Blood 112:461
2. Ljunggren HG, Karre K (1990) Immunol Today 11:237
3. Liu WR, Kim J, Nwankwo C, Ashworth LK, Arm JP (2000) Immunogenetics 51:659
4. Trowsdale J, Barten R, Haude A, et al. (2001) Immunol Rev 181:2
5. Wende H, Colonna M, Ziegler A, Volz A (1999) Mamm Genome 10:154
6. Biron CA, Nguyen KB, Pien GC, Salazar-Mather TP (1999) Annu Rev Immunol 17:189
7. Baxter AG, Smyth MJ (2002) Autoimmunity 35:1
8. Shi F, Ljunggren H, Sarvetnick N (2001) Trends Immunol 22:97

9. Williams F, Meenagh A, Sleator C et al (2005) Hum Immunol 66:836
10. Hiby SE, Walker JJ, O'Shaughnessy KM et al (2004) J Exp Med 200:957
11. Ruggeri L, Capanni M, Mancusi A, et al. (2004) Bailliere's Best Pract Res Clin Haematol 17:427
12. Cooley S, Trachtenberg E, Bergmann TL, et al. (2009) Blood 113:726
13. Moretta L, Biassoni R, Bottino C, Mingari MC, Moratta A (2000) Immunol Today 9:20
14. Vilches C, Parham P (2002) Annu Rev Immunol 20:217
15. Martin AM, Freitas EM, Witt CS, Christiansen FT (2000) Immunogenetics 5l:268
16. Wilson MJ, Torkar M, Haude A et al (2000) Proc Natl Acad Sci USA 97:4778
17. Norman PJ, Stephens HA, Verity DH, Chandanyingyong D, Vaughan RW (2001) Immunogenetics 52:195
18. Middleton D, Curran M, Maxwell L (2002) Transplant Immunol 10:147
19. Marsh SGE, Parham P, Dupont B et al (2003) Hum Immunol 64:648
20. Gomez-Lozano N, Gardiner CM, Parham P, Vilches C (2002) Immunogenetics 54:314
21. Rajagopalan S, Fu J, Long EO (2001) J Immunol 167:1877
22. Moesta AK, Norman PJ, Yawata M et al (2008) J Immunol 180:3969
23. Norman PJ, Abi-Rached L, Gendzekhadze K, et al (2007) Nat Genet 39:1092
24. Uhrberg M, Valiante NM, Shum BP et al (1997) Immunity 7:753
25. Robinson J, Waller MJ, Stoehr P, Marsh SGE (2005) Nucl Acids Res 331:D523
26. Horton R, Coggill P, Miretti MM et al (2006) Tissue Antigens 68:450
27. Vilches C, Castano J, Gomez-Lozano N, Estefania E (2007) Tissue Antigens 70:415
28. Gomez-Lozano N, Vilches C (2002) Tissue Antigens 59:184
29. Norman PJ, Cook MA, Carey BS et al (2004) Immunogenetics 56:225
30. Middleton D, Williams F, Halfpenny IA (2005) Transplant Immunol 14:135
31. De Santis D, Witt C, Gomez-Lozano N, et al (2006) In Hansen JA (ed) Immunobiology of the Human MHC. IHWG, Seattle, p 1233
32. Cook MA, Norman PJ, Curran MD et al (2003) Hum Immunol 64:567
33. http://.hla.ucla.edu./cellDna.htm
34. Middleton D, Menchaca L, Rood H, Komerofsky R (2003) Tissue Antigens 61:403
35. Turino C, Scavello G, Ferriola D, Kunkel M, Dapprich J (2006) Hum Immunol 67(Suppl 1):S41
36. Du Z, Sharma SK, Spellman S, Reed EF, Rajalingham R (2008) Genes Immun 9:470
37. Gomez-Lozano N, de Pablo R, Puente S, Vilches C (2003) Eur J Immunol 33:639
38. Norman PJ, Carrington CVF, Byng M et al (2002) Genes Immun 3:86
39. Middleton D, Meenagh A, Gourrand PA (2007) Immunogenetics 59:145
40. Rajalingam R, Du Z, Meenagh A et al (2008) Immunogenetics 60:207
41. Toneva M, Lepage V, Lafay G et al (2001) Tissue Antigens 57:358
42. Rajalingam R, Krausa P, Shilling HG et al (2002) Immunogenetics 53:1009
43. Parham P (2003) Tissue Antigens 62:194
44. Yawata M, Yawata N, McQueen KL et al (2002) Immunogenetics 54:543
45. Flores AC, Marcos CY, Paladino N et al (2007) Tissue Antigens 69:568
46. Gendzekhadze K, Norman PJ, Abi-Rached L, Layrisse Z, Parham P (2006) Immunogenetics 58:474
47. Kulkarni S, Single RM, Martin MP et al (2008) Immunogenetics 60:121
48. Gardiner CM, Guethlein LA, Shilling HG et al (2001) J Immunol 166:2992
49. Middleton D, Meenagh A, Moscoso J, Arnaiz-Villena A (2007) Tissue Antigens 71:105
50. Middleton D, Meenagh A, Serrano-Vela JI, Moscoso J, Arnaiz-Villena A (2008) Open Immunol J 1:42
51. Shilling HG, Lienert-Weidenbach K, Valiante NM, Uhrberg M, Parham P (1998) Immunogenetics 48:413
52. Rajalingam R, Parham P, Rached L (2004) J Immunol 172:356–369
53. Khakoo SI, Rajalingam R, Shum BP et al (2000) Immunity 12:687
54. Kelley J, Walter L, Trowsdale J (2005) Plos Genet 1:129
55. Parham P (2005) Nat Rev Immunol 5:201

56. Williams F, Maxwell LD, Halfpenny IA et al (2003) Hum Immunol 64:729
57. Martin MP, Bashirova A, Traherne J, Trowsdale J, Carrington M (2003) J Immunol 171:2192
58. Gomez-Lozano N, Estefania E, Williams F et al (2004) Eur J Immunol 35:16
59. Shilling HG, Guethlein LA, Cheng NW et al (2002) J Immunol 168:2307
60. Williams F, Meenagh A, Sleator C, Middleton D (2004) Hum Immunol 65:31
61. Maxwell LD, Williams F, Gilmore P, Meenagh A, Middleton D (2004) Hum Immunol 65:613
62. Keaney L, Williams F, Meenagh A, Sleator C, Middleton D (2004) Tissue Antigens 64:188
63. Halfpenny IA, Middleton D, Barnett YA, Williams F (2004) Hum Immunol 65:602
64. Meenagh A, Williams F, Sleator C, Halfpenny IA, Middleton D (2004) Tissue Antigens 64:226
65. Meenagh A, Gonzalez A, Sleator C, McQuaid S, Middleton D (2008) Tissue Antigens 72:383
66. Gonzalez A, Meenagh A, Sleator C, Middleton D (2008) Tissue Antigens 72:11
67. Yawata M, Draghi M, Little AM, Partheniou F, Parham P (2006) J Exp Med 203:633
68. Abi-Rached L, Parham P (2005) J Exp Med 201:1319
69. Maxwell LD, Wallace A, Middleton D, Curran MD (2002) Tissue Antigens 60:254
70. Middleton D, Gonzalez A, Gilmore PM (2007) Hum Immunol 68:128
71. Ordonez D, Meenagh A, Gomez-Lozano N et al (2008) Genes Immun 9:431
72. Shilling HG, Young N, Guethlein LA et al (2002) J Immunol 169:239
73. Carrington M, Martin MP (2006) Curr Top Microbiol Immunol 298:225
74. Single RM, Martin MP, Gao X et al (2007) Nat Genet 39:1114
75. Fauriat C, Andersson S, Bjorklund AT et al (2008) J Immunol 181:6010
76. Valiante NM, Uhrberg M, Shilling H et al (1997) Immunity 7:739
77. Santourlidis S, Trompeter HI, Weinhold S et al (2002) J Immunol 169:4253
78. Van Bergen J, Stewart CA, van Des Elsen PJ, Trowsdale J (2005) Eur J Immunol 35:2191
79. Davies GE, Locke SM, Wright PW et al (2007) Genes Immun 8:245
80. Martin MP, Pascal V, Yeager M et al (2007) Immunogenetics 59:823
81. Pando MJ, Gardiner CM, Gleimer M, McQueen KL, Parham P (2003) J Immunol 171:6640
82. Martin MP, Qi Y, Gao X et al (2007) Nat Genet 39:733
83. Sanjanwala B, Draghi M, Norman PJ, Guethlein LA, Parham P (2008) J Immunol 181:6293
84. Cella M, Longo A, Ferrara GB, Strominger JL, Colonna M (1994) J Exp Med 180:1235
85. Foley BA, De Santis D, Van Beelen E et al (2008) Blood 112:435
86. Carr WH, Pando MJ, Parham P (2005) J Immunol 175:5222
87. Dempster AP (1977) J Roy Stat Soc B 39:921
88. Gourraud PA, Genin E, Cambon-Thomsen A (2004) Eur J Hum Genet 12:805
89. Gourraud PA, Gagne K, Bignon JD, et al (2007) Tissue Antigens 69(Suppl 1):96
90. Single RM, Martin MP, Meyer D, Gao X, Carrington M (2008) Immunogenetics 60:711

NK Cell Education and *CIS* Interaction Between Inhibitory NK Cell Receptors and Their Ligands

Jacques Zimmer, François Hentges, Emmanuel Andrès, and Anick Chalifour

1 Introduction

Natural killer (NK) cells are major players of the innate immunity. Their capacity to synthesize cytokines and chemokines, to lyse various cells and to allow crosstalk between innate and adaptive immunity, make them important angular effector cells in the global immune system.

NK cells have to be "educated" to correctly fulfill their functions. The expression of inhibitory receptors (IR) must be regulated in a way such that NK cells remain tolerant towards normal autologous cells while recognizing and eliminating cells that have lost, in part or in total, the expression of autologous major histocompatibility complex class I (MHC-I) molecules. This loss frequently reflects tumor transformations or viral infections. In other words, NK cells are "educated" or "selected" for sparing normal autologous cells (normal presence of "self") and for detecting abnormal autologous cells (abnormal absence of "self," in other words "missing-self") which leads to the elimination of diseased cells. The mechanisms of this education/selection process have not yet been completely elucidated, although dramatic progress has been made in recent years.

J. Zimmer (✉) and F. Hentges
Laboratoire d'Immunogénétique-Allergologie, Centre de Recherche Public de la Santé (CRP-Santé), 84 Val Fleuri, L-1526, Luxembourg
e-mail: jacques.zimmer@crp-sante.lu

E. Andrès
Department of Internal Medicine, Clinique Médicale B, Hôpitaux Universitaires de Strasbourg, Strasbourg, France

A. Chalifour
Mymetics, 4 route de la Corniche, 1066, Epalinges, Switzerland
e-mail: anick.chalifour@mymetics.com

Historically, three different but not mutually exclusive models have been proposed to explain the phenomenon:

1. The "at least one model" postulates that the processes governing NK cell development and education only allow for the emergence and survival of NK cells bearing at least one IR specific for autologous MHC-I molecules. The sequential acquisition of IR by developing NK cells would stop, once the interactions between IR and the autologous MHC-I molecules are sufficiently strong to prevent the killing of normal cells. This model does not take into account a second potential "missing-self" recognition system based on the interactions of the NK cell IR NKRP-1 and the Ocil/Clr-b molecules on surrounding cells.
2. The "anergy model" claims that NK cells that either express no or not enough self-specific IR are not eliminated but conserved. However, they would be "anergic" in order to avoid a break of tolerance.
3. The "receptor calibration model" proposes that the expression levels of NK cell IR would be adapted to the level of expression of autologous MHC-I molecules. The hypothesis is based on the observation that in mice expressing a MHC-I ligand for a given Ly49 IR, the expression level of the latter goes down in contrast to mice that do not express such a ligand. In this context, we have published data (see below) showing that it is rather the accessibility of tetramers of ligand and antibodies against the Ly49 receptor that diminishes in the presence of a ligand *in cis*. Thus, if the explanation of the observed receptor's down modulation changes, the concept of receptor calibration in itself is still valid.

2 NK Cell Education

The "modern" concept of NK cell education started in 2005, when Raulet's group published a work showing that, in contrast to previously held beliefs, a subpopulation of NK cells lacked the expression of every known self-specific IR (SPIR) [1]. The size of this subset in C57BL/6 (B6) mice was between 10–13% of all spleen NK cells. Known SPIR in this mouse strain are the C-type lectins Ly49C, Ly49I, and NKG2A.

Similar to MHC-I-deficient NK cells, the SPIR− NK cells poorly responded to MHC-I-deficient lymphoblasts, the readout being IFN-γ production after a brief target cell exposure. The same hypo responsiveness was observed in cytotoxicity assays against various types of targets and after antibody (Ab)-mediated cross linking of activating receptors (AR, reverse ADCC). This population thus functionally mimics the self-tolerant NK cells from MHC-I-deficient individuals. Importantly, despite the lack of SPIR, these NK cells were self-tolerant as they did not kill syngeneic B6 lymphoblasts [1].

The data, in particular the strongly reduced reverse ADCC, suggest an attenuated stimulatory signaling. However, the observed hypo responsiveness is not due to a general inability of the cells to execute an activation program, as they produced IFN-γ to the same extent as SPIR+ NK cells after in vivo infection with *Listeria monocytogenes* or stimulation with PMA/ionomycin [1].

Phenotypically, apart from the presence or absence of SPIR, both NK cell subsets were similar with regard to the presence and expression levels of other IR and AR. In addition, the SPIR— NK cells did not display an immature phenotype. The only marker that was clearly different was the MHC-I-independent IR KLRG1, which was expressed at a much higher percentage by SPIR+ than by SPIR— NK cells. While the significance of this differential expression is unknown, it is unlikely to play a role in the hypo responsiveness and the self-tolerance, as it is expressed only by a small subpopulation of SPIR— cells. The same reduced expression is observed in MHC-I-deficient mice and thus seems to correlate with hypo responsiveness of NK cells in general. Finally, SPIR—, in contrast to SPIR+ NK cells, did not play any role in the rejection of β2m-deficient bone marrow grafts, which is consistent with their in vitro hypo responsiveness [1].

On the basis of their results, the authors develop a model of NK cell education which is currently known as the "disarming" model [2, 3]. In summary, this model claims that the balance of stimulatory and inhibitory signaling generated by the encounter of developing NK cells and surrounding self cells determines the subsequent responsiveness of the mature NK cells. In the absence of MHC-dependent inhibition, more hypo responsiveness is observed, whereas less hypo responsiveness appears in the presence of such an inhibition. In other words, too much stimulation would lead to inhibition, by, for example, activating signals exhaustion. This hypothesis would explain both the hypo responsiveness of SPIR— NK cells in normal mice and that of all NK cells from MHC-I-deficient animals.

Reduced stimulatory signaling could explain self-tolerance of SPIR— NK cells, as signal transduction by self-specific AR would be down regulated to such an extent that no auto reactivity would appear.

The next milestone in the field of NK cell education was the paper by Kim et al. [4] which introduced the "licensing" concept. In this work, it was shown that NK cells expressing a SPIR ("licensed" cells) produced much more IFN-γ than NK cells without such a receptor ("unlicensed" cells). The former are inhibited by the same MHC-I molecule that originally conferred "licensing," whereas the latter do not need to be inhibited by MHC-I as they are not functionally competent.

Anfossi et al. [5] confirmed the data in humans: approximately 13% of circulating human NK cells do not express a SPIR (KIR or NKG2A), and these cells are hypo responsive in terms of cytotoxic activity and cytokine production. In contrast, NK cells with a self-specific IR were functionally competent.

Clearly, the last two papers lead to the conclusion that there must be opposite signaling capacities of SPIR according to the stage of NK cell development. They transmit activating signals during NK cell development but inhibitory messages in mature NK cells. In the meantime, the Yokoyama group has confirmed the existence of the "licensing" mechanism in human NK cells [6].

In summary, it appears that NK cells have to express a SPIR (an inhibitory molecule) to become correctly activated.

Additional work by other groups has abundantly confirmed the previously presented data [7–10], without being able to demonstrate which one of the two models ("arming" versus "disarming") is right.

Interestingly, Yokoyama's group has recently published a paper [11] showing that the continuous engagement of a self-specific AR also induces NK cell self-tolerance, but they continue to claim that this is not in favor of the "disarming" model.

Detailed review of both the "disarming" and the "licensing" (or "arming") models have been published by the Raulet group [2, 3] – the reader might refer to it for an in-depth analysis of the topic. Recent reviews, also introducing new ideas) have been presented by Brodin and Höglund [12] as well as by Held [13].

One topic that has recently emerged and which is indirectly related to NK cells education is the discovery of so-called "memory" NK cells. Indeed, two groups [14, 15] have shown that after an initial activation, NK cells respond more strongly to the same stimulus provided several weeks later. The increased response only concerns proliferation capacity and cytokine production but not cytotoxicity. Further work is needed to completely understand this memory-like behavior.

3 Cis Interaction Between Ly49A and Its Ligand H–2D^d

As mentioned, "arming," "disarming" and "licensing" models are not able to explain all observed functions of NK cells. For instance, a significant subset of NK cells either do not express inhibitory receptors (or express one still unknown) and/or the correspondent ligand (or yet unidentified ligand) and do not show any fault in their self tolerance behavior. What is the in vivo relevance of these NK cells? Are they immature NK cells? How to explain the reactivity/functional maturity of activated NK cells from MHC-I deficient individuals that show "normal" NK cell functions: cytokines/chemokines synthesis and cytotoxicity? Such examples are too numerous to correspond to exceptional or erratic acting subsets. They suggest by their number that yet unidentified mechanisms are behind the global NK cells behaviors. Furthermore, it is still puzzling to understand/ envisage how the same receptor can be inhibiting or activating depending of the maturation stage of the NK cell, as postulated above.

A still unappreciated and solidly demonstrated property of IR is their ability to interact in *cis* with their ligand, per se, IR and its ligand(s) present on the same NK cell membrane. We will now describe in detail the experimental strategy that ended up with this demonstration. Secondly, we will discuss the implications of these findings for NK cell education.

The work started with the analysis of double transgenic H–2D^d (D^d) × Mx–Cre mice on the B6 background. A D^d-transgenic mouse in which the transgene is flanked by loxP sites, recognized by the viral recombinase Cre, had been previously developed. In Mx–Cre mice, the Cre gene is under the control of the promoter Mx which responds to IFN-α and IFN-β with the result that Cre expression is inducible after administration of these cytokines. However, in double transgenic mice, the system is leaky to some extent as approximately 30% of hematopoietic cells are D^d-, 30% D^d intermediate and the resting 30% D^d bright (same expression level as in B6 × D^d single transgenic mice), despite the absence of any exogenous induction. Thus, the double transgenic animals display a mosaic expression of the MHC-I molecule D^d. The expression pattern is similar in NK, NKT, conventional T and B cells from spleen, bone marrow and peripheral blood. The percentage of D^d- cells increases with the age of the animals [16].

As NK cells likewise display a mosaic expression of D^d, we wanted to check if the IR Ly49A, specific for the MHC-I molecule D^d, was homogeneously expressed in the three subpopulations defined by D^d expression. Surprisingly, this is not the case: whereas the percentage of Ly49A+ NK cells among the D^d bright population is normal (20%), it is increased to 32% among D^d intermediate NK cells and strongly reduced among D^d- NK cells (6%). The global percentage of Ly49A+ NK cells is normal (18%) and thus similar to the one observed in B6 mice and B6D^d single transgenic animals. This might suggest a better resistance of Ly49A+ NK cells (compared to Ly49A− NK cells) to the total loss of D^d, for instance due to a selective inactivity of the Mx promoter [16].

The mean fluorescence intensities (MFI) of Ly49A, which reflect the expression levels, diminish in parallel to the expression of D^d, such that they are weakest in the D^d- population. This finding is in total contrast to what is observed in B6 mice (genetically D^d-) with a MFI that is the double of that from B6D^d mice in which the D^d ligand is present.

In vivo Cre expression was induced through the intraperitoneal injection of the IFN-inducer polyI/C. Ten days after the last injection, more than 86% of total splenocytes and also of Ly49A− NK cells have become D^d-. In contrast, only 54% of the Ly49A+ NK cells are now D^d-. However, after 10 days of culture in the presence of IL-2, virtually all splenocytes, including the Ly49A+ NK cells, have become D^d-. Thus, the latter cells have also successfully recombined their transgene, although the disappearance of D^d seems to take longer than in other cells.

In combination with this, the low percentage of Ly49A+D^d- NK cells might suggest that most of the Ly49A+ NK cells have actually acquired D^d molecules from the environment. To elucidate this possibility, we performed a mixed chimera experiment: lethally irradiated B6D^d recipients are grafted with a 1/1 mixture of bone marrow from B6 and B6D^d donors. In order to distinguish between both types of cells independently from D^d expression, we used B6 mice expressing the CD45.1 isoform of the pan-leukocyte marker CD45, whereas B6D^d mice are CD45.2+.

Chimerism was investigated a few months after bone marrow transplantation by staining with anti-CD45.1 and anti-CD45.2 monoclonal Ab. Approximately 33% of splenocytes, bone marrow cells and peripheral blood cells are CD45.1+ and thus of B6 origin. A normal percentage of Ly49A+ NK cells (18–20% of all NK cells) is found both among CD45.1+ and CD45.2+ splenocytes. However, in contrast to Ly49A−CD45.1+ NK cells which are D^d-, Ly49A+CD45.1+ NK cells appear as D^d intermediate. As they are of B6 origin and therefore genetically D^d-, our observation implies that these cells have necessarily acquired D^d molecules from the surrounding D^d+ cells. In the same way, it is conceivable that in the case of $D^d \times$ Mx–Cre mice, Ly49A+ NK cells that have recombined the transgene and have lost D^d expression appear as D^d intermediate cells after capturing these molecules from the environment [16].

MFI of Ly49A of B6 origin (in the chimeras) are much lower than on Ly49A+ NK cells of B6D^d origin, which reproduces the similar correlation between absence of D^d and low MFI of Ly49A in mosaic mice (in contrast, as we already mentioned, to the high MFI of Ly49A in non chimeric B6 mice).

To strengthen the idea of acquisition of D^d molecules from surrounding D^d+ cells, we performed in vitro experiments by mixing in equal parts fresh splenocytes from B6 and B6 $\times$ D^d (transgenic expression of D^d) or B10.D2 (endogenous expression of D^d) mice. Cells are cultured for 2 h at 37°C, then stained and analyzed by flow cytometry. Indeed, Ly49A+ NK cells (but not Ly49A− NK cells) from B6 mice appear as D^d intermediate and, as they are genetically D^d-, this demonstrates that the D^d molecules were provided by the surrounding D^d+ cells. The transfer of D^d is completely abolished by anti-D^d and by anti-Ly49A Ab, showing that it is specifically dependent on Ly49A.

In order to determine the specificity of acquisition by Ly49A+ NK cells, we included two other MHC-I molecules, H–$2D^k$ (D^k) and H–$2D^b$ (D^b). The former is a known ligand for Ly49A [17], although with a slightly lower affinity than D^d, whereas the latter is not a ligand for this receptor. The "donors" of MHC-I molecules are B10.D2 (D^d), B10.BR (D^k) and B6 (D^b) mice, respectively, the "recipients" being splenocytes from β2m KO mice (MHC-I-). Ly49A+ NK cells from recipient mice very efficiently acquire D^d, significantly but to a lesser extent D^k, and not at all D^b. Thus, the acquisition capacity reflects precisely the affinity of Ly49A for its MHC-I ligands.

Next, we wanted to check what happens if the Ly49A+ NK cells already express themselves a ligand for Ly49A. For that purpose, we mixed splenocytes from B10.D2 (D^d) and B10.BR (D^k) animals. Under these conditions, Ly49A+ D^d+ splenocytes do not capture D^k molecules, and Ly49A+ D^k+ NK cells do not acquire D^d molecules. Therefore, the presence of a Ly49A ligand on the NK cells themselves precludes the acquisition of a second ligand.

This result could be confirmed in vivo after crossing of $D^d \times$ Mx–Cre mice with B10.BR mice. In contrast to the former (only very few Ly49A+D^d- NK cells), D^d- NK cells were present in high and, importantly, similar percentages of Ly49A+ and Ly49A− populations of ($D^d \times$ Mx–Cre) $\times$ B10.BR mice, suggesting that D^d capture is also blocked in vivo under these conditions [16].

These different results might lead to consider the possibility that the expression of a MHC-I ligand by Ly49A+ NK cells influences the structure and/or the accessibility of Ly49A. We chose, as an approach to elucidate this question, to stain NK cells from B6 and B10.D2 mice with three different monoclonal anti-Ly49A Ab: JR9-318 (JR9), YE1-48 and A1. In both types of mice, the three Ab stain approximately 20% of NK cells, which perfectly corresponds to the percentages usually observed in both strains. MFI obtained on B6 NK cells are arbitrarily considered as 100%. In B10.D2 mice, with the first two Ab, a reduction to 40–50% of the B6 levels is observed, in accordance with prior work by others [18, 19]. In contrast, MFI obtained with the A1 Ab are only 18% of B6 mice. This suggests that the fixation of this Ab to its epitope is strongly and selectively reduced in the presence of a Ly49A ligand on NK cells.

The next step was to analyze if the receptor/ligand co expression also had functional consequences, i.e., if the inhibitory properties of Ly49A are modified. Activated NK cells from Ly49A single transgenic and Ly49A × D^d double transgenic mice were used as effectors in cytotoxicity assays. Both types of NK cells do not lyse syngeneic lymphoblasts, in accordance with the known self tolerance of NK cells towards normal autologous cells. However, the tumor target C1498D^d is efficiently killed by double transgenic NK cells but completely resistant to single transgenic effectors. The parental target cell line C1498 (prototype sensitive target cell line) is equally well killed by the two activated NK cell populations, showing that the cytotoxic capacity of Ly49A single transgenic NK cells is intact.

All this confirms and extends prior research by others and shows that the inhibitory capacity of Ly49A is reduced at least tenfold on B6D^d compared to B6 NK cells. Structural constraints seem to strongly limit the interaction of Ly49A with its ligands on other cells when one ligand is already expressed by the Ly49A+ NK cell itself.

A major part of this work was based on the observation of the acquisition of MHC-I molecules by their corresponding IR expressed by NK cells. This is an example of the so-called "trogocytosis" phenomenon. It is not specific at all to our system but has been described in parallel (1) in another mouse model [21], (2) for human NK cells, whose KIR can acquire HLA-I ligands [22], and (3) for other types of lymphocytes . Thus, peptide – MHC-I complexes are transferred from antigen-presenting cells to CD8+ T cells [23–25]. The capture depends on the TCR and is only observed for peptide – MHC complexes specifically recognized by this TCR. Acquisition of MHC-II molecules by CD4+ T cells has likewise been described [26]. Furthermore, CD8+ and CD4+ T lymphocytes acquire, in a specifically CD28-dependent manner, CD80 from the surface of antigen-presenting cells [25, 27]. B lymphocytes can capture antigens coupled to cell membranes through specific interactions with the BCR [28]. Usually, the transferred molecules are rapidly internalized and degraded by the "receiving" cell [23, 25, 28]. NK cell trogocytosis and its functional consequences are reviewed in detail in this book by HoWangYin et al.

Vanherberghen et al. [29] demonstrated that the opposite direction "trogocytosis" is also possible: human and mouse IR are transferred from NK cells to MHC-I surrounding cells expressing the correspondent ligand.

Surprisingly however, we could not observe the transfer of D^d molecules to Ly49A+ NK cells when the NK cells already expressed a Ly49A ligand themselves. It is interesting to consider this result in the context of the crystal structure of Ly49A bound to D^d [30]. Indeed, according to this structure, one Ly49A molecule can interact with D^d through two different binding sites. Site 1 implicates the N-terminal residues of the α1 helix and the C-terminal residues of the α2 helix of D^d. Regarding site 2, it is located under the peptide-binding groove and involves residues of domains $\alpha 2$ and $\alpha 3$ as well as of $\beta 2m$. Both sites allow an interaction *in trans* between Ly49A and D^d, each molecule being expressed by a different cell. However, site 2 could also mediate an interaction *in cis*, Ly49A and its ligand D^d being expressed by the same cell. Therefore we proposed at this stage of the work that under physiological conditions (NK cells expressing self MHC-I as well as surrounding cells), the capture of D^d from surrounding cells is rendered very difficult or even impossible due to a *cis* interaction between Ly49A and D^d.

What are the arguments for this *cis* interaction? The A1 Ab only weakly stains Ly49A when a ligand is coexpressed *in cis*. It is interesting to mention that the epitope recognized by this Ab is located within a region of Ly49A that is in contact with the interaction site 2 of the D^d molecule [30] which might be masked by receptor/ligand *cis* interaction on Ly49A+D^d+ NK cells. It had also been reported [17, 31] that D^d tetramers, although binding very efficiently to Ly49A+ NK cells from B6 mice, stain much less the same receptor in mice expressing D^d. Finally, the inhibitory capacity of Ly49A on D^d+ cells is strongly reduced compared to D^d− NK cells.

Altogether, this suggests that the expression of D^d by NK cells, which means the presence of D^d *in cis*, reduces the accessibility of Ly49A for ligands *in trans*. This lower accessibility allows Ly49A to be particularly sensitive to even minor modifications in the expression level of D^d on target cells and consequently to be very efficient in the detection of diseased cells and in immune surveillance.

The next logical step was to try to directly demonstrate the existence of a *cis* interaction between Ly49A and D^d [32].

For that purpose, we investigated IFN-γ production by Ly49A+ compared to Ly49A− NK cells from non Ly49A transgenic, B6 and B6D^d, mice. After 4 days of culture in the presence of IL-2, NK cells were exposed overnight to C1498 or C1498D^d, and IFN-γ production was measured the following day by intracellular flow cytometry. C1498 induced cytokine production by a substantial fraction of both Ly49A+ and Ly49A− NK cells from both strains. In contrast, IFN-γ production by Ly49A+ NK cells from B6 mice was nearly completely abolished in the presence of C1498D^d cells, while it remained high in Ly49A+ NK cells from B6 × D^d mice. Under these conditions again, Ly49A did not confer a significant inhibition to NK cells even in the presence of its ligand D^d on the target cells.

It was important to discriminate between a potential role of D^d molecules expressed by surrounding cells (*in trans*) compared to D^d molecules expressed by the Ly49A+ NK cells (*in cis*). For this purpose, we used Ly49A × D^d × Mx–Cre triple transgenic mice, in which Ly49A+D^d+ and Ly49A+D^d− NK cells coexist and develop in the same environment characterized by a mosaic expression of D^d. By flow cytometry cell sorting, NK cells from triple transgenic mice were separated into D^d+ and D^d− populations and separately cultured in the presence of IL-2. After 4 days, cytotoxic activity was checked. Both cell types were equally activated as shown by the comparably efficient lysis of C1498 targets. With C1498D^d cells, a strong inhibition of lysis mediated by D^d− NK cells contrasts with a very minor inhibition of lysis by D^d+ NK cells. Thus, it is indeed the presence of D^d *in cis*, but not *in trans*, that governs the inhibitory capacity of Ly49A on NK cells.

In order to check if the inhibitory function of Ly49A depends on its capacity of ligand binding, we stained NK cells with D^k tetramers complexed with mouse β2m. These tetramers brightly stained 20% of B6 NK cells and all NK cells from Ly49A single transgenic animals. In contrast, they hardly stained NK cells from B6 × D^d, Ly49A × D^d and B10.BR mice. This shows that D^k tetramers cannot efficiently bind to Ly49A in the presence of D^d or D^k ligands on the same NK cell.

This low tetramer binding could be due to a steric hindrance resulting from the physical association between Ly49A and D^d. Alternatively, it could be the consequence of a structural modification of Ly49A due to the co expression of D^d, as has been described for the posttranscriptional modifications of CD8 [33].

To investigate these two possibilities, we repeated the tetramer binding experiments after destruction of MHC-I complexes with soft acid treatment. This leads to a very weak detection of β2m and of the heavy chain D^d on double transgenic Ly49A × D^d NK cells. However, D^k tetramers bind ten times better to Ly49A+D^d + NK cells than in the absence of acid treatment, whereas nothing changes at the level of Ly49A+ NK cells of B6 origin. Thus, tetramer binding does not seem to be diminished due to covalent modifications of Ly49A. On the contrary, the results obtained after acid treatment rather suggest that a non covalent complex between Ly49A and D^d *in cis* limits ligand binding *in trans*.

The next step consisted in searching for a potential co localization between Ly49A and D^d on cell membranes of Ly49A × D^d double transfectants of the tumor cell line C1498. Such a co localization was indeed evidenced by confocal microscopy.

The final direct proof of the *cis* interaction was possible due to immunoprecipitation experiments. In order to allow for the detection of Ly49A and D^d in cellular lysates, we added intracellular tags to both molecules: FLAG to D^d and vesicular stomatitis virus (VSV) to Ly49A. D^d–FLAG and Ly49A–VSV were then transfected alone or together into the cell line C1498. In double transfectants, immunoprecipitation of Ly49A with an anti-VSV Ab allowed the detection of D^d–FLAG in the immunoblots. Likewise, Ly49A is found in immunoprecipitates with an anti-FLAG Ab. The same results are obtained with T cells from Ly49A/D^d double transgenic mice, demonstrating the existence of the physical receptor/ligand *cis* interaction which was strongly suggested by our various functional assays.

Next, we wanted to know which one of the two interaction sites of D^d with Ly49A was important for *cis* interaction. Site 2 contains, among others, several residues of β2m, and other groups [34, 35] had previously demonstrated the importance of murine β2m for *trans* interactions. Based on these observations, we transfected the β2m-negative cell line C4.4.25 (endogenous expression of Ly49A) with D^d and with either human or murine β2m. D^k tetramers bind well to the human, but not the murine transfectants, which suggests the importance of murine β2m residues and of site 2 for *cis* interaction.

In the same context, we [32] mutated several residues of both interaction sites (1 and 2). Tetramer binding selectively increases for three of the five mutants of site 2, showing that it is this site that mediates the *cis* interaction.

The following conclusions can be drawn from these studies:

1. If D^d is present only on target cells (*in trans*), the inhibition transmitted to the NK cell by Ly49A is very strong, but this does not occur under physiological conditions. In contrast, if D^d and Ly49A are co expressed by the same NK cell, the inhibition transmitted by Ly49A is much weaker, although it remains sufficient to protect normal cells. Cancer cells remain sensitive and the presence of D^d on NK cells permits the "calibration" of Ly49A for an optimal detection of diseased cells. If the NK cell would be regulated like it is in the absence of D^d *in cis*, it would be permanently inhibited via Ly49A and would thus not be efficient in immunosurveillance.

2. In previous work by other groups, the lower staining intensity of Ly49A with specific Ab has been attributed to a lower expression level of Ly49A induced by its ligand *in trans*, to allow the NK cell to adapt to this presence while remaining capable of discriminating normal and diseased cells [18, 36]. This is the "receptor calibration model" that we mentioned above. Our results do not go against this model, but we rather suggest that the calibration is not due to a reduced expression level of Ly49A in the presence of D^d, but to the *cis* interaction between Ly49A and D^d which negatively influences the fixation of available reagents (Ab and tetramers) to Ly49A.

3. We confirmed and extended previous data demonstrating that site 2 is the most important one for the inhibitory activity of Ly49A in the presence of D^d *in trans*. We demonstrated in addition that site 2 also mediates *cis* interaction. As one homodimer of Ly49A binds to one single D^d molecule in *cis* [30], *cis* interaction precludes *trans* interaction of the same Ly49A molecule. This leads to our proposed model explaining, on the basis of available structural data, both the *trans* and the *cis* interactions. The model predicts an important flexibility of the stalk region of the receptor which is quite long.

Further work has clearly demonstrated that the same *cis* interaction also exists between the NK cell IR Ly49C and its ligands K^b and D^b as well as for the IR Ly49Q and D^d, on the basis of an experimental strategy largely using tetramer binding studies [37]. Importantly, these results suggest that receptor/ligand *cis* interaction is widely present in the NK cell population.

The confocal microscopy has allowed us to investigate the NK/target cell synapse composition for Ly49A and its ligand D^d according to the presence or absence of D^d *in cis* [38]. The results showed that in the absence of D^d on NK cells (*in cis*), Ly49A significantly accumulates at immunologic synapses with D^d+ target cells. D^d *in cis* however prevents accumulation of a large part of the Ly49A receptors at the synapse and in fact sequesters Ly49A outside of it. These observations fit well with the concept of reduced inhibitory function of Ly49A+D^d+ NK cells as reduced numbers of Ly49A receptors are available to interact *in trans* and consequently transmit inhibitory signals. These results further extend the previous ones as they suggest that receptor/ligand interaction tunes the receptor's inhibitory function to the inherited self-MHC-I, therefore/hence a cell-autonomous phenomenon.

So far, NK cell education models did not take into account the fact that some receptors as Ly49A notably, are constitutively associated *in cis* with their ligand(s). Our most recent experiments directly tried to investigate the role of this receptor/ligand *cis* interaction for NK cell education [39].

As a starting point, it has been confirmed that it's the Ly49A+ NK cells from D^d background that are responsible of the missing-self recognition, as upon their removal from the NK cell population, this property does no longer exist. This means, that Ly49A+ NK cells from D^d mice are educated and not Ly49A− NK cells.

In order to check if the inhibitory receptor Ly49A itself is responsible of this education, B6 and B6 × D^d mice transgenic for the Balb/c allele of Ly49A (Ly49A$^{Balb/c}$) were generated. In these mice, NK cells expressing only the transgene Ly49A$^{Balb/c}$ (JR9+A1-)(100%) can be distinguished from NK cells expressing both the transgene Ly49A$^{Balb/c}$ and the endogenous receptor (Ly49A^{B6}) (JR9+A1+) (20%) through staining with the anti-Ly49A Ab JR9 (both Ly49A$^{Balb/c}$ and Ly49A^{B6} alleles-specific) and A1 (only Ly49A^{B6} allele-specific). On the B6 background, although to a slightly lesser extent than the B6 allele, the transgenic Ly49A$^{Balb/c}$ bound D^d multimers and inhibited NK cells in the presence of D^d on target cells. This transgenic allele educated D^d NK cells as they killed B6 targets (missing-D^d recognition), even if those Ly49A$^{Balb/c}$ × D^d-derived NK cells are depleted from the endogenous Ly49A^{B6}-expressing cells. Those results demonstrated that the receptor Ly49A is responsible of the education acquisition by D^d NK cells which express it.

Next, the question if *cis* interaction between Ly49A and D^d was important for NK cell education was addressed. We have previously presented (see Fig. 1) a model hypothesizing that the stalk region of Ly49A has to be flexible to allow *cis* interaction. In order to assess this model, this stalk region was replaced by the stalk region of CD72 which is predicted to adopt a rigid conformation because of its α-helical coiled-coil structure. The resulting Ly49A$^{Balb/c}$/CD72 chimera (Ly49A·CD72) was expected not to be able to undergo *cis* interaction with D^d but rather to point away from the NK cell membrane and thereby being able only of *trans* interaction with D^d expressed on apposed membranes.

A Ly49A·CD72 transgenic mouse was generated. On the B6 background, NK cells were efficiently stained by D^d multimers and more importantly, were inhibited upon exposure to D^d+ targets; essential evidence demonstrating that the chimeric receptor is functional. A double Ly49A·CD72 × D^d mouse showed that D^d expression *in cis* did not significantly reduce multimer binding to this receptor and that this fixation was not increased after acid treatment. Furthermore, immunoprecipitation of the chimeric receptor did not reveal D^d in the lysates, as previously reported for the wild-type Ly49A receptor. All this suggests that the receptor cannot interact with D^d *in cis* but is able to mediate functional *trans* interaction with its ligand.

The obvious next step was to test if chimeric receptor-expressing NK cells were educated. As expected, B6 × D^d NK cells expressing chimeric Ly49A·CD72 receptor could not lyse B6 targets (missing-D^d recognition) as opposed to B6 × D^d NK cells expressing the wild-type Ly49A. Thus, the chimeric receptor can

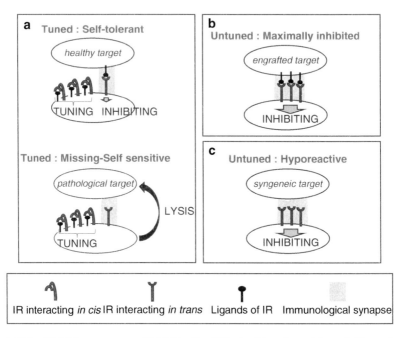

Fig. 1 NK cell inhibitory receptors: "Inhibiting" and "Tuning" functions. Schematic illustration of both inhibitory receptor (IR) functions, per se, "inhibiting" and "tuning", based on compiling data here described in details and recently reviewed by Held and Mariuzza [40].NK cells are illustrated in different contexts: (**a**) physiological conditions: MHC-I expressed both on NK cells and on their environment. (*Upper part*) healthy targets and (*lower part*) abnormal/pathological targets; (**b**) artificial conditions like grafts; (**c**) minor conditions like NK cell subpopulations expressing IR without known ligand expressed. IR/ligand *cis* interaction allows NK cells to be efficiently tolerant towards cells expressing correctly "self" and as opposite, efficiently untolerant towards cells with "missing-self" phenotype. Thus, *cis* tuned IR in order to allow NK cells to be properly inhibited. In the absence of *cis*, engagement of IR with its ligand *in trans* causes the substantial recruitment of IR at the immunological synapse in an uncontrolled way resulting with massive inhibition of NK cells. The latter are totally useless as they will not be able to sense improper "self" expression, as retrieved upon infection or tumorization. In fact, even if *trans* ligands get down regulated, a substantial quantity of IR is still engaged and inhibits the NK cell. This recruitment (uncounteracted by absent *cis* ligand) upon syngeneic encounters has been also demonstrated for NK cells expressing IR without *cis* ligand and *trans* ligand, explaining their hyporeactivity phenotype

inhibit NK cells but cannot instruct them missing-self recognition, which demonstrates the crucial role of *cis* interaction for NK cell education.

Ly49A expression without D^d (in B6 mice) actually dampened the functional properties of NK cells even in the absence of a ligand *in trans*, as was shown by stimulation experiments with plate-bound anti-NK1.1 Ab. Indeed, a lower percentage of Ly49A transgenic than of nontransgenic B6 NK cells produced IFN-γ upon this activation. The same transgenic Ly49A × B6 NK cells also did not, as opposed to B6 NK cells, kill efficiently MHC-I- target cells (β2m KO lymphoblasts), whereas D^d expression by those NK cells (and therefore *cis* interaction) restored this lysis. This was further confirmed by the observation that IFN-γ production was not restored to normal by the presence of D^d when Ly49A·CD72 transgenic NK cells (no *cis* interaction with D^d) were tested.

Crucially, all results obtained with transgenic mice were confirmed with (1) two independent transgenic lines and with (2) endogenous IR expressed on B6, Balb/c and β2m KO backgrounds.

A more in-depth experimentation has led to an important observation. Indeed, after NK1.1-mediated activation of wild-type B6 and B6 × D^d NK cells, a significant fraction of the Ly49A−Ly49C+ population produced IFN-γ. This percentage was significantly reduced (by more than 50%) in Ly49A+Ly49C+ cells from B6, showing that even the endogenous Ly49A dampens NK cell activation in the absence of ligand. Cytokine production was restored in Ly49A+Ly49C+ NK cells from B6 × D^d mice. Thus, the engagement of two IR by high affinity ligands, D^d and K^b respectively, leads to optimal NK cell education. This observation is not inclusive of the "disarming" model which stipulates that the presence of Ly49C alone should fulfill the requirements to ensure proper NK cell education.

Further work was necessary to investigate the mechanism behind the hypo responsiveness of NK cells expressing unengaged Ly49A. Yet another transgenic mouse allowed demonstrating that the dampening of these NK cell functions is inhibitory function-dependent and is mediated by their cytoplasmic ITIM motifs. For this transgenic a critical tyrosine residue of the ITIM sequence which is essential for the inhibitory function of Ly49A has been mutated to phenylalanine. Although they displayed normal numbers of NK cells with a normal phenotype, NK cells from these mice were not inhibited, as expected, by D^d-expressing tumor cells. On the other hand, they killed β2m lymphoblasts as efficiently as wild-type B6 NK cells in contrary to Ly49A transgenic B6 NK cells. It was then shown by confocal microscopy that Ly49A accumulated significantly at the NK cell synapse in the absence, but not in the presence, of D^d on the NK cell surface, and this even when the targets were MHC-I-deficient.

Finally, cocrosslinking experiments with anti-NK1.1 and anti-Ly49A Ab revealed that the cocrosslinking inhibits, whereas Ly49A sequestration (artificially induced by the Ab) improves NK1.1-mediated stimulation. This sequestration outside of the immunological synapse, physiologically realized through *cis* interaction with D^d, is thus crucial for NK cell education and function. Therefore, hypo responsiveness of NK cells expressing an unengaged IR can be counteracted by the expression of its ligand(s) *in cis*.

The following conclusions can be drawn from this second series of studies:

1. *Cis* interaction continuously sequesters a considerable fraction of Ly49 receptors, limits their redistribution at the synapse as well as their activity as negative regulators of NK cell functions
2. *Cis* interaction adjusts Ly49 receptor functions to the inherited self-MHC-I environment. In fact, the NK cells functional adaptation is based on the fact that the MHC-I expressed by NK cells corresponds to that of their environment
3. *Trans* interaction is not sufficient to educate NK cells
4. *Cis* interaction is necessary for NK cell education

5. Importantly, our data suggest two roles for Ly49; inhibiting and tuning. This model reconciles the fact that IR behaves as instinctively expected; as an inhibitory receptor (see Figure)

Our model hypothesizes that unassociated Ly49A, both *in cis* and *in trans*, would interfere in some instance, either by interacting directly or indirectly, with a component of the activation signaling cascade (receptor, adhesion molecule or signaling molecule) or with the proper synapse constitution. Importantly, this function is ITIM-dependent, suggesting an inhibitory-dependent mechanism. The presence of its ligand *in trans*, saves its owner by strong inhibitory signal to the NK cell. Tumorization for example, will not be sensed by those "only *trans*" interacting Ly49A, such as C1498D^d that will not be lysed. That said, the presence of its ligand *in cis*, will "tune" the activation threshold of the NK cell accordingly as they will be now able to recognize and eliminate tumorized cells despite the expression of the ligand (*in trans*), as is the case for Ly49A $\times$ D^d NK cells in face to C1498D^d targets.

Several important questions still remain to be answered, notably: are AR able of *cis* interaction? Does human and nonhuman primate IR possess *cis* ability? What is the functional relevance of this absence or existence?

The path to fully understand the mechanisms behind the diverse functions of NK cells, such as their education, is still going on, but major advances have been achieved in recent years. All models presented above cannot be entirely excluded and would probably be finally mutually nonexclusive. The picture is certainly complex, as is the level of the importance of the functions presumed for these cells.

References

1. Fernandez NC, Treiner E, Vance RE, Jamieson AM, Lemieux S, Raulet DH (2005) A subset of natural killer cells achieves self-tolerance without expressing inhibitory receptors specific for self-MHC molecules. Blood 105:4416–4423
2. Joncker N, Raulet DH (2008) Regulation of NK cell responsiveness to achieve maximal self-tolerance and maximal responses to diseased target cells. Immunol Rev 224:85–97
3. Raulet DH, Vance RE (2006) Self-tolerance of natural killer cells. Nat Rev Immunol 6: 520–531
4. Kim S, Poursine-Laurent J, Truscott SM, Lyberger L, Song YS, Yang L, French AR, Sunwoo JB, Lemieux S, Hansen TH et al (2005) Licensing of natural killer cells by host major histocompatibility complex class I molecules. Nature 436:709–713
5. Anfossi N, André P, Guia S, Falk CS, Roetynck S, Stewart CA, Breso V, Frassati C, Reviron D, Middleton D et al (2006) Human NK cell education by inhibitory receptors for MHC class I. Immunity 25:331–342
6. Kim S, Sunwoo JB, Yang L, Choi T, Song YJ, French AR, Vlahiotis A, Piccirillo JF, Cella M, Colonna M et al (2008) HLA alleles determine differences in human natural killer cell responsiveness and potency. Proc Natl Acad Sci USA 105:3053–3058
7. Cooley S, Xiao F, Pitt M, Gleason M, McCullar V, Bergemann TL, McQueen KL, Guethlein LA, Parham P, Miller JS (2007) A subpopulation of human peripheral blood NK cells that lacks inhibitory receptors for self-MHC is developmentally immature. Blood 110:578–586

8. Das A, Saxena RK (2004) Role of interaction between Ly49 inhibitory receptors and cognate MHC I molecules in IL2-induced development of NK cells in murine bone marrow cell cultures. Immunol Lett 94:209–214
9. Hasenkamp J, Borgerding A, Uhrberg M, Falk C, Chapuy B, Wulf G, Jung W, Trümper L, Glass B (2008) Self-tolerance of human natural killer cells lacking self-HLA-specific inhibitory receptors. Scand J Immunol 67:218–229
10. Yu J, Herller G, Chewning J, Kim S, Yokoyama WM, Hsu KC (2007) Hierarchy of the human natural killer cell response is determined by class and quantity of inhibitory receptors for self-HLA-B and HLA-C ligands. J Immunol 179:5977–5989
11. Tripathy SK, Keyel PA, Yang L, Pingel JT, Cheng TP, Schneeberger A, Yokoyama WM (2008) Continuous engagement of a self-specific activation receptor induces NK cell tolerance. J Exp Med 205:1829–1841
12. Brodin P, Höglund P (2008) Beyond licensing and disarming: a quantitative view on NK-cell education. Eur J Immunol 38:2934–2937
13. Held W (2008) NK cell education: licensing, arming, disarming, tuning? Which model? Eur J Immunol 38:2930–2933
14. Cooper MA, Elliott JM, Keyel PA, Yang L, Carrero JA, Yokoyama WM (2009) Cytokine-induced memory-like natural killer cells. Proc Natl Acad Sci USA 106:1915–1919
15. Sun JC, Beilke JN, Lanier LL (2009) Adaptive immune features of natural killer cells. Nature 457:557–561
16. Zimmer J, Ioannidis V, Held W (2001) H-2D ligand expression by Ly49A$^+$ natural killer (NK) cells precludes ligand uptake from environmental cells. Implications for NK cell functions. J Exp Med 194:1531–1539
17. Hanke T, Takizawa H, McMahon CW, Busch DH, Pamer EG, Miller JD, Altman JD, Liu Y, Cado FA, Lemonnier F et al (1999) Direct assessment of MHC class I binding by seven Ly49 inhibitory NK cell receptors. Immunity 11:67–77
18. Höglund P, Sundbäck J, Olsson-Ahlheim MY, Johansson M, Salcedo M, Ohlen C, Ljunggren HG, Sentman CL, Kärre K (1997) Host MHC class I gene control of NK-cell specificity in the mouse. Immunol Rev 155:11–28
19. Olsson MY, Kärre K, Sentman CL (1995) Altered phenotype and function of natural killer cells expressing the major histocompatibility complex receptor Ly-49 in mice transgenic for its ligand. Proc Natl Acad Sci USA 92:1649–1653
20. Ohlsson-Ahlheim MY, Salcedo M, Ljunggren HG, Kärre K, Sentman CL (1997) NK cell receptor calibration. Effects of MHC class I induction on killing by Ly49A^{high} and Ly49A^{low} NK cells. J Immunol 159:3189–3194
21. Sjöström A, Eriksson M, Cerboni C, Johansson MH, Sentman CL, Kärre K, Höglund P (2001) Acquisition of external major histocompatibility complex class I molecules by natural killer cells expressing inhibitory Ly49 receptors. J Exp Med 194:1519–1530
22. Carlin LM, Eleme K, McCann FE, Davis DM (2001) Intercellular transfer and supramolecular organization of human leukocyte antigen C at inhibitory natural killer cell immune synapses. J Exp Med 194:1507–1517
23. Huang JF, Yang Y, Sepulveda H, Shi W, Hwang I, Peterson PA, Jackson MR, Sprent J, Cai Z (1999) TCR-mediated internalization of peptide-MHC complexes by T cells. Science 286:952–954
24. Hudrisier D, Riond J, Mazarguil H, Gairin JE, Joly E (2001) CTLs rapidly capture membrane fragments from target cells in a TCR signaling-dependent manner. J Immunol 166:3645–3649
25. Hwang I, Huang JF, Kishimoto H, Brunmark A, Peterson PA, Jackson MR, Surh CD, Cai Z, Sprent J (2000) T cells can use either T cell receptor or CD28 receptor to absorb and internalize cell surface molecules derived from antigen-presenting cells. J Exp Med 191:1137–1148
26. Patel DM, Arnold PY, White GA, Nardella JP, Mannie D (1999) Class II MHC/peptide complexes are released from APC and are acquired by T cell responders during specific antigen recognition. J Immunol 163:5201–5210

27. Sabzevari H, Kantor J, Jaigirdar A, Tagya Y, Naramura M, Hodge JW, Bernon J, Schlom J (2001) Acquisition of CD80 (B7-1) by T cells. J Immunol 166:2505–2513
28. Batista FD, Iber D, Neuberger MS (2001) B cells acquire antigen from target cells after synapse formation. Nature 411:489–494
29. Vanherberghen B, Anderson K, Carlin LM, Nolte-t'Hoen EN, Williams GS, Höglund P, Davis DM (2004) Human and murine inhibitory natural killer cell receptors transfer from natural killer cells to target cells. Proc Natl Acad Sci USA 101:16873–16878
30. Tormo J, Natarajan K, Margulies DH, Mariuzza RA (1999) Crystal structure of a lectin-like natural killer cell receptor bound to its MHC class I ligand. Nature 402:623–631
31. Michaelsson J, Achour A, Salcedo M, Kase-Sjöström A, Sudback J, Harris RA, Kärre K (2000) Visualization of inhibitory Ly49 receptor specificity with soluble major histocompatibility complex class I tetramers. Eur J Immunol 30:300–307
32. Doucey MA, Scarpellino L, Zimmer J, Guillaume P, Luescher IF, Bron C, Held W (2004) *Cis* association with MHC class I restricts natural killer cell inhibition. Nat Immunol 5:328–336
33. Moody AM, Chui D, Reche PA, Priatel JJ, Marth JD, Reinherz EL (2001) Developmentally regulated glycosylation of the $CD8\alpha\beta$ coreceptor stalk modulates ligand binding. Cell 107:501–512
34. Matsumoto N, Mitsuki M, Tajima K, Yokoyama WM, Yamamoto K (2001) The functional binding site for the C-type lectin-like natural killer cell receptor Ly49A spans three domains of its major histocompatibility complex class I ligand. J Exp Med 193:147–157
35. Wang J, Whitman MC, Natarajan K, Tormo J, Mariuzza RA, Margulies DH (2002) Binding of the natural killer cell inhibitory receptor Ly49A to its major histocompatibility complex class I ligand. Crucial contacts include both $H-2D^d$ and $\beta 2$-microglobulin. J Biol Chem 277:1433–1442
36. Sentman CL, Olsson MY, Kärre K (1995) Missing self recognition by natural killer cells in MHC class I transgenic mice: a "receptor calibration model" for how effector cells adapt to self. Semin Immunol 7:109–119
37. Scarpellino L, Oeschger F, Guillaume P, Coudert JD, Lévy F, Leclercq G, Held W (2007) Interactions of Ly49 family receptors with MHC class I ligands in *trans* and *cis*. J Immunol 178:1277–1284
38. Back J, Chalifour A, Scarpellino L;, Held W (2007) Stable masking by $H-2D^d$ cis ligand limits Ly49A relocalization to the site of NK cell/target cell contact. Proc Natl Acad Sci USA 104:3978–3983
39. Chalifour A, Scarpellino L, Back J, Brodin P, Devèvre E, Gros F, Lévy F, Leclercq G, Höglund P, Beermann F, Held W (2009) A role for *cis* interaction between the inhibitory Ly49A receptor and MHC class I for natural killer cell education. Immunity 30:337–347
40. Held W, Mariuzza RA (2008) *Cis* interactions of immunoreceptors with MHC and non-MHC ligands. Nat Rev Immunol 8:269–278

Trogocytosis and NK Cells in Mouse and Man

Kiave-Yune HoWangYin, Edgardo D. Carosella, and Joel LeMaoult

Abstract The intercellular exchange of membrane-associated proteins between interacting cells is a mechanism which has recently become the subject of intense investigations, whether at the phenomenological, mechanistic, functional, or even conceptual levels. Even though the most prominent examples of intercellular transfer of membranes concern antigen presenting cell–T cell interactions, they have also been described between NK cells and target cells. In this chapter, we will focus on the characteristics of these mechanisms in the context of NK/target cell communication, and discuss the functions associated with intercellular transfers of membrane-bound proteins and their possible relevance.

1 Introduction

NK cells may serve as a first line of defense against tumors and viruses, thanks to their capability to kill MHC class I deficient cells. The function of NK cells is regulated by the balance between activating and inhibitory signals that are sent through triggering and inhibitory receptors.

Activating receptors recognize ligands usually expressed at low levels on healthy cells but that are often up regulated on tumor cells or virus-infected cells. Examples are the activating complex NKG2D/DAP10 which recognizes polymorphic MHC class I chain-related molecules (MIC) MICA and MICB, several toll-like receptors (TLRs) and the low affinity Fc receptor CD16. On the other hand, NK inhibitory receptors recognize MHC class I molecules that are constitutively expressed by most healthy cells but that are often down regulated in tumor cells. In humans, the main inhibitory receptors are killer cell immunoglobulin-like

---

K.-Y. HoWangYin, E.D. Carosella and J. LeMaoult (✉)
Institut Universitaire d'Hématologie, CEA, I^{2}BM, Service de Recherches en Hémato-Immunologie, Hôpital Saint Louis, 1 Avenue Claude Vellefaux, F-75475 Paris, France
e-mail: Joel.LeMaoult@cea.fr

receptors (KIR) and the lectin-like CD94/NKG2A heterodimers. In mice, the main inhibitory receptors are the lectin-like Ly49 dimers. These inhibitory receptors contain one or two intracytoplasmic inhibitory signaling domains called immunoreceptor tyrosine-based inhibition motifs (ITIMs) and recruit phosphatases such as SHP-1 to switch off activating signals. Hence, NK cells do not react to cells which express low levels of activating ligands and high levels of MHC class I molecules, but kill cells such as tumor cells which express higher levels of activating ligands and low levels of MHC class I molecules (for review, see [1]).

Activating and inhibitory signals are sent through a specific structure called the *immunological synapse* (IS), i.e., highly organized supramolecular clusters of specific molecules recruited to the interface between the NK cell and its target [2]. It is now known that inhibitory and activating receptors are spatially segregated in two different structures called cytotoxic NK IS (cNKIS) and inhibitory NK IS (iNKIS). It is the integration of the signals received through these structures which dictates whether or not NK cells will lyse their targets.

Thus, the function of NK cells seems to be controlled by the signals they receive from target cells through their activating and inhibitory receptors. However, NK cells are constantly proving to be more complex than previously thought, and their function to be ever more complex than the basic activating ligand versus inhibitory ligand dichotomy. In particular, it was recently shown that NK cells exchange membranes and proteins with their surroundings, and that this has potentially dramatic consequences on their function. Intercell antigen exchanges do not fit well into the basic model of the immune reaction because they imply that cells can act through molecules they do not express, react to molecules they express no receptor for, and perform functions unrelated to their nature. Yet, intercellular exchange of materials might be part of the normal behavior of NK cells and thus bear high significance.

Intercell antigen exchange was first demonstrated in 1973 in the context of LPS transfer between lymphocytes [3], and since then, various mechanisms have been described for it. One can cite uptake of shed antigens, uptake of apoptotic cell antigens, and uptake of exosomes, nanotube formation, and trogocytosis (for a review on each of these various mechanisms, see [4]). In the present chapter, we will focus on intercellular exchanges through trogocytosis, although we shall incorporate data concerning intercellular exchanges through nanotubes. Indeed, it is only recently that the differences between these two mechanisms have been highlighted. Hence, most of the available literature only deals with antigen exchanges taken as a whole and does not distinguish between nanotubes and trogocytosis. In truth, these two mechanisms do share common features and might happen concomitantly [4].

Membrane nanotubes (reviewed in [5]) are thin structures composed of F-actin that connect two cells. Nanotubes can either be actin-driven protrusions extending from one cell to another [6], or tethers which form when two cells dissociate after cell-to-cell interaction [7]. Depending on whether or not membrane fusion subsequently occurs, either an open-ended or closed membrane nanotube may be formed. Membrane nanotubes have been observed in a wide variety of cells, including NK, B, T cells and myeloid cells; they represent a novel route for intercellular communication [7–9]. Nanotubes may mediate the intercellular transfer of material through several mechanisms: (1) cargo such as vesicles or larger organelles might traffic between cells within open-ended nanotubes [6], or (2) cargo might traffic along the surface of nanotubular connections, as demonstrated by intercellular trafficking of bacteria [10] or viral particles [11]. (3) Cell-surface proteins and/or patches of surface membrane might also transfer via nanotubular connections [4, 6, 7]. (4) Finally, activation signals such as calcium signals may be transmitted from one cell to another through nanotubular connections [12]. In this chapter, we will only consider the transfer of membranes or membrane-bound molecules via nanotubular connections as these may resemble trogocytic transfer.

Trogocytosis (reviewed in [4, 13]) is a mechanism of fast, cell-to-cell contact-dependent uptake of membranes and associated molecules from one cell by another. Trogocytosis has been documented in α/β T cells [14–16] γ/δ T cells [17], B cells [18], NK cells [19], antigen presenting cells (APC) [20] and tumor cells [21]. Molecules whose transfers have been studied the most, include but are not restricted to MHC-I, MHC-II, CD54, CD80, CD86, and NK receptors. Regardless of the cell types involved, the main parameters of trogocytosis are (1) a dependence on cell-to-cell contact, (2) fast transfer kinetics in the order of minutes, (3) transfer of membrane patches containing intact membrane-bound molecules and proper orientation on the new cell, and (4) limited half-life of the acquired proteins at the surface of the acquirer cell.

The strict dependence of trogocytosis on cell-to-cell contact means that disruption of the initial cross-talk process might be sufficient to abrogate membrane transfer. For example, it was shown that $CD4^+$ and $CD8^+$ T cell acquisition of APC MHC Class II and MHC Class I molecules, respectively, was antigen-specific and consequently, could be blocked by disrupting TCR–MHC interaction [16]. This also implies that interactions between membrane-bound ligands and receptors drive the trogocytic transfer. In the example cited, MHC–TCR interaction was shown to be one of these driving interactions, but this cannot be generalized. In other systems and even in some APC-T trogocytosis experiments, transfers could be blocked by disrupting other ligand–receptor interactions [22], or could not be blocked at all [23], or happened in all-autologous antigen-free conditions [24–26]. Hence, it is commonly accepted that the mechanisms which underlie trogocytic transfers are either multiple/redundant, and/or incompletely understood.

Nevertheless, trogocytosis is not a transfer of individual molecules, but one of the entire membrane patches that contain intra- and trans-membrane proteins. Consequently, since disrupting only one receptor–ligand interaction is sometimes sufficient to block trogocytosis [16, 26, 27], it is clear that all molecules other than those absolutely necessary to trogocytosis are transferring passively. This is well illustrated by the fact that $CD8^+$ T cells can non-specifically acquire MHC Class II molecules along with the MHC-Class I:peptide complexes they are specific for [24, 28], the reverse being true for $CD4^+$ T cells [28].

What makes these transfers important is that the transferred membrane patches may temporarily endow the acceptor cell with some functions of the donor cells. Indeed, (1) $CD8^+$ T cells which acquired their cognate MHC Class I:peptide ligands became susceptible to "fratricide" antigen-specific cytolysis [16, 29], (2) T cells which acquired HLA-DR and CD80 could stimulate resting T cells in an antigen-specific manner, and thus behave as APC themselves [25, 30, 31], and (3) CD4[+] T cells and NK cells which acquired HLA-G behaved as suppressor cells [23, 32].

Even though the most prominent examples of intercellular transfer of membranes concern APC–T cell interactions, trogocytic transfers have also been described between NK cells and target cells. In this chapter, we will focus on the characteristics of these mechanisms in the context of NK/target cell

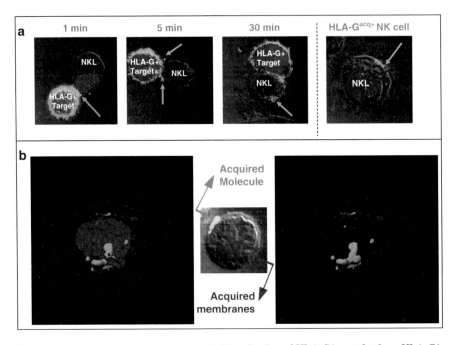

Fig. 1 Confocal visualization of trogocytosis Visualization of HLA-G1 transfer from HLA-G1-positive melanoma cells to the NK cell line NKL by confocal microscopy. (**a**): Kinetics of intercellular protein exchanges between target and NK cells. *Red*: cytoplasmic labeling of target cells, *Blue*: cytoplasmic labeling of NKL cells, *Green*: membrane bound HLA-G molecule. The pictures were taken at the indicated times after the beginning of the target-NK cell co incubation. After 1 min of co incubation, contact areas between target and NK cells were clearly visible. After 5 min and even more so after 30 min of co incubation, multiple HLA-G-positive areas can be seen on the surface of the NKL cells, some of which located outside of the contact area with target cells. The latter patches might indicate multiple contacts between the NK cell and HLA-G-positive target cells, and/or membrane movements. An isolated NKL cell which acquired HLA-G from target cells by trogocytosis is shown on the right. Arrows indicate areas of interest. (**b**): Three-dimensional reconstruction of an NK cell which acquired two membrane patches of different composition from a target cell: the transfer of lipids of target cell origin is evident in both cases (*red labeling*), but only one patch contains the HLA-G molecule (*green staining*). *Blue*: nuclear staining of NKL cells, *Red*: stained membranes from the target cells: prior to co incubation with NK cells, the target cells were labeled with the lipophilic PKH26 dye. *Green*: membrane bound HLA-G molecule originally expressed by the target cells

communication, and discuss the functions associated with these transfers and their possible relevance. For the sake of clarity, one example of target-to-NK cell transfer of membranes and associated molecules (including HLA-G, in this case) are shown in Fig. 1.

2 Intercellular Transfer Between Target Cells and NK Cells: Characteristics and Mechanisms

2.1 *Phenomenology*

The transfer of molecules between NK cells and target cells has now been described in several contexts. Arguably, the most complete demonstration was made in two independent reports published at the same time, which described in depth the acquisition of target MHC molecules by NK cells expressing Ly49 receptors at the iNKIS [33, 34]. Ly49A is a C-type lectin-like Ly49 receptor that recognizes H-2D^d and H-2D^k strongly but H-2D^b very weakly. In vitro experiments from both articles show that Ly49A$^+$ NK cells from C57BL/6 mice (H-2^b) acquire Ly49A ligand H-2D^d from the splenocytes of H-2^d mice. In this system, the transfer of H-2D^d was directly dependent of the H-2D^d:Ly49A interaction and could be blocked by anti-Ly49A or anti-H-2D^d antibodies. Of note, the transfer of membrane patches was not demonstrated in these reports but by using C-terminus GFP-tagged H2-D^d it was recently shown that the transferred MHC molecules were whole and included the intact intracellular part [35]. It seems therefore logical to assume that at least some donor cell membrane transferred along with H2-D^d in this system as it was shown in others [9, 19, 21, 36]; whether trogocytosis or nanotubes are involved, remains an open issue. In these two reports, it was further established that H2-D^d transfer to Ly49A$^+$ NK cells occurred in vivo. For instance, H-2^{b}D^d mice were lethally irradiated and then reconstituted with a mixture of bone marrow cells from B6 (H-2^b) and H-2^{b}D^d donors, and in this system, H2-D^d-negative NK cells from B6 origin stained weakly positive for H2-D^d molecules which they continuously acquired from H2-D^d-expressing cells in their environment [33].

Trogocytosis by NK cells is an active process. Indeed, it was shown at the cNKIS, that intercellular membrane exchanges are regulated by Src kinases and relied upon ATP, PKC, calcium, actin cytoskeleton and activation of NK cells at the cNKIS [37]. Interestingly, the transfer of HLA-C from target cells to NK cells at the iNKIS requires actin cytoskeleton and NK cell activation [19] but in this system Src kinases were not involved [9].

Several reports further showed that following NK cell–target cell interaction, proteins also transferred from NK cells to target cells. One can cite the NK-to-target transfer of KIR2DL1 [9, 21], NKG2D [38, 39], 2B4 [40], and in the mouse,

Ly49A [21]. These transfers have been comparatively less studied than target-to-NK transfers, but it seems they depend on the same parameters.

The existence of NK-to-target transfers asks the question of whether target-to-NK and NK-to-target transfers which have been described within the same experimental system occur independently as concomitant monodirectional transfers or in the same time as bidirectional transfers. This question is difficult to answer because it requires an analysis at the level of the structure where transfers occur with a possible requirement for kinetics analysis. Furthermore, the results obtained may depend on the nature of the individual transfers that are investigated. Nevertheless, it was shown that KIR2DL1 and its ligand HLA-C transferred from within the same synapse to target and to NK cell, respectively [9]. These results proved that bidirectional transfers of membrane-bound molecules are possible.

Thus, intercellular protein transfers occur between NK cells and target cells but open questions remain on the basic parameters of the intercellular protein exchanges, in particular: (1) which are the mechanisms involved (trogocytosis or nanotubes), are the proteins integrated or not in the recipient membrane, (2) where does the transfer take place (cNKIS or iNKIS), is it killing dependent and (3) is it specific of a particular interaction? In the following paragraphs, we will focus on these different issues.

2.2 Mechanisms Involved in the Transfer of Membrane-Anchored Proteins, and Integration of Acquired Proteins Within the Recipient Cell Membrane

Trogocytosis is a transfer of membrane proteins contained within a membrane patch, whereas nanotubes might cargo membrane-anchored proteins from one cell to another in the absence of lipid transfer [4, 5]. Several situations have been observed for protein exchanges between target cells and NK cells.

First, membrane-bound proteins may transfer even though membranes do not. It has been hypothesized that H-2 molecules which are specifically acquired by Ly49 expressing NK cells may transfer in the absence of lipid exchange [34]. This hypothesis would resemble the cargo of MHC molecules through nanotubular connections, but definite proof and characterization of the mechanism involved remain elusive.

Second, membrane-bound proteins may transfer within membrane patches but may not integrate within the membrane of the acceptor cell. In this case, the transferred structures may remain affixed onto the acceptor cell. This was particularly well illustrated in the context of HLA-C transfer from target cells to NK cells, for which the transferred proteins were shown to remain inside long thin membranous structures at the surface of the recipient cells [9]. The same mechanism seemed to drive the transfer of KIR2DL1 from NK cells to target cells [21] and in both configurations, transferred proteins and membranes could be removed by a brief acid wash. It was postulated that transferred HLA-C was contained within nanotubes which remained stamped on the NK cells after they dissociated from their targets. This may occur not only at the iNKIS, but also at the cNKIS, as shown for the transfer of the activating receptor 2B4 and its associated signaling adaptor molecule SAP from NK cells to target cells [40].

Third, membrane and proteins may integrate within the membrane of the acceptor cell. Although there is no formal demonstration that acquired membranes integrate properly and in a way that allows the acquired molecules to function within this new environment, evidence accumulates that this may be happening. For instance, in one of the systems described in the previous paragraph, only a fraction of the transferred KIR2DL1 could be removed from the acceptor cell surface [21], indicating that some membranes and associated molecules may have been more than just affixed on acceptor cells. In a recent report [41], it was shown that recombinant H-Ras, a signal transducing protein that is attached to the inner leaflet of the plasma membrane integrated into the plasma membrane of the adopting NK cell in a way that seemed to allow free diffusion. There is some debate on whether or not definite proof of free diffusion was provided in this report, but still, after transfer, H-Ras was found evenly redistributed within the plasma membrane of the adopting NK cell, and not clustered within a membrane patch of donor cell origin, which suggests proper integration.

2.3 Do Intercellular Transfers Depend on the Cytotoxic Activity of NK Cells?

NK cells are mainly characterized by their killing activity on target cells, i.e., on tumor cells or class I deficient cell lines. As trogocytosis and nanotubes are dependent of the activation of NK cells, one of the main questions of the exchange of proteins is whether they depend on the lytic activity of the NK cells.

In some cases, the inhibition of lysis prevented membrane uptake by NK cells, thus demonstrating that intercellular transfers may be dependent on cytolytic function. Indeed, Tabiasco et al. reported that cytolytic NK cells acquired membranes from their targets, but that when target cells expressed HLA-B27, the interaction between this molecule and ILT2 blocked both killing of target cells and membrane uptake [37]. In another study, the same authors demonstrated that CD21 acquisition by NK cells was inhibited when target cells were protected from lysis [42]. Recently, a study demonstrated that IL-2 activated NK cells could acquire membrane fragments from autologous monocytes and that this was correlated with the lytic activity of the NK cells towards the monocytes [43]. These results demonstrate that intercellular transfers may be linked to the cytolytic activity of NK cells. Yet, it was also shown that NK cells acquired the activating molecule MICA at the cNKIS rapidly after intercellular contact, and before the lysis could begin [38]. This interesting observation does not question the possible link between cytolysis and membrane transfers, but challenges the dependence of transfers on lysis.

In other contexts, intercellular protein transfers occurred in the complete absence of lysis. For instance, NK cells from a perforin-deficient patient acquired the inhibitory molecule carcinoembryonic antigen (CEA) from target cells even though they had no cytolytic function [44]. Similarly, NK cells and NK lines efficiently acquired the nonclassical HLA class I molecule, HLA-G from target cells even though this very molecule completely inhibited the cytolytic function of NK cells through interaction with its receptor on NK cells, ILT2 [32].

All data considered, there is not one definite answer to the question of whether or not intercellular transfer from/to NK cells depends on the cytotoxic activity of the NK cells and it all seems to depend on the interactive context. As far as our knowledge stands, it is safest to state that NK-target antigen exchanges seem to depend on NK activation, but may or may not be linked to NK cytolytic activity. The reason for this vague statement is that since we do not fully understand the mechanisms that drive NK-target antigen exchanges it is likely that one key parameter eludes us, which prevents us from discriminating between various exchange situations.

2.4 Do Intercellular Transfers Depend on a Specific Ligand–Receptor Interaction?

One of the key questions which remain unanswered is what are the molecules which actually mediate the intercellular transfer? All studies carried on so far emphasize the absolute requirement for cell-to-cell contact, and most hypothesize that antigen exchange occurs at the NKIS, since this is mainly where cell-to-cell contact occurs. Thus, it seems logical to conclude that some ligand–receptor interaction is required for antigen transfer. Most studies have investigated this point and some have identified a ligand–receptor pair that, in the system being considered, mediated the transfer.

For instance, in the murine system, the uptake of $H-2D^d$ by NK cells depended on the presence of Ly49A receptors at the surface of these NK cells, and blocking of the $H-2D^d$:Ly49A interaction prevented $H-2D^d$ uptake. This demonstrates that the $H-2D^d$:Ly49A interaction was directly responsible for antigen transfer in this system [33, 34]. Similarly transfer of MICA and MICB molecules to NK cells was shown to depend on recognition by NKG2D [38, 39]. In these systems, the interaction being considered seemed to play a crucial role. In other systems, specific interactions between receptor and ligand could only enhance protein transfers. This is the case of HLA-Cw6 uptake by NK cells, which is enhanced by KIR2DL1 recognition but does not require it [9]. This likely means that another ligand–receptor interaction drove HLA-Cw6 uptake by NK cells, and that several ligand–receptor interactions may participate in antigen transfer.

Many other examples exist showing that the specific interaction between the protein being studied and its receptor or ligand is not responsible for the transfer of that very protein. To cite only one: it was shown that the very efficient transfer of HLA-G from target cells to activated human NK cells bearing only one HLA-G receptor, ILT2, does not involve the HLA-G:ILT2 interaction. Indeed, the blocking of HLA-G:ILT2 interaction using anti-HLA-G or anti-ILT2 antibodies had no effect on HLA-G uptake by NK cells [32].

Thus, it seems that results which concern the requirement for a specific receptor–ligand interaction for antigen exchange to occur between NK cells and target cells are conflicting. This is however not the case. Indeed, one has to keep in mind that trogocytosis is the transfer of membrane patches, and not of individual molecules, which means that many molecules transfer at the same time. The fact that only few (sometimes even one) receptor–ligand interactions are sufficient to drive this transfer means that most of the proteins transfer passively. A passive transfer of a protein is one which does not require the interaction between this protein and one of its specific ligand/receptor. Consequently, a protein which transfers passively follows the molecules that really drive membrane and protein uptake [21]. Eventually, it all comes down to what is the cellular system and what is the protein followed by the investigator who performs the experiment. In some cases the protein that is looked at happens to be the one which is important for transfer and in other cases not. However, we believe it is safe to state that in all systems, a specific ligand–receptor interaction is indeed required for trogocytic transfer.

Thus, transfer of membrane-associated proteins such as trogocytosis is a process that is easily observable, but the mechanisms of which remain largely unknown. Some parameters are clear, such as activation of the NK cell, and the need for cell-to-cell contact, but others remain largely unknown, and in particular what is the molecular event that is the key to intercellular transfer. It is very possible that many types of intercellular exchanges exist, which take place differentially or concomitantly, depending on the cell–cell interaction considered, and also on the purpose of these interactions. As long as the various types of cell-to-cell antigen exchanges are not clearly identified, it will remain difficult to draw a clear picture of what are intercellular protein exchanges and/or what is their real contribution to immune responses.

3 Functions and Physiological Significance of Protein Transfers

Once it is admitted that proteins transfer between NK and target cells, the next obvious question is "so what?," in other words: how does an acquired molecule impact an immune response, and is this impact significant?

As far as the functions of the acquired molecules are concerned, one line must be drawn to separate transferred molecules with a ligand-type or a receptor-type function. Indeed, a ligand-type function is that of a proteic structure which acts through a receptor expressed on another cell. In this configuration nothing is asked of the transferred molecule other than to retain its original receptor-binding properties, and the actual functional changes are observed on the cell that bears the receptor. On the other hand, a receptor-type function is that of a proteic structure which signals to the cell that bears it after engagement with its ligand. In this configuration, the transferred molecule must not only retain its ligand-binding properties, but also its capability to send a signal that will yield an observable function. Indeed, in this case, the functional changes will be observed on the cell that has acquired a new set of molecules. A proper regrafting of the acquired receptor within the biochemical machinery of the new host cell is therefore mandatory.

To the best of our knowledge, the function of transferred molecules with ligand-type functions has been demonstrated numerous times, whereas the function of transferred molecules with receptor-type functions is still to be proven.

3.1 Functional Impact of the Intercellular Transfer of Molecules with Ligand-Type Functions

It has been demonstrated that many proteins with ligand-type functions are still functional after transfer to NK cells. For instance, $H-2D^d$ molecules acquired by $Ly49A^+$ murine NK cells retained their Ly49A-binding capabilities. As a consequence, $H-2D^d$-negative but Ly49A-positive NK cells which had acquired exogenous $H-2D^d$ molecules showed a reduced killing activity against MHC class I negative target cells compared to control NK cells which did not acquire exogenous $H-2D^d$ [34]. This could be the result of acquired-$H-2D^d$:endogenous-Ly49 interaction either in *cis* or in *trans* between NK cells.

In human beings, one study reports a lack of inhibitory functions for transferred HLA-C molecules. However, it is possible that no function was observed because no acquired molecule was left on the NK cell surface due to rapid internalization of the acquired molecules, along with the receptors they bind [19]. On the contrary, another study reports that HLA-G1 acquired by NK cells from tumor cells retains its full inhibitory function and has a dramatic impact on the function of the cells which acquired it [32]. Indeed, after acquisition of HLA-G1, NK cells stopped proliferating, became unresponsive to IL-2 stimulation, no longer lysed their targets, and behaved as regulatory NK cells capable of blocking the functions of other NK cells. All these inhibitions were directly due to HLA-G1 since blocking HLA-G1 or its receptor ILT2 completely restored NK cell functions. Thus, $HLA-G1^+$ NK cells were able to inhibit each other as well as other NK cells through the interaction, at least in *trans*, of acquired HLA-G1 with ILT2. The consequence of HLA-G1 acquisition by NK cells was the complete inhibition of the NK cell population's lytic capabilities and the transformation of NK cells supposed to lyse the targets into NK cells which protected that same target. It is interesting to note that this suppressive function, acquired through membrane transfers was temporary. Indeed, if HLA-G1-donor cells were removed, NK cells which had acquired HLA-G1 but no longer had access to it lost HLA-G1 cell-surface expression in a matter of hours, lost their inhibitory function along with it, and concomitantly recovered their original lytic capabilities.

These examples are demonstrative but are not the only ones. Indeed, it has been shown that (1) MICA that had been acquired by NK cells from tumor cells caused the degranulation of NK cells [38], that (2) NK cells which had acquired MICB from tumor cells exhibited a reduced NKG2D-dependent cytotoxicity [39], and that (3) transfer of CD21 on NK cells facilitated binding of and possibly infection by EBV [42].

Thus, molecules with ligand-type functions seem to keep their receptor-binding capabilities after intercellular transfer to NK cells. Depending on the molecule transferred, this may endow the acceptor NK cells with a new function, one that may alter dramatically its cross-talk capabilities and/or its natural function, albeit temporarily.

3.2 Functional Impact of the Intercellular Transfer of Molecules with Receptor-Type Functions

The transfer of proteins with receptor-type functions has been observed in both target-to-NK (H-Ras [41]) and NK-to-target direction (Ly49A [21], 2B4 [40], NKG2D [38, 39], KIR2DL1 [9]). Furthermore, the transferred receptors seem to be correctly oriented, in a manner that should allow binding in *trans*. In some cases, activating receptors transferred from NK cells to target cells along with costimulatory or adaptor molecules, which means that these receptors could potentially keep their functions and still signal inside their new host cells. For instance, (1) NKG2D transferred from NK cells to target cells along with DAP10, a transmembrane adaptor molecule containing a YINM motif that binds and activates via phosphatidylinositol 3-kinase and Grb2 [39]; (2) 2B4, a human activating co receptor also transferred from NK cells to target cells in association with SLAM associated protein (SAP), a signaling adaptor molecule. With the clear exception of transfers through apposition of nanotubes onto the acquirer cell (e.g., KIR2DL1 [9]), it seems that the conditions could allow signaling of the acquired receptor to its new cell host. However, with the possible exception of H-Ras transfer [41] no such signaling and no function of the transferred molecules with receptor-type ligands have been reported to date. There are, of course, many possible explanations to this fact, including the fact that such a function may not have been investigated, but we favor one simple interpretation: as stated before, the function of an acquired receptor is strictly dependent on, at least, (1) a proper insertion in the plasma membrane of the new host cell, (2) the existence of a compatible biochemical machinery within the host cell that will allow the acquired receptor to get its signal through, and (3) an outcome of this signaling that one can predict and measure. These are the simplest and the most basic requirements for acquired receptor function, and there must be many additional ones. Assuming a proper orientation and integration of the transferred molecules, in the systems which have been investigated so far, the transferred receptors may not function at all because donor cells and acceptor cells may be biochemically/metabolically too different (tumor line versus NK cells), causing a clear break in the progression of a signal originating from the transferred receptor. Furthermore, even if the donor cell and acquirer cells were compatible biochemically with respect to the signaling of the transferred receptor, were the experimental conditions right for the observation of the transferred receptor's function? For instance: how does one observe the effect of an inhibitory receptor transferred to a tumor cell, what is this receptor going to inhibit, is the acquirer cell in a state at which inhibition is observable? Thus, it seems to us that no function has been reported for transferred receptors because the experimental designs were set-up to focus on ligand transfers and not receptor transfers, with one exception. Indeed, a recent paper reports the transfer of the oncogenic molecule H-Ras from transfected B cell lines to lymphocytes that include NK cells [41]. In this configuration, donor and acquirer cells were both of lymphocytic origin, even if the possible biochemical compatibility is significantly weakened by the fact that one was a tumor line. Furthermore, the experiments were specifically set up to focus on the function of the transferred receptor. In this report, the authors make a strong case for a function of the transferred receptor and argue that it induced ERK phosphorylation, cytokine secretion and enhanced proliferation. Even though we believe that undisputable proof was still lacking, this was the first time the case for a functional regrafting of an acquired molecule with receptor-type functions was so strongly made.

The possibility of receptor-type molecules functional transfer is a fascinating one and fully applies to NK cells. For instance, in the simplest of configurations, the acquisition of a functional receptor which the acceptor cell does not endogenously produce would temporarily allow this acceptor cell to sense and react to molecules (membrane-bound or soluble) it was insensitive to before. In more complex situations, the acquired receptor might signal and trigger pathways which, in the context of the new host cell, may differ from the one triggered by the same stimulus in the original host cell. Variations of this are of course open to speculation, but regardless of the details, the demonstration that transferred receptor-type molecules may be functional opens a window on an entirely new way of looking at the regulation of immune responses, one that would fully integrate the contribution of environmental cells, be microenvironment- and situation-dependent, and highly adaptable, even though it would also be a very unpredictable one.

It is now evident that NK cells exchange antigens with their targets upon cell-to-cell contact, and in particular membrane-bound proteins. It is also clear that in the case of proteins with ligand-type functions, functions may be transferred from cell to cell, and that this may have a dramatic impact on the eventual behavior of NK cells, implying that the biology of NK cells is really more complicated than originally thought. The requirements of antigen transfers between NK cells and target cells are so that it is difficult not to consider the idea that antigen (and function?) transfers are an integral part of immune responses. This is already a great achievement: if intercellular transfers are part of the normal course of immune responses, then they constitute one potent regulation mechanism that is clinically completely unexploited. However, the parameters which dictate which molecules transfer from which to which cell, and with what possible impact are less evident. When all the various types of membrane-bound antigen exchanges have been clearly identified, the picture may get clearer and the consequences more predictable, but for the moment, only possibilities are certain.

References

1. Lanier LL (2005) NK cell recognition. Annu Rev Immunol 23:225
2. Orange JS, Harris KE, Andzelm MM, Valter MM, Geha RS, Strominger JL (2003) The mature activating natural killer cell immunologic synapse is formed in distinct stages. Proc Natl Acad Sci USA 100(24):14151
3. Bona C, Robineaux R, Anteunis A, Heuclin C, Astesano A (1973) Transfer of antigen from macrophages to lymphocytes. II. Immunological significance of the transfer of lipopolysaccharide. Immunology 5(24):831
4. Davis DM (2007) Intercellular transfer of cell-surface proteins is common and can affect many stages of an immune response. Nat Rev Immunol 7(3):238
5. Davis DM, Sowinski S (2008) Membrane nanotubes: dynamic long-distance connections between animal cells. Nat Rev Mol Cell Biol 9(6):431
6. Rustom A, Saffrich R, Markovic I, Walther P, Gerdes H-H (2004) Nanotubular highways for intercellular organelle transport. Science 303(5660):1007
7. Onfelt B, Nedvetzki S, Yanagi K, Davis DM (2004) Cutting edge: membrane nanotubes connect immune cells. J Immunol 173(3):1511
8. Sowinski S, Jolly C, Berninghausen O, Purbhoo MA, Chauveau A, Kohler K, Oddos S, Eissmann P, Brodsky FM, Hopkins C, Onfelt B, Sattentau Q, Davis DM (2008) Membrane nanotubes physically connect T cells over long distances presenting a novel route for HIV-1 transmission. Traffic 10(2):211
9. Williams GS, Collinson LM, Brzostek J, Eissmann P, Almeida CR, McCann FE, Burshtyn D, Davis DM (2007) Membranous structures transfer cell surface proteins across NK cell immune synapses. Traffic 8(9):1190
10. Onfelt B, Nedvetzki S, Benninger RKP, Purbhoo MA, Sowinski S, Hume AN, Seabra MC, Neil MAA, French PMW, Davis DM (2006) Structurally distinct membrane nanotubes between human macrophages support long-distance vesicular traffic or surfing of bacteria. J Immunol 177(12):8476
11. Sherer NM, Lehmann MJ, Jimenez-Soto LF, Horensavitz C, Pypaert M, Mothes W (2007) Retroviruses can establish filopodial bridges for efficient cell-to-cell transmission. Nat Cell Biol 9(3):310
12. Watkins SC, Salter RD (2005) Functional connectivity between immune cells mediated by tunneling nanotubules. Immunity 23(3):309
13. LeMaoult J, Caumartin J, Carosella ED (2007) Exchanges of membrane patches (Trogocytosis) split theoretical and actual functions of immune cells. Hum Immunol 68(4):240
14. Arnold PY, Davidian DK, Mannie MD (1997) Antigen presentation by T cells: T cell receptor ligation promotes antigen acquisition from professional antigen-presenting cells. Eur J Immunol 27(12):3198
15. Huang J-F, Yang Y, Sepulveda H, Shi W, Hwang I, Peterson PA, Jackson MR, Sprent J, Cai Z (1999) TCR-mediated internalization of peptide-MHC complexes acquired by T cells. Science 286(5441):952

16. Hudrisier D, Riond J, Mazarguil H, Gairin JE, Joly E (2001) Cutting edge: CTLs rapidly capture membrane fragments from target cells in a TCR signaling-dependent manner. J Immunol 166(6):3645
17. Espinosa E, Tabiasco J, Hudrisier D, Fournie J-J (2002) Synaptic transfer by human gamma}{delta T cells stimulated with soluble or cellular antigens. J Immunol 168(12):6336
18. Batista FD, Iber D, Neuberger MS (2001) B cells acquire antigen from target cells after synapse formation. Nature 411(6836):489
19. Carlin LM, Eleme K, McCann FE, Davis DM (2001) Intercellular transfer and supramolecular organization of human leukocyte antigen C at inhibitory natural killer cell immune synapses. J Exp Med 194(10):1507
20. Herrera OB, Golshayan D, Tibbott R, Ochoa FS, James MJ, Marelli-Berg FM, Lechler RI (2004) A novel pathway of alloantigen presentation by dendritic cells. J Immunol 173(8):4828
21. Vanherberghen B, Andersson K, Carlin LM, Nolte-'t Hoen EN, Williams GS, Hoglund P, Davis DM (2004) Human and murine inhibitory natural killer cell receptors transfer from natural killer cells to target cells. Proc Natl Acad Sci USA 101(48):16873
22. Hudrisier D, Aucher A, Puaux A-L, Bordier C, Joly E (2007) Capture of target cell membrane components via trogocytosis is triggered by a selected set of surface molecules on T or B cells. J Immunol 178(6):3637
23. LeMaoult J, Caumartin J, Daouya M, Favier B, Le Rond S, Gonzalez A, Carosella E (2007) Immune regulation by pretenders: cell-to-cell transfers of HLA-G make effector T cells act as regulatory cells. Blood 109(5):2040
24. Hwang I, Huang JF, Kishimoto H, Brunmark A, Peterson PA, Jackson MR, Surh CD, Cai Z, Sprent J (2000) T cells can use either T cell receptor or CD28 receptors to absorb and internalize cell surface molecules derived from antigen-presenting cells. J Exp Med 191(7):1137
25. Tatari-Calderone Z, Semnani RT, Nutman TB, Schlom J, Sabzevari H (2002) Acquisition of CD80 by human T cells at early stages of activation: functional involvement of CD80 acquisition in T cell to T cell interaction. J Immunol 169(11):6162
26. Xia D, Hao S, Xiang J (2006) CD8+ cytotoxic T-APC stimulate central memory CD8+ T cell responses via acquired peptide-MHC class I complexes and CD80 costimulation, and IL-2 secretion. J Immunol 177(5):2976
27. Hwang I, Shen X, Sprent J (2003) Direct stimulation of naive T cells by membrane vesicles from antigen-presenting cells: distinct roles for CD54 and B7 molecules. Proc Natl Acad Sci USA 100(11):6670
28. Lorber M, Loken M, Stall A, Fitch F (1982) I-A antigens on cloned alloreactive murine T lymphocytes are acquired passively. J Immunol 128(6):2798
29. Hudrisier D, Riond J, Garidou L, Duthoit C, Joly E (2005) T cell activation correlates with an increased proportion of antigen among the materials acquired from target cells. Eur J Immunol 35(8):2284
30. Xiang J, Huang H, Liu Y (2005) A new dynamic model of CD8+ T effector cell responses via CD4+ T helper-antigen-presenting cells. J Immunol 174(12):7497
31. Game DS, Rogers NJ, Lechler RI (2005) Acquisition of HLA-DR and costimulatory molecules by T cells from allogeneic antigen presenting cells. Am J Transplant 7(5):1614
32. Caumartin J, Favier B, Daouya M, Guillard C, Moreau P, Carosella E, LeMaoult J (2007) Trogocytosis-based generation of suppressive NK cells. EMBO J 26:1423
33. Zimmer J, Ioannidis V, Held W (2001) H-2D ligand expression by Ly49A+ natural killer (NK) cells precludes ligand uptake from environmental cells: implications for NK cell function. J Exp Med 194(10):1531
34. Sjostrom A, Eriksson M, Cerboni C, Johansson MH, Sentman CL, Karre K, Hoglund P (2001) Acquisition of external major histocompatibility complex class I molecules by natural killer cells expressing inhibitory Ly49 receptors. J Exp Med 194(10):1519
35. Andersson KE, Williams GS, Davis DM, Hoglund P (2007) Quantifying the reduction in accessibility of the inhibitory NK cell receptor Ly49A caused by binding MHC class I proteins in cis. Eur J Immunol 37(2):516

36. Caumartin J, Favier B, Daouya M, Guillard C, Moreau P, Carosella ED, LeMaoult J (2007) Trogocytosis-based generation of suppressive NK cells. EMBO J 26(5):1423
37. Tabiasco J, Espinosa E, Hudrisier D, Joly E, Fournié J-J, Vercellone A (2002) Active trans-synaptic capture of membrane fragments by natural killer cells. Eur J Immunol 32(5):1502
38. McCann FE, Eissmann P, Onfelt B, Leung R, Davis DM (2007) The activating NKG2D ligand MHC class I-related chain A transfers from target cells to NK cells in a manner that allows functional consequences. J Immunol 178(6):3418
39. Roda-Navarro P, Vales-Gomez M, Chisholm SE, Reyburn HT (2006) Transfer of NKG2D and MICB at the cytotoxic NK cell immune synapse correlates with a reduction in NK cell cytotoxic function. Proc Natl Acad Sci USA 103(30):11258
40. Roda-Navarro P, Mittelbrunn M, Ortega M, Howie D, Terhorst C, Sanchez-Madrid F, Fernandez-Ruiz E (2004) Dynamic redistribution of the activating 2B4/SAP complex at the cytotoxic NK cell immune synapse. J Immunol 173(6):3640
41. Rechavi O, Goldstein I, Vernitsky H, Rotblat B, Kloog Y (2007) Intercellular transfer of oncogenic H-Ras at the immunological synapse. PLoS ONE 2(11):e1204
42. Tabiasco J, Vercellone A, Meggetto F, Hudrisier D, Brousset P, Fournie J-J (2003) Acquisition of viral receptor by NK cells through immunological synapse. J Immunol 170(12):5993
43. Poupot M, Fournie J-J, Poupot R (2008) Trogocytosis and killing of IL-4-polarized monocytes by autologous NK cells. J Leukoc Biol 84(5):1298
44. Stern-Ginossar N, Nedvetzki S, Markel G, Gazit R, Betser-Cohen G, Achdout H, Aker M, Blumberg RS, Davis DM, Appelmelk B, Mandelboim O (2007) Intercellular transfer of carcinoembryonic antigen from tumor cells to NK cells. J Immunol 179(7):4424

Virus Interactions with NK Cell Receptors

Vanda Juranić Lisnić, Iva Gašparović, Astrid Krmpotić, and Stipan Jonjić

Abstract Natural killer cells are among the first cells of the immune response to recognize and react to threats. They do so by surveying other cells for aberrant behavior such as altered expression of MHC class I, and molecules produced or induced by pathogens. As such, they are very important in host resistance to viral infection. Various unrelated viruses have evolved numerous evasion techniques in order to avoid detection by NK cells. The many immunoevasive techniques may be roughly divided into two main groups: camouflage of infected cells aimed at inhibitory receptors and obstruction of activating receptors. By differential down-modulation of MHC class I molecules and production of MHC class I homologues, viruses prevent CTL recognition and camouflage their presence from NK cells. Additionally, viruses have directed even greater attention towards preventing the engagement of activating receptors by interfering with the receptors per se or by down modulating their ligands and coactivating molecules, providing soluble competitors, modification and interference with translation of ligands.

1 Introduction to NK Cells and Their Receptors

Natural killer (NK) cells are a subset of bone-marrow derived lymphocytes initially identified by their ability to lyse cancer cells without prior sensitization. Nowadays, it is known that this ability, termed natural killing, plays a major role not only in the rejection of tumors but also in the early defense against pathogenic organisms, especially viruses, while the adaptive immunity is still being mounted. NK cell activity is primarily regulated by a balance of inhibitory and activating signals

---

V.J. Lisnić, I. Gašparović, A. Krmpotić, and S. Jonjić (✉)
Department of Histology and Embryology, Faculty of Medicine, Braće Branchetta 20, 51000, Rijeka, Croatia
e-mail: jstipan@medri.hr

coming through receptor structures on the cell surface [1]. Probably, the most important role of inhibitory receptors is to prevent NK cells from engaging self antigens, a situation which could lead to autoimmune disorders. In general, inhibitory receptors monitor the presence of MHC class I molecules on the cell surface as explained by the "missing self" hypothesis [2]. It was noticed that NK cells preferentially kill tumor cells that do not express MHC class I molecules on their surface. Lately, it has been shown that upon ligation of their respective ligands, inhibitory receptors confer silencing signals thus preventing the release of cytotoxic molecules and the secretion of cytokines and chemokines.

All mature NK cells express at least one self MHC class I-specific inhibitory NK cell receptor (the "at least one" rule), ensuring NK cell tolerance to self [3, 4]. This implies that inhibitory receptors are needed for the proper maturation of NK cells in order for the NK cells to be able to recognize the "self" and become tolerant to it ("licensing" model) [5, 6] or to prevent hyporesponsiveness ("disarming" model) [7, 8]).

Inhibitory receptors signal through a characteristic amino-acid sequence motif located in their cytoplasmic tail: the immunoreceptor tyrosine-based inhibitory motif (ITIM). Upon ligand binding, the tyrosines located in the ITIM motif become phosphorylated leading to a signaling cascade. In turn, this results in the inhibition of NK cells through dephosphorylation of substrates for tyrosine kinases linked to activating NK cell receptors (reviewed in [9]). This interesting feature may explain why inhibitory signals often dominate over activating ones.

Aside from surveying other cells for any sign of transformation or viral infection through the expression of MHC class I molecules via inhibitory receptors, NK cells also rely on a vast and very diverse array of activating and coactivating receptors. Quite a few are structurally related and belong to families of inhibitory NK receptors. As with their receptors, activating receptor ligands are equally diverse and still not a sufficiently understood group of molecules. Some of them are constitutively expressed (like MHC class I), others are inducible as a consequence of cellular stress or viral infection while some are encoded by viruses. Despite their diversity, activating receptors use common signaling pathways, many of which are employed by T and B cells. However, unlike inhibitory receptors, cytoplasmic domains of most activating receptors do not contain signaling motifs. Rather, they transmit their signals through their interaction with signaling motif-bearing adaptor proteins [10].

2 NK Cell Receptors

Activating and inhibitory NK cell receptors can be classified into two families based on the chemical structure of their extracellular domains: receptors containing Ig-like ectodomains and receptors containing C-type lectin-like domains [11]. Most families of NK cell receptors contain both activating and inhibitory members. It is important to mention that many NK cell receptors are shared by other cells of the immune system. However, since numerous inhibitory and activating receptors and their signaling pathways far exceed the scope of this chapter, we will focus only on those most often targeted by viral evasion techniques.

2.1 The Ly49 Family

The first among many of MHC class I specific receptors discovered was the *Ly49 family*. Expressed in mice but lacking in humans, genes encoding this family of C-type lectins are highly polygenic and polymorphic and encode both inhibitory and activating receptors which are expressed on the cell surface as disulphide homodimers (reviewed in detail in [10, 12, 13]). Although the majority recognize MHC class I molecules, the most studied activating member of this family, Ly49H, adds an interesting twist to the story – it recognizes mouse cytomegalovirus (MCMV)-encoded protein m157 [14–19] conferring NK dependent resistance to Ly49H$^+$ mice. Recently it was shown that Ly49P plays a similar role: expressed in MA/My mice, which show a partial resistance to MCMV, it recognizes the H2-D^k molecule in complex with the MCMV m04 protein. Similar to the infection with the m157 deletion mutant in C57BL/6 mice [20], the infection of MA/My mice with the m04 deletion mutant abrogates the resistance of MA/My mice to MCMV infection [21].

Yet another interesting feature of Ly49 receptors is the fact that each NK cell expressing a particular inhibitory receptor will also express at least one activating, a feature shared by their human functional homologues belonging to the KIR (killer cell immunoglobulin-like receptors) family [4]. The potential problem of two receptors with opposing activity sharing the same or similar ligands is explained by the fact that most inhibitory receptors dominate over the activating ones, as mentioned above. Two or more activating signals are needed to rescue NK cells from inhibition. An important role in providing the additional positive signal is played by various cytokines [22].

2.2 The KIR Family

Human functional homologues of Ly49, *KIR*, belong to the immunoglobulin (Ig) superfamily of receptors. Unlike Ly49, KIR display a greater degree of complexity and diversity in ligand binding [23]. Various members of the KIR family can be distinguished by the number of their extracellular Ig domains (two or three), while their function depends on the length of their cytoplasmic tails [24]. KIR bind peptide loaded HLA-A, HLA-B, HLA-C and HLA-G by specifically recognizing amino acids belonging to the C-terminal portion of the MHC class I α1 helix [25].

2.3 The ILT Family

Ig-like domains in the extracellular region are also a common feature of another family of human receptors named ILT (Ig-like transcripts; also known as LILR and MIR), which map close to KIR genes. ILT receptors are expressed on the surface of many cells including monocytes, macrophages, dendritic cells, B, T and NK cells [26]. Only one of these, the leukocyte immunoglobulin receptor (LIR-1, ILT2, MIR7, CD85j) [27, 28], is expressed on a subset of NK cells where it acts as an inhibitory receptor. Like KIR and Ly49, this receptor also surveys the cell surface for the presence of MHC class I by binding to a broad spectrum of HLA class I molecules (A, B, C, G alleles and nonclassical HLA-F molecules) [29, 30]. When bound by a ligand, LIR-1 blocks killing and antigen dependent cellular cytotoxicity (ADCC) by NK cells, TCR induced cytotoxicity of T cells and the activation of B cells and monocytes [27].

2.4 The CD94/NKG2 Family

Aside from directly monitoring the expression of MHC class I molecules on the cell surface, NK cells also survey the expression of these molecules indirectly. This is achieved by CD94/NKG2, a C-type lectin family of receptors, which recognizes nonclassical MHC type I molecules (HLA-E in humans and Qa-1 in mice). Though structurally similar to classical MHC class I molecules, HLA-E and Qa-1 display only peptides derived from a leader sequence of classical class I molecules. A nine amino acid peptide sequence, contained within the presented leader peptide, is recognized by CD94/NKG2 [31]. The members of this family include the inhibitory receptor CD94/NKG2A and the activating receptors CD94/NKG2C and CD94/NKG2E.

2.5 NKG2D

Despite its name, NKG2D, a C-type lectin-like glycoprotein expressed as a transmembrane homodimer, is only distantly related to other NKG2 molecules [32]. The many evasive tactics employed by viruses, which target NKG2D, underline its importance in innate and possibly adaptive immunity.

NKG2D comes in two isoforms; long and short which differ not only in the length of their cytoplasmic tails but also in adaptor molecules they can interact with. NKG2D-L (long) splice variant only interacts with DAP10, whereas NKG2D-S (short) can transmit its signaling through the interaction with either DAP10 or DAP12 molecules [33]. In mice, both isoforms are present while humans only possess the long variant. However, signaling through DAP10 is sufficient to induce cell-mediated cytotoxicity in both humans and mice. Moreover, unlike most other activating receptors, signaling through NKG2D can overcome inhibitory signals and can therefore activate NK cells even if the target expresses sufficient amounts of self-MHC class I molecules to avoid killing via the "missing self" axis [34].

Another remarkable feature that distinguishes NKG2D from other activating NK cell receptors is its promiscuity: so far seven ligands have been identified in humans and nine in mice. Human NKG2D ligands fall into two families, the MHC-class-I-polypeptide-related sequence A (MICA) and B (MICB) family and CMV UL16-binding protein (ULBP) family (five members; ULBP1-4 and RAET1g) (reviewed in [35]), whereas mice express five retinoic acid early transcripts 1 (RAE-1α to ε) [36–38], minor histocompatibility protein 60 (H60 a to c [39, 40]) and murine UL16-binding-protein-like transcript 1 (MULT-1) [41]. All NKG2D ligands are characterized by a structural homology to MHC class I molecules and are inducible by cell stress, rather than constitutively expressed on the cell surface [35]. They show a great structural variability and are either expressed as GPI-anchored or as transmembrane proteins. Many are polymorphic in human populations or among various mouse strains and different alleles can show different susceptibility to viral evasion [42].

2.6 The NKR-P1 Family

Another group of C-type lectin-like receptors, *Klrb1* genes encoding NKR-P1 (natural killer cell receptor protein 1) molecules were originally identified in rats and are expressed on NK cells and T cell subsets [43, 44]. So far, five distinct murine Nkr-p1 genes and one pseudogene have been described encoding for both activating (NKR-P1A, NKR-P1C, and NKR-P1F) and inhibitory (NKR-P1B and NKR-P1D) isoforms. The *Klrb1* genes in rodents are characterized by a high number of copies and also allelic variability, whereas in humans only one ortholog has been found (encoding NKR-P1A) (reviewed in [45]).

2.7 NCR1

All receptors mentioned so far are not expressed on NK cells exclusively but are also shared by other cells of the immune system. However, natural cytotoxicity receptors (NCR) seem to be an exception to this rule [46, 47]. These members of the Ig superfamily have been found in both humans and mice. Human NK cells express three types of NCR1 receptors – NKp30, NKp44 and NKp46 [48], while mouse cells express only one – NKp46, although NKp30 exists in the mouse genome as a pseudogene [49, 50]. Although their binding to several viral hemagglutinins has been shown, their cellular ligands are still unknown [51, 52]. Despite this, the importance of NCR1 receptors is evident since it has been shown that their surface density correlates well with the ability of NK cells to lyse tumors [53], while a reduced or defective NCR expression during HIV or HCV infection has been associated with a decrease in NK-cell mediated killing [54, 55].

3 Viral Evasion Techniques

3.1 Camouflage of Virally Infected Cells

The immune system is a complex collection of different biological mechanisms involving numerous, highly specialized cell types. In order to effectively cope with various threats, two specialized "arms" of the immune response have evolved: less specialized but quick to act innate immune system, and a slower but highly specific adaptive immune system. The innate immune system plays a crucial role during the early times after viral infection, a time during which the more specific adaptive immune response is still not properly mounted. Since the cells of the adaptive immune response, specialized in eliminating intracellular pathogens such as viruses, can only recognize the threat in the context of MHC class I molecules, many viruses use down-modulation of MHC class I molecules as an immunoevasive strategy [56]. However, this renders the infected cells susceptible to NK cells which survey other cells for aberrant behavior including diminished levels of MHC class I and expression of ligands for NK activating receptors. In response to this, in order to evade the detection and control by the innate immune system, viruses have developed various strategies aimed at compromising NK cell function (Table 1). Production of MHC class I homologues and the upregulation of nonclassical MHC class I molecules not recognized by T lymphocytes as well as the production of ligands for the inhibitory NK receptors are commonly used evasion strategies.

3.1.1 Production of MHC Class I Homologues

The most obvious and straightforward approach in camouflaging the viral presence is by producing MHC class I homologues. In order to efficiently perform their role, these homologues must not be capable of inducing a T cell response, while still retaining their ability to engage the inhibitory NK cell receptors. Cytomegaloviruses (CMV) are especially successful in employing this strategy [57]. Interestingly, although all CMV-encoded homologues are structurally very similar to MHC class I and some can even bind peptides, they share only a moderate degree of sequence homology with MHC class I. As a perfect example of this the human CMV (HCMV) gpUL18 shares only 21% of sequence homology with MHC class I, while still retaining a structural similarity to MHC class I [58], the ability to

Table 1 Common viral camouflaging techniques

Virus	Immunoevasin	Receptor	Mode of action	References
MHC I homologues				
HCMV	gpUL18	LIR-1	Binds to inhibitory receptor LIR-1	[28, 57]
MCMV	m144	Unknown	Possible ligand for inhibitory receptors	[58]
RCMV	ORF r144	Unknown	Unknown	[59, 60]
MCV	MC080R	Unknown	Unknown	[61]
Regulators of MHC I expression				
HCMV	US6	–	Prevents peptide loading into MHC I molecules	[62–64]
HCMV	US2, US11	–	Differentially target MHC I for degradation	[62, 65–68]
HCMV	US3, US10	–	Retain HLA molecules in ER	[69–71]
HCMV	gpUL40	–	Maintains HLA-E expression on the cell surface by providing leader sequence for loading to HLA-E	[72, 73]
MCMV	m04	Ly49P	Binds to MHC I in ER and rescues their expression on the cell surface	[21, 74–76]
MCMV	m06	–	Targets MHC I for lysosomal degradation	[77]
MCMV	m152	–	Causes retention of MHC I in ER-cis-Golgi intermediate compartment	[78]
HIV	p24 aa14-22a	–	Binds to HLA-E and stabilizes it	[79, 80]
HIV	Nef	–	Downmodulates HLA-A and HLA-B by accelerating their endocytosis but spares HLA-C and HLA-E	[81–83]
KSHV	K3	–	Downregulates all four allotypes (A,B,C and E) by rapid endocytosis	[84]
KSHV	K5	–	Downregulates HLA-A and HLA-B by rapid endocytosis, HLA-C is affected only weakly, while HLA-E is unaffected	[84]
Homologues of non-classical MHC I molecules				
RCMV	RCTL	NKR-P1B	Ocil (Clr-b) homologue, binds to inhibitory NKR-P1B receptor	[45, 85]

associate with β2-microglobulin and the ability to bind endogenous peptides [59, 60]. An MHC class I-deficient cell line endogenously expressing gpUL18 was shown to express an increased level of cell surface β2-microglobulin and was resistant to NK cell lysis [61]. Moreover, it has been shown that the receptor recognizing this immunoevasin – LIR-1, binds UL18 with a 1,000-fold higher affinity than other HLA class I molecules [28, 62]; therefore, even a small amount of UL18 is sufficient for a successful immune evasion. In addition, UL18 has another interesting feature – it is heavily glycosylated. It appears that glycosylation stabilizes the protein [63] and protects it against degradation. Also it was speculated that glycosylation may play a role in the prevention of its down-modulation by HCMV encoded immunoevasins [64], such as US2, US3, US6, US10 and US11 which down-modulate MHC class I (reviewed in [65]). Glycosylation may also play a role in the prevention of UL18 engagement by other receptors or coreceptors recognizing MHC class I [63]. Recently, it has been shown that despite a high structural homology to MHC class I, UL18 is capable of avoiding down-modulation by US6 by restoring TAP function [66]. Subsequent reappearance of MHC class I on the cell surface is prevented by a coordinated action of UL18 and US6 where UL18 interferes with peptide loading by preventing the physical association between MHC class I and TAP. Therefore, the virus has not only ensured its survival by camouflaging MHC class I deficiency through a homologue with a higher affinity for the inhibitory ligand, but also by ensuring its long life on the cell surface (Fig. 1a).

Interestingly, despite the apparent successful strategy of UL18 as viral immunoevasin, it has been shown that the cellular receptor recognizing UL18, LIR-1, is expressed only on a minor subset of NK cells [29]. The controversy surrounding this receptor has been broadened by the finding that UL18 not only inhibits NK cells but it can also in some cases activate NK cells not expressing the LIR-1 receptor [28, 67, 68]. Since LIR-1 appears also on monocytes and dendritic cells much earlier in their development than during NK cell maturation it can be speculated that this viral tactic is targeted mostly at monocytes and dendritic cells [67].

An MCMV encoded MHC I homologue, m144, like UL18 uses glycosylation to ensure stability and avoid down-modulation [69]. However, unlike UL18 it does not bind peptide. Although its receptor is still unknown, its importance is underscored by the finding that deletion mutants showed increased susceptibility to NK cell control in vivo [70].

Other structural MHC homologues like ORF r144 [71] in rat CMV or a poxvirus gene MC080R [72] are still subjects of studies.

3.1.2 Differential Modulation of MHC Class I Molecules

Viruses are pretty successful at subverting the cell to do their bidding. A perfect example is differential modulation of different MHC molecules. Not all MHC class I molecules are capable of presenting viral peptides and alerting T cells of viral infection. Such MHC molecules, like HLA-E in humans and mouse Qa-1, present only peptides derived from leader sequences of other HLA/MHC class I molecules and thus serve as an indirect indicator of MHC class I production. NK cells survey the presence of these nonclassical MHC molecules through the CD94/NKG2 family of receptors. Since the binding of HLA-E to inhibitory CD94/NKG2A prevents NK cell mediated killing, even when other MHC class I molecules are down-

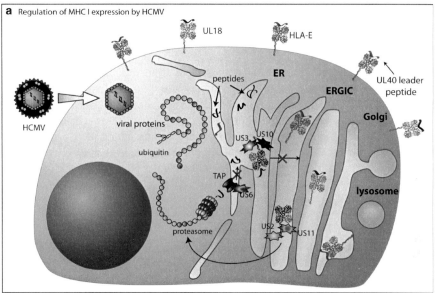

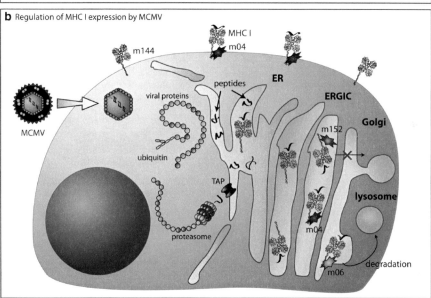

Fig. 1 Camouflaging techniques employed by cytomegaloviruses. (**a**) Regulation of MHC class I expression by HCMV. HCMV encodes several proteins which selectively interfere with MHC I expression on the cell surface. US6 prevents peptide loading, US2 and US11 target HLA-A for degradation, while US3 and US10 retain MHC class I in the ER. In order to evade recognition by NK cells, HLA-E is spared from down-modulation, while HCMV-encoded UL40 provides leader peptide for loading into HLA-E. Additionally, HCMV encodes MHC class I homologue UL18. (**b**) Regulation of MHC class I expression by MCMV. Like HCMV, MCMV also encodes several proteins which interfere with MHC class I expression. m152 causes retention of MHC class I in ERGIC compartment while m06 targets MHC class I for degradation. Interestingly, m04 does not prevent MHC class I expression on the cell surface. Rather, a complex of m04 and MHC class I appears on the cell surface. Like HCMV, MCMV also encodes one MHC homologue – m144

modulated, it should not come as a surprise that HLA-E molecules are common targets for viral camouflaging techniques.

HCMV encodes several proteins that down-modulate various MHC class I molecules with different efficacy. US6 protein prevents peptide loading into MHC class I molecules and thus prevents them from reaching the cell surface. US2 and US11 target HLA-A for degradation, while sparing HLA-E expression on the cell surface [65]. The activation of NK cells is prevented by gpUL40 which has a leader sequence identical to other MHC class I leader sequences [73, 74]. Expression of this protein results in a maintained expression of HLA-E molecules on the cell surface and consequent inhibition of NK cell-dependent lysis even in the case of diminished expression of other MHC molecules.

Like HCMV, MCMV also down-modulates MHC class I molecules in order to prevent the destruction of infected cells by CD8+ T cells (Fig. 1b). Among three regulators of MHC class I molecules, m152 arrests the maturation of MHC molecules at the level of ERGIC compartment [75], whereas the product of m06 redirects MHC class I molecules to lysosomes for degradation [76]. While the role of m152 and m06 in MHC class I presentation is well established, the role of the third MCMV regulator of MHC class I molecules, m04 is more complex. Contrary to the other two, m04 does not prevent MHC class I molecules from reaching the cell surface. Instead, it binds to MHC class I, forming a complex that is expressed on the cell surface [77, 78]. Although it has been shown that m04 may inhibit MHC class I presentation [79], it has also been shown that it may actually antagonize the function of m152 and enhance the recognition of infected cells by virus specific CD8+ T cells [80]. Since m04 brings class I molecules to the cell surface, it has originally been proposed that this function serves to inhibit NK cells by providing ligands for its inhibitory receptors. Although it has not been so far demonstrated that m04 does indeed serve as an NK cell inhibitor, recent results showing that m04 is essential for the recognition of infected cells by the activating Ly49P receptor [21] suggest that inhibitory Ly49 receptors may, in a similar way, recognize host MHC class I in complex with m04 in order to prevent NK cell activation. Therefore we speculate that m04 did actually evolve during the coevolution of the host and the virus to serve as a NK cell inhibitor, since the downregulation of MHC class I in order to prevent CD8 recognition could make infected cells susceptible to NK cells via the "missing self" mechanism.

HIV-1 virus also takes advantage of this mechanism by expressing p24 aa14-22a, a HIV T cell epitope that binds to HLA-E and stabilizes its presence on the cell surface [81, 82], while Nef protein downregulates HLA-A and HLA-B molecules, but has no effect on HLA-C or HLA-E [83]. Kaposi's Sarcoma Associated Herpes Virus (KSHV) K3 and K5 proteins show similar selectivity [84, 85].

An interesting twist comes from the finding that HLA-E molecules are ligands for the activating CD94/NKG2 receptors as well. An increase of NK cells expressing activating CD94/NKG2C receptor was observed in HCMV infected and HCMV and HIV-1 coinfected humans [86, 87]. Moreover, these NK cells outnumbered those expressing inhibitory CD94/NKG2A. The increased proliferation of CD94/NKG2C+ NK cells was independent of the UL16, UL18, and UL40

HCMV genes, but was impaired upon infection with a mutant lacking the US2-11 gene region responsible for MHC class I inhibition [88].

3.1.3 Viral Homologues of Nonclassical MHC Molecules

The original "missing self" hypothesis assumed that only the altered expression of classical MHC class I molecules prompts NK cells to reaction. However, the finding that tumors expressing the non-MHC molecule osteoclast inhibitory lectin (Ocil, also known as Clr-b) are protected from NK cell lysis, indicates that "missing self" recognition can also be mediated through other molecules [89].

Infection with rat CMV (RCMV) causes the down-modulation of Ocil, most probably due to cellular stress. The subsequent NK cell activation is prevented by RCMV-encoded ortholog RCTL (rat C-type lectin-like product), which binds to an inhibitory receptor present on NK cells, NKR-P1B. RCTL sequence closely resembles that of rat Ocil indicating that virus might have acquired it through an act of "molecular piracy" [45, 90]. Interestingly, some rat strains counteracted this mechanism by evolving an activating NKR-P1A receptor capable of recognizing RCTL, while others have evolved inhibitory NKR-P1B variants which fail to bind the decoy while still binding Ocil; this shows that viruses can be an important factor in shaping the host evolution [90].

3.2 Obstruction of Activating Receptors

A second line of viral defense against the immune system utilizes a more direct approach by obstructing activating receptors. As inhibitory signals prevail over activating ones, obstruction of activating receptors ensures NK cell evasion even when inhibitory signals are weakened. Immunoevasive tactics by obstruction are many and can target activating receptors directly or circumspectly by down-modulation of their ligands or coactivating molecules, providing competition for the receptor or by modulation of host gene expression (Table 2).

3.2.1 Interference with Activating NK Cell Receptors

This most direct of all approaches has been observed in HIV and Hepatitis C virus (HCV) infections. NK cells of chronically infected patients are reduced in numbers and functionally impaired, displaying reduced cytotoxicity [91]. Additionally, NK cells of HIV patients show altered expression of activating receptors [92, 93].

The first mechanism providing a possible explanation of this phenomenon came with the discovery that binding of HCMV tegument protein pp65 causes a general inhibition of NK cell cytotoxicity towards a wide variety of targets [94]. Tegument molecules are part of an HCMV virion positioned between the lipid envelope and

Table 2 Common viral obstruction techniques

Virus	Immunoevasin	Receptor	Mode of action	References
Interference with activating NK cell receptors				
HCMV	pp65	NKp30	Causes dissociation of CD3ξ chain from NKp30 receptor complex	[104]
Downmodulation of ligands for activating NK cell receptors				
HCMV	UL16	NKG2D	Causes intracellular retention of NKG2D ligands MICB, ULBP-1 and ULBP-2	[105, 106]
HCMV	UL141	CD96, CD226	Retains CD155 in an immature form associated with ER	[107]
HCMV	UL142	NKG2D	Downregulates full-length MICA	[108]
MCMV	m138 (fcr-1)	NKG2D	Downregulates MULT-1, H60 and RAE-1ε, interferes with clathrin-dependent endocytosis	[109, 110]
MCMV	m145	NKG2D	Intracellular retention of MULT-1	[111]
MCMV	m152	NKG2D	Intracellular retention of RAE-1	[112]
MCMV	m155	NKG2D	Intracellular retention of H60	[113, 114]
KSHV	K5	NKG2D, NKp80	Ubiquitylates MICA and targets to unknown compartment, downmodulates MICB and targets NKp80 ligand AICL for lysosomal degradation	[115]
HIV	Nef	NKG2D	Downmodulates MICA, ULBP-1 and ULBP-2	[116]
Ligand modification				
Influenza virus	HA	NKp46	Viral attachment and fusion protein, prevents recognition by NKp46 by addition of new glycosilation motives	[117]
Competition for receptor				
zoonotic orthopoxvirus	OCMP	NKG2D	Soluble antagonist of NKG2D; prevents recognition and binding of host NKG2D ligands	[118]
Downmodulation of co-activating and accessory molecules				
KSHV	K5	LFA-1	Inhibits ICAM-1 expression by promoting its ubiquitylation and lysosomal degradation	[99, 119, 120]

(*continued*)

Table 2 (continued)

Virus	Immunoevasin	Receptor	Mode of action	References
HTLV-1	p12^I	–	Downmodulates ICAM-1 and ICAM-2 thus interfering with NK cell adhesion to infected cells	[121]
HCV	E2	CD81	CD81 crosslinking generates inhibitory signals	[122, 123]
Micro RNAs				
HCMV	miR-UL112	NKG2D	Completely prevents translation of MICB and to some extent full length MICA, also targets viral IE1	[124–126]
Other				
MCMV	m157	Ly49H; Ly49I	Binds to inhibitory Ly49I and activating Ly49H receptors	[15, 18]

capsid and, as such, are among the first molecules to be delivered into the host cell upon infection. While pp65 is a dominant antigen for CTL, it exerts just the opposite effect on NK cells. Direct binding of pp65 to NKp30 causes the dissociation of the CD3ξ chain from the receptor complex, thus preventing the transduction of activating signals (Fig. 2a). By inhibiting NK mediated lysis via NKp30, HCMV does not only protect infected cells from lysis but also modulates the adaptive immune response via dendritic cells. Mainly, NKp30 is the only activating NK receptor involved in the maturation and lysis of normal, self cells – primarily immature dendritic cells (DC). Additionally, NK cell inhibition may impair DC-NK cross talk which is considered important in DC priming as well as in the maturation and consequent CD4+ Th1 cell induction and CD8+ CTL development. Therefore, long-term inhibition of NK cells could impair the DC maturation process and thus have an indirect effect on the efficacy of T cell responses [95]. Also, recent findings suggest that DC progenitor cells are a major hiding place for the virus during latency, while differentiation to mature DC has been linked with reactivation of infectious virus particles (reviewed in [96]). Therefore, pp65-mediated inhibition of NK cell killing may serve to protect this hiding place [94]. Taking all this into account, it is not surprising that HCMV infection often comes accompanied by general immunosuppression and an increased susceptibility to various infections.

3.2.2 Downmodulation of Ligands for Activating Receptors

The most widespread of all obstruction techniques, the down-modulation of ligands for activating receptors offers a wide variety of targets. Taking into account its title

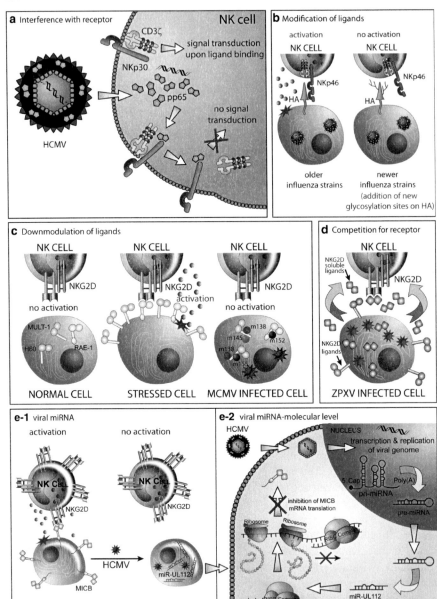

Fig. 2 *Viral obstruction techniques.* (**a**) Interference with receptor. Influenza virus protein pp65, a component of tegument, enters the cell upon infection and by binding to NKp30 receptor causes the dissociation of CD3ξ from the receptor complex thus preventing transduction of activating signals. (**b**) Modification of ligands. Major Influenza virus protein, hemagglutinin (HA) is recognized by NKp46. However, in newer influenza virus strains, recognition by NKp46 is prevented by the addition of new glycosylation motifs in HA. (**c**) Downmodulation of ligands. Murine NKG2D ligands are not normally expressed on healthy cells. Rather, their expression is triggered by cellular stress such is transformation, irradiation or infection leading to NK cell activation.

of the promiscuous activating receptor [35], as well as the nature of its ligands whose expression is connected to cellular stress, infection or transformation, it does not come as a surprise that NKG2D is the most common target of this evasion technique. Moreover, the importance of NKG2D in early control of viral infection is underscored by the number of viruses targeting this particular receptor, as well as the redundancy of the immunoregulatory genes targeting it.

Although many viruses target this versatile receptor, the most advanced arsenal against NKG2D belongs to CMV. The first protein shown to negatively regulate NKG2D ligand is HCMV UL16. This 50 kDa transmembrane glycoprotein causes intracellular retention of NKG2D ligands MICB, ULBP-1 and ULBP-2 [97–99].

MCMV also encodes negative regulators of NKG2D ligands: surface expression of MULT-1 is affected by m145, RAE-1 family by m152 and H60 by m155 [100–104] (Fig. 2c). Deletion of either of these genes from the viral genome is followed by the reappearance of their target NKG2D ligands on the infected cell surface and results in enhanced sensitivity of the mutant viruses to NK-mediated killing (reviewed in [105]). MULT-1 and H60 are, in addition to their own specific down-regulators, both targeted by *m138*/fcr-1 [106], a protein previously described as MCMV FcγR and also implicated in down-modulation of T cell costimulatory molecule B7-1 and some RAE-1 proteins [105, 107]. As was the case with other downregulators of NKG2D ligands, deletion of this gene resulted in virus attenuation in vivo. Recently, we have observed that various members of the RAE-1 family are not equally susceptible to down-modulation by MCMV. We have shown that m152 acts on RAE-1 molecules in a similar fashion to MHC I molecules – blocking newly synthesized molecules in the ERGIC compartment. Therefore, since m152 cannot affect mature RAE-1 molecules, another viral immunoevasin, *m138*/fcr-1, is needed for down-modulation of mature RAE-1ε [108].

The broad range of ligand modulators for just one receptor can be explained by the differential expression of viral and/or cellular genes and protein localization in different tissues, and implies a strong selective pressure on the virus exerted by the NKG2D-mediated immune responses. Additionally, unlike the majority of signals from NK cell activating receptors, the engagement of NKG2D can override inhibitory signals [109, 110]. Also, the numbers and nature of NKG2D receptor ligands, which are induced by the infection, might play a significant role. In this light, it does not seem so unlikely that it would take a coordinated action of two or more immunomodulators to bring down a resilient or versatile ligand (reviewed in detail in [105]).

←

Fig. 2 (Continued) MCMV encodes multiple regulators of NKG2D ligand expression. RAE-1 proteins are targeted by m152, MULT-1 by m145 and m138 and H-60 by m155 and m138. (**d**) Competition for receptor. Zoonotic orthopoxviruses secrete soluble NKG2D ligands which, by competitive binding, prevent binding of NKG2D ligands expressed on the surface of the cell as a consequence of viral infection. (**e**) Viral miRNA. HCMV regulates MICB by RNA interference. It encodes miRNA miR-UL112 which is homologous to 3′-UTR of MICB (and to some extent, full-length MICA). Binding of miR-UL112 to MICB mRNA initiates RNA silencing and results in the blockade of MICB translation

Another accomplished NKG2D obstruction specialist is KSHV. Its most powerful weapon is K5 – a member of the E3 family of membrane bound ubiquitin ligases encoded by some herpes and pox viruses and often associated with the downmodulation of MHC class I molecules [111]. K5 reduces cell surface expression of NKG2D ligands MICA and MICB and the newly defined ligand for NKp80 AICL (activation-induced C-type lectin) by ubiquitylation of three lysine residues positioned in the cytoplasmic tail near the plasma membrane. Ubiquitylated MICA is targeted to a currently undefined intracellular compartment, while AICL is degraded in endolysosomes. However, the coevolution of host and virus resulted in a widespread MICA 008 allele which has a frameshift mutation leading to a premature stop codon (thus lacking lysine residues in cytoplasmic tail) and its resistance to K5 down-modulation. This allele is also resistant to down-modulation by HCMV [112]. Full length MICA is targeted by HCMV UL142 [113]. Interestingly, the most commonly used HCMV laboratory strain AD169 lacks the Ulb' region which contains the UL142 gene but is still capable of MICA down-modulation by an undefined gene product [114].

NKG2D ligands are also targets for HIV Nef protein. Unjustly termed negative function, this protein exerts a majority of HIV's immunomodulatory functions. It selectively down-modulates HLA-A and HLA-B molecules thus protecting the cells from CTL attack while avoiding NK cell activation through the engagement of inhibitory receptors through HLA-C and HLA-E molecules [83]. However, since not all NK cells express inhibitory receptors specific for HLA-C and HLA-E molecules, this could still render virus-infected cells vulnerable. Nef counters that by down-modulation of NKG2D ligands MICA, ULBP-1 and ULBP-2 thus adding another, more general layer of protection from NK cells and enabling HIV to establish chronic infection [115]. The ability of Nef to downregulate MICA and ULBP-2 was shown to be conserved among patient-derived HIV strains indicating an importance of NK cells in anti-HIV defense.

3.2.3 Modification of Activating Ligands

Another approach employed by viruses is chemical modification of ligands on the cell surface. Zheng et.al [116] have shown that retroviruses HTLV-I, HTLV-II (Human T-cell Leukemia Virus) and HIV prevent lysis of infected cells by the addition of sialic acid residues on cell-surface molecules, presumably including activating receptor ligands. In contrast to this, sialylation of NKp44 and NKp46 is crucial for the recognition of hemagglutinin (HA) of influenza virus and hemagglutinin-neuroaminidase of Sendai virus [51, 52]. HA serves as viral attachment and fusion protein which, by binding to sialic acid expressed on the surface of target cells, ensures infection. It seems that NK cells have outsmarted the virus by sialylating their own activating receptors. However, it has been shown that recent, more virulent influenza virus strains owe their virulence in part to weaker recognition of HA by NKp46 receptor expressed on NK cells [117]. This weaker recognition is due to the acquisition of new glycosilation motives in HA which alter its affinity for NK cell receptors (Fig. 2b). Experiments showing severe sensitivity of NKp46 knockout mice to influenza virus infection emphasize the importance of NK cells in the early control of viral infection.

3.2.4 Competition for Receptor

Rather than targeting its many ligands, zoonotic orthopoxviruses (ZPXV) have opted for a less selective approach in dealing with the NKG2D receptor. In contrast to CMV, ZPXV have adapted the class-I domain to create a soluble competitive antagonist of NKG2D [118]. Binding of soluble antagonist prevents recognition of host ligands, NKG2D receptor cross-linking and blunts activation signals (Fig. 2d). It also attenuates NKG2D-driven immune responses to any NKG2D-controlled pathogen which may compromise host resistance to other infections. A similar mode of action can be seen in some tumors which shed soluble NKG2D ligand MICA inducing internalization and degradation of the receptor [119]. Taking into account that NKG2D expression is not reserved to NK cells, but that it also plays an important role as a coactivating molecule on T cells, especially under conditions of suboptimal activation signals [120], this immunoevasive strategy might seriously compromise the immunological fitness of the host.

3.2.5 Down-Modulation of Coactivating and Accessory Molecules

NK cell cytotoxicity is a complex mechanism involving various processes like adhesion, synapse formation, polarization and degranulation which are regulated by a variety of receptors and accessory molecules. Such a complex mechanism offers a wide variety of targets for viral obstruction.

Certain strains of HCMV increase the resistance of their infected host cells to NK cell cytotoxicity by down-regulating LFA-3, a member of the integrin family necessary for efficient cytotoxicity [121]. This downregulation possibly interferes with the binding of LFA-3 to the NK cell activating receptor CD2 [122].

KSHV K5, among its many other roles, inhibits ICAM-1 expression by promoting its ubiquitylation and lysosomal degradation [85, 123, 124]. ICAM-1 is a cell-surface ligand for the leukocyte receptor LFA-1, an integrin which is believed to play a dominant role in target cell lysis. Expressed in low amounts on all leukocytes and endothelial cells, its concentrations increase upon cytokine stimulation and it is required for neutrophil migration into inflamed tissue. Antibody blocking of LFA-1–ICAM interactions impairs ADCC and natural cytotoxicity by human NK cells [125, 126]. Strong adhesion to target cells, mediated by integrins such as LFA-1 is involved in the formation of NK cell–target cell immune synapse and therefore critical in triggering NK cell mediated cytotoxicity [127].

ICAM-1, along with ICAM-2, is also a target for the HTLV-1 p12^I protein [127]. HTLV-1, agent of adult T-cell leukemia, an aggressive and fatal T cell malignancy, infects CD4+ T cells and establishes life-long persistence. Its T cell evasion strategy through MHC I down-modulation leaves this virus susceptible to NK cell mediated attack. By down-modulating ICAM-1 and -2 on infected CD4+ T cells, HTLV-1 reduces the ability of NK cells to adhere to infected cells.

HCV envelope protein E2 directly inhibits NK cell activity by cross-linking CD81, a component of large molecular complexes that act as cell-surface organizers (coupling different cellular functions) through an as of yet undefined, negative signaling pathway [128]. Interestingly, CD81 is a potent costimulator of T and B cells, so the engagement of CD81 by HCV's E2 has just the opposite effect on T and B cell activity. This suggests that the inhibition of NK cells early in infection allows the virus a replicative advantage prior to the induction of adaptive immunity [95]. This obstruction tactic could have even graver consequences since the inhibition of NK cells during early times of infection could affect NK-DC cross-talk. As previously mentioned, the impairment of NK-DC cross-talk can affect T cell maturation through impairment of DC maturation, thus leading to lowered efficacy of T cell responses even in the presence of CD81-mediated costimulatory signals [95].

3.2.6 Targeting the NK Cell Receptor Ligands by Various Herpesviral microRNAs

Recently, it has been shown that many viruses, primarily herpesviruses (HCMV; EBV, HSV-1, MCMV, KSHV, but also simian polyomaviruses, HIV-1 and human adenovirus) encode several micro(mi) RNAs (reviewed in [129]). miRNAs are members of a large family of small, 22 nucleotides long, noncoding RNAs whose function is posttranscriptional regulation of gene expression. They are transcribed from the genomes of all multicellular organisms and some viruses. Although the exact function of viral miRNAs has not yet been fully determined, it has been predicted that more than 30% of animal genes may be subject to regulation by miRNAs. Due to their small size, lack of immunogenicity and the fact that they do not encode proteins, they cannot be targeted by known effectors of the immune system. This fact makes miRNAs a very attractive mechanism for viral immunoevasion [111, 130]. The importance of miRNA in viral evasion is perhaps underscored best by the HCMV cDNA library generated by Zhang et al. [131]; surprisingly, 45% of all transcripts analyzed were derived from genomic regions predicted to be noncoding and more than half were at least partially antisense to known or predicted HCMV genes.

The first direct evidence for miRNA-related immunoevasion mechanism was discovered in SV40 virus infection. miRNA, encoded by this virus, regulates expression of its own early transcripts thus reducing the expression of viral T cell antigens and resulting in CTL evasion [132].

HCMV encodes miR-UL112 which, by binding to 3'-UTR of MICB (and to a lesser extent MICA) miRNA prevents their translation [133] (Fig. 2e).

Interestingly, miR-UL112 also targets one of the virus's own proteins – a major immediate early protein IE1 [134, 135].

Considering the fact that immune regulation/evasion by the usage of miRNA has many advantages including an evolutionary advantage (it is faster to produce a small antisense molecule than to develop a whole protein and it costs less to do it), and the fact that antisense transcripts seem to constitute a large part of the viral genome, it will not come as a surprise should many more similar immune evasion strategies be discovered in the future.

3.3 Evolutionary Arms Race

The plethora of receptors and the robust activation of NK cells during viral infection have put selective pressure on the viruses to evade such recognition making NK cells an important force in shaping viral evolution. Alterations in the structure of one of the influenza virus' main proteins – hemagglutinin have led to its weaker recognition by NK cells and consequently enhanced virulence, thus showing that the necessity to avoid NK cells plays an important role in driving the evolution of viruses [51, 52]. The importance of NK cells in shaping the outcome of viral infections is perhaps best pointed out by the vast numbers of unrelated pathogens and various NK evasion techniques employed by them. However, the host has not turned a blind eye to constantly emerging viral tactics, thus leading to a fierce and ongoing evolutionary arms race between viruses and the immune system.

One of the most well known examples of the evolutionary race between hosts and viruses is Ly49H-mediated resistance to MCMV in the C57BL/6 mouse strain [14, 16, 17]. Unlike MCMV susceptible strains, C57BLL/6 mice effectively control the infection with the wild type MCMV. The resistance is conferred by a single dominant locus named *Cmv1* located in the NK-cell complex (NKC) on mouse chromosome 6 [136] which encodes many NK cell receptors, including NKG2D and the members of the Ly49 family. In C57BL/6 mice, resistance is provided by the activating receptor Ly49H, which specifically recognizes viral MHC-like protein m157 [17, 19]. In susceptible mouse strains, such as 129/J, m157 is recognized by an inhibitory Ly49I receptor [18]. This and the extensive homology to MHC class I molecules of m157 indicate that the virus might have obtained m157 from its host for the purpose of evading NK-mediated control. Such an act of "molecular piracy" is not unheard of in the viral world – KSHV obtained many of its genes by an act of theft from its host and is using them to shape the cell cycle, apoptosis and immune responses [111]. Indeed, multiple passages of MCMV through resistant C57BL/6 mice lead to loss-of-function mutations in the *m157* gene [137] while a strong selective pressure of Ly49H+ B6-SCID mice selected escape variants in just one passage [138] showing NK cell-mediated selective pressure against the expression of this gene in Ly49H+ strains.

A comparison of extracellular domains of Ly49H, Ly49I and Ly49U showed they all share extensive homology and have probably arisen from a common ancestral gene [139]. Activating Ly49 receptors (and their human analogs KIR) have in fact evolved from inhibitory receptors by losing their ITIM and through gene duplication and genetic recombination [140].

Cmv1 is not the only genetic locus conferring NK cell-dependant resistance to MCMV infection. In MA/My mice resistance is conferred by an alternative, the *Cmv3* locus encoding Ly49P activating receptor recognizing H-2D^k MHC molecules coupled with viral m04 protein [21]. Additionally, in wild derived PWK mouse strains resistance to MCMV is conferred through *Cmv4* locus encoding a still undefined receptor [141].

Virus driven evolution and diversification of receptor genes is also present in the rat NKR-P1 family of receptors. Selective pressure by RCMV decoy protein RCTL resulted in the appearance of inhibitory NKR-P1B receptor variants with no specificity for the decoy ligand and appearance of activating NKRP-1A receptor variants with mild specificity for the decoy [45, 90]. Similar allelic diversity exists among mNKR-P1B and mNKR-P1C gene products from the C57BL/6, SJL, BALB/c, and 129 strains [142].

Recent findings that differential expression of activating NK cell receptor KIR3DS1 can lead to a better control of HIV virus by NK cells and slower progression of the disease [143], along with similar effect in individuals possessing KIR2DL3 allele coupled with HLA-C I alleles, who are able to clear HCV infection [144], imply that selective pressure exerted by the viruses may also play an important role in shaping human evolution at the population level [145].

Additionally, viral activity directed against NK cells influences other cells and may lead to weaker response from the second, adaptive arm of the immune response. Although NK cells are considered to be a part of the innate immunity, their interactions with other cells indicate that they may also act as a bridge between the two arms of the immune response. The line between adaptive and innate immunity is blurred even more in the light of the fact that many NK cell receptors are present in cells of the adaptive immunity. Thus, viral evasion of those receptors may not be exclusively targeted at NK cells.

4 Conclusion

Since their discovery in early 1970s, NK cells have been the subject of intensive research, and new data about their importance is being generated every day. However, there are many questions still left unanswered. In addition, with each new discovery new questions are emerging. So far we know that the importance of NK cells in the early control of viral infection is unquestionable, but there are still a lot of problems to address. For example, what is the purpose of NK immunoevasion for viruses, which regardless of NK control, still enter latency? Why is there a need for so many possibly redundant viral immunoevasins directed at NK cells? What is their role in the context of other immune cells bearing the receptor being targeted? In order to answer these and other questions yet to arise in the future we feel that new, more complex and more sensitive approaches will be needed. Therefore, these are challenging but interesting times to be involved in NK cell and viral evasion research.

References

1. Kirwan SE, Burshtyn DN (2007) Regulation of natural killer cell activity. Curr Opin Immunol 19(1):46–54
2. Karre K et al (1986) Selective rejection of H-2-deficient lymphoma variants suggests alternative immune defence strategy. Nature 319(6055):675–678
3. Valiante NM et al (1997) Killer cell receptors: keeping pace with MHC class I evolution. Immunol Rev 155:155–164
4. Raulet DH et al (1997) Specificity, tolerance and developmental regulation of natural killer cells defined by expression of class I-specific Ly49 receptors. Immunol Rev 155:41–52
5. Yokoyama WM, Kim S (2006) How do natural killer cells find self to achieve tolerance? Immunity 24(3):249–257
6. Kim S et al (2005) Licensing of natural killer cells by host major histocompatibility complex class I molecules. Nature 436(7051):709–713
7. Raulet DH, Vance RE (2006) Self-tolerance of natural killer cells. Nat Rev Immunol 6(7):520–531
8. Fernandez NC et al (2005) A subset of natural killer cells achieves self-tolerance without expressing inhibitory receptors specific for self-MHC molecules. Blood 105(11):4416–4423
9. Lanier LL (2003) Natural killer cell receptor signaling. Curr Opin Immunol 15(3):308–314
10. Lanier LL (2008) Up on the tightrope: natural killer cell activation and inhibition. Nat Immunol 9(5):495–502
11. Raulet DH (2003) Natural killer cells. In: Paul WE (ed) Fundamental immunology. Lippincott Williams & Wilkins, Philadelphia, pp 365–391
12. Dimasi N, Biassoni R (2005) Structural and functional aspects of the Ly49 natural killer cell receptors. Immunol Cell Biol 83(1):1–8
13. Dimasi N, Moretta L, Biassoni R (2004) Structure of the Ly49 family of natural killer (NK) cell receptors and their interaction with MHC class I molecules. Immunol Res 30(1):95–104
14. Daniels KA et al (2001) Murine cytomegalovirus is regulated by a discrete subset of natural killer cells reactive with monoclonal antibody to Ly49H. J Exp Med 194(1):29–44
15. Smith HR et al (2002) Recognition of a virus-encoded ligand by a natural killer cell activation receptor. Proc Natl Acad Sci USA 99(13):8826–8831
16. Lee SH et al (2001) Susceptibility to mouse cytomegalovirus is associated with deletion of an activating natural killer cell receptor of the C-type lectin superfamily. Nat Genet 28(1):42–45
17. Brown MG et al (2001) Vital involvement of a natural killer cell activation receptor in resistance to viral infection. Science 292(5518):934–937
18. Arase H et al (2002) Direct recognition of cytomegalovirus by activating and inhibitory NK cell receptors. Science 296(5571):1323–1326
19. Adams EJ et al (2007) Structural elucidation of the m157 mouse cytomegalovirus ligand for Ly49 natural killer cell receptors. Proc Natl Acad Sci USA 104(24):10128–10133
20. Bubic I et al (2004) Gain of virulence caused by loss of a gene in murine cytomegalovirus. J Virol 78(14):7536–7544

21. Kielczewska, A., et al. (2009) Ly49P recognition of cytomegalovirus-infected cells expressing H2-Dk and the virally encoded glycoprotein m04 is associated with NK cell-mediated resistance to infection. J Exp Med 206(3):515–523
22. Ortaldo JR, Young HA (2005) Mouse Ly49 NK receptors: balancing activation and inhibition. Mol Immunol 42(4):445–450
23. Trowsdale J et al (2001) The genomic context of natural killer receptor extended gene families. Immunol Rev 181:20–38
24. Vilches C, Parham P (2002) KIR: diverse, rapidly evolving receptors of innate and adaptive immunity. Annu Rev Immunol 20:217–251
25. Boyington JC, Sun PD (2002) A structural perspective on MHC class I recognition by killer cell immunoglobulin-like receptors. Mol Immunol 38(14):1007–1021
26. Colonna M et al (1999) A novel family of Ig-like receptors for HLA class I molecules that modulate function of lymphoid and myeloid cells. J Leukoc Biol 66(3):375–381
27. Colonna M et al (1997) A common inhibitory receptor for major histocompatibility complex class I molecules on human lymphoid and myelomonocytic cells. J Exp Med 186(11):1809–1818
28. Cosman D et al (1997) A novel immunoglobulin superfamily receptor for cellular and viral MHC class I molecules. Immunity 7(2):273–282
29. Vitale M et al (1999) The leukocyte Ig-like receptor (LIR)-1 for the cytomegalovirus UL18 protein displays a broad specificity for different HLA class I alleles: analysis of LIR-1 + NK cell clones. Int Immunol 11(1):29–35
30. Lepin EJ et al (2000) Functional characterization of HLA-F and binding of HLA-F tetramers to ILT2 and ILT4 receptors. Eur J Immunol 30(12):3552–3561
31. Yokoyama WM, Plougastel BF (2003) Immune functions encoded by the natural killer gene complex. Nat Rev Immunol 3(4):304–316
32. Raulet DH (2003) Roles of the NKG2D immunoreceptor and its ligands. Nat Rev Immunol 3(10):781–790
33. Diefenbach A et al (2002) Selective associations with signaling proteins determine stimulatory versus costimulatory activity of NKG2D. Nat Immunol 3(12):1142–1149
34. Mistry AR, O'Callaghan CA (2007) Regulation of ligands for the activating receptor NKG2D. Immunology 121(4):439–447
35. Eagle RA, Trowsdale J (2007) Promiscuity and the single receptor: NKG2D. Nat Rev Immunol 7(9):737–744
36. Cerwenka A et al (2000) Retinoic acid early inducible genes define a ligand family for the activating NKG2D receptor in mice. Immunity 12(6):721–727
37. Girardi M et al (2001) Regulation of cutaneous malignancy by gammadelta T cells. Science 294(5542):605–609
38. Carayannopoulos LN et al (2002) Ligands for murine NKG2D display heterogeneous binding behavior. Eur J Immunol 32(3):597–605
39. Takada A et al (2008) Two novel NKG2D ligands of the mouse H60 family with differential expression patterns and binding affinities to NKG2D. J Immunol 180(3):1678–1685
40. Diefenbach A et al (2000) Ligands for the murine NKG2D receptor: expression by tumor cells and activation of NK cells and macrophages. Nat Immunol 1(2):119–126
41. Carayannopoulos LN et al (2002) Cutting edge: murine UL16-binding protein-like transcript 1: a newly described transcript encoding a high-affinity ligand for murine NKG2D. J Immunol 169(8):4079–4083
42. Ogasawara K, Lanier LL (2005) NKG2D in NK and T cell-mediated immunity. J Clin Immunol 25(6):534–540
43. Giorda R et al (1990) NKR-P1, a signal transduction molecule on natural killer cells. Science 249(4974):1298–1300
44. Giorda R, Trucco M (1991) Mouse NKR-P1: a family of genes selectively coexpressed in adherent lymphokine-activated killer cells. J Immunol 147(5):1701–1708

45. Mesci A et al (2006) NKR-P1 biology: from prototype to missing self. Immunol Res 35(1–2):13–26
46. Walzer T et al (2007) Identification, activation, and selective in vivo ablation of mouse NK cells via NKp46. Proc Natl Acad Sci U S A 104(9):3384–3389
47. Gazit R et al (2006) Lethal influenza infection in the absence of the natural killer cell receptor gene Ncr1. Nat Immunol 7(5):517–523
48. Bottino C et al (2000) The human natural cytotoxicity receptors (NCR) that induce HLA class I-independent NK cell triggering. Hum Immunol 61(1):1–6
49. Hollyoake M, Campbell RD, Aguado B (2005) NKp30 (NCR3) is a pseudogene in 12 inbred and wild mouse strains, but an expressed gene in Mus caroli. Mol Biol Evol 22(8):1661–1672
50. Biassoni R et al (2003) Human natural killer cell receptors: insights into their molecular function and structure. J Cell Mol Med 7(4):376–387
51. Mandelboim O et al (2001) Recognition of haemagglutinins on virus-infected cells by NKp46 activates lysis by human NK cells. Nature 409(6823):1055–1060
52. Arnon TI et al (2001) Recognition of viral hemagglutinins by NKp44 but not by NKp30. Eur J Immunol 31(9):2680–2689
53. Moretta A et al (2001) Activating receptors and coreceptors involved in human natural killer cell-mediated cytolysis. Annu Rev Immunol 19:197–223
54. Nattermann J et al (2006) Surface expression and cytolytic function of natural killer cell receptors is altered in chronic hepatitis C. Gut 55(6):869–877
55. De Maria A et al (2003) The impaired NK cell cytolytic function in viremic HIV-1 infection is associated with a reduced surface expression of natural cytotoxicity receptors (NKp46, NKp30 and NKp44). Eur J Immunol 33(9):2410–2418
56. Alcami A, Koszinowski UH (2000) Viral mechanisms of immune evasion. Immunol Today 21(9):447–455
57. Farrell HE et al (2002) Function of CMV-encoded MHC class I homologues. Curr Top Microbiol Immunol 269:131–151
58. Chapman TL, Bjorkman PJ (1998) Characterization of a murine cytomegalovirus class I major histocompatibility complex (MHC) homolog: comparison to MHC molecules and to the human cytomegalovirus MHC homolog. J Virol 72(1):460–466
59. Browne H et al (1990) A complex between the MHC class I homologue encoded by human cytomegalovirus and beta 2 microglobulin. Nature 347(6295):770–772
60. Fahnestock ML et al (1995) The MHC class I homolog encoded by human cytomegalovirus binds endogenous peptides. Immunity 3(5):583–590
61. Reyburn HT et al (1997) The class I MHC homologue of human cytomegalovirus inhibits attack by natural killer cells. Nature 386(6624):514–517
62. Willcox BE, Thomas LM, Bjorkman PJ (2003) Crystal structure of HLA-A2 bound to LIR-1, a host and viral major histocompatibility complex receptor. Nat Immunol 4(9):913–919
63. Rudd PM et al (2001) Glycosylation and the immune system. Science 291(5512):2370–2376
64. Park B et al (2002) The MHC class I homolog of human cytomegalovirus is resistant to down-regulation mediated by the unique short region protein (US)2, US3, US6, and US11 gene products. J Immunol 168(7):3464–3469
65. Lin A, Xu H, Yan W (2007) Modulation of HLA expression in human cytomegalovirus immune evasion. Cell Mol Immunol 4(2):91–98
66. Kim Y et al (2008) Human cytomegalovirus UL18 utilizes US6 for evading the NK and T-cell responses. PLoS Pathog 4(8):e1000123
67. Leong CC et al (1998) Modulation of natural killer cell cytotoxicity in human cytomegalovirus infection: the role of endogenous class I major histocompatibility complex and a viral class I homolog. J Exp Med 187(10):1681–1687
68. Prod'homme V et al (2007) The human cytomegalovirus MHC class I homolog UL18 inhibits LIR-1+ but activates LIR-1- NK cells. J Immunol 178(7):4473–4481

69. Natarajan K et al (2006) Crystal structure of the murine cytomegalovirus MHC-I homolog m144. J Mol Biol 358(1):157–171
70. Farrell HE et al (1997) Inhibition of natural killer cells by a cytomegalovirus MHC class I homologue in vivo. Nature 386(6624):510–514
71. Beisser PS et al (2000) The r144 major histocompatibility complex class I-like gene of rat cytomegalovirus is dispensable for both acute and long-term infection in the immunocompromised host. J Virol 74(2):1045–1050
72. Senkevich TG et al (1996) Genome sequence of a human tumorigenic poxvirus: prediction of specific host response-evasion genes. Science 273(5276):813–816
73. Tomasec P et al (2000) Surface expression of HLA-E, an inhibitor of natural killer cells, enhanced by human cytomegalovirus gpUL40. Science 287(5455):1031
74. Ulbrecht M et al (2000) Cutting edge: the human cytomegalovirus UL40 gene product contains a ligand for HLA-E and prevents NK cell-mediated lysis. J Immunol 164(10):5019–5022
75. Ziegler H et al (1997) A mouse cytomegalovirus glycoprotein retains MHC class I complexes in the ERGIC/cis-Golgi compartments. Immunity 6(1):57–66
76. Reusch U et al (1999) A cytomegalovirus glycoprotein re-routes MHC class I complexes to lysosomes for degradation. EMBO J 18(4):1081–1091
77. Kavanagh DG, Koszinowski UH, Hill AB (2001) The murine cytomegalovirus immune evasion protein m4/gp34 forms biochemically distinct complexes with class I MHC at the cell surface and in a pre-Golgi compartment. J Immunol 167(7):3894–3902
78. Kleijnen MF et al (1997) A mouse cytomegalovirus glycoprotein, gp34, forms a complex with folded class I MHC molecules in the ER which is not retained but is transported to the cell surface. EMBO J 16(4):685–694
79. Pinto AK et al (2006) Coordinated function of murine cytomegalovirus genes completely inhibits CTL lysis. J Immunol 177(5):3225–3234
80. Holtappels R et al (2006) Cytomegalovirus encodes a positive regulator of antigen presentation. J Virol 80(15):7613–7624
81. Nattermann J et al (2005) HIV-1 infection leads to increased HLA-E expression resulting in impaired function of natural killer cells. Antivir Ther 10(1):95–107
82. Martini F et al (2005) HLA-E up-regulation induced by HIV infection may directly contribute to CD94-mediated impairment of NK cells. Int J Immunopathol Pharmacol 18(2):269–276
83. Cohen GB et al (1999) The selective downregulation of class I major histocompatibility complex proteins by HIV-1 protects HIV-infected cells from NK cells. Immunity 10(6):661–671
84. Ishido S et al (2000) Downregulation of major histocompatibility complex class I molecules by Kaposi's sarcoma-associated herpesvirus K3 and K5 proteins. J Virol 74(11):5300–5309
85. Ishido S et al (2000) Inhibition of natural killer cell-mediated cytotoxicity by Kaposi's sarcoma-associated herpesvirus K5 protein. Immunity 13(3):365–374
86. Guma M et al (2004) Imprint of human cytomegalovirus infection on the NK cell receptor repertoire. Blood 104(12):3664–3671
87. Guma M et al (2006) Human cytomegalovirus infection is associated with increased proportions of NK cells that express the CD94/NKG2C receptor in aviremic HIV-1-positive patients. J Infect Dis 194(1):38–41
88. Guma M et al (2006) Expansion of CD94/NKG2C+ NK cells in response to human cytomegalovirus-infected fibroblasts. Blood 107(9):3624–3631
89. Carlyle JR et al (2004) Missing self-recognition of Ocil/Clr-b by inhibitory NKR-P1 natural killer cell receptors. Proc Natl Acad Sci USA 101(10):3527–3532
90. Voigt S et al (2007) Cytomegalovirus evasion of innate immunity by subversion of the NKR-P1B:Clr-b missing-self axis. Immunity 26(5):617–627

91. Meier UC et al (2005) Shared alterations in NK cell frequency, phenotype, and function in chronic human immunodeficiency virus and hepatitis C virus infections. J Virol 79 (19):12365–12374
92. Titanji K et al (2008) Altered distribution of natural killer cell subsets identified by CD56, CD27 and CD70 in primary and chronic human immunodeficiency virus-1 infection. Immunology 123(2):164–170
93. O'Connor GM et al (2007) Natural Killer cells from long-term non-progressor HIV patients are characterized by altered phenotype and function. Clin Immunol 124(3):277–283
94. Arnon TI et al (2005) Inhibition of the NKp30 activating receptor by pp 65 of human cytomegalovirus. Nat Immunol 6(5):515–523
95. Golden-Mason L, Rosen HR (2006) Natural killer cells: primary target for hepatitis C virus immune evasion strategies? Liver Transpl 12(3):363–372
96. Sinclair J (2008) Human cytomegalovirus: Latency and reactivation in the myeloid lineage. J Clin Virol 41(3):180–185
97. Dunn C et al (2003) Human cytomegalovirus glycoprotein UL16 causes intracellular sequestration of NKG2D ligands, protecting against natural killer cell cytotoxicity. J Exp Med 197(11):1427–1439
98. Cosman D et al (2001) ULBPs, novel MHC class I-related molecules, bind to CMV glycoprotein UL16 and stimulate NK cytotoxicity through the NKG2D receptor. Immunity 14(2):123–133
99. Kubin M et al (2001) ULBP1, 2, 3: novel MHC class I-related molecules that bind to human cytomegalovirus glycoprotein UL16, activate NK cells. Eur J Immunol 31 (5):1428–1437
100. Krmpotic A et al (2005) NK cell activation through the NKG2D ligand MULT-1 is selectively prevented by the glycoprotein encoded by mouse cytomegalovirus gene m145. J Exp Med 201(2):211–220
101. Lodoen M et al (2003) NKG2D-mediated natural killer cell protection against cytomegalovirus is impaired by viral gp40 modulation of retinoic acid early inducible 1 gene molecules. J Exp Med 197(10):1245–1253
102. Hasan M et al (2005) Selective down-regulation of the NKG2D ligand H60 by mouse cytomegalovirus m155 glycoprotein. J Virol 79(5):2920–2930
103. Lodoen MB et al (2004) The cytomegalovirus m155 gene product subverts natural killer cell antiviral protection by disruption of H60-NKG2D interactions. J Exp Med 200(8):1075–1081
104. Krmpotic A et al (2002) MCMV glycoprotein gp40 confers virus resistance to CD8+ T cells and NK cells in vivo. Nat Immunol 3(6):529–535
105. Lenac T et al (2008) Murine cytomegalovirus regulation of NKG2D ligands. Med Microbiol Immunol 197(2):159–166
106. Lenac T et al (2006) The herpesviral Fc receptor fcr-1 down-regulates the NKG2D ligands MULT-1 and H60. J Exp Med 203(8):1843–1850
107. Mintern JD et al (2006) Viral interference with B7–1 costimulation: a new role for murine cytomegalovirus fc receptor-1. J Immunol 177(12):8422–8431
108. Arapovic J et al (2009) Promiscuity of MCMV immunoevasin of NKG2D: m138/fcr-1 down-modulates RAE-1ε in addition to MULT-1 and H60. Mol Immunol
109. Cerwenka A, Baron JL, Lanier LL (2001) Ectopic expression of retinoic acid early inducible-1 gene (RAE-1) permits natural killer cell-mediated rejection of a MHC class I-bearing tumor in vivo. Proc Natl Acad Sci USA 98(20):11521–11526
110. Regunathan J et al (2005) NKG2D receptor-mediated NK cell function is regulated by inhibitory Ly49 receptors. Blood 105(1):233–240
111. Coscoy L (2007) Immune evasion by Kaposi's sarcoma-associated herpesvirus. Nat Rev Immunol 7(5):391–401

112. Thomas M et al (2008) Down-regulation of NKG2D and NKp80 ligands by Kaposi's sarcoma-associated herpesvirus K5 protects against NK cell cytotoxicity. Proc Natl Acad Sci USA 105(5):1656–1661
113. Chalupny NJ et al (2006) Down-regulation of the NKG2D ligand MICA by the human cytomegalovirus glycoprotein UL142. Biochem Biophys Res Commun 346(1):175–181
114. Zou Y et al (2005) Effect of human cytomegalovirus on expression of MHC class I-related chains A. J Immunol 174(5):3098–3104
115. Cerboni C et al (2007) Human immunodeficiency virus 1 Nef protein downmodulates the ligands of the activating receptor NKG2D and inhibits natural killer cell-mediated cytotoxicity. J Gen Virol 88(Pt 1):242–250
116. Zheng ZY, Zucker-Franklin D (1992) Apparent ineffectiveness of natural killer cells vis-a-vis retrovirus-infected targets. J Immunol 148(11):3679–3685
117. Owen RE et al (2007) Alterations in receptor binding properties of recent human influenza H3N2 viruses are associated with reduced natural killer cell lysis of infected cells. J Virol 81(20):11170–11178
118. Campbell JA et al (2007) Zoonotic orthopoxviruses encode a high-affinity antagonist of NKG2D. J Exp Med 204(6):1311–1317
119. Groh V et al (2002) Tumour-derived soluble MIC ligands impair expression of NKG2D and T-cell activation. Nature 419(6908):734–738
120. Groh V et al (2001) Costimulation of CD8alphabeta T cells by NKG2D via engagement by MIC induced on virus-infected cells. Nat Immunol 2(3):255–260
121. Fletcher JM, Prentice HG, Grundy JE (1998) Natural killer cell lysis of cytomegalovirus (CMV)-infected cells correlates with virally induced changes in cell surface lymphocyte function-associated antigen-3 (LFA-3) expression and not with the CMV-induced down-regulation of cell surface class I HLA. J Immunol 161(5):2365–2374
122. Orange JS et al (2002) Viral evasion of natural killer cells. Nat Immunol 3(11):1006–1012
123. Coscoy L, Sanchez DJ, Ganem D (2001) A novel class of herpesvirus-encoded membrane-bound E3 ubiquitin ligases regulates endocytosis of proteins involved in immune recognition. J Cell Biol 155(7):1265–1273
124. Coscoy L, Ganem D (2001) A viral protein that selectively downregulates ICAM-1 and B7–2 and modulates T cell costimulation. J Clin Invest 107(12):1599–1606
125. Hildreth JE et al (1983) A human lymphocyte-associated antigen involved in cell-mediated lympholysis. Eur J Immunol 13(3):202–208
126. Miedema F et al (1984) Both Fc receptors and lymphocyte-function-associated antigen 1 on human T gamma lymphocytes are required for antibody-dependent cellular cytotoxicity (killer cell activity). Eur J Immunol 14(6):518–523
127. Banerjee P, Feuer G, Barker E (2007) Human T-cell leukemia virus type 1 (HTLV-1) p12I down-modulates ICAM-1 and -2 and reduces adherence of natural killer cells, thereby protecting HTLV-1-infected primary CD4+ T cells from autologous natural killer cell-mediated cytotoxicity despite the reduction of major histocompatibility complex class I molecules on infected cells. J Virol 81(18):9707–9717
128. Crotta S et al (2002) Inhibition of natural killer cells through engagement of CD81 by the major hepatitis C virus envelope protein. J Exp Med 195(1):35–41
129. Pedersen I, David M (2008) MicroRNAs in the immune response. Cytokine 43(3):391–394
130. Gottwein E, Cullen BR (2008) Viral and cellular microRNAs as determinants of viral pathogenesis and immunity. Cell Host Microbe 3(6):375–387
131. Zhang G et al (2007) Antisense transcription in the human cytomegalovirus transcriptome. J Virol 81(20):11267–11281
132. Sullivan CS et al (2005) SV40-encoded microRNAs regulate viral gene expression and reduce susceptibility to cytotoxic T cells. Nature 435(7042):682–686
133. Stern-Ginossar N et al (2007) Host immune system gene targeting by a viral miRNA. Science 317(5836):376–381

134. Murphy E et al (2008) Suppression of immediate-early viral gene expression by herpesvirus-coded microRNAs: implications for latency. Proc Natl Acad Sci USA 105(14):5453–5458
135. Grey F et al (2007) A human cytomegalovirus-encoded microRNA regulates expression of multiple viral genes involved in replication. PLoS Pathog 3(11):e163
136. Scalzo AA et al (1992) The effect of the Cmv-1 resistance gene, which is linked to the natural killer cell gene complex, is mediated by natural killer cells. J Immunol 149(2):581–589
137. Voigt V et al (2003) Murine cytomegalovirus m157 mutation and variation leads to immune evasion of natural killer cells. Proc Natl Acad Sci USA 100(23):13483–13488
138. French AR et al (2004) Escape of mutant double-stranded DNA virus from innate immune control. Immunity 20(6):747–756
139. Makrigiannis AP et al (2001) Class I MHC-binding characteristics of the 129/J Ly49 repertoire. J Immunol 166(8):5034–5043
140. Abi-Rached L, Parham P (2005) Natural selection drives recurrent formation of activating killer cell immunoglobulin-like receptor and Ly49 from inhibitory homologues. J Exp Med 201(8):1319–1332
141. Adam SG et al (2006) Cmv4, a new locus linked to the NK cell gene complex, controls innate resistance to cytomegalovirus in wild-derived mice. J Immunol 176(9):5478–5485
142. Carlyle JR et al (2006) Molecular and genetic basis for strain-dependent NK1.1 alloreactivity of mouse NK cells. J Immunol 176(12):7511–7524
143. Alter G et al (2007) Differential natural killer cell-mediated inhibition of HIV-1 replication based on distinct KIR/HLA subtypes. J Exp Med 204(12):3027–3036
144. Khakoo SI et al (2004) HLA and NK cell inhibitory receptor genes in resolving hepatitis C virus infection. Science 305(5685):872–874
145. Lanier LL (2008) Evolutionary struggles between NK cells and viruses. Nat Rev Immunol 8(4):259–268
146. Kloover JS et al (2002) A rat cytomegalovirus strain with a disruption of the r144 MHC class I-like gene is attenuated in the acute phase of infection in neonatal rats. Arch Virol 147(4):813–824
147. Llano M et al (2003) Differential effects of US2, US6 and US11 human cytomegalovirus proteins on HLA class Ia and HLA-E expression: impact on target susceptibility to NK cell subsets. Eur J Immunol 33(10):2744–2754
148. Hengel H et al (1997) A viral ER-resident glycoprotein inactivates the MHC-encoded peptide transporter. Immunity 6(5):623–632
149. Ahn K et al (1997) The ER-luminal domain of the HCMV glycoprotein US6 inhibits peptide translocation by TAP. Immunity 6(5):613–621
150. Barel MT et al (2003) Human cytomegalovirus-encoded US2 differentially affects surface expression of MHC class I locus products and targets membrane-bound, but not soluble HLA-G1 for degradation. J Immunol 171(12):6757–6765
151. Schust DJ et al (1998) Trophoblast class I major histocompatibility complex (MHC) products are resistant to rapid degradation imposed by the human cytomegalovirus (HCMV) gene products US2 and US11. J Exp Med 188(3):497–503
152. Wiertz EJ et al (1996) The human cytomegalovirus US11 gene product dislocates MHC class I heavy chains from the endoplasmic reticulum to the cytosol. Cell 84(5):769–779
153. Wiertz EJ et al (1996) Sec61-mediated transfer of a membrane protein from the endoplasmic reticulum to the proteasome for destruction. Nature 384(6608):432–438
154. Jones TR et al (1996) Human cytomegalovirus US3 impairs transport and maturation of major histocompatibility complex class I heavy chains. Proc Natl Acad Sci USA 93(21):11327–11333
155. Misaghi S et al (2004) Structural and functional analysis of human cytomegalovirus US3 protein. J Virol 78(1):413–423

156. Furman MH et al (2002) The human cytomegalovirus US10 gene product delays trafficking of major histocompatibility complex class I molecules. J Virol 76(22):11753–11756
157. Greenberg ME, Iafrate AJ, Skowronski J (1998) The SH3 domain-binding surface and an acidic motif in HIV-1 Nef regulate trafficking of class I MHC complexes. EMBO J 17(10):2777–2789
158. Le Gall S et al (1998) Nef interacts with the mu subunit of clathrin adaptor complexes and reveals a cryptic sorting signal in MHC I molecules. Immunity 8(4):483–495
159. Tomasec P et al (2005) Downregulation of natural killer cell-activating ligand CD155 by human cytomegalovirus UL141. Nat Immunol 6(2):181–188

The Role of NK Cells in Bacterial Infections

Brian P. McSharry and Clair M. Gardiner

Abstract Natural killer (NK) cells are best known for their ability to kill virally infected and transformed cells. Evidence of the contributions they make to other immune functions is, however, growing. This review will focus on the important role that NK cells play in the immune response to bacterial pathogens. We will present the experimental evidence defining recent advances in our understanding of NK cell receptor recognition of bacteria. In particular, we now appreciate the fact that NK cells can recognize and respond to bacteria directly through pattern recognition receptors including Toll-like receptors and Nod-2. They can also be activated indirectly by accessory cells that have responded to pathogen infection. Stimulated accessory cells produce cytokines including IL-12 and IL-18 which activate NK cell functions. They also upregulate NK cell ligands leading to NK cell activation through engagement of receptors on their surface. In addition to their antiviral defences, we will describe the impressive arsenal of antimicrobial defences that NK cells employ, including cytokine production, cytotoxicity, production of antimicrobial peptides, and immuno-regulation.

1 Introduction

Natural killer (NK) cells are lymphocytes which are best known for their ability to kill virally infected and cancer cells [1–3]. Indeed, it is because freshly isolated NK cells demonstrated their lack of need for priming (in contrast to T cells) to mediate cytotoxicity that they received their provocative name. Although their name inspired significant research on their cytotoxic functions over the last 30 years or

---

B.P. McSharry
Immunology Research Centre, School of Biochemistry and Immunology, Trinity College, Dublin 2, Ireland

C.M. Gardiner (✉)
NK cell group, School of Biochemistry and Immunology, Trinity College, Dublin 2, Ireland

so, it probably biased scientists in terms of their research strategies and delayed the broader potential of NK cells as immune effector cells from being appreciated. Recently however, additional important roles for NK cells in regulating immune responses and as effector cells in bacterial and parasitic infections have been defined. In contrast to the historical dogma surrounding NK cells, cytotoxicity is probably not the primary effector mechanism involved. It is the production of cytokines, in particular IFN-γ, by which NK cells contribute most to the immune response controlling non-viral infections. Monocytes produce IL-12 early in bacterial infection and it is extremely potent and important in stimulating NK cells to produce IFN-γ [4–6]. Cells that provide "help" to another cell to stimulate a functional response are termed accessory cells and although almost all antigen presenting cells can act as such, monocytes are the most common accessory cells to provide help to NK cells in peripheral blood [7]. In fact, extremely limited numbers of monocytes can dramatically affect NK cell functions. This complicates interpretation of scientific data using "purified" NK cells. In the human system, <1% contaminating $CD14^+$ monocytes in a preparation of magnetic bead purified NK cells (>97% CD56+CD3− cells), can still have profound effects on NK cells, mainly through cytokine production [7]. Cell sorting is currently the only definitive way to investigate the effects of particular molecules/pathogens directly on NK cells without the confounding presence of accessory cell populations. This needs to be considered when reading the literature defining a role for NK and accessory cells in response to bacterial infection.

2 Evidence of a Role for NK Cells in the Immune Response to Bacterial Infections

Much of the evidence of a role for NK cells in bacterial infections comes from murine models that allow defined deletion or depletion of NK cells. Pulmonary infections with the extracellular bacteria *Staphylococcus aureus* induces markedly increased numbers of TNF-α^+, activated NK cells in the airway lumen of wild type mice [8]. IL-15−/− (knock-out (KO)) mice that are NK-cell deficient, and NK cell depleted wild type mice also have significantly increased bacterial burdens in the lung and spleen compared to control [8]. Similarly, pulmonary *Bordetella pertussis* challenge is associated with the recruitment of activated NK cells to the lung, and depletion of NK cells with an anti-asialo GM1 antibody renders mice more susceptible to higher bacterial loads and accelerated mortality [9].

IFN-γ is a key cytokine involved in the resolution of *Legionella pnemophilia* infection with NK cells identified as the major source of early IFN-γ secretion in response to challenge [10]. NK cell-depleted or IFN-γ receptor KO mice also demonstrated increased bacterial loads and failure to control infection. MyD88, an adaptor molecule in Toll-like receptor (TLR) signaling, KO mice have a severely diminished cytokine response to legionella infection suggesting an essential role for

TLR signaling in driving NK cell IFN-γ production [10]; however no specific TLR was identified as the principal receptor promoting the bacterial induced response. Another TLR signaling molecule, IRAK-4, has been implicated in the immune response to bacterial pathogens in humans, with children deficient in IRAK-4 highly susceptible to *Streptococcus pnemoniae* and *S. aureus* infections. In addition to other effector cells, NK cells isolated from these individuals failed to produce IFN-γ in response to TLR stimulation suggesting that defective NK cell function could play a role in the symptoms seen [11].

A significant body of literature exists, investigating the role of NK cells in mycobacterial infections. Tuberculosis, primarily caused by *Mycobacterium tuberculosis,* is one of the most serious bacterial infections in the world. Although depletion of NK cells from T-cell sufficient wild type mice infected with mycobacterial species has been reported to have a minimal effect on bacterial loads and pathogenesis [12, 13], most evidence supports a role for NK cells in the control of mycobacterial infections: infection with mycobacterial species is known to activate NK cells in vivo [14–16] and NK cell depleted mice exhibit elevated bacterial loads suggesting a protective role for NK cells [17]. Experiments from model systems using mice lacking functional T-cells have implicated NK cells as playing a vital role in the response to mycobacterial infection. Depletion of NK cells from SCID mice, which lack mature T and B cells, was associated with decreased granuloma formation following *Mycobacterium avium* infection [18]. In a more recent study using a Rag−/− model (again lacking mature T and B cells), depletion of NK cells with antibody or use of specific genetic mutants that inhibit IFN-γ production/function indicated that NK cells play an important role in inhibiting mycobacterial replication and pathogenesis [19].

A number of studies have examined NK cell activity following infection with *Listeria monocytogenes*, a facultative intracellular bacteria associated with food borne infections. There is conflicting evidence in the literature regarding the role that NK cells play in *L. monocytogenes* infections. In support of a protective role for NK cells, IFN-γ deficient mice are extremely sensitive to listeria infection [20, 21] and listeria is a potent activator of NK cells in vivo [22, 23]. Furthermore, Dunn and North 1991, described that NK cell depletion prior to infection with *L. monocytogenes* via the mouse footpad was associated with enhanced bacterial loads in draining lymph nodes [24]. However, other studies suggest that depleting NK cells appears to be primarily beneficial to the host and is associated with enhanced bacterial clearance [23, 25, 26]. This phenomenon is not unique to listeria infections as depletion of NK cells has also been shown to be associated with protection from challenge with other bacteria, in particular in models of sepsis, including *Escherichia coli* [27], *S. pneumoniae* [28], and *Pseudomonas aeruginosa* [29]. The mechanism underpinning this enhanced replication and reduced pathogenesis is not well understood. A more recent report indicates a potentially important role for the Natural Killer dendritic cell subset (NKDC) in listeria infections. NKDC have attributes of both NK cells (cytotoxicity) and dendritic cells (DC; antigen presentation). Adoptive transfer of NKDC into IFN-γ deficient mice was associated with diminished bacterial loads and protection from listeria challenge that was dependent on IFN-γ production [30]. Further study will be necessary to confirm the role of this specific cellular subset on listeria infection in wild type mice.

It is important to note that the majority of these reports use one of two antibodies to deplete NK cells (anti-NK1.1 or anti-asialo GM1). Both these antibodies can target the depletion of additional cell subsets including invariant natural killer T-cells (anti-NK1.1), myeloid and T-cell subsets (anti-asialo GM1). The use of murine models that allow more specific deletion of the NK cell subset [31] may help resolve some of the conflicting literature that exists on the role of NK cells in a number of bacterial infections. Differences in the model systems used, routes of administration, infecting inoculum, and specific bacterial strains could all contribute to the different responses reported and suggest that the role of NK cells in a specific bacterial infection cannot be predicted a priori but must be determined empirically. It is clear however that NK cells play an important role following bacterial infection in murine models with NK cell deficiencies or depletion associated with both positive and negative effects on bacterial replication and pathogenesis.

There is also a growing body of literature demonstrating activation of human NK cells in bacterial infections, and NK cells isolated from the blood and pleural effusions of patients with tuberculosis pleuritis demonstrate an activated phenotype [14, 15, 32]. Stimulation of peripheral blood mononuclear cells (PBMC) and/or purified NK cells in vitro with a wide range of bacteria leads to increased cytotoxic potential and cytokine release from human NK cells. Such studies have also confirmed the important role of accessory cells in promoting NK activity (see later section). Additional evidence of the importance of NK cells in bacterial infection has also come from studies of individuals with natural mutations in different components of the immune system. While humans with defects in their NK cell compartment or effector functions are best known for deficiencies in their immune response to viral infections, they are also highly susceptible to bacterial infections and often present with severe disease. Wendland et al. described an NK cell deficient patient (lacking all $CD56^+$ cells) who had no cytotoxic activity against a classical NK cell target - the myelogenous leukemia cell line, K562 [33]. The patient also had low numbers of monocytes and B-cells but could produce and respond to both IL-12 and IFN-γ. This individual presented with a disseminated infection of the facultative intracellular bacterium *M. avium* and was also susceptible to infection from another intracellular bacterium, *Salmonella enteritidis* [33]. A number of case studies reported in the literature describe increased susceptibility to bacterial infections in patients presenting with immune defects including decreased or absent NK cell function (reviewed by Orange, 2002) but almost all are complicated due to the common occurrence of additional immune defects. This makes it difficult to specifically implicate defective NK cell responses as the primary immune deficiency in these patients [34]. The importance of cytokine secretion as an effector mechanism in response to bacterial infections is also exhibited by patients presenting with mutations in genes important in the production/activity of IFN-γ including IL-12 p40, IL-12Rβ1, IFN-γR1, and IFN-γR2 genes as they have increased susceptibility to intracellular bacterial infections, in particular, mycobacteria and salmonella species [35–39]. NK cells can play a crucial role in such responses, e.g., individuals lacking IL-12Rβ1 expression had virtually no IFN-γ secretion by unstimulated NK cells in vitro [36]. Interestingly, basal NK cell cytotoxicity in these patients was normal which supports the tenet that cytokine production and not cytotoxicity is a more important effector mechanism in the NK cell response to bacterial pathogen.

The capacity of NK cells to be activated by bacterial infection has been utilized therapeutically in developing a number of anticancer therapies. The administration of the live vaccine strain *Mycobacterium bovis* bacillus Calmette Guerin (BCG) via catheter has been used as a successful immunotherapy for superficial bladder cancer. Such treatment induces an inflammatory immune response associated with the recruitment and activation of NK cells [40]. In a murine model of bladder cancer, depletion of NK cells or the use of NK cell function deficient beige mice was associated with increased tumor load and enhanced morbidity [41]. In addition, depletion of NK cells in vitro from spleen cells of BCG vaccinated mice significantly reduced the ability of such cells to lyse bladder cancer cells [42]. *L. monocytogenes* infection is also associated with potent NK cell activation in vivo with NK cells isolated following challenge showing enhanced cytotoxicity against the YAC-1 tumor cell line [43, 44]. This immune activation has also been investigated as a potential treatment for hepatic tumors due to the natural tropism of the bacteria for liver cells. Administration of attenuated *L. monocytogenes* strains to mice provokes antitumor responses against hepatic metastases that are dependent on NK cell activation and NK cell depletion is associated with accelerated mortality [45].

3 NK Cells Have a Diverse Repertoire of Antibacterial Effector Mechanisms

3.1 Secretion of Cytokines

NK cells secrete a number of Th1-type cytokines – in particular IFN-γ and TNF-α. IFN-γ plays a key role in controlling bacterial infections, exemplified by the exquisite sensitivity of IFN-γ or IFN-γR deficient patients or genetically mutated mice to infection with a range of bacteria [20, 38, 39]. These cytokines and IFN-γ in particular, have a wide range of antibacterial activities. IFN-γ and TNF-α secreted by NK cells play a vital role in controlling DC function; they promote DC cell maturation through upregulation of MHC Class I molecules and costimulatory molecules in addition to stimulating secretion of IL-12 [46]. These activated mature DC in turn control the T-cell effector responses that are vital for an effective immune response to bacterial infection. IFN-γ secretion also stimulates phagocytosis of bacteria by macrophages and facilitates their elimination via a number of mechanisms including the generation of reactive oxygen and nitrogen species via the NADPH phagocyte oxidase and iNOS pathways [47–50]. IFN-γ also limits the availability of iron, an essential nutrient for bacterial replication, in *Salmonella typhimurium* infected macrophages, by modulating iron efflux and uptake pathways [51]. NK cell secretion of the cytokines TNF-α and IFN-γ are known to play a crucial role in granuloma formation following challenge with intracellular bacteria, including *M. avium* and *Francisella tularensis* [18, 52]. Granulomas help protect the host from bacterial dissemination by isolating infectious foci.

3.2 NK Cell Cytotoxicity

NK cells mediate target cell death by one of two major mechanisms that require direct contact between NK cells and target cells – secretion of cytotoxic molecules or engagement of death receptors on the target cell. In the first pathway, cytoplasmic granule contents are secreted by exocytosis. The main cytotoxic effector molecules include a membrane pore inducing protein, perforin, and granzymes - a family of serine proteases that can promote both caspase-dependent and independent apoptosis. The second pathway of NK cell cytotoxicity involves the engagement of death receptors (e.g., Fas/CD95, TRAIL-R) on target cells by their appropriate ligand (e.g., FasL, TRAIL) on NK cells, resulting in caspase-dependent apoptosis of the target cell. NK cells have been shown to directly lyse cells infected with intracellular bacteria [46, 53–55]. However, the specific mechanisms involved in this process are as yet unknown. Brill et al. demonstrated that NK cells induced apoptosis of *M. tuberculosis* infected monocytes and could directly target intracellular bacteria [53]. This apoptosis was not inhibited by either blocking Fas–FasL interactions or blocking cytotoxic granule release from NK cells.

3.3 Secretion of Antimicrobial Proteins

NK cells can also secrete molecules that are directly cytotoxic to intracellular bacteria. Granulysin, a membrane damaging peptide with homology to the antimicrobial defensin proteins, is secreted by NK cells from their cytolytic granules [56]. Granulysin induces discrete lesions and distortions in the membrane of bacteria when viewed by electron microscopy [57] and has demonstrated potent antimicrobial activity against both gram-positive and gram-negative bacteria including *M. tuberculosis, S. typhimurium, L. monocytogenes, E. coli*, and *S. aureus*. Importantly, these include intracellular bacterial infections, e.g., mycobacteria and listeria species. The ability of granulysin to target such intracellular pathogens is at least partially dependent on perforin activity. Stenger et al.

demonstrated that granulysin alone was incapable of lysing intracellular mycobacteria in the absence of perforin; the authors suggested that this was due to the inability of granulysin to access mycobacteria containing phagosomes [57]. However, Walch et al. reported a perforin-independent pathway of granulysin-mediated lysis of *Listeria innocua* in infected human DC [58]. In this system, granulysin was bound and endocytosed into lipid rafts before transfer into *L. innocua* containing phagosomes. However, perforin can increase the anti-listeria efficacy of granulysin: coadministration of perforin and granulysin promoted colocalization of granulysin and listeria by inducing endosome–phagosome fusion (triggered by Ca^{2+} flux), leading to enhanced levels of bacteriolysis compared to granulysin alone [59].

Defensins are a family of small cysteine rich antimicrobial peptides, which are primarily known to be secreted by activated neutrophils [60]. These molecules also mediate their antimicrobial activity through disruption of the bacterial cell membrane. Defensin mRNA was first shown in IL-2 stimulated NK cells [61] and more recent reports have indicated that freshly isolated NK cells exhibit constitutive expression of α-defensins 1–3 [62]. NK cells can be activated to secrete defensins following stimulation with bacterially derived molecules and higher levels of secretion are noted in combination with IL-2 [62]. In addition to their direct antimicrobial effects, defensins also have immunomodulatory activity promoting proinflammatory cytokine secretion including IL-8 and MCP-1 [63]. Cathelicidins are precursor proteins that release antimicrobial peptide after proteolytic cleavage; such peptides have been demonstrated to play an important role in the innate immune response to tuberculosis infection [64]. LL-37 is thus far, the only cathelicidin-derived antimicrobial peptide described in humans. Although it is primarily associated with neutrophils, its expression has also been reported in freshly isolated NK cells and NK cell clones [61].

3.4 Regulation of Other Immune Cells

NK cell cytotoxic activity is also important in regulating the cellular immune response to bacteria. T-regulatory cells (T-regs) are a population of suppressor, regulatory cells which are defined as $CD4^+CD25^+Foxp3^+$ T-cells that play an important role in controlling the immune response to intracellular pathogens. T-regs suppress Th1 immune responses through the secretion of immunosuppressive cytokines including IL-10 and TGF-β, and direct cell–cell interactions. T-regs induced to proliferate in response to *M. tuberculosis* lysate stimulated monocytes, are susceptible to NK cell mediated cytolysis via NKG2D and NKp46 activating receptors [65]. T-regs exhibited higher surface expression of the NKG2D ligand, ULBP1, which was at least partially responsible for the enhanced recognition, reported. These data suggest that NK cells could promote an enhanced Th1 response to mycobacterial infection through elimination of a suppressor cell population. It has also been shown that NK cells directly activated by BCG in combination with IL-12 can lyse immature DC [66]. This process where NK cells selectively target immature DC while sparing mature DC, is known as NK cell-mediated DC editing. Cytolysis of immature DC was fully or partially blocked using antibodies against the NK activating receptors, NKp30 and DNAM-1 respectively [66]. It has also been reported that the protection of mature DC is due to their enhanced expression of NK cell inhibitory ligands including classical HLA Class I molecules and the non-classical Class I molecule, HLA-E [67]. Specific elimination of immature DC may allow the selection of the most potent antigen presenting cells expressing higher levels of MHC Class I and costimulatory molecules, thereby promoting a more robust T-cell response to the invading pathogen [46].

4 Receptors Involved in NK Cell Sensing of Bacterial Pathogen

4.1 Pattern Recognition Receptors

It has been known for many years that the immune system has receptors that recognize foreign pathogen, e.g., mannose receptor on macrophages which can distinguish the specific orientation and spacing of microbial sugars. In recent years, major advances have been made in the identification of novel families of pattern recognition receptors (PRR) that have evolved to recognize conserved microbial moieties termed pathogen associated molecular patterns (PAMPs). PRRs benefit the immune system by allowing a relatively small number of receptors to recognize many diverse microbes. In addition, to avoid microbes evading immune recognition, these PRRs target conserved molecules that are essential for microbial growth. Among the PRRs recently identified are the TLRs, the Nod-like receptors (NLRs), and the RIG-I like receptors [68–70].

TLRs are a family of highly conserved proteins expressed by cells of the innate immune system [68]. There are ten human family members and each receptor has a different specificity for a conserved microbial product. Some of them (TLR3, TLR7, TLR8, and TLR9) recognize products of viral/bacterial replication while others (TLR2, TLR4, TLR5, TLR6, and TLR9) are more specific for bacterial structural components [68, 71, 72]. TLRs, as one of the first receptor systems to recognize and respond to invading pathogen, play a key role in initiating an adaptive immune response. TLR signaling on DC increases antigen processing, activates DC functions by upregulating costimulatory molecules, and induces their migration to lymph nodes facilitating antigen presentation to naïve T cells [71, 72]. TLRs on monocytes and macrophages function to orchestrate an appropriate innate immune response by directly activating antimicrobial effector functions and by stimulating cytokine production which can in turn regulate other cells of the immune system.

4.2 TLRs on NK Cells

Given the importance of NK cells in the early innate immune response, we and others have examined TLR expression in NK cell lines and purified primary NK cells. We have found mRNA for TLR2, TLR3, TLR4, TLR7, and TLR8 in the human NKL cell line [7 and unpublished observations]. Human, primary NK cells have been shown to express TLR2 [73], TLR3 [7, 74, 75], TLR4 [3, 73], TLR5 [62], TLR7 [7, 73], TLR8 [7, 73], and TLR9 [75]. However, data regarding the functional relevance of TLRs on NK cells is only starting to emerge and it will be some time before we fully understand their role in the overall NK cell response. Although TLR signaling on accessory cells can potently affect NK cell responses, TLRs on NK cells themselves can directly sense pathogen and transduce positive activating signals to NK cells (see Fig. 1).

TLR2 recognizes lipoproteins and lipopeptides from a wide range of microbial pathogens including parasites and bacteria [76]. It can also form heterodimers with TLR1 and TLR6 which increases its ligand range. mRNA for TLR2 has been found in purified primary human NK cells and cell surface expression of the receptor, albeit at low levels, has been described by a couple of groups [62, 66]. TLR2 provides a receptor on NK cells that allows them to directly sense and respond to bacterial pathogen. Experiments with the TLR2 ligand, KpOmpA (outer membrane protein A of *Klebsiella pneumoniae*), demonstrated that it could directly activate human NK cells [62]. Importantly, these experiments were performed using sorted NK cells as it is known that the presence of even tiny numbers of accessory cells (<1% monocyte contamination after purification from a PBMC preparation) are sufficient to provide accessory cell help in the form of cytokines for NK cell activation [7]. The authors supported this result by demonstrating, using NK cells purified from TLR2$-/-$ mice, that KpOmpA mediated activation of NK cells was TLR2 dependent. Cytokines known to be produced locally at the sites of bacterial entry (IL-1β, IL-12, IL-15 and IFN-α) all synergized with KpOmpA to increase the level of NK cell activation and functional responses. In particular, NK cells produced IFN-γ and the classical antibacterial defense molecule, α-defensin. More recently, blocking antibody experiments were used to identify TLR2 as a receptor by which human NK cells directly recognized *M. bovis* (BCG). Stimulation of human NK cells with live BCG activated NK cell functions, including upregulation of activation antigens, production of cytokines (IFN-γ and TNF-α), and cytotoxicity against target cells including immature DC [66]. Although the NK cells were not purified by cell sorting, the authors did include a blocking anti-IL-12 antibody in experiments to exclude possible confounding effects of IL-12 (IL-12 is the most potent accessory cell derived cytokine which promotes NK cell activities that is likely to be present in their experiments). Exogenous addition of IL-12 in experiments greatly enhanced the NK cell responses to BCG which was similar to the response observed for KpOmpA. Indeed, this is a repeating theme seen for NK cell activation with PAMPs and supports the emerging paradigm that NK cells can and do directly respond to pathogen but that this response is greatly enhanced and regulated by the presence of accessory cell cytokines [3, 7, 75]. It is also likely that

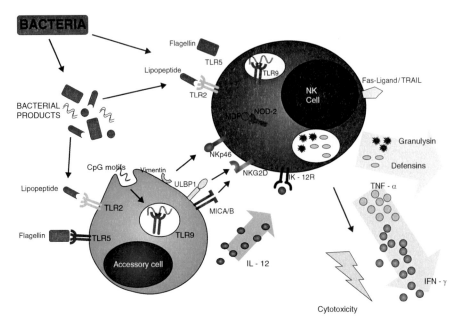

Fig. 1 NK cells in the immune response to bacterial infection. Upon infection, bacteria and bacterial components activate cells of the immune system. TLRs and NLRs interact with bacterial pathogen associated molecular patterns (PAMPs) which results in immune signaling. Natural killer (NK) cells can be activated directly by whole bacteria and bacterial components or indirectly through accessory cell activation. NK cells express a range of pathogen recognition receptors (PRR) that allow direct interaction with bacterial PAMPs including TLR2 (lipopeptide), TLR5 (flagellin), TLR9 (CpG motifs), and Nod-2 (muramyl dipeptide, MDP). Accessory cells, including monocytes, are activated very early in infection by recognition of PAMPs through specific PRRs. Activated accessory cells produce cytokines including IL-12, IL-18, and IL-15 which activate NK cell functions. Interaction with pathogen also changes surface molecule expression on accessory cells and alters the repertoire of NK cell ligands they express. Stress molecules, e.g., ULBP1 and MICA/B are upregulated in response to infection; both these provide ligands for the activating NK cell receptor, NKG2D. Vimentin has recently been reported to be expressed on mycobacterial infected monocytes, where it provides a ligand for the NKp46 activating receptor on NK cells. All of the above mechanisms signal the presence of bacterial pathogen in the NK cell. It becomes activated and deploys a range of antimicrobial effector mechanisms. NK cells produce IFN-γ and TNF-α which have inherent antimicrobial activities. NK cells can also induce cytotoxicity of infected cells using Fas-ligand and conventional perforin/granzyme pathways. Finally, it has recently been appreciated that NK cells can also secrete bactericidal proteins including defensins and granulysin

accessory cells and NK cells will be activated by the same PAMP, e.g., TLR2 is present on monocytes and also on NK cells; this facilitates a rapid and coordinated response to invading pathogen. It is also interesting that the authors of the BCG study found heterogeneity in terms of TLR2 expression by NK cells [66]. Not all donors tested had detectable TLR2 expression and NK cell clones from a single donor varied in their TLR2 expression patterns. However, the authors did not investigate if variability in TLR2 expression by NK cells was responsible for the heterogeneity of functional responses seen in NK cells responding to BCG stimulation.

TLR5 is the receptor for the bacterial PAMP, flagellin, a component of flagella involved in bacterial motility [68, 76]. mRNA for TLR5 is expressed by human NK cells and highly purified NK cells (isolated by cell sorting) respond directly to flagellin by upregulating activation antigens, increasing cytokine production (IFN-γ), and producing α-defensins [62]. Although KpOmpA was more potent in terms of direct NK cell activation, cytokines including IL-1β, IL-12, IL-15, and IL-2 synergized with flagellin to activate NK cell functions to a level seen with KpOmpA and cytokine. IL-10 was defined as a potent inhibitor of flagellin and KpOmpA induced responses [62]. Tsujimuto et al. reported that NK cell numbers and activation status were also increased after in vivo administration of flagellin to C57/BL6 mice [77]. The authors report that flagellin directly activated murine NK cells in vitro, inducing proliferation and CD69 antigen expression (but not cytokine production). Flagellin was also demonstrated to indirectly activate NK cells by inducing DC maturation; however, as the purities of the preparations were only approximately 60%, it is difficult to determine the relative effect of flagellin directly on NK cells versus indirect effects from flagellin-activated accessory cells.

TLR9 is a receptor which recognizes unmethylated CpG motifs that are characteristic of microbial (bacteria and some viruses), but not vertebrate DNA [78, 79]. It is highly expressed by B cells and plasmacytoid DC. Although, there have been some reports to the contrary [80, 81], the general consensus is that human NK cells do express TLR9 [73, 75]. It is less clear however, whether they respond directly to TLR9 ligand or if they are activated indirectly through TLR9 activation of accessory cells (or a combination of both). Using NK cell clones, Sivori et al. demonstrated direct activation of NK cells using the TLR9 agonist, oligodeoxynucleotide (ODN)-A/B [75]. However, NK cell clones and freshly isolated cells only responded to TLR9 agonists in the presence of IL-12 which suggests priming of NK cells as a prerequisite for TLR9 signaling. We have previously reported similar findings where accessory cell derived IL-12 was essential for NK cell cytokine production in response to the TLR7/8 agonist, R848 [7]. Subclasses of ODN differentially modulate immune responses with ODN-A and ODN-C most effective at activating human NK cells [82]. ODN induced expression of the activation antigen, CD69, and increased both cytotoxic activity and cytokine secretion by human NK cells. It is unclear what proportion of NK cells express TLR9 as almost all freshly isolated NK cells expressed CD69 and CD25 in response to ODN (suggesting that most NK cells are responsive to TLR agonists); however, similar to findings for TLR2 [66], there was heterogeneity in the responses to ODN among the NK cell clones examined [82].

4.3 Nod-Like Receptors

Unlike TLRs which recognize a wide spectrum of pathogens (bacteria, virus, parasite), NLRs are best known for their ability to respond to bacterial PAMPs [69, 83]. The best characterized NLRs, Nod-1, and Nod-2, are activated by

D-γ-glutamyl-meso-DAP and muramyl dipeptide (MDP) peptidoglycan subunits of gram positive and negative bacteria respectively. Both receptors signal through RIP-2 leading to NFκB activation [83]. Nod-2 plays an essential role in detecting invasive bacteria in the gut and mutations in the protein are associated with the development of Crohn's disease [84, 85]. Until recently, Nod-2 expression was only associated with monocytes and DC in humans [86, 87]. However, we have recently demonstrated a role for Nod-2 in human NK cells [88]. mRNA for Nod-2 was found in NK cell lines and in purified primary NK cells, and western blotting demonstrated constitutive expression of Nod-2 protein [88]. Stimulation of NK cells with MDP resulted in NK cell activation, as measured by CD69 antigen expression and IFN-γ production. In the absence of exogenous cytokine, the ability of MDP to activate CD69 expression on NK cells was highly variable and modest increases in CD69 expression were seen in only five out of fifteen donors. However, in support of the emerging theme of NK cells requiring cytokine priming for an effective response to PAMPs [3, 89], MDP synergised with IL-12 to increase IFN-γ production, and IFN-α to increase CD69 expression by NK cells. Furthermore, and in contrast to NK cell activation through TLRs which results in the activation of global NK cell responses (cytokine production, cytotoxicity, proliferation etc.), MDP did not induce cytotoxicity in NK cells, even in the presence of IL-12 or IFN-α. This provides further supporting evidence for the relative importance of cytokine production, and not cytotoxicity, as an NK cell effector function important in bacterial infection. Our data support a scenario in which NK cells function as effector cells against bacterial pathogen: they can respond directly to bacterial products (e.g., MDP) but function optimally in the presence of cytokine secreted from accessory cells. It is likely that different pathogens will stimulate different cytokine secretion profiles depending on the cells of the immune system that they first encounter (and the PRR they express) and their constitutive PAMPs [90]. These different cytokine milieus will prime NK cells for an appropriate effector response when they encounter pathogen. While cytotoxicity may be important in the context of a viral infection, it is not as critical in bacterial infections. NK cell production of cytokines and possibly defensins (as seen in NK cells in response to TLR2 and TLR5 agonists), may be more appropriate.

4.4 Natural Cytotoxicity Receptors

Natural cytotoxic receptors (NCRs) (NKp30, NKp44 and NKp46) are receptors that are exclusively expressed by NK cells [91]. The cellular ligands for NKp44 and NKp46 have not been conclusively identified, but both NKp44 and NKp46 are known to be engaged by haemagglutinin of influenza virus [92, 93]. Recently a putative cellular ligand for NKp30, BAT-3, has been identified and demonstrated to promote NK cell recognition of immature DC [94, 95]. pp65 of human cytomegalovirus (HCMV) also interacts with NKp30 to prevent its activating activity [96].

There is emerging evidence in the literature of a role for NCRs in the NK cell response to bacterial pathogen.

The NKp44 receptor is poorly expressed on freshly isolated NK cells and is associated with an activated NK cell phenotype [97]. Esin et al. have suggested that NKp44 is involved in the direct recognition of BCG by human NK cells [98]. They reported increased expression of NKp44 in response to BCG stimulation of NK cells, in the absence of accessory cells (although NK cells were not purified by cell sorting). The authors ruled out a role for NKp46 in this response; however, the dot-plot shown in the paper would suggest that NKp46 also increases on NK cells and while the data do not reach statistical significance, the MFI values also suggest a possible increasing trend. Following up on their more dramatic findings, the authors used a fusion protein to NKp44 to demonstrate binding to bacterial BCG cells using both flow cytometry, and immunogold staining followed by analysis with transmission electron microscopy. Interestingly, they did not find any binding of BCG using a fusion protein to NKp46 (see below). Given that expression is low and that blocking the NKp44 receptor did not inhibit BCG induced CD69 expression on NK cells, the extent of importance of NKp44 for the primary detection of BCG by NK cells is unclear.

NKp46 is found uniquely on NK cells (it is not present on NKT cells) [99]. It has also been implicated as a receptor involved in NK cell recognition of *M. tuberculosis* [54]. Vankayalapati et al. found an increase in two NCRs (NKp30 and NKp46, but not NKp44) and also NKG2D on NK cells in response to stimulation with monocytes infected with *M. tuberculosis* [54, 55]. This contrasts with the data from Esin et al. (no change in NKp30 or NKp46) although one study was using live bacteria and the other bacterially infected monocytes to activate NK cells [98]. In this second body of work [55], important roles for both NKp46 and NKG2D were described. In addition to blocking NKp46 to demonstrate functional importance, the authors have biochemically identified and characterized vimentin, expressed on the surface of *M. tuberculosis* infected monocytes, as a putative ligand for NKp46 on NK cells [100]. The main receptors involved in direct recognition of bacterial PAMPs by NK cells are detailed in Table 1.

4.5 Additional Receptors on NK Cells Activated by Bacterial Infection

A number of ligands for the activating receptor NKG2D have been identified – the MHC I chain related proteins A and B (MICA/B) and the UL16 binding proteins ULBP (RAET1) in man, and the RAE-1 molecules, MULT-1, and H60 in mice [101]. The activating NK cell receptor DNAM-1 (CD226) binds to the nectin and nectin-like molecules, CD112 and CD155. NKp80, another C-type lectin activating receptor, has recently shown to bind the AICL molecule and stimulate cytokine production by NK cells [102]. The 2B4 receptor, expressed on NK cells, can be

Table 1 Direct activation of NK cells by bacteria/ bacterial derived agonists

Agonist/ pathogen	Receptor mediating recognition	Accessory cytokine	NK cell function	Reference
KpOmpA	TLR2	IL-1β, IL-12, IL-15, IL-2, IFN-α	IFN-γ secretion Defensin production NK cell proliferation CD69 expression	[62]
Flagellin	TLR5			
M. bovis (BCG)	TLR2	IL-12	IFN-γ, TNF-α secretion Cytotoxicity (CD107a, ^{51}Cr release) CD69, CD25 expression	[66]
ODN	TLR9	IL-12, IL-8, IFN-α, IL-2	IFN-γ, TNF-α secretion Cytotoxicity (CD107a, ^{51}Cr release) CD69 expression	[75]
Muramyl dipeptide (MDP)	NOD-2	IFN-αIL-12	IFN-γ secretion CD69 expression	[88]
M. tuberculosis	TLR2, TLR4	IL-12	IFN-γ secretion CD69 expression	[15]
H. pylori lysate	Unknown	IL-12	IFN-γ secretion CD69, CD25 expression	[122]

ligated by its ligand CD48 to drive both NK cell proliferation and cytotoxicity [103]. In contrast, NKR-P1A (CD161) is an inhibitory C-type lectin receptor expressed by NK cells which interacts with the lectin-like transcript-1 (LLT-1) [104]. Bacterial infection and stimulation with bacterially derived or synthetic ligands has been shown to modulate the surface expression of these ligands on accessory cells that can help control NK cell function (see Fig. 1).

5 Activation of NK Cells by Accessory Cells

5.1 Infection Induced Changes in NK Cell Ligand Expression

As described above, NK cells have a range of receptors that allow them to recognize pathogen directly. However, we now know that accessory cells are fundamentally important in regulating NK cell responses to a diverse array of pathogens [3, 89]. This can occur either by the release of cytokine which activates NK cell functions, or by the altered expression of receptors on the surface of the accessory cell, e.g., pathogen induced expression of NK cell ligands, which affects the ability of accessory cells to interact with and modulate NK cell functions (see Fig. 1). A number of studies have examined the expression of NK ligands following infection with the intracellular bacteria *M. tuberculosis*. It has been known for many years that infection of monocytes with mycobacterial species renders such cells susceptible to NK cell lysis. NK cells also appear to limit intracellular bacterial growth and/or target mycobacteria for lysis [53, 105]. More recently the specific interactions governing the recognition/activation of such cells have begun to be characterized. Infection of monocytes or alveolar macrophages with *M. tuberculosis* rendered cells susceptible to NK cell lysis that was inhibited by both anti-NKG2D and anti-NKp46 antibodies [54, 55]. Co-culture of NK cells with mycobacteria infected cells also stimulated the upregulation of NKG2D, NKp30, and NKp46 on the surface of the effector cell. The NKG2D ligand, ULBP1, was upregulated in both *M. tuberculosis* infected monocytes and alveolar macrophages, and blocking with an anti-ULBP1 antibody indicated this was the specific NKG2D ligand involved in NK cell recognition of infected cells [55]. The upregulation of surface ULBP1 by *M. tuberculosis* was blocked by anti-TLR2 antibody. As there is a known TLR2 ligand (19 kDa lipoprotein of *M. tuberculosis*), known to drive ULBP1 expression, it is likely that this may be the specific bacterial derived molecule promoting NKG2D ligand expression in infection. A previous report also demonstrated increased surface expression of another NKG2D ligand, MICA, on *M. tuberculosis* infected monocyte derived DCs; however a very high multiplicity of infection (2000:1) was used in this study and the expression of other NK cell ligands was not tested [106]. As mentioned above, infection of monocytes with the H37Ra strain of *M. tuberculosis* (or *L. monocytogenes*), also upregulated surface expression of vimentin, a putative ligand for the activating NKp46 receptor [100]; however further studies will be necessary to confirm the direct interaction between NKp46 and vimentin.

It is not just for *M. tuberculosis* for which evidence exists for bacterial modulation of NK cell ligands on accessory cells. Binding of diffusely adherent *E. coli* via the adhesion AfaE to CD55 on epithelial cell lines leads to an upregulation of the NKG2D activating receptor MICA on the cell surface [107]. Co-culture of lipopolysaccharide (LPS, *Salmonella minnesota*) stimulated macrophages with autologous NK cells induced NK cell proliferation, IFN-γ production, and cytotoxicity. LPS stimulation of macrophages was associated with increased surface expression of the NKG2D ligands ULBP1, 2, 3 and MICA/B, and cytolysis of stimulated macrophages was blocked by anti-NKG2D antibodies. However NK cell proliferation and IFN-γ secretion were not modulated by anti-NKG2D blocking antibody but were inhibited by blocking the interaction between 2B4 on the NK cell, and CD48 on the macrophage [108].

TLR ligands which activate NK cells also dramatically activate other cells of the immune system and can thus lead to indirect activation of NK cells. Treatment of human monocytes with the TLR ligands LPS (TLR4), Poly (I:C) (TLR3), or Pam$_2$Cys SK4 (TLR2) stimulated the surface expression of the NK cell activating ligand, AICL. AICL upregulation sensitized treated cells to NK cell-mediated cytolysis [102]. Although monocytes are generally considered to be key accessory cells, other cell types can also modulate NK cell activities, e.g., activation of TLR signaling by synthetic ligands or heat killed *E. coli* and *L. monocytogenes* increased surface expression of NKG2D ligands on the surface of murine peritoneal macrophages [109]. In particular, the interplay between NK cells and DC has become a significant area of recent scientific research and there seems to be an important two-way communication (and subsequent regulation) between these cells occurring early in infection and which affects the development of a downstream adaptive

Table 2 Regulation of NK cell ligands by bacteria/bacterial derived agonists

Pathogen/agonist	Receptor mediating recognition	Accessory cell	NK cell ligand upregulated	Reference
M. tuberculosis	TLR2	Human alveolar macrophage/Monocytes	ULBP1	[54]
E. coli AfaE	CD55	Human epithelial cells (Hela)	MICA	[107]
LPS (*S. minnesota*)	TLR4	Human macrophages	MICA/B, ULBP1-3 CD48	[108]
M. tuberculosis, L. monocytogenes	Not shown	Human moncytes	NKp46 Ligand (vimentin)	[100]
E. coli L. monocytogenes LPS	TLR4 (LPS)	Murine peritoneal macrophages	RAE-1 proteins	[109]
LPS (*E. coli*)	TLR4	Human immature DC	ULBP1, ULBP2	[111]
Pam$_2$Cys SK4 LPS CpG DNA	TLR2 TLR4 TLR9	Human monocytes	AICL	[102]
LPS	TLR4	Monocyte derived DC	LLT1	[112]

immune response [46, 110]. Stimulation of immature DC with specific TLR ligands resulted in differential cell surface expression of NKG2D ligands. Treatment with LPS from *E. coli* induced both mRNA and surface expression of ULBP1 and ULBP2 but did not induce a significant change in surface expression of other NKG2D ligands. In contrast, Pam3 (TLR2) and Malp2 (TLR2/6) did not modulate surface expression of any NKG2D ligands indicating differential effects depending on the specific TLR activated [111]. While most reports to date have demonstrated increased expression of activating ligands following stimulation with PAMPs, Rosen et al. recently demonstrated induction of the NK cell inhibitory ligand LLT1 following stimulation with TLR3, TLR4, TLR7, TLR8, and TLR9 agonists. Interestingly, specific cell types demonstrated alternative activation patterns: plasmacytoid DC stimulated via TLR7 and TLR9 exhibited increased LLT1 expression while monocyte derived DC was activated in response to TLR3, TLR4, and TLR8 agonists [112]. Thus, there is a growing range of microbial ligands (summarized in Table 2) which, in addition to directly activating NK cells, can indirectly regulate NK cell functions by inducing changes in receptor expression on accessory cells.

5.2 *Infection Induced Changes in the Cytokine Environment*

The importance of cytokine production in NK cell activation in vitro has been demonstrated in studies that separate purified NK cells from accessory cells via semipermeable membranes, the incubation of NK cells with supernatants from pathogen stimulated accessory cells, and the demonstration that cytokine neutralizing antibodies inhibit NK cell activation. In particular, IFN-γ plays a central role in the NK cell response to a wide range of bacterial pathogens. Both IL-12 and IL-18 are potent cytokines capable of driving IFN-γ production from NK cells. The stimulation of IFN-γ secretion from NK cells in vivo by M. tuberculosis was inhibited in Rag−/− mice that lacked IL-12 expression (both p35−/− and p40−/−) and such mice were susceptible to higher bacterial loads and accelerated mortality [19]. In humans, in vitro mycobacterial stimulation of pleural fluid lymphocytes isolated from patients with tuberculosis pleuritis indicated that the production of IFN-γ following bacterial stimulation was partially dependent on the production of IL-12 from antigen presenting cells, and was blocked by treatment with anti-TLR2 and anti-TLR4 antibodies [15]. Interestingly, NK cells were most potently activated by the whole bacterium rather than purified TLR2 and TLR4 agonists. Treatment of human monocytes with UV irradiated gram positive bacteria (including staphylococccus, streptococcus, and lactobacillus species) potently induced IL-12 secretion [113]. Pompei et al. examined IL-12 production from specific cellular subsets following TLR stimulation and demonstrated higher levels of IL-12 production from murine bone marrow derived DC than macrophages, with the enhanced IL-12 secretion associated with the triggering of TLR9 in DC rather than TLR2 in macrophages [114]. This may provide another level of control in determining levels of NK cell activation in response to different pathogens.

Listeriolysin O (LLO), the major cytolytic virulence factor from *L. monocytogenes*, is a potent inducer of IL-12 and IL-18 both in vivo and in vitro. Treatment of murine spleen cells with LLO purified from L. monocytogenes induced the expression of IFN-γ in vitro and was blocked by both anti-IL-12 and IL-18 antibodies [115]. Depletion of cellular subsets from spleen cells implicated NK cells as the major IFN-γ producing cell type and CD11b$^+$ cells (mainly macrophages) as the primary source of both biologically active IL-12 and IL-18 [115]. Activation of IL-12 expression by listeria appears primarily to be a TLR-dependent mechanism with KO mice lacking TLR2, TLR4, and MyD88 expression all exhibiting reduced capacity to drive efficient IL-12 secretion [116, 117]. In contrast, the secretion of IL-18 in listeria infection appears primarily to be TLR independent but rather involves the NLRs. IL-18 is produced in an inactive proform that must be cleaved to produce the biologically active cytokine. Initiation of this processing is controlled by a multiprotein complex known as the inflammasome in which NLRs activate caspase-1. Infection of murine macrophages with *L. monocytogenes* or *S. aureus* induced the release of IL-18 that was dependent on the presence of an intact inflammasome complex and caspase-1 activity [118]. A more recent study using a LLO listeria mutant implicated LLO as the listeria factor driving the production of biologically active IL-18 via caspase-1 [116]. However, a number of other reports have implicated flagellin, from a number of bacteria including *L. monocytogenes, S. typhimurium,* or *L. pneumophilia*, as being responsible for inflammasome complex activation and IL-18 secretion [119–121]. It has been suggested that the deletion of LLO from the bacterial genome inhibits the access of the bacteria to the cytosol and thus limits access to the inflammasome complex but this remains controversial [116]. Bacteria can also modulate IL-12 dependent NK cell functions indirectly. Pulsing of purified NK cells with *Helicobacter pylori* lysate induced an upregulation of IL-12βR2 expression rendering NK cells more sensitive to activation with IL-12 [122].

Type 1 interferons, primarily known for their potent antiviral effects are also capable of activating NK cells. A large number of bacteria or bacterially derived molecules are capable of inducing the production of type 1 IFNs (reviewed in Bogdan et al. [123]) and studies have shown that IFN production induced by stimulation with bacteria/TLR agonists is important in activating NK cells both in vivo and in vitro [124, 125]. Infection of DC with *M. tuberculosis* induced secretion of both IL-12 and type 1 IFNs [126, 127]. This production of 1 IFN was also associated with an increased secretion of the chemokine CXCL10 from infected cells, which was shown to enhance the chemotactic ability of purified NK cells [124].

IL-15 plays a vital role in the development and expansion of NK cells and can induce increased cytotoxic potential. IL-15 is a soluble molecule that appears to be bound by the IL-15 receptor on the surface of the cytokine secreting cell and it is presented to the NK cell in a cell contact dependent manner (*trans* presentation) and so straddles the crossover between soluble/cell mediated interactions [128]. Lucas et al. described the importance of *trans* presentation of IL-15 in priming NK cells in vivo following TLR agonist stimulation or *L. monocytogenes* infection [125]. The production of IL-15 from CD11c^{high} DC was dependent on type 1 IFN signaling and NK cell effector function was markedly reduced following specific deletion of such DC. Stimulation of human DC with TLR agonists alone or in combination with NOD1/2 agonists in vitro also induced surface expression of IL-15; however, soluble secreted cytokine was undetectable [129].

6 Bacterial Modulation of NK Cell Functions

It is well known that pathogenic organisms manipulate the host immune system to enhance their own survival. They employ a wide range of mechanisms to evade and interfere with host immunity. While evasion from NK cells has best been described for viruses [130, 131], it seems likely that similar type mechanisms will also be discovered that help bacteria escape from NK cell immune responses. Indeed some examples have already been described. *Yersinia pestis*, the causative agent of plague, secretes Yop proteins into host cells. One of these, YopM, has been implicated in targeting innate immunity by inducing targeted depletion of NK cells from the host [132]. This depletion was associated with reduced transcription of IL-15 mRNA from spleen cells and IL-15Rα mRNA expression specifically from NK cells. In mycobacterial infections, ManLAM (a component of the bacterial cell wall), inhibits NK cell activation and IFN-γ production. One possible mechanism for this is through bacterial interference of DC maturation. ManLAM binds to

DC-SIGN, interferes with DC maturation, and inhibits its ability to secrete NK cell activating cytokines, e.g., IL-12 and also promotes expression of the immunosuppressive cytokine IL-10 [133]. More controversial is the idea of bacterial activation of NK cells as a pathogen virulence factor. Humann et al. suggest that *L. monocytogenes* secretes a protein - p60 autolysin - which activates NK cell cytokine production, thereby allowing bacterial expansion in the host [134]. Indeed, other reports, which describe NK cell depletion as beneficial to the host in certain bacterial infections (see section 2 of this review) also exist in the literature. However, as it has also been reported that IFN-γ is critical in the control of listeria, further experimentation by a number of groups is required for clarification.

In summary, significant progress has been made towards identifying the role played by NK cells in bacterial infections. As with any field in its infancy, conflicting reports abound and much needs to be done to resolve the issues. The beneficial versus detrimental effects of NK cells reported following in vivo challenge must be rigorously tested in numerous, well-defined models of infection. More clinical studies on the role of NK cells in human bacterial infections are also required. In recent years, heterogeneity of response has become a defining characteristic of NK cells and it is possible that NK cell subsets may differ in their antibacterial activities. This will be elucidated through more detailed investigations of NK cell function at both cellular and molecular levels. While significant advances in innate immunology have been made with the discovery of novel immune receptor families, in particular the TLRs and NLRs, their role in NK cell biology has only begun to be appreciated. While these receptors are highly conserved between species, humans and mice differ significantly in other NK cell receptors, particularly those for MHC class I (KIR in humans and Ly49 in mice); the significance of this in terms of NK cell responses during bacterial infection awaits discovery. Finally, the identification of novel pathogen molecules that can activate NK cells directly and indirectly also adds to the growing body of evidence which supports an important role for NK cells in bacterial infections. These recent discoveries may provide new therapeutic targets for modulation of NK cell activities in infection and disease. It seems that the time is right for the importance of NK cells to be appreciated, not just as an antiviral effector cell, but as a significant and potent effector cell in the immune struggle against bacterial pathogens.

References

1. Hamerman JA, Ogasawara K, Lanier LL (2005) Curr Opin Immunol 17:29
2. Moretta L, Bottino C, Pende D, Vitale M, Mingari MC, Moretta A (2005) Immunol Lett 100:7
3. O'Connor GM, Hart OM, Gardiner CM (2006) Immunology 117:1
4. Fehniger TA, Carson WE, Caligiuri MA (1999) Transplant Proc 31:1476
5. Fehniger TA, Shah MH, Turner MJ, VanDeusen JB, Whitman SP, Cooper MA, Suzuki K, Wechser M, Goodsaid F, Caligiuri MA (1999) J Immunol 162:4511
6. Orange JS, Biron CA (1996) J Immunol 156:1138

7. Hart OM, Athie-Morales V, O'Connor GM, Gardiner CM (2005) J Immunol 175:1636
8. Small CL, McCormick S, Gill N, Kugathasan K, Santosuosso M, Donaldson N, Heinrichs DE, Ashkar A, Xing Z (2008) J Immunol 180:5558
9. Byrne P, McGuirk P, Todryk S, Mills KH (2004) Eur J Immunol 34:2579
10. Sporri R, Joller N, Albers U, Hilbi H, Oxenius A (2006) J Immunol 176:6162
11. Ku CL, von Bernuth H, Picard C, Zhang SY, Chang HH, Yang K, Chrabieh M, Issekutz AC, Cunningham CK, Gallin J, Holland SM, Roifman C, Ehl S, Smart J, Tang M, Barrat FJ, Levy O, McDonald D, Day-Good NK, Miller R, Takada H, Hara T, Al-Hajjar S, Al-Ghonaium A, Speert D, Sanlaville D, Li X, Geissmann F, Vivier E, Marodi L, Garty BZ, Chapel H, Rodriguez-Gallego C, Bossuyt X, Abel L, Puel A, Casanova JL (2007) J Exp Med 204:2407
12. Florido M, Correia-Neves M, Cooper AM, Appelberg R (2003) Int Immunol 15:895
13. Junqueira-Kipnis AP, Kipnis A, Jamieson A, Juarrero MG, Diefenbach A, Raulet DH, Turner J, Orme IM (2003) J Immunol 171:6039
14. Lorgat F, Keraan MM, Ress SR (1992) Clin Exp Immunol 90:215
15. Schierloh P, Yokobori N, Aleman M, Landoni V, Geffner L, Musella RM, Castagnino J, Baldini M, Abbate E, de la Barrera SS, Sasiain MC (2007) Infect Immun 75:5325
16. Wolfe SA, Tracey DE, Henney CS (1976) Nature 262:584
17. Harshan KV, Gangadharam PR (1991) Infect Immun 59:2818
18. Smith D, Hansch H, Bancroft G, Ehlers S (1997) Immunology 92:413
19. Feng CG, Kaviratne M, Rothfuchs AG, Cheever A, Hieny S, Young HA, Wynn TA, Sher A (2006) J Immunol 177:7086
20. Harty JT, Bevan MJ (1995) Immunity 3:109
21. Huang S, Hendriks W, Althage A, Hemmi S, Bluethmann H, Kamijo R, Vilcek J, Zinkernagel RM, Aguet M (1993) Science 259:1742
22. Bancroft GJ, Sheehan KC, Schreiber RD, Unanue ER (1989) J Immunol 143:127
23. Teixeira HC, Kaufmann SH (1994) J Immunol 152:1873
24. Dunn PL, North RJ (1991) Infect Immun 59:2892
25. Schultheis RJ, Kearns RJ (1990) Nat Immun Cell Growth Regul 9:376
26. Takada H, Matsuzaki G, Hiromatsu K, Nomoto K (1994) Immunology 82:106
27. Badgwell B, Parihar R, Magro C, Dierksheide J, Russo T, Carson WE 3rd (2002) Surgery 132:205
28. Kerr AR, Kirkham LA, Kadioglu A, Andrew PW, Garside P, Thompson H, Mitchell TJ (2005) Microb Infect 7:845
29. Newton DW Jr, Runnels HA, Kearns RJ (1992) Nat Immunol 11:335
30. Plitas G, Chaudhry UI, Kingham TP, Raab JR, DeMatteo RP (2007) J Immunol 178:4411
31. Walzer T, Blery M, Chaix J, Fuseri N, Chasson L, Robbins SH, Jaeger S, Andre P, Gauthier L, Daniel L, Chemin K, Morel Y, Dalod M, Imbert J, Pierres M, Moretta A, Romagne F, Vivier E (2007) Proc Natl Acad Sci USA 104:3384
32. Ota T, Okubo Y, Sekiguchi M (1990) Am Rev Respir Dis 142:29
33. Wendland T, Herren S, Yawalkar N, Cerny A, Pichler WJ (2000) Immunol Lett 72:75
34. Orange JS (2002) Microb Infect 4:1545
35. Altare F, Durandy A, Lammas D, Emile JF, Lamhamedi S, Le Deist F, Drysdale P, Jouanguy E, Doffinger R, Bernaudin F, Jeppsson O, Gollob JA, Meinl E, Segal AW, Fischer A, Kumararatne D, Casanova JL (1998) Science 280:1432
36. Altare F, Lammas D, Revy P, Jouanguy E, Doffinger R, Lamhamedi S, Drysdale P, Scheel-Toellner D, Girdlestone J, Darbyshire P, Wadhwa M, Dockrell H, Salmon M, Fischer A, Durandy A, Casanova JL, Kumararatne DS (1998) J Clin Invest 102:2035
37. de Jong R, Altare F, Haagen IA, Elferink DG, Boer T, van Breda Vriesman PJ, Kabel PJ, Draaisma JM, van Dissel JT, Kroon FP, Casanova JL, Ottenhoff TH (1998) Science 280:1435
38. Dupuis S, Doffinger R, Picard C, Fieschi C, Altare F, Jouanguy E, Abel L, Casanova JL (2000) Immunol Rev 178:129
39. Newport MJ, Huxley CM, Huston S, Hawrylowicz CM, Oostra BA, Williamson R, Levin M (1996) N Engl J Med 335:1941

40. Koga S, Kiyohara T, Taniguchi K, Nishikido M, Kubota S, Sakuragi T, Shindo K, Saitoh Y (1988) Urol Res 16:351
41. Brandau S, Riemensberger J, Jacobsen M, Kemp D, Zhao W, Zhao X, Jocham D, Ratliff TL, Bohle A (2001) Int J Cancer 92:697
42. Sonoda T, Sugimura K, Ikemoto S, Kawashima H, Nakatani T (2007) Oncol Rep 17:1469
43. Holmberg LA, Ault KA (1984) Cell Immunol 89:151
44. Kearns RJ, Leu RW (1984) Cell Immunol 84:361
45. Yoshimura K, Laird LS, Chia CY, Meckel KF, Slansky JE, Thompson JM, Jain A, Pardoll DM, Schulick RD (2007) Cancer Res 67:10058
46. Moretta A, Marcenaro E, Sivori S, Della Chiesa M, Vitale M, Moretta L (2005) Trends Immunol 26:668
47. Brennan RE, Russell K, Zhang G, Samuel JE (2004) Infect Immun 72:6666
48. Foster N, Hulme SD, Barrow PA (2003) Infect Immun 71:4733
49. Gordon MA, Jack DL, Dockrell DH, Lee ME, Read RC (2005) Infect Immun 73:3445
50. Portnoy DA, Schreiber RD, Connelly P, Tilney LG (1989) J Exp Med 170:2141
51. Nairz M, Fritsche G, Brunner P, Talasz H, Hantke K, Weiss G (2008) Eur J Immunol 38:1923
52. Bokhari SM, Kim KJ, Pinson DM, Slusser J, Yeh HW, Parmely MJ (2008) Infect Immun 76:1379
53. Brill KJ, Li Q, Larkin R, Canaday DH, Kaplan DR, Boom WH, Silver RF (2001) Infect Immun 69:1755
54. Vankayalapati R, Wizel B, Weis SE, Safi H, Lakey DL, Mandelboim O, Samten B, Porgador A, Barnes PF (2002) J Immunol 168:3451
55. Vankayalapati R, Garg A, Porgador A, Griffith DE, Klucar P, Safi H, Girard WM, Cosman D, Spies T, Barnes PF (2005) J Immunol 175:4611
56. Krensky AM, Clayberger C (2005) Am J Transplant 5:1789
57. Stenger S, Hanson DA, Teitelbaum R, Dewan P, Niazi KR, Froelich CJ, Ganz T, Thoma-Uszynski S, Melian A, Bogdan C, Porcelli SA, Bloom BR, Krensky AM, Modlin RL (1998) Science 282:121
58. Walch M, Eppler E, Dumrese C, Barman H, Groscurth P, Ziegler U (2005) J Immunol 174:4220
59. Walch M, Latinovic-Golic S, Velic A, Sundstrom H, Dumrese C, Wagner CA, Groscurth P, Ziegler U (2007) BMC Immunol 8:14
60. Menendez A, Brett Finlay B (2007) Curr Opin Immunol 19:385
61. Agerberth B, Charo J, Werr J, Olsson B, Idali F, Lindbom L, Kiessling R, Jornvall H, Wigzell H, Gudmundsson GH (2000) Blood 96:3086
62. Chalifour A, Jeannin P, Gauchat J-F, Blaecke A, Malissard M, N'Guyen T, Thieblemont N, Delneste Y (2004) Blood 104:1778
63. Liu CY, Lin HC, Yu CT, Lin SM, Lee KY, Chen HC, Chou CL, Huang CD, Chou PC, Liu WT, Wang CH, Kuo HP (2007) Life Sci 80:749
64. Martineau AR, Newton SM, Wilkinson KA, Kampmann B, Hall BM, Nawroly N, Packe GE, Davidson RN, Griffiths CJ, Wilkinson RJ (2007) J Clin Invest 117:1988
65. Roy S, Barnes PF, Garg A, Wu S, Cosman D, Vankayalapati R (2008) J Immunol 180:1729
66. Marcenaro E, Ferranti B, Falco M, Moretta L, Moretta A (2008) Int Immunol 20:1155
67. Della Chiesa M, Vitale M, Carlomagno S, Ferlazzo G, Moretta L, Moretta A (2003) Eur J Immunol 33:1657
68. Akira S, Sato S (2003) Scand J Infect Dis 35:555
69. Creagh EM, O'Neill LA (2006) Trends Immunol 27:352
70. Koyama S, Ishii KJ, Coban C, Akira S (2008) Cytokine 43:336
71. Pearce EJ, Kane CM, Sun J (2006) Chem Immunol Allergy 90:82
72. Steinman RM, Hemmi H (2006) Curr Top Microbiol Immunol 311:17
73. Lauzon NM, Mian F, MacKenzie R, Ashkar AA (2006) Cell Immunol 241:102

74. Schmidt KN, Leung B, Kwong M, Zarember KA, Satyal S, Navas TA, Wang F, Godowski PJ (2004) J Immunol 172:138
75. Sivori S, Falco M, Della Chiesa M, Carlomagno S, Vitale M, Moretta L, Moretta A (2004) Proc Natl Acad Sci USA 101:10116
76. Carpenter S, O'Neill LA (2007) Cell Microbiol 9:1891
77. Tsujimoto H, Uchida T, Efron PA, Scumpia PO, Verma A, Matsumoto T, Tschoeke SK, Ungaro RF, Ono S, Seki S, Clare-Salzler MJ, Baker HV, Mochizuki H, Ramphal R, Moldawer LL (2005) J Leukoc Biol 78:888
78. Bauer S, Kirschning CJ, Hacker H, Redecke V, Hausmann S, Akira S, Wagner H, Lipford GB (2001) Proc Natl Acad Sci USA 98:9237
79. Kumagai Y, Takeuchi O, Akira S (2008) Adv Drug Deliv Rev 60:795
80. Ballas ZK (2007) Immunol Res 39:15
81. Hornung V, Rothenfusser S, Britsch S, Krug A, Jahrsdorfer B, Giese T, Endres S, Hartmann G (2002) J Immunol 168:4531
82. Sivori S, Carlomagno S, Moretta L, Moretta A (2006) Eur J Immunol 36:961
83. Chen G, Shaw MH, Kim YG, Nunez G (2009) Annu Rev Pathol 4:365–398
84. Inohara N, Ogura Y, Fontalba A, Gutierrez O, Pons F, Crespo J, Fukase K, Inamura S, Kusumoto S, Hashimoto M, Foster SJ, Moran AP, Fernandez-Luna JL, Nunez G (2003) J Biol Chem 278:5509
85. Watanabe T, Kitani A, Strober W (2005) Gut 54:1515
86. Fritz JH, Girardin SE, Fitting C, Werts C, Mengin-Lecreulx D, Caroff M, Cavaillon JM, Philpott DJ, Adib-Conquy M (2005) Eur J Immunol 35:2459
87. Ogura Y, Inohara N, Benito A, Chen FF, Yamaoka S, Nunez G (2001) J Biol Chem 276:4812
88. Athie-Morales V, O'Connor GM, Gardiner CM (2008) J Immunol 180:4082
89. Newman KC, Riley EM (2007) Nat Rev Immunol 7:279
90. Kadowaki N, Ho S, Antonenko S, Malefyt RW, Kastelein RA, Bazan F, Liu YJ (2001) J Exp Med 194:863
91. Moretta L, Moretta A (2004) EMBO J 23:255
92. Arnon TI, Lev M, Katz G, Chernobrov Y, Porgador A, Mandelboim O (2001) Eur J Immunol 31:2680
93. Mandelboim O, Lieberman N, Lev M, Paul L, Arnon TI, Bushkin Y, Davis DM, Strominger JL, Yewdell JW, Porgador A (2001) Nature 409:1055
94. Pogge von Strandmann E, Simhadri VR, von Tresckow B, Sasse S, Reiners KS, Hansen HP, Rothe A, Boll B, Simhadri VL, Borchmann P, McKinnon PJ, Hallek M, Engert A (2007) Immunity 27:965
95. Simhadri VR, Reiners KS, Hansen HP, Topolar D, Simhadri VL, Nohroudi K, Kufer TA, Engert A, Pogge von Strandmann E (2008) PLoS ONE 3:e3377
96. Arnon TI, Achdout H, Levi O, Markel G, Saleh N, Katz G, Gazit R, Gonen-Gross T, Hanna J, Nahari E, Porgador A, Honigman A, Plachter B, Mevorach D, Wolf DG, Mandelboim O (2005) Nat Immunol 6:515
97. Cantoni C, Bottino C, Vitale M, Pessino A, Augugliaro R, Malaspina A, Parolini S, Moretta L, Moretta A, Biassoni R (1999) J Exp Med 189:787
98. Esin S, Batoni G, Counoupas C, Stringaro A, Brancatisano FL, Colone M, Maisetta G, Florio W, Arancia G, Campa M (2008) Infect Immun 76:1719
99. Pessino A, Sivori S, Bottino C, Malaspina A, Morelli L, Moretta L, Biassoni R, Moretta A (1998) J Exp Med 188:953
100. Garg A, Barnes PF, Porgador A, Roy S, Wu S, Nanda JS, Griffith DE, Girard WM, Rawal N, Shetty S, Vankayalapati R (2006) J Immunol 177:6192
101. Lanier LL (2005) Annu Rev Immunol 23:225
102. Welte S, Kuttruff S, Waldhauer I, Steinle A (2006) Nat Immunol 7:1334
103. Vaidya SV, Mathew PA (2006) Immunol Lett 105:180
104. Rosen DB, Bettadapura J, Alsharifi M, Mathew PA, Warren HS, Lanier LL (2005) J Immunol 175:7796

105. Millman AC, Salman M, Dayaram YK, Connell ND, Venketaraman V (2008) J Interferon Cytokine Res 28:153
106. Das H, Groh V, Kuijl C, Sugita M, Morita CT, Spies T, Bukowski JF (2001) Immunity 15:83
107. Tieng V, Le Bouguenec C, du Merle L, Bertheau P, Desreumaux P, Janin A, Charron D, Toubert A (2002) Proc Natl Acad Sci USA 99:2977
108. Nedvetzki S, Sowinski S, Eagle RA, Harris J, Vely F, Pende D, Trowsdale J, Vivier E, Gordon S, Davis DM (2007) Blood 109:3776
109. Hamerman JA, Ogasawara K, Lanier LL (2004) J Immunol 172:2001
110. Moretta L, Ferlazzo G, Bottino C, Vitale M, Pende D, Mingari MC, Moretta A (2006) Immunol Rev 214:219
111. Ebihara T, Masuda H, Akazawa T, Shingai M, Kikuta H, Ariga T, Matsumoto M, Seya T (2007) Int Immunol 19:1145
112. Rosen DB, Cao W, Avery DT, Tangye SG, Liu YJ, Houchins JP, Lanier LL (2008) J Immunol 180:6508
113. Hessle C, Andersson B, Wold AE (2000) Infect Immun 68:3581
114. Pompei L, Jang S, Zamlynny B, Ravikumar S, McBride A, Hickman SP, Salgame P (2007) J Immunol 178:5192
115. Nomura T, Kawamura I, Tsuchiya K, Kohda C, Baba H, Ito Y, Kimoto T, Watanabe I, Mitsuyama M (2002) Infect Immun 70:1049
116. Hara H, Tsuchiya K, Nomura T, Kawamura I, Shoma S, Mitsuyama M (2008) J Immunol 180:7859
117. Seki E, Tsutsui H, Tsuji NM, Hayashi N, Adachi K, Nakano H, Futatsugi-Yumikura S, Takeuchi O, Hoshino K, Akira S, Fujimoto J, Nakanishi K (2002) J Immunol 169:3863
118. Mariathasan S, Weiss DS, Newton K, McBride J, O'Rourke K, Roose-Girma M, Lee WP, Weinrauch Y, Monack DM, Dixit VM (2006) Nature 440:228
119. Franchi L, Amer A, Body-Malapel M, Kanneganti TD, Ozoren N, Jagirdar R, Inohara N, Vandenabeele P, Bertin J, Coyle A, Grant EP, Nunez G (2006) Nat Immunol 7:576
120. Lightfield KL, Persson J, Brubaker SW, Witte CE, von Moltke J, Dunipace EA, Henry T, Sun YH, Cado D, Dietrich WF, Monack DM, Tsolis RM, Vance RE (2008) Nat Immunol 9:1171
121. Warren SE, Mao DP, Rodriguez AE, Miao EA, Aderem A (2008) J Immunol 180:7558
122. Yun CH, Lundgren A, Azem J, Sjoling A, Holmgren J, Svennerholm AM, Lundin BS (2005) Infect Immun 73:1482
123. Bogdan C, Mattner J, Schleicher U (2004) Immunol Rev 202:33
124. Lande R, Giacomini E, Grassi T, Remoli ME, Iona E, Miettinen M, Julkunen I, Coccia EM (2003) J Immunol 170:1174
125. Lucas M, Schachterle W, Oberle K, Aichele P, Diefenbach A (2007) Immunity 26:503
126. Giacomini E, Iona E, Ferroni L, Miettinen M, Fattorini L, Orefici G, Julkunen I, Coccia EM (2001) J Immunol 166:7033
127. Remoli ME, Giacomini E, Lutfalla G, Dondi E, Orefici G, Battistini A, Uze G, Pellegrini S, Coccia EM (2002) J Immunol 169:366
128. Kobayashi H, Dubois S, Sato N, Sabzevari H, Sakai Y, Waldmann TA, Tagaya Y (2005) Blood 105:721
129. Tada H, Aiba S, Shibata K, Ohteki T, Takada H (2005) Infect Immun 73:7967
130. Lanier LL (2008) Nat Rev Immunol 8:259
131. Wilkinson GW, Tomasec P, Stanton RJ, Armstrong M, Prod'homme V, Aicheler R, McSharry BP, Rickards CR, Cochrane D, Llewellyn-Lacey S, Wang EC, Griffin CA, Davison AJ (2008) J Clin Virol 41:206
132. Kerschen EJ, Cohen DA, Kaplan AM, Straley SC (2004) Infect Immun 72:4589
133. Geijtenbeek TB, Van Vliet SJ, Koppel EA, Sanchez-Hernandez M, Vandenbroucke-Grauls CM, Appelmelk B, Van Kooyk Y (2003) J Exp Med 197:7
134. Humann J, Bjordahl R, Andreasen K, Lenz LL (2007) J Immunol 178:2407

NK Cells and Autoimmunity

Hanna Brauner and Petter Höglund

Abstract Most, if not all, individuals carry immune cells in their bodies which display inherent reactivity to self cells. For example, all T lymphocytes are self-reactive to some extent as a result of positive selection on self MHC in the thymus. This reactivity is required for self–nonself discrimination by T cells and is therefore necessary for proper function of T cells. However, carrying self-reactive T cells in the circulation is potentially dangerous and multiple mechanisms have developed to ensure that self-reactive T cells are kept in check under normal circumstances. Despite all the control, there are situations in which self-tolerance fails and autoimmune diseases result. T cells are critical players in autoimmunity, either as primary effector cells or as helper cells promoting autoreactive B cell responses and autoantibody production. Many cell types can modulate T cell responses, however, and may act as critical determinants whether or not autoimmune diseases will prevail. In this chapter, we discuss the role for natural killer (NK) cells in autoimmunity. Most evidence suggest that NK cells participate in disease pathogenesis, primarily as cytokine-producing cells, but it is also possible that cytotoxicity may be of importance. Furthermore, while most data so far suggest a disease-promoting role for NK cells, there are also evidence for a downregulatory role for NK cells in some models. We here review data from individual autoimmune diseases and give suggestions for possible future directions for the field.

H. Brauner and P. Höglund (✉)
Department of Microbiology Tumor and Cell Biology (MTC), Karolinska Institutet, Box 280, 171 77, Stockholm, Sweden
e-mail: petter.hoglund@ki.se

1 Autoimmunity

Autoimmune diseases result from breakdown of self-tolerance and an immune-mediated attack on self cells. The disease can be either systemic or affect a specific organ, such as the thyroid, the pancreas or the brain (Table 1). Autoimmune diseases are mostly polygenic and complex, meaning that they are controlled by multiple genetic loci and usually modified by a multitude of unknown environmental factors. However, autoimmune manifestations may also be part of genetic diseases with monogenic traits. Studies of such monogenic diseases have demonstrated the role of specific genes controlling autoimmunity. For example, APECED patients have a mutation in the *AIRE* gene, which controls the presentation of self antigens in the thymus [1]. Without AIRE, self-reactive T cells escape clonal deletion and cause organ-specific damage. IPEX patients have a mutation in the gene *FoxP3*, which results in deficiency of regulatory T cells, leading to autoimmunity [2]. Complex autoimmune diseases are not possible to link to single genes but result from an interplay between functional variants of several genes that act in concert to create an autoimmune phenotype. Just a few years ago, disappointingly few genes had been identified besides the MHC, but powerful techniques that allow genome-wide scans of common polymorphisms in human cohorts and animal models are now generating exciting results [3].

Consistent with the complexity of this group of diseases, the immunological effector mechanisms vary between different autoimmune disorders. For example, deposition of autoantibodies is a central pathogenic mechanism of systemic autoimmune diseases, such as Systemic Lupus Erythematosis (SLE) and Rheumatoid Arthritis (RA). In organ-specific autoimmune diseases, such as Type 1 diabetes (T1D) and Addison's disease, T cells are instead key players in the disease development, and autoantibodies, are considered a consequence, rather than a cause, of disease.

Specific treatments that prevent autoreactive responses in autoimmune diseases are rare. Systemic immunosuppressive treatments, for example corticosteroids, cytostatic drugs and antibodies against TNF-α are powerful and efficient in many patients [4]. However, they sometimes cause unwanted side effects and do not act in a very selective way, and more specific therapies are therefore needed. Better knowledge regarding the role of individual cell subsets in autoimmunity will help this development. In this chapter, we discuss the role of one specific innate cell type, the natural killer (NK) cell, in autoimmune diseases.

2 NK Cells as Possible Players in Autoimmunity

NK cells protect against viruses and cancer [5, 6], playing roles both as cytotoxic cells and as cells that affect other cells via cytokines [7]. NK cells are abundant in hematopoietic organs, including lymph nodes [8, 9], and are surprisingly frequent

Table 1 Examples of autoimmune diseases, the primary cells and organs that are attacked by the immune system, the most important immune players and some examples of the self antigens that are recognized

Target organ	Disease	Target cell/tissue	Suggested effector mechanisms	Examples of autoantigens (when known)
Endocrine glands	Type 1 diabetes	Pancreatic β cells	T cells, Monokines	Insulin, GAD, IC69
	Hashimoto's thyreoditis	Thyroid cells	T cells	Thyroid enzymes
	Grave's disease	Thyroid cells	Stimulatory antibodies	Receptor for TSH
	Autoimmune adrenocortical failure	Adrenal cortex	T cells	17-α and 21-hydroxylase
Gastrointestinal tract	Crohn's disease/ Ulcerative cholitis	Gut epithelium	T cells, Cytokines, Monokines	Tropomyosine isoform, other gut antigens
	Primary biliary cirrhosis	Intrahepatic bile ducts	Immune complexes, Cytokines	Mitochondrial enzymes
	Goodpasture's disease	Basal membranes in the kidney	Antibodies	Collagen type III
Nervous system	Multiple sclerosis	Myelin	T cells	Myelin basic protein, Proteolipid protein, Myeline oligodendrocyte glycoprotein
	Myasthenia gravis	Neuromuscular junctions	Blocking antibodies	Acetylcholine receptor
Blood	Autoimmune hemolytic anemia	Erythrocytes	Antibodies, Complement	I antigen, Rh antigen
	Autoimmune thrombocytopenia	Platelets	Antibodies	Platelet glycoproteins GPIIb – GPIIIa, GPIb – GPIX
Skin	Psoriasis	Skin epithelium	T cells	Skin antigens
	Vitiligo	Melanocytes	Antibodies	Melanocyte antigens
	Pemfigus Vulgaris	Skin epithelium	Antibodies, Complement	Proteins joining cells together
Eye	Sympathetic ophthalmia	Retinal cells	T cells, Antibodies	Eye proteins
Systemic	Rheumatoid arthritis	Joints	T cells, Monokines	Collagen type II, cartilage
	Systemic lupus erythematosis (SLE)	Kidney, joints, arterial walls	Immune complexes, Complement	DNA, histones, ribosomes

also in various nonhematopoietic organs such as liver, lung, uterus, thymus and intestine [10–15]. NK cells enter inflamed lymph nodes in a CXCR3-dependent fashion [16, 17], where they can influence the outcome of T cell activation via interactions with dendritic cells (DC) [8, 18]. NK cells can also be recruited across endothelium to nonlymphoid inflamed tissue where they could participate in inflammatory reactions [19, 20]. In humans, the minor population of $CD56^{bright}$ NK cells and the major population of $CD56^{dim}$ NK cells express overlapping as well as specific chemokine receptors [20] implying heterogeneity in migratory properties.

The location of NK cells in most normal organs suggests a possible role of those cells as sentinels, perhaps surveilling the body for altered cells or danger, in a similar fashion as DC. The widespread location also raises the question if the NK cell population is more heterogenous than previously thought. In fact, growing evidence supports the existence and importance of environmental niches for NK cells in vivo, with an increasing awareness of the influence of the microenvironment in shaping NK cell responses [21]. This notion is greatly relevant for the understanding of how NK cells partake in autoimmune diseases, especially organ-specific diseases.

In addition to their localization, the increasing array of identified effector functions displayed by NK cells [22] has opened up multiple possible roles in autoimmune development [23]. The effector functions of NK cells include cellular cytotoxicity mediated by perforin and granzymes, as well as production of Th1 and Th2 cytokines [24]. Via their cytokine production, NK cells are able to modulate the innate and adaptive immunity, for example, by affecting DC maturation and T cell polarization [17, 18, 25, 26]. In addition to DC, NK cells have also been shown to influence the function of several other cells in the immune system, such as regulatory T cells [27], NKT cells [28], T cells [29, 30] and macrophages [31], suggesting that NK cells may be capable of influencing the outcome of autoimmune inflammation in many different ways [22, 23].

3 Evidence for the Involvement of NK Cells in Autoimmune Diseases

3.1 Myasthenia Gravis

In an experimental model of Myasthenia Gravis (EAMG), NK cells promoted the development of the disease by influencing Th1 responses [32]. Consequently, NK cell depletion skewed the cytokine response of $CD4^+$ T cell towards Th2 dominance and reduced levels of pathogenic antibodies against the acetylcholine receptor [32]. By the use of gene deletion strains lacking NKT cells, a role for IFN-γ production in NK cells was demonstrated in an unusually rigorous way. The regulatory role of NK cells in EAMG seemed to take place during the priming phase of the disease, since depletion of NK cells during later stages of the disease had no effect. The role of NK cells in patients with Myastenia Gravis is poorly studied and is limited to analyses of NK cell frequencies, which seem to be lowered in patients with disease [33, 34].

3.2 Type 1 Diabetes

Conflicting results have been generated as to the role of NK cells in type 1 diabetes, leaving open the question as to whether or not NK cells are protective cells or accelerate disease. Poirot et al. compared two transgenic mouse strains: BDC2.5/NOD, which develop a mild insulitis but no diabetes, and BDC2.5/B6^{g7}, which rapidly develop both an aggressive form of insulitis and diabetes [35, 36]. A correlation between expression of NK genes in the infiltrating cells and aggressive insulitis was shown, as well as a higher frequency of NK cells in aggressive infiltrates compared to the mild ones. When NK cells were depleted in these models of diabetes, disease incidence was decreased, suggesting a role for NK cells in the effector phase also in these models of diabetes. However, the antibodies used to deplete NK cells also recognize other immune cells, which must be taken into consideration when interpreting the data. A similar disease-accelerating role for NK cells was found in transgenic NOD mice expressing IFN-β in the β cells [37], which led to recruitment of NK cells to the inflamed islets. Yet another example for an accelerating role of NK cells came from Flodström et al., who suggested that NK cells were pathogenic during later disease stages in a virally induced model of autoimmune diabetes [38].

Conversely, a protective role of NK cells was reported in NOD mice in which diabetes was prevented by administration of complete Freund's adjuvant (CFA) [39, 40]. CFA induced NK cell trafficking to the blood and spleen, induced IFN-γ production by NK cells and decreased activation of β cell-specific T cells. Depletion of NK cells abrogated the protective effect of CFA and addition of sorted NK cells to the depleted mice restored the protective effect. NK cell IFN-γ was recently been found to be responsible for the protective effect [40], which is surprising given results from other autoimmune models in which IFN-γ was the accelerating cytokine. It is possible that the use of CFA to induce diabetes protection, which is triggered at a location distant from the pancreas (skin), enhances a special type of immune response with immunosuppressive functions and prevention of diabetes development. More work is required to explore the interesting difference in the role of NK cell IFN-γ in diabetes pathogenesis in these various models.

Several studies have proposed a reduced NK cell activity in NOD mice [41, 42] and in patients with T1D [43]. It will be of importance to further clarify the role of this impairment for diabetes pathogenesis. It must also be remembered that few studies so far thoroughly address the numbers, phenotypes and functions of NK cells from the target organ itself, in diabetes or in any other autoimmune disease. This remains a critical development of autoimmunity studies in order to reach a better understanding of organ-specific autoimmunity. One interesting report in this aspect is a recent study where NK cells and coxackie virus were found in the pancreas of 3/6 newly diagnosed type 1 diabetic patients, suggesting involvement of NK cells in human virus-induced diabetes [44]. Another study has analyzed 29 pancreases from patients that died from recent-onset diabetes [45] and found evidence for infiltration with many cell types. NK cells were found in some islets but did not seem to be very abundant. In our ongoing work in NOD mice, NK cells are present in the islet infiltrate of prediabetic mice. As in humans, they are less abundant that T cells, but are nevertheless part of the islet infiltrate, where they may execute effector functions.

3.3 Rheumatoid Arthritis

RA is a chronic systemic autoimmune disease characterized by joint inflammation and immune-mediated cartilage destruction. Cytokine production and cytotoxicity mediated by NK cells are deficient in RA patients [46], which may be related to a macrophage activation syndrome similar to the abnormalities observed in patients with hemophagocytic lymphohistocytosis [46]. The inverse relationship between NK cells and macrophages suggest a possibility that NK cells actively regulate macrophages [31]. Furthermore, NK cells accumulate within inflammatory joint lesions [47, 48], where they could influence monocytes to differentiate into DC. Thus, NK cells are actively participating in the initiation of the joint pathology and perhaps determine the intensity of local inflammation by engaging other lymphocytes.

It is of interest that joint NK cells are phenotypically different from blood NK cells. They are almost all $CD56^{bright}$, are less likely to express KIR receptors and rely on CD96/NKG2 receptors for inhibition [47, 48]. It is unclear if this difference results from selective recruitment of a small population of blood NK cells to the joint niche [21], or from local differentiation from stem cells of a unique NK cell subset in the joint. This finding further emphasizes the need for a more in depth exploration of NK cell in the target organ of the autoimmune attack.

3.4 Multiple Sclerosis

Multiple Sclerosis (MS) has been shown to be associated with low NK cell activity [49–51]. A protective role for NK cells in humans with MS was proposed [52]. Protection was associated with secretion of Th2 cytokines such as IL-5 and IL-13 [52, 53]. Takahashi et al. suggest that those NK cells may control IFN-γ secretion in memory T cells, as depletion of NK cells in ex vivo PBMC increased IFN-γ responses in T cells after stimulation with myelin basic protein [52]. A regulatory role for NK cells was further suggested by the fact that, before disease relapse, NK cells lost their NK2 phenotype, and thus presumably their regulatory role.

Experimental autoimmune encephalomyelitis (EAE) is an animal model of MS that can be induced in susceptible strains of rats and mice. NK cells are present in the central nervous system of rats at the early stages of EAE and when NK cells were depleted, the disease was aggravated [54]. Zhang et al. found a similar protective role of NK cells during the effector phase of EAE [55]. Interestingly, NK cell depletion led to an in vivo increase in production of Th1 cytokines by CD4 T cells and an increased T cell proliferation in vitro when NK depleted spleen cells were used as antigen presenting cells. The regulatory role of NK cells in rodent models of EAE is further strengthened by a study of Smeltz et al. showing that NK cells inhibit proliferation of autoreactive T cells from DA rat in vitro [56].

3.5 Systemic Lupus Erythematosis

In SLE, which is another relapsing-remitting disease, it has been shown that low NK cell numbers in the blood are associated with relapses, while the NK cell number is restored during remission. Some studies have also shown that a low NK cell activity on a per cell basis is sustained throughout the disease cycle in these patients [49, 50, 57]. This low NK cell activity has been explained either by a genetic polymorphism in the FcγIIIR (CD16), resulting in a low avidity binding of IgG antibodies [58], or by an abnormality in the expression of the signaling adapter molecule DAP12 [59–61]. A recent paper did not replicate the low number of NK cells in SLE patients, but instead found an increased frequency of cytokine-producing $CD56^{bright}$ NK cells [62]. No functional assessments were done.

Lpr mice, harboring a mutation affecting Fas expression, display an SLE-like phenotype [63] and have low NK activity, especially in aging mice [64]. An association in time has been found between disease development and ceasing NK activity [65]. In this model, the disease process could also be accelerated or decelerated by depleting or transferring $NK1.1^+$ cells, respectively. A regulatory role of $NK1.1^+$ cells was seen in vitro, and NK cells from nude mice were also efficient in this respect, making an effect of contaminating NKT cells less likely.

3.6 Some Other Autoimmune Diseases

NK cells have been suggested to play a role also in several other autoimmune disorders. In Pemphigus Vulgaris, a blistering autoimmune disease affecting the skin and multiple mucosal membranes, an increased percentage of circulating NK cells was observed [66]. Interestingly, these NK cells showed an activated phenotype and secreted Th2-associated cytokines [66]. In another recent study, NK cells were suggested to act as antigen presenting cells, on the basis of their expression of MHC class II and the costimulatory molecule B7-H3. In cocultures, they stimulated proliferation of $CD4^+$ T cells after addition of the autoimmunogenic peptide desmoglein 3 [67].

In Sjögren's syndrome, studies on the number of NK cells in the blood have given conflicting results, ranging from no difference [68], increased frequencies [69] and decreased frequencies [70]. The latter study also showed alterations in the expression pattern of NK cell activating receptors and a higher percentage of apoptotic Annexin V positive cells compared to controls [70]. It cannot be discerned if these changes are primary or secondary changes in relation to disease. Nevertheless, they imply an altered peripheral NK cell compartment in Sjögren syndrome. In the salivary gland, less is known about the presence and function of NK cells. In an early study, it was shown that affected salivary glands from patients with Sjögren's disease lacked the antigens Leu-7 and Leu-11 (which were used as NK cell markers at the time) and lacked NK cell activity [71]. These interesting data suggest a lack of infiltrating NK cells, or a change in phenotype and function of NK cells, in the autoimmune target organ. In our studies of NK cells infiltrating in the pancreas of NOD mice, we have come to the conclusion that NK cells do infiltrate this target organ, but that they display a completely different phenotype and have a decreased function (our unpublished data), possibly consistent with the latter idea regarding NK cells in Sjögren's syndrome [71, 72].

4 Synthesis

The role of NK cells in immunoregulation needs further study. This appears especially important given the recent notion that NK cells not only patrol the blood and hematopoietic organs, but also occupy many different niches in the

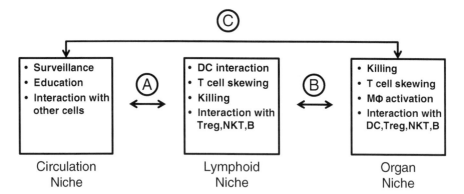

Fig. 1 Schematic drawing of NK cell niches and possible functions of NK cells in each niche. The source of NK cells in human studies is the peripheral blood. In the mouse, spleen NK cells are normally used. These niches are thought to be in equilibrium, but this is not well studied. NK cells likely migrate freely between blood and lymphoid organs (A). Traffic from organs to the blood and spleen is less clear (B, C), but is suggested by the possibility to transfer autoimmune diseases with spleen cells in diseased animals. Migration from blood and the lymphoid niche to organs (B and C, other direction) is unclear; NK cells residing in organs may derive from the blood or represent entirely different subsets. Suggested functions of NK cells in each niche are indicated

body such as the gut, liver, decidua and pancreas. Figure 1 depicts possible role of NK cells in various locations in the body (Fig. 1). Regarding the role of NK cells in autoimmunity, the current picture contains evidence for both disease-promoting and disease-preventing roles. A closer and comparative dissection of diseases models in which NK cells perform different functions will be important to identify potential differences in NK cell biology that may be responsible for these opposing roles. In humans, low NK cell numbers in the blood are frequently observed but this is also difficult to interpret, both because autoimmunity may be associated with systemic inflammation that may deplete or affect NK cells, and because it is uncertain, at least in organ-specific autoimmunity, which roles are played by circulating NK cells in disease pathogenesis.

Difficulties in understanding discrepancies in the role of NK cells in autoimmunity may also lie in our incomplete understanding of the in vivo biology of NK cells as well as the role of different functional subsets. In order to dissect the role of NK cells in autoimmune diseases, the following aspects of NK cell biology must be studied in greater detail.

4.1 Exploring NK Cell Niches in Relation to Autoimmune Disease

It is becoming increasingly clear that there exist several subsets of NK cells, with distinct phenotype and/or function and/or tissue distribution. A more thorough look at NK cells in tissues and target organs during different phases of disease must therefore be undertaken. In the target organ, NK cells may change their phenotype and function compared to when in the blood. They may even represent different subsets with different ontogenies. The balance between NK cell functions at different anatomical sites may thus be different in different autoimmune conditions and may explain why NK cells play different roles depending on which disease that is studied. This problem includes the question of how NK cell circulate and whether organ-residing NK cells and NK cells in the circulation are the same cell type or distinct subsets.

4.2 Characterizing Functional NK Cell Subsets

There certainly exists several subsets of NK cells that mediate different effector functions depending on how, where and when they may be stimulated. Just as for Th1 and Th2 T cells, such subset compositions are likely to be partly genetically determined but may also be a consequence of the local microenvironment to which they home and become activated. An important question will be how functionally distinct NK cell subsets overlap with subsets distinguished by their expression of activating and inhibitory MHC class I-specific receptors, the balance of which is important in autoimmunity.

4.3 Study the Genetic Control of Autoimmunity with NK Cell Eyes

The genetic analyses of autoimmunity may reveal NK cell-related genes. A particular gene family, killer immunoglobulin-like (KIR) receptors, show interesting polymorphisms that have been associated to disease suseceptibility. Also ligands for other NK cell receptors are polymorphic with some variants being associated with disease, suggesting a genetic control of autoimmunity at the level of both NK cell receptors and their ligands. Studying these genes in more detail should give novel insights into possible roles for a balance between activating and inhibitory signals in autoimmunity. When genetic profiles for different functional subsets are generated, new markers suitable for disease predictions could potentially be developed.

4.4 A Kinetic look of NK Cell Responses in Different Diseases

Cells and molecules of the immune system may play varying and sometime opposing roles at different ages and different stages of disease. For example, several cytokines, TNF-α being one example, play opposite roles at early and late stages of autoimmunity. Thus, differences in time kinetics of critical NK-related events in induction, progression and final stages of different autoimmune diseases may hold clues as to why NK cells could play both protective and disease-promoting roles in different diseases.

4.5 Intensify Studies of Human Autoimmunity

Much of the studies on NK cells in autoimmunity are performed in animal models. Although such model systems have proven to be very useful tools, several differences exist relative to humans that need further exploration. One example is the apparently different patterns of lymphocyte infiltration in the pancreas preceding diabetes. Greater efforts to perform studies on human subjects will therefore be of utmost importance to fully understand the human autoimmune pathogenesis and ultimately be able to design new treatments.

References

1. Villasenor J, Benoist C, Mathis D (2005) AIRE and APECED: molecular insights into an autoimmune disease. Immunol Rev 204:156–164
2. van der Vliet HJ, Nieuwenhuis EE (2007) IPEX as a result of mutations in FOXP3. Clin Dev Immunol 2007:89017

3. Gregersen PK, Olsson LM (2009) Recent advances in the genetics of autoimmune disease. Annu Rev Immunol 27:363–391
4. Wong M, Ziring D, Korin Y et al (2008) TNFalpha blockade in human diseases: mechanisms and future directions. Clin Immunol 126:121–136
5. Yokoyama WM, Kim S, French AR (2004) The dynamic life of natural killer cells. Annu Rev Immunol 22:405–429
6. Smyth MJ, Hayakawa Y, Takeda K et al (2002) New aspects of natural-killer-cell surveillance and therapy of cancer. Nat Rev Cancer 2:850–861
7. Raulet DH (2004) Interplay of natural killer cells and their receptors with the adaptive immune response. Nat Immunol 5:996–1002
8. Fehniger TA, Cooper MA, Nuovo GJ et al (2003) CD56bright natural killer cells are present in human lymph nodes and are activated by T cell-derived IL-2: a potential new link between adaptive and innate immunity. Blood 101:3052–3057
9. Ferlazzo G, Thomas D, Lin SL et al (2004) The abundant NK cells in human secondary lymphoid tissues require activation to express killer cell Ig-like receptors and become cytolytic. J Immunol 172:1455–1462
10. Basse PH, Hokland P, Gundersen HJ et al (1992) Enumeration of organ-associated natural killer cells in mice: application of a new stereological method. APMIS 100:202–208
11. Stein-Streilein J, Bennett M, Mann D et al (1983) Natural killer cells in mouse lung: surface phenotype, target preference, and response to local influenza virus infection. J Immunol 131:2699–2704
12. Wiltrout RH, Mathieson BJ, Talmadge JE et al (1984) Augmentation of organ-associated natural killer activity by biological response modifiers. Isolation and characterization of large granular lymphocytes from the liver. J Exp Med 160:1431–1449
13. Moffett A, Regan L, Braude P (2004) Natural killer cells, miscarriage, and infertility. BMJ 329:1283–1285
14. Vosshenrich CA, Garcia-Ojeda ME, Samson-Villeger SI et al (2006) A thymic pathway of mouse natural killer cell development characterized by expression of GATA-3 and CD127. Nat Immunol 7:1217–1224
15. Vivier E, Spits H, Cupedo T (2009) Interleukin-22-producing innate immune cells: new players in mucosal immunity and tissue repair? Nat Rev Immunol 9:229–234
16. Rot A, von Andrian UH (2004) Chemokines in innate and adaptive host defense: basic chemokinese grammar for immune cells. Annu Rev Immunol 22:891–928
17. Martin-Fontecha A, Thomsen LL, Brett S et al (2004) Induced recruitment of NK cells to lymph nodes provides IFN-gamma for T(H)1 priming. Nat Immunol 5:1260–1265
18. Ferlazzo G, Pack M, Thomas D et al (2004) Distinct roles of IL-12 and IL-15 in human natural killer cell activation by dendritic cells from secondary lymphoid organs. Proc Natl Acad Sci USA 101:16606–16611
19. Fogler WE, Volker K, McCormick KL et al (1996) NK cell infiltration into lung, liver, and subcutaneous B16 melanoma is mediated by VCAM-1/VLA-4 interaction. J Immunol 156:4707–4714
20. Campbell JJ, Qin S, Unutmaz D et al (2001) Unique subpopulations of CD56+ NK and NK-T peripheral blood lymphocytes identified by chemokine receptor expression repertoire. J Immunol 166:6477–6482
21. Di Santo JP (2008) Natural killer cells: diversity in search of a niche. Nat Immunol 9:473–475
22. Vivier E, Tomasello E, Baratin M et al (2008) Functions of natural killer cells. Nat Immunol 9:503–510
23. Johansson S, Berg L, Hall H et al (2005) NK cells: elusive players in autoimmunity. Trends Immunol 26:613–618
24. Bryceson YT, Long EO (2008) Line of attack: NK cell specificity and integration of signals. Curr Opin Immunol 20:344–352

25. Ferlazzo G, Tsang ML, Moretta L et al (2002) Human dendritic cells activate resting natural killer (NK) cells and are recognized via the NKp30 receptor by activated NK cells. J Exp Med 195:343–351
26. Gerosa F, Baldani-Guerra B, Nisii C et al (2002) Reciprocal activating interaction between natural killer cells and dendritic cells. J Exp Med 195:327–333
27. Ghiringhelli F, Menard C, Martin F et al (2006) The role of regulatory T cells in the control of natural killer cells: relevance during tumor progression. Immunol Rev 214:229–238
28. Carnaud C, Lee D, Donnars O et al (1999) Cutting edge: cross-talk between cells of the innate immune system: NKT cells rapidly activate NK cells. J Immunol 163:4647–4650
29. Assarsson E, Kambayashi T, Schatzle JD et al (2004) NK cells stimulate proliferation of T and NK cells through 2B4/CD48 interactions. J Immunol 173:174–180
30. Shanker A, Verdeil G, Buferne M et al (2007) CD8 T cell help for innate antitumor immunity. J Immunol 179:6651–6662
31. Nedvetzki S, Sowinski S, Eagle RA et al (2007) Reciprocal regulation of human natural killer cells and macrophages associated with distinct immune synapses. Blood 109:3776–3785
32. Shi FD, Wang HB, Li H et al (2000) Natural killer cells determine the outcome of B cell-mediated autoimmunity. Nat Immunol 1:245–251
33. Suzuki Y, Onodera H, Tago H et al (2005) Altered populations of natural killer cell and natural killer T cell subclasses in myasthenia gravis. J Neuroimmunol 167:186–189
34. Nguyen S, Morel V, Le Garff-Tavernier M et al (2006) Persistence of CD16+/CD56-/2B4+ natural killer cells: a highly dysfunctional NK subset expanded in ocular myasthenia gravis. J Neuroimmunol 179:117–125
35. Poirot L, Benoist C, Mathis D (2004) Natural killer cells distinguish innocuous and destructive forms of pancreatic islet autoimmunity. Proc Natl Acad Sci U S A 101:8102–8107
36. Gonzalez A, Katz JD, Mattei MG et al (1997) Genetic control of diabetes progression. Immunity 7:873–883
37. Alba A, Planas R, Clemente X et al (2008) Natural killer cells are required for accelerated type 1 diabetes driven by interferon-beta. Clin Exp Immunol 151:467–475
38. Flodström M, Maday A, Balakrishna D et al (2002) Target cell defense prevents the development of diabetes after viral infection. Nat Immunol 3:373–382
39. Lee IF, Qin H, Trudeau J et al (2004) Regulation of autoimmune diabetes by complete Freund's adjuvant is mediated by NK cells. J Immunol 172:937–942
40. Lee IF, Qin H, Priatel JJ et al (2008) Critical role for IFN-gamma in natural killer cell-mediated protection from diabetes. Eur J Immunol 38:82–89
41. Poulton LD, Smyth MJ, Hawke CG et al (2001) Cytometric and functional analyses of NK and NKT cell deficiencies in NOD mice. Int Immunol 13:887–896
42. Johansson SE, Hall H, Bjorklund J et al (2004) Broadly impaired NK cell function in non-obese diabetic mice is partially restored by NK cell activation in vivo and by IL-12/IL-18 in vitro. Int Immunol 16:1–11
43. Rodacki M, Svoren B, Butty V et al (2007) Altered natural killer cells in type 1 diabetic patients. Diabetes 56:177–185
44. Dotta F, Censini S, van Halteren AG et al (2007) Coxsackie B4 virus infection of beta cells and natural killer cell insulitis in recent-onset type 1 diabetic patients. Proc Natl Acad Sci USA 104:5115–5120
45. Willcox A, Richardson SJ, Bone AJ et al (2009) Analysis of islet inflammation in human type 1 diabetes. Clin Exp Immunol 155:173–181
46. Villanueva J, Lee S, Giannini EH et al (2005) Natural killer cell dysfunction is a distinguishing feature of systemic onset juvenile rheumatoid arthritis and macrophage activation syndrome. Arthritis Res Ther 7:R30–R37
47. Dalbeth N, Gundle R, Davies RJ et al (2004) CD56bright NK cells are enriched at inflammatory sites and can engage with monocytes in a reciprocal program of activation. J Immunol 173:6418–6426

48. de Matos CT, Berg L, Michaelsson J et al (2007) Activating and inhibitory receptors on synovial fluid natural killer cells of arthritis patients: role of CD94/NKG2A in control of cytokine secretion. Immunology 122:291–301
49. Erkeller-Yusel F, Hulstaart F, Hannet I et al (1993) Lymphocyte subsets in a large cohort of patients with systemic lupus erythematosus. Lupus 2:227–231
50. Yabuhara A, Yang FC, Nakazawa T et al (1996) A killing defect of natural killer cells as an underlying immunologic abnormality in childhood systemic lupus erythematosus. J Rheumatol 23:171–177
51. Loza MJ, Zamai L, Azzoni L et al (2002) Expression of type 1 (interferon gamma) and type 2 (interleukin-13, interleukin-5) cytokines at distinct stages of natural killer cell differentiation from progenitor cells. Blood 99:1273–1281
52. Takahashi K, Aranami T, Endoh M et al (2004) The regulatory role of natural killer cells in multiple sclerosis. Brain 127:1917–1927
53. Takahashi K, Miyake S, Kondo T et al (2001) Natural killer type 2 bias in remission of multiple sclerosis. J Clin Invest 107:R23–29
54. Matsumoto Y, Kohyama K, Aikawa Y et al (1998) Role of natural killer cells and TCR gamma delta T cells in acute autoimmune encephalomyelitis. Eur J Immunol 28:1681–1688
55. Zhang B, Yamamura T, Kondo T et al (1997) Regulation of experimental autoimmune encephalomyelitis by natural killer (NK) cells. J Exp Med 186:1677–1687
56. Smeltz RB, Wolf NA, Swanborg RH (1999) Inhibition of autoimmune T cell responses in the DA rat by bone marrow-derived NK cells in vitro: implications for autoimmunity. J Immunol 163:1390–1397
57. Green MR, Kennell AS, Larche MJ et al (2005) Natural killer cell activity in families of patients with systemic lupus erythematosus: demonstration of a killing defect in patients. Clin Exp Immunol 141:165–173
58. Wu J, Edberg JC, Redecha PB et al (1997) A novel polymorphism of FcgammaRIIIa (CD16) alters receptor function and predisposes to autoimmune disease. J Clin Invest 100:1059–1070
59. Toyabe SI, Kaneko U, Uchiyama M (2004) Decreased DAP12 expression in natural killer lymphocytes from patients with systemic lupus erythematosus is associated with increased transcript mutations. J Autoimmun 23:371–378
60. Djeu JY, Jiang K, Wei S (2002) A view to a kill: signals triggering cytotoxicity. Clin Cancer Res 8:636–640
61. Colonna M (2003) TREMs in the immune system and beyond. Nat Rev Immunol 3:445–453
62. Schepis D, Gunnarsson I, Eloranta ML et al (2009) Increased proportion of CD56bright natural killer cells in active and inactive systemic lupus erythematosus. Immunology 126:140–146
63. Watanabe-Fukunaga R, Brannan CI, Copeland NG et al (1992) Lymphoproliferation disorder in mice explained by defects in Fas antigen that mediates apoptosis. Nature 356:314–317
64. Scribner CL, Steinberg AD (1988) The role of splenic colony-forming units in autoimmune disease. Clin Immunol Immunopathol 49:133–142
65. Takeda K, Dennert G (1993) The development of autoimmunity in C57BL/6 lpr mice correlates with the disappearance of natural killer type 1-positive cells: evidence for their suppressive action on bone marrow stem cell proliferation, B cell immunoglobulin secretion, and autoimmune symptoms. J Exp Med 177:155–164
66. Takahashi H, Amagai M, Tanikawa A et al (2007) T helper type 2-biased natural killer cell phenotype in patients with pemphigus vulgaris. J Invest Dermatol 127:324–330
67. Stern JN, Keskin DB, Barteneva N et al (2008) Possible role of natural killer cells in pemphigus vulgaris – preliminary observations. Clin Exp Immunol 152:472–481
68. Markeljevic J, Marusic M, Uzarevic B et al (1991) Natural killer cell number and activity in remission phase of systemic connective tissue diseases. J Clin Lab Immunol 35:133–138
69. Szodoray P, Gal I, Barath S et al (2008) Immunological alterations in newly diagnosed primary Sjogren's syndrome characterized by skewed peripheral T-cell subsets and inflammatory cytokines. Scand J Rheumatol 37:205–212

70. Izumi Y, Ida H, Huang M et al (2006) Characterization of peripheral natural killer cells in primary Sjogren's syndrome: impaired NK cell activity and low NK cell number. J Lab Clin Med 147:242–249
71. Fox RI, Hugli TE, Lanier LL et al (1985) Salivary gland lymphocytes in primary Sjogren's syndrome lack lymphocyte subsets defined by Leu-7 and Leu-11 antigens. J Immunol 135:207–214
72. Schrambach S, Ardizzone M, Leymarie V et al (2007) In vivo expression pattern of MICA and MICB and its relevance to auto-immunity and cancer. PLoS ONE 2:e518

NK Cells and Allergy

Tatiana Michel, Maud Thérésine, Aurélie Poli, François Hentges, and Jacques Zimmer

Abstract NK cells are an important component of the innate immune response. They play a significant role in defense against tumor cells and viral infections, but also during human pregnancy. Until recently, few things were known about a possible role of NK cells in allergic diseases. Studies in animal models of asthma but also of peripheral blood from allergic patients suggest the possibility that NK cells also have a key function in inflammatory airway diseases. This hypothesis will be examined in this chapter.

1 Introduction

Immediate hypersensitivity is the most widespread immune disease in humans. In some countries, allergies represent a prevalent chronic health problem among individuals over 15 years of age. In the United States, allergic diseases affect more than 50 million people [1].

Common allergic reactions include eczema, hives, hay fever, asthma, food allergies, and reactions to the venom of stinging insects such as wasps and bees. Environmental substances called allergens, especially of animal, mould, pollen and mite, can trigger allergic reactions. These reactions are dependent upon previous exposure and sensitization to specific allergens that ultimately result in the development of antigen-specific IgE antibodies. Allergenic cross-linking of IgE bound to the surface of mast cells and basophils will release numerous preformed or secondary mediators from these cells. These mediators induce constriction of smooth

---

T. Michel (✉), M. Thérésine, A. Poli, F. Hentges and J. Zimmer
Laboratory of Immunogenetics and Allergology, Centre de Recherche Public (CRP)-Santé, 84 Val Fleuri, L-1526, Luxembourg
e-mail: tatiana.michel@crp-sante.lu

muscle in the airways and edema in the tissues. They increase mucosal secretions and stimulation of nerves. Macrophages, neutrophils, eosinophils and lymphocytes also play a significant role. The T lymphocyte response in allergic diseases is characterized by a Th2-biased respiratory response. Allergen exposure in sensitized individuals leads to chronic airway inflammation and asthma. The asthmatic airways are characterized by infiltrates of eosinophils and T lymphocytes secreting type 2 cytokines: interleukin (IL)-4, IL-5 and IL-13 [2]. The production of IL-5 results in maturation and enhances the recruitment of eosinophils in airways mucosa. In certain sensitized people, acute allergic triggers may result in life-threatening anaphylactic reactions. To identify significant therapeutic targets in these frequently disabling diseases, great efforts have been made in these last years to characterize the regulatory events involved in the balance of Th1/Th2 cytokines. The factors responsible for initial induction of the Th2-like response in vivo remain incompletely understood.

Recent data suggest that NK cells play a role in the initiation of allergen-specific T cell response and hence, in the subsequent development of allergic airway inflammation [3, 4]. The early appearance of NK cells at the site of immunization in animal models and their capacity to produce a panel of cytokines suggest that these cells play a critical role at several steps in the development of the acquired immune response to allergens. NK cells influence the adaptive immune response by the secretion of cytokines, but also through other mechanisms like lysing dendritic cells (DC) or macrophages, and thereby modulating antigen presentation. The NK cell cytokine production rather than cytolytic activity contributes to the resistance against infectious agents. NK cells can rapidly produce IFNγ. In addition, they produce a variety of other immunoregulatory mediators, like several members of the IL-family: IL-1, IL-3, IL-5, IL-8, and IL-10, as well as TGFβ, TNFα, TNFβ, GM-CSF, macrophage inflammatory protein (MIP)-1α, [5–7].

NK cells are divided in two functionally different subsets NK1 and NK2, analogous to T cell subsets Th1 and Th2. NK1 cells produce IFN-γ and IL-10, whereas NK2 cells produce IL-4, IL-5 and IL-13 [8]. NK cells are influenced by the nature of cytokines present in the microenvironment. NK cells stimulated in vitro by IL-12 produce increased levels of IFNγ and decreased levels of IL-4. On the contrary, the presence of IL-4 in the medium inhibits IFNγ and increases the secretion of IL-13 [9]. However, NK1 and NK2 subsets show similar cytotoxicity to K562 cells [10]. It is suggested that depending on the cytokine profile of individuals, NK cells may display different inflammatory properties.

In human, NK cells are subdivided in two major functional subtypes according to the level of CD56 expression ($CD56^{bright}$ and $CD56^{dim}$) [11]. $CD56^{bright}$ NK cells are capable to produce large quantities of type 1 and type 2 cytokines. Therefore, they are considered immunoregulatory cells, whereas $CD56^{dim}$ NK cells produce less cytokines but are cytotoxic.

NK cells are present in human lung interstitium, and their profile is comparable to that of NK cells found in peripheral blood. The involvement of these cells in pulmonary immunity [12] and also in allergic diseases is a matter of great interest.

2 NK Cells Involved in Human Allergy

Timonen et al. were among the first to describe the role of NK cells in asthma. They demonstrated that patients with atopic asthma had significantly stronger NK cell cytotoxicity activity against the leukemic cell line K562 [13]. Since then, it has been confirmed that patients with asthma (for instance seasonal asthma due to grass pollen allergy) show increased numbers of NK cells and elevated NK cell cytotoxicity in peripheral blood compared to healthy persons [14–16]. Recent data suggest that cells of innate immunity like NK cells are important regulators of the allergen-specific T cell response and possibly enhance allergic airway inflammation. The increased NK cell activity in asthma could reflect a predisposition of individuals with high NK cell activity to develop exaggerated T cell responses to allergens and hence to be at risk of developing asthma.

NK cells from healthy individuals produce more type 1 cytokines compared to those of atopic asthmatic patients whose NK cells secrete more type 2 cytokines. The ratio of IL-4+CD56+ NK2 cells in the blood of asthmatic patients is higher than that in healthy individuals, and the level of IFNγ+CD56+ NK1 cells is lower in these asthmatic patients [17, 18]. Furthermore, STAT6 (signal transducer and activator of transcription 6), a key regulator for type 2 cytokines [19], is constitutively activated in NK cells from asthmatic patients [17]. Aktas et al. have shown that in polyallergic atopic dermatitis (AD) patients, (the patients are polyallergic to food and/or airborne allergens, some of them show allergic rhinoconjunctivitis and asthma), NK cells produce higher amounts of IL-4, IL-5, IL-13 and IFNγ than do NK cells from healthy individuals [10]. By increasing these cytokines, the NK cell might influence the overall inflammatory response in allergy. Furthermore, these author show that the expression of killer inhibitory receptors and costimulatory molecules in NK cells differs between polyallergic and healthy subjects. But different phenotypes of asthma or allergy were associated with receptor expression. For example, the CD95 (fas) receptor on NK cells of AD patients is expressed at the same level than in healthy donors [10]. On the contrary, CD95 is significantly more expressed on NK cells from asthmatics compared to NK cells from normal persons in the study of Wingett et al. [20]. The consequence of the CD95 over-expression may be a signal of inducing apoptosis by cell-to-cell contact between NK cells and T cells. The expression of CD16 and CD56 on peripheral blood NK cells demonstrates heterogeneity in immune profile in patients with polyallergic atopic diseases. Lower percentages of CD56+ CD16+ cells are found in patients with AD compared to patients with allergic rhinitis and healthy controls [21]. Krejsek et al. showed that the number of CD56+ NK cells is higher in patients with difficult-to-control asthma in comparison to healthy controls [22]. Recently, a significant decrease in NK cell number was observed in asthmatic patients with acute asthma episodes in comparison to patients with stable asthma or healthy subjects [23]. A controversial situation is described by Scordamaglia et al. [18], who showed that peripheral blood lymphocytes from allergic rhinitis patients or healthy controls have similar percentages of CD56+ NK cells. But after analysis of the NK cell subsets, they found that

$CD56^{bright}$ $CD16^{dim}$ IFNγ NK cells are reduced in patients with allergy. $CD56^{bright}$ $CD16^{dim}$ NK cells give rise to a complex NK/DC cross-talk that would help Th1 responses. A reduction of this NK cell subset in most allergic patients may therefore affect the capacity to produce IFNγ after interaction with DC, and these NK cells are less efficient in promoting DC maturation and/or in killing immature DC. These data confirm a role for NK1/NK2 imbalance in allergy.

Different NK cell subsets and different NK cell functions may be affected in allergy. One hypothesis is that the particular cytokine milieu characteristic of patients with allergy may play a role [17, 18]. A hypothesis from Korsgren [24] is that individuals with high NK cell activity have a predisposition to develop high T cell responses to antigen and consequently an asthma disease. It could be also correlated with the elevated level of total serum IgE [25].

3 NK Cells in Animal Models of Asthma

To develop new therapeutic strategies and elucidate the pathogenesis of allergic asthma, animal models help to investigate allergic airway inflammatory mechanisms. Although these models are not perfect replicas of clinical asthma, they have exposed a number of clinical and pathological correlates to the human disease. These include a strong Th2-driven inflammation involving lymphocytes, eosinophils and mast cells and the creation of a proinflammatory pulmonary milieu involving cytokines such as IL-4, IL-5 and IL-13, and growth factors.

Bogen et al. [26] have demonstrated that after a subcutaneous administration of the protein antigen ovalbumin (OVA) in mouse, the first cells to appear at the site of immunization were IFNγ-producing-NK1.1+ cells (NK and NKT cells). Moreover in OVA-sensitized and challenged rats, NK cells are increased in lung parenchyma [27].

In a mouse model of allergic peritonitis, after immunization and challenge with a short ragweed antigen extract, eosinophils were analyzed in peritoneal lavage fluid. Intracellular staining shows IL-5 production by NK cells, which contributes to eosinophil infiltration. IL-5 regulates the production, activation of eosinophils, their survival and their degranulation. Depletion of NK1.1 cells results in reduced peritoneal eosinophilia and a complete loss of IL-5 producing NK cells. This strongly suggests that significant amounts of the IL-5 detectable at the site of inflammation are derived from NK cells [28].

Depletion of NK1.1 cells before immunization inhibits pulmonary eosinophil and CD3+ T cell infiltration and increases the level of IL-4, IL-5, IL-12 in bronchoalveolar lavage fluid in a murine model of asthma. That this is mediated by NK cells and not by NKT cells is suggested by the fact that NKT cell-deficient CD1d1 mutant mice developed airway disease like wild-type mice [24].

On the contrary, Wang et al. [29] investigated the role of NK and NKT cells in the initial induction of immune responses after immunization with OVA allergen in rat, and they did not observe any effect of NK1.1+ cell depletion on OVA-specific

IgG2a and IgE levels or on cytokine production. But compared to Korsgen et al. [24], the NK1.1+ cell depletion was not there during the early stage of immunization protocol, which could explain the different results. This suggests that NK cells play a role during the immunization stage and not during the challenge period. NK cells might influence the initial antigen presentation to T cells. Hence the role of NK cells in the regulation of allergic airway disease remains controversial. Recently, Matsubara et al. [30] studied the effect of exogenous IL-2 combined with IL-18 on asthma disease in a mouse model. The injection of these cytokines just before each of OVA allergen challenges, suppressed allergen-induced AHR (airway hyperresponsiveness), airway eosinophilia and goblet cell metaplasia. However, combined IL-2/IL-18 administration increased the number of NK cells and IFNγ secretion in the asthmatic mouse lung. These different results suggest that NK cells could have different roles on airway inflammation depending on the nature of cytokines present in the environment.

Doganci et al. [31] have shown the effect induced by local blockade of the common IL-2R/IL-15R β-chain (CD122) in a Balb/c murine model of asthma. IL-2 and IL-15 influence the survival of NK cells thereby CD122 is important for signal transduction since it acts as a common chain for IL-2R and IL-15R. This group has observed that CD49b+ cells (NK and NKT cells) represented the majority of cells in the lung expressing the CD122. The induction of asthma did not influence CD122 expression on these cells but after an anti-IL-2Rβ-chain antibody treatment, the CD49b+ cell number was reduced in the lung.

4 The Environment Influences NK Cells in Asthma

Genetic and environmental factors influence susceptibility to asthma. When rats are postnatally exposed to different stress factors, they present different clinical patterns of asthma symptoms. Krutschinski et al. [32] were interested in the maternal factors in early postnatal period as being relevant for the course of asthma in adulthood. A rat model of asthma was used after maternal separation, deprivation or short separation and the development of endocrine and immune regulation was analyzed after OVA sensitization and challenge. They have shown that ex vivo NK cell cytotoxicity is increased in the handling stimulation (deprivation or short separation) compared to prolonged maternal separation or undisturbed rats. Thereby specific postnatal life events lead to a modification of allergic airway inflammation in adult individuals.

Subjects with asthma or allergic rhinitis exhibit increased pathophysiological effects after a rhinovirus infection compared to nonatopic, nonasthmatic patients. NK cells are implicated in the elimination of virus-infected cells and the modulation of adaptive immunity against viruses. During airway inflammation the environment is rich in type 2 cytokines, consequently the NK1 function for antiviral activity might be inhibited. Moreover NK2 cytokines produced by NK cells in response to viral infection could be responsible for increasing the allergic inflammation [33].

Chlamydial infection before OVA allergen challenge in a mouse model of asthma caused an inhibition of the Th2 cytokine response and an increase of the IFNγ production by NK cells. In that case, depletion of NK cells partially abolished the inhibitory effect of chlamydial infection. NK-depleted mice infected by *Chlamydia*, showed higher levels of eosinophilia and mucus production after asthma induction than infected mice without NK cell depletion [34]. This result confirms that NK cells play a major role in airway eosinophilia and mucus overproduction caused by allergen exposure.

5 Conclusion

NK cells, components of the innate immune response, could play an important role during the sensitization phase of an allergen-specific adaptive immune response. How and whether NK cells are activated by allergens themselves or by other factors like viral infection remains to be established. By secretion of the cytokines IFNγ or IL-5 [28], NK cells could influence the antigen presenting cells, DC maturation [18] and could regulate the allergic eosinophilic airway disease (Fig. 1). The NK cells could also play a role by cytolytic mechanisms, by lysing of DC or macrophages and thereby influence the T cell response.

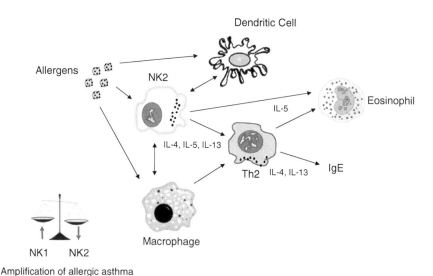

Fig. 1 Possible roles of NK cells in asthmatic airway inflammation. In asthmatic patients, the NK1/NK2 balance is modified in favor of an NK2 response. They present lower levels of IFNγ+NK1 cells and lower interactions with DC compared to healthy controls. One hypothesis is that individuals with high NK cell activity have a predisposition to develop high T cell responses to antigen and therefore to have a risk to induce asthma

References

1. Gergen PJ, Turkeltaub PC, Kovar MG (1987) The prevalence of allergic skin test reactivity to eight common aeroallergens in the U.S. population: results from the second National Health and Nutrition Examination Survey. J Allergy Clin Immunol 80:669–679
2. Robinson DS et al (1992) Predominant TH2-like bronchoalveolar T-lymphocyte population in atopic asthma. N Engl J Med 326:298–304
3. Orange JS, Ballas ZK (2006) Natural killer cells in human health and disease. Clin Immunol 118:1–10
4. Korsgren M (2002) NK cells and asthma. Curr Pharm Des 8:1871–1876
5. Biron CA (1997) Activation and function of natural killer cell responses during viral infections. Curr Opin Immunol 9:24–34
6. Scharton-Kersten TM, Sher A (1997) Role of natural killer cells in innate resistance to protozoan infections. Curr Opin Immunol 9:44–51
7. Warren HS, Kinnear BF, Phillips JH, Lanier LL (1995) Production of IL-5 by human NK cells and regulation of IL-5 secretion by IL-4, IL-10, and IL-12. J Immunol 154:5144–5152
8. Kimura MY, Nakayama T (2005) Differentiation of NK1 and NK2 cells. Crit Rev Immunol 25:361–374
9. Erten G, Aktas E, Deniz G (2008) Natural killer cells in allergic inflammation. Chem Immunol Allergy 94:48–57
10. Aktas E et al (2005) Different natural killer (NK) receptor expression and immunoglobulin E (IgE) regulation by NK1 and NK2 cells. Clin Exp Immunol 140:301–309
11. Cooper MA, Fehniger TA, Caligiuri MA (2001) The biology of human natural killer-cell subsets. Trends Immunol 22:633–640
12. Weissler JC, Nicod LP, Lipscomb MF, Toews GB (1987) Natural killer cell function in human lung is compartmentalized. Am Rev Respir Dis 135:941–949
13. Timonen T, Stenius-Aarniala B (1985) Natural killer cell activity in asthma. Clin Exp Immunol 59:85–90
14. Jira M et al (1988) Natural killer and interleukin-2 induced cytotoxicity in asthmatics. I. Effect of acute antigen-specific challenge. Allergy 43:294–298
15. Lin SJ et al (2003) Decreased intercellular adhesion molecule-1 (CD54) and L-selectin (CD62L) expression on peripheral blood natural killer cells in asthmatic children with acute exacerbation. Allergy 58:67–71
16. Di Lorenzo G et al (2001) Effects of in vitro treatment with fluticasone propionate on natural killer and lymphokine-induced killer activity in asthmatic and healthy individuals. Allergy 56:323–327
17. Wei H et al (2005) Involvement of human natural killer cells in asthma pathogenesis: natural killer 2 cells in type 2 cytokine predominance. J Allergy Clin Immunol 115:841–847
18. Scordamaglia F et al (2008) Perturbations of natural killer cell regulatory functions in respiratory allergic diseases. J Allergy Clin Immunol 121:479–485
19. Katsumoto T et al (2004) STAT6-dependent differentiation and production of IL-5 and IL-13 in murine NK2 cells. J Immunol 173:4967–4975
20. Wingett D, Nielson CP (2003) Divergence in NK cell and cyclic AMP regulation of T cell CD40L expression in asthmatic subjects. J Leukocyt Biol 74:531–541
21. Wehrmann W et al (1990) Selective alterations in natural killer cell subsets in patients with atopic dermatitis. Int Arch Allergy Appl Immunol 92:318–322
22. Krejsek J et al (1998) Decreased peripheral blood gamma delta T cells in patients with bronchial asthma. Allergy 53:73–77
23. Liu F, Luo YL, Wu YX, Kong QY (2008) Peripheral natural killer cell counting and its clinical significance blood in patients with bronchial asthma. Nan Fang Yi Ke Da Xue Xue Bao 28:780–782

24. Korsgren M et al (1999) Natural killer cells determine development of allergen-induced eosinophilic airway inflammation in mice. J Exp Med 189:553–562
25. Kusaka Y et al (1997) Association of natural killer cell activity with serum IgE. Int Arch Allergy Immunol 112:331–335
26. Bogen SA, Fogelman I, Abbas AK (1993) Analysis of IL-2, IL-4, and IFN-gamma-producing cells in situ during immune responses to protein antigens. J Immunol 150:4197–4205
27. Schuster M, Tschernig T, Krug N, Pabst R (2000) Lymphocytes migrate from the blood into the bronchoalveolar lavage and lung parenchyma in the asthma model of the brown Norway rat. Am J Respir Crit Care Med 161:558–566
28. Walker C, Checkel J, Cammisuli S, Leibson PJ, Gleich GJ (1998) IL-5 production by NK cells contributes to eosinophil infiltration in a mouse model of allergic inflammation. J Immunol 161:1962–1969
29. Wang M, Ellison CA, Gartner JG, HayGlass KT (1998) Natural killer cell depletion fails to influence initial CD4 T cell commitment in vivo in exogenous antigen-stimulated cytokine and antibody responses. J Immunol 160:1098–1105
30. Matsubara S et al (2007) IL-2 and IL-18 attenuation of airway hyperresponsiveness requires STAT4, IFN-gamma, and natural killer cells. Am J Respir Cell Mol Biol 36:324–332
31. Doganci A et al (2008) IL-2 receptor beta-chain signaling controls immunosuppressive CD4+ T cells in the draining lymph nodes and lung during allergic airway inflammation in vivo. J Immunol 181:1917–1926
32. Kruschinski C et al (2008) Postnatal life events affect the severity of asthmatic airway inflammation in the adult rat. J Immunol 180:3919–3925
33. Message SD, Johnston SL (2001) The immunology of virus infection in asthma. Eur Respir J 18:1013–1025
34. Han X et al (2008) NK cells contribute to intracellular bacterial infection-mediated inhibition of allergic responses. J Immunol 180:4621–4628

Natural Killer Cells and Their Role in Hematopoietic Stem Cell Transplantation

Deborah L.S. Goetz and William J. Murphy

Abstract Hematopoietic stem cell transplantation (HSCT) has been used for the treatment of malignant and nonmalignant disorders, including leukemias, lymphomas, renal cell carcinomas and melanomas. Limitations of allogeneic HSCT are numerous and include a dearth of suitable donors leading to graft versus host disease (GvHD), graft rejection, immune deficiency and relapse when used for cancer. Despite the challenges, the goal of allogeneic HSCT for cancer patients is to eliminate the tumor through enhancing the antitumor effects of the donor cells. Natural killer (NK) cells are innate effector cells capable of mediating MHC unrestricted killing of neoplastic and virally infected cells. The use of NK cells in HSCT using preclinical models has been found to increase antitumor effects without increasing GvHD. NK cells have also been found to enhance reconstitution of the donor hematopoietic stem cells. This chapter will highlight current research on NK cells used in HSCT including the role that alloreactive NK cells play in HSCT, NK reconstitution following HSCT, NK involvement in graft versus leukemia (GvL) and GvHD, and NK adoptive immunotherapy.

1 Introduction

Hematopoietic stem cell transplantation (HSCT) has been used for the treatment of many types of malignant and nonmalignant disorders, including leukemias, lymphomas, renal cell carcinomas, melanomas as well as thalassemia and other primary

---

D.L.S. Goetz
Whittemore Peterson Institute for Neuro Immune Diseases, University of Nevada, Reno, Nevada, USA

W.J. Murphy (✉)
Department of Dermatology, Sacramento, University of California, Davis School of Medicine, California, USA
e-mail: william.murphy@ucdmc.ucdavis.edu

immunodeficiency diseases. HSCT has also been applied to solid tumors such as breast cancer [1–4]. The use of allogeneic HSCT is limited, however, due to several significant issues. The search for a suitable donor can take months, allowing time for disease progression. Conditioning of the recipient with cytoreductive therapies leaves the recipient immunocompromised and subject to opportunistic infections. Graft rejection, relapse from the original cancer, and the specter of acute or chronic Graft versus host disease (GvHD) pose serious complications for the patient [5–9]. Despite these challenges, the goal of allogeneic HSCT for cancer patients is to eliminate the tumor through enhancing the antitumor effects of the donor cells while reducing the severity of GvHD. Adoptive cellular immune therapy using NK cells represents an attractive means to circumvent many of these problems. This review will highlight current research on NK cells used in HSCT including the role that alloreactive NK cells play in HSCT, NK reconstitution following HSCT, NK involvement in graft versus leukemia (GvL) and GvHD, and NK adoptive immunotherapy.

2 Hematopoietic Stem Cell Transplantation: Overview

Hematopoietic stem cells (HSC) are derived from one of two sources: autologous and allogeneic sources. Autologous stem cells are extracted from the patient and transplanted back into the same person as a rescue after cytoreductive therapy. Without histocompatibility barriers, autologous transplants show a higher degree of engraftment and reconstitution. Autologous transplants, however, have a higher rate of cancer relapse since the transplanted immune cells lack ability to kill the tumor cells. By contrast, allogeneic stem cells are derived from a different donor and transplanted into an HLA-matched (as much as possible) recipient. The donor immune cells can attack the host tumor cells exhibiting graft versus tumor (GvT) effects, but the allogeneic donor cells can also react to and attack normal host cells damaging normal tissues in the host causing GvHD [4, 7–9], which is a major cause of morbidity.

For patients in need of allogeneic HSCT, HLA-matched sibling donors are optimal. Without an HLA-matched sibling as HSC donor, the use of partial HLA-matches and even unrelated HLA-matches may be necessary. The incidence of GvHD, however, increases dramatically among patients who receive HSCT from such unrelated matches. Currently, some transplantation centers manipulate donor HSC graft by depleting donor T cells in order to reduce GvHD. Depleting T and/or NK cells can, however, provide a greater chance for opportunistic infection including EBV lymphoproliferative disease, greater graft failure and a greater relapse rate. Other transplantation centers choose to use the T and NK replete donor graft, opting for a greater GvT effect, and treating the GvHD post transplant with systemic immunosuppression [4, 8, 10].

The tissues used to generate HSC generally include bone marrow, cytokine-mobilized peripheral blood and umbilical cord blood. Bone marrow was typically the only source until the 1990s, when utilizing peripheral blood as a source of HSC became more efficacious. Using growth factors, such as GM-CSF or G-CSF, hematopoietic precursors could be mobilized into peripheral blood, yielding a rich source of HSC, and a faster neutrophil and platelet recovery [9, 11]. Currently, peripheral blood is used for most HSCT, although the use of umbilical cord blood is becoming more widespread. Umbilical cord blood has been investigated since the mid-1990s as a rich source of HSCs [12, 13]. Initially, the limited size of the cord blood graft restricted use to pediatric patients, but with the use of two unrelated, partially matched grafts, cord blood transplantation is being investigated in adults [14–24]. As compared with bone marrow transplantation, cord blood transplantation provides greater ease and safety of the collection of HSC, a lower risk of viral contamination, a reduced incidence of GvHD due to immunologically naïve T cells, and greater availability from established public cord blood banks [9, 19, 21]. Immune reconstitution and the limited HSC number obtained from cord blood, however, limit the use in adults.

The allogeneic HSCT process includes HLA matching of donor to recipient, conditioning or preparative regimen, graft manipulation, actual transplant of HSC, and post-transplant treatments. The successful process will suppress or eliminate host immunity to prevent graft rejection, eliminate tumor cells, minimize or preferably avoid GvHD without compromising engraftment or GvT effects and to minimize toxicity to other tissues. Suppression or elimination of the host immune system is critical to the success of the HSCT. Without a successful conditioning regimen, host effector cells (NK and T cells) can prevent engraftment by the donor cells, exerting a host versus graft effect, thus rejecting the graft [5, 25]. Generally this involves either radiation or chemotherapy, and can either be myeloablative or be reduced intensity. Reduced intensity conditioning uses less cytoreductive therapies but requires extensive immunosuppression. Following the conditioning regimen, the unfractionated or fractionated bone marrow, peripheral blood or cord blood cells are given to the host to repopulate the immune system and generate GVT effects. This has allowed allogeneic HSCT to be used in elderly patients but delayed immune recovery remains a significant hurdle. Full engraftment by the various cell types can take years. The host can be subject to acute or chronic GvHD, yet can also be subject to opportunistic infection due to the extensive immunosuppression. Finally, relapse of the original tumor can occur [8, 9]. NK cells represent a means to promote GvT without inducing GvHD.

3 Natural Killer Cells – Overview

NK cells were functionally described by Cudkowizc and Bennett in 1971 when they found that lethally irradiated mice were capable of rejecting allogeneic or parental bone marrow cell grafts [26, 27]. It was later found that the cells mediating the cytotoxicity were lymphoid cells that were radioresistant. In vitro studies in human cells found a population of lymphoid cells that mediated cytotoxicity in an MHC-unrestricted fashion. This unique cell type was termed Natural Killer cells [28–32]. NK cells are a subset of circulating lymphocytes comprising between 10% and 20% of all peripheral blood lymphocytes, with as many as 2 billion in the human adult [33–37]. While their lifespan is considered short, they serve primary cytotoxic and cytostatic functions in the innate immune system and serve as a bridge between the innate and adaptive immune systems [28, 33, 34, 38–40]. NK cells are morphologically classified as a large granular lymphocyte, yet NK cells share many of the same characteristics as other lymphocytes, including T and B cells. Similar to T cells, NK cells exert their cytotoxicity through secretion of perforin and granzyme molecules, thus lysing virally infected cells and tumor cells, as well as secrete other cytotoxic and immunoregulatory cytokines. Phenotypically, NK cells also share some of the same cell surface markers as T and B cells that define each as a subset of lymphocytes.

Despite similarities with T cells, NK cells are not antigen-specific, do not clonally expand and do not exhibit classical immune memory responses. NK cells do not have receptors which undergo rearrangement such as T (TCR) and B (Immunoglobulin) cell receptors. Also, NK cytotoxicity involves a specific balance of the NK cell's activating and inhibitory receptors discussed later in this chapter. Whereas T cells require prior sensitization by an antigen presenting cell, and presentation of the same antigen in the target cell's major histocompatibility complex (MHC) molecule on the cell surface, NK cells do not require antigen presentation or "self" presentation by the cell's MHC in order to induce cytotoxicity. NK cells have receptors which interact with the MHC molecule directly. Many virally transformed or malignant cells can downregulate MHC molecules in order to escape immune surveillance by T cells; but NK cells detect downregulated or modified MHC and induce cell lysis. Furthermore, NK cells can also mediate cell death through death receptor (e.g., Fas, TRAIL) pathways and are potent mediators of antibody-dependent cell cytotoxicity [34, 41, 42]. NK cells also perform cytotoxic, cytostatic and immunoregulatory functions through the production of cytokines including IL-1, IL-5, IL-8, IFN-γ, TNF-α, TGF-β and GM-CSF, among others.

Although NK cells share some cell surface markers with other lymphocytes, the combination of cell surface markers label NK cells as unique. Among the many cell markers in humans that NK cells share with other cell types, CD56 and CD16 have characteristically been used to define NK cells. Since NK cells do not express CD3, the typical NK signature has been CD56$^+$ CD16$^{+/-}$ CD3$^-$ lymphocytes. The NK cell subset that demonstrates a cytotoxic function has been shown as CD56^{DIM}CD16$^+$ NK cells in peripheral blood, whereas cytokine generation has been known as CD56^{BRIGHT}CD16$^-$ NK cells which predominate in lymph nodes in man, and represents a significant difference between man and mouse because a specific cytokine generating NK subset has not be found in mice [43–45]. Unfortunately, not only do other cell types express CD56 and CD16, some NK cells do not express either CD56 or CD16. Other phenotypic markers that have been used to delineate NK cells include NK activating and inhibitory receptors such as NKG2D, CD161, CD244 and NKp46, as well as killer immunoglobulin receptors (KIR) in humans and the Ly49 family of activating and inhibitory receptors in mice [34, 43]. These markers also are not exclusively found on NK cells and may not be expressed consistently or at all on NK cells [43, 46, 47]. In the murine system, NK cells are phenotypically defined as $NK1.1^+CD3^-$ lymphocytes in C57BL/6 mice or $DX5^+CD3^-$ lymphocytes in Balb/c mice which are negative for NK1.1. These phenotypic markers can also be found infrequently on murine T cell subsets. AsialoGM1 is also expressed on murine NK cells, but like NK1.1 and DX5, this can also be expressed on other cell types including T cells and macrophages [43, 48]. CD122, the IL-2/IL-15 β/γ receptor chain, has been proposed as a defining surface marker for NK cells in both humans and mice [34, 43]. CD122 expression is essential for IL-15 to be functional, and has been found as an early marker of developing murine NK cells [49, 50]. Through studies with IL-15 knock-out mice, IL-15 has been shown to be absolutely critical to the development and survival of NK cells [51–55]. CD122 expression, together with expression of CD56 and CD16 in humans, and NK1.1 or DX5 in mice, together with PRF1 (perforin) in humans and the nonexpression of CD3, may be the most reliable primary markers for NK cells.

4 Natural Killer Cells and Their Roles in Allogeneic HSCT

Natural killer cells can serve both positive and negative functions in allogeneic HSCT depending on whether they are of donor or host origin. Donor NK cells have the potential to enhance engraftment, promote GvT and reduce or suppress GvHD. Host NK cells affect GvHD by lysing donor HSC, and can cause graft rejection [5, 25, 56–63].

The cytotoxic function of the NK cell is regulated by the balance of inhibitory and activating receptors. Preclinical studies with mice have found that humans and mice share similarities and differences in the NK inhibitory and activating receptors.

Discovery of the first MHC class I specific receptors, the Ly49 family in mice paved the way for the discovery of the KIR family in humans [64]. The Ly49 receptors are C-type lectin receptors whereas the human KIR receptors differ structurally, in that they are characterized by extracellular immunoglobulin domains [64]. The murine Ly49 family recognizes the MHC Class Ia molecules H-2D and/or H-2K [64–69]. In preclinical studies with mice, evidence has shown that NK cells can serve as not only a barrier to engraftment, but can mediate graft rejection [70–73]. We have clearly shown that severe combined immune deficient (SCID) mice which lack T and B cells can reject allogeneic bone marrow cells [70]. In humans, graft rejection by recipient NK cells varies with the cytoreductive conditioning, as well as the balance and makeup of the recipient KIR and donor HLA repertoire [5, 74–76]. Generally, some of this can be overcome by the volume and makeup of the graft. Increasing the number of HSC in a graft will overcome the potential rejection by NK cells [77]. There are, however, numerous conditions which can affect rejection by NK cells. Subclinical infections prior to transplant may have an effect upon rejection as agents such as poly I:C can promote NK cell-mediated HSC rejection.

In preclinical studies with mice, Blazar and Bennett found that the conditions of the mouse colony can also markedly affect rejection and transplant outcomes [78]. Furthermore, murine studies have unlocked a key role of the Ly49 family members in NK cell-mediated rejection. Different strains of mice have a specific subset of activating and inhibitory Ly49 receptors. Depending upon the MHC expression on the HSC, a particular NK subset can reject an allogeneic graft. For example, rejection of HSC will occur by NK cells without inhibitory Ly49C/I receptors for MHC Class I molecule H-2K^d and with activating Ly49D receptors for H-2K^d. However, if the NK cells possess both inhibitory Ly49C/I receptors and activating Ly49D receptors, the NK cell will not lyse a HSC with the MHC Class I molecule for H-2K^d. Our earlier studies have also indicated that depletion of specific and "opposite" inhibitory Ly49 subsets increase rejection [79–81]. The idea of licensing, which will be discussed later in this chapter, can also possibly play a role in rejection.

The distribution of human MHC Class I molecules for inhibitory KIR are relatively generic. KIR2DL2 and/or KIR2DL3 are the receptors for HLA-C group 1, which is ubiquitous among humans. KIR2DL1 is the receptor for HLA-C group 2 present in about 97% of humans [82]. HLA-Bw4 binds to the KIR3DL1 receptor which is found in about 90% of humans [83]. In contrast, KIR activating receptors display extensive genetic variation. This leads to heterogeneity in the human population. In contrast to inhibitory KIR, activating KIR may be absent from large percentages of ethnic groups [84].

In addition to the murine Ly49 and human KIR inhibitory receptors, the C-type lectin molecule, CD94 covalently bound to an inhibitory member of the NKG2 family can be found on both mice and humans. CD94/NKG2A or CD94/NKG2B binds to nonclassical HLA-E molecules in humans. Human allele specificity on the CD94/NKG2 receptors is less well defined than KIR receptors. In mice CD94/NKG2A binds to the Qa-1 molecule on the target cell [47, 85, 86, 87].

Recently, another family of NK cell receptors consisting of type II transmembrane molecules with an extracellular C type lectin motif has been characterized in mice and rats. The killer cell lectin-like receptors (KLRE/I1 and KLRE/I2) have been shown to regulate NK cell cytotoxicity. KLRE/I1 forms heterodimers with NK cell receptor chains KLRI1 and KLRI2. KLRI1 contains two immunoreceptor tyrosine based motifs (ITIM). When KLRI1 and KLRI2 associate with KLRE/I1, NK cell-mediated cytotoxicity is inhibited. KLRE/I2, conversely, has been found to function as an activating heterodimeric receptor and functions in an inhibitory manner, whereas KLRE/I2 has been found to function in an activating manner [88, 89].

Despite the divergent receptor types, human and murine NK inhibitory receptors essentially inhibit the NK cell in a similar manner. They possess ITIM that are phosphorylated when the receptor engages its respective ligand. Tyrosine phosphatases SHP-1 and SHP-2 are recruited to the junction of the NK cell and its target cell, where they suppress functions associated with NK cell activating receptors, including calcium influx, cytokine production, proliferation and degranulation, thus inhibiting the NK cell from lysing the target cell [47]. Furthermore, inhibitory KIR have a higher affinity for their ligands than activating receptors. When both inhibitory and activating receptors are engaged, the inhibitory signal predominates, preventing NK cell-mediated cytotoxicity [90].

In recent studies, Yokoyama et al. demonstrated that in both mouse and human, the activation of NK cells as measured by IFN-γ release depends upon the NK receptor's early recognition of self-MHC. In a functional maturation process, NK cells with specific receptors for "self" MHC Class I become activated, undergoing a "licensing" process, whereas NK cells lacking that specific MHC Class I receptor are more likely to be less or nonfunctional or anergic [83, 91, 92]. In a 2008 study of NK cells from 39 normal, healthy volunteers, researchers found increased functional responsiveness from NK cells with the KIR3DL1 receptor from donors possessing the HLA-Bw4/4 group, a lesser functional responsiveness in donors with the HLA-Bw4/6, and a low responsiveness in donors with the HLA-Bw6/6 allele [92]. These data indicate a licensing effect on human NK cells by self-HLA molecules. While further research in this area is ongoing, the impact for HSCT is enormous. Specific licensed NK subsets could be isolated, enhanced and incorporated with the HSCT graft for the greatest GvT effect.

5 NK Alloreactivity in HSCT

NK cells can exhibit a beneficial GvL or GvT effect which T cells cannot achieve. This is particularly the case when NK cells are KIR ligand-mismatched in the donor–to-host direction. KIR ligand-mismatching takes advantage of a fundamental attribute of alloreactive NK cells – the "missing self" recognition. The "missing self" hypothesis was first proposed by Kärre, and has been observed in both mice and humans [77, 85–87, 93–100]. In this case, donor NK cells express a KIR for an HLA Class I molecule that is absent in the host as their only inhibitory receptor for self. In this situation, the donor NK cells react to the missing expression of the self HLA Class I molecule on the host target cells and lyse the host cancer cells.

In 2005, Ruggeri et al. reported the use of T cell-depleted peripheral blood transplants from haplotype-mismatched (haploidentical) family members [63]. The recipients of these transplants suffered from high-risk acute myeloid leukemia (AML) and did not have matched sibling donors. Of the 57 AML patients who received these haploidentical transplants, 20 of them received these transplants from NK alloreactive donors. Of the 57 patients, 90.2% achieved engraftment, 26% achieved event free survival, and only 8.6% acquired GvHD. Interestingly, transplants from NK alloreactive donors totally protected the recipients from rejection, GvHD and relapse. The Ruggeri group later updated their findings. Including the 2002 findings, primary engraftment was obtained in 103 of 112 patients. Recipients from NK alloreactive donors experienced a 6% rejection versus the 10% rejection of recipients from non-NK alloreactive donors. Grade 2 acute GvHD occurred in 10% of the NK alloreactive transplants versus 11% in non-NK alloreactive transplants. Relapse was also substantially lower in transplants from NK alloreactive donors (3% versus 47%) [100]. Despite the lack of difference in rejection and GvHD, transplantation from NK alloreactive donors in a haploidentical transplantation setting did improve overall event free survival. In a similar study of KIR ligand mismatched HSCT from unrelated donors, conducted by three transplantation centers, 90% of the patients with KIR ligand mismatch showed event free survival [101]. Of this group, only one patient, with acute lymphoblastic leukemia (ALL), relapsed. This patient later died from chronic GvHD. None of the KIR ligand mismatched patients experienced graft rejection. The incidence of GvHD, however, was similar between the KIR ligand mismatched patients and the KIR ligand matched patients, although the cumulative GvHD related mortality was 6% versus 29% respectively. Other studies, however, indicated that KIR ligand mismatch HSCT does not produce the GvL effect associated with a survival advantage, particularly with unrelated donors. In a recent retrospective study of 1,571 unrelated HSCT, no differences were seen in disease reoccurrence between KIR matched and KIR mismatched groups [102]. This may be due to differences in the HSCT conditioning and immunosuppression used by the different institutions. In contrast to the "missing self" mechanism conveying the advantages reported of donor alloreactive NK cells, other researchers have proposed the "missing ligand" or "receptor ligand" mechanism that conveys the effectiveness of donor alloreactive NK cells. The "receptor ligand" model hypothesizes that some donors have an extra inhibitory KIR that neither donor nor host have the matching HLA ligand for. This extra KIR is considered to be anergic or hyporesponsive because it is not licensed. The unlicensed state is not permanent and can be overcome by proinflammatory cytokines [103–106]. This would account for control or reduction of relapse and improved survival. Leung et al. in 2004 proposed that relapse rates could be better predicted using the receptor ligand model. They used this model to predict outcomes in 36 pediatric patients receiving HLA-haploidentical peripheral blood HSCT [107]. They found that while the KIR ligand–ligand model proposed by Ruggeri et al. was able to adequately predict those at low risk for relapse, the KIR ligand–ligand model misclassified those at high risk for relapse.

In a 2005 study, Leung et al. found that some inhibitory KIR are expressed at very low levels, thus impacting outcomes predicted by the KIR ligand–ligand model [108]. Further, in a 2007 study, they applied the KIR receptor–ligand mismatch model to 16 patients with lymphoma or solid tumors who received an autologous HSCT [106]. Fifty percent of the patients experienced disease progression. Of this group, 83% had no receptor–ligand mismatch and 50% had one receptor–ligand mismatch. None of the patients with two mismatched receptor–ligand pairs experienced disease progression. Hsu et al. conducted a retrospective study looking at HLA and KIR genotyping to predict clinical outcomes of T cell depleted HSCT [109]. They hypothesized that patients who lacked HLA ligands for their donor inhibitory KIR and who received HLA identical sibling transplants would show differences in outcomes. The 178 patients who received HSCT suffered from AML, chronic myelogenous leukemia (CML), ALL or myelodysplastic syndrome (MDS). In patients with CML and ALL, they found that a lack of HLA ligand (missing ligand) for the donor inhibitory KIR had no effect on survival or relapse. There was, however, a significant missing ligand effect on survival and lower incidence of relapse for patients with AML and MDS. Furthermore, those AML and MDS patients who lacked two HLA ligands for donor inhibitory KIR had the highest survival. These findings were further substantiated by a recent study showing that even highly effective NK cells expressing just one inhibitory KIR to self class I ligands will be enough to terminate a cytotoxic response, indicating that the HLA ligands are the key to the alloreactive NK response in HSCT [110]. To substantiate the finding that the lack of KIR ligand in patients receiving HLA-mismatched HSCT may be a predictor for relapse, researchers for the International Histocompatibility Working Group reported their findings of 1,770 patients receiving unrelated HLA matched or mismatched HSCT [111]. These patients were being treated for AML MDS, CML and ALL, and each received myeloablative conditioning together with T-replete donor stem cells. Among the HLA mismatched transplants, decreased probability of relapse was associated with recipient homozygosity for the KIR epitopes HLA-B and HLA-C. In addition, recipients with AML were predicted to have a reduced hazard of relapse similar to ALL and CML. The missing ligand effect was not seen in unrelated transplants.

Ruggeri et al. applied the missing ligand model to 112 patients who received haploidentical HSCT [100]. Their results indicated that among non-NK alloreactive patients, event-free survival did not differ between the group with no missing ligands versus the group having fewer than three missing ligands. Further, after pooling the patients into missing ligand or KIR ligand-mismatch groups, they found that survival was lower in the missing ligand cohort.

All of these studies suggest HSCT outcome based upon inhibitory and activating KIR together with their accompanying ligands is complex and affected by treatment procedures, graft composition, conditioning and post transplantation immunosuppression regimens, not to mention the patient population status. All of these factors may explain the disparities between the studies discussed above. Additional preclinical studies with mice may shed further light upon this controversy and provide mechanistic insights and how best to exploit this.

6 NK Reconstitution in HSCT

Elimination or strong suppression of host hematopoiesis is the goal of preconditioning in HSCT. The importance for this is multifaceted, but includes providing space for engraftment of donor derived hematopoietic cells. Alloreactive NK and T cells from the donor help eliminate remaining host immune cells that attack the donor graft. However, prior to and during donor cell engraftment, the host is susceptible to opportunistic infection. As a result, the members of the innate immune system are the first to recover, generally within weeks following transplantation. The members of the adaptive immune system are much slower, taking months to years to normalize depending on the age of the recipient [112, 113].

Within the first few weeks following HSCT, NK cells increase substantially, but they decrease, many times to below original levels, between 100 and 200 days following HSCT [21, 114–119]. This allows the innate immune system to defend against opportunistic infection including viral infections [120]. Because of this, the NK cell subset, $CD56^{BRIGHT}$ $CD16^-$, that secretes proinflammatory cytokines, such as IFN-γ emerge first [115, 116, 121]. The cytotoxic $CD56^{DIM}$ $CD16^-$ subset also emerges and has been noted to increase in number more rapidly [116]. Schulze et al. recently characterized the emerging NK cell populations from eight patients who received a myeloablative PBMC $CD34^+$ haploidentical mismatched HSCT [121]. They found the characteristic shift towards the $CD56^{BRIGHT}$ $CD16^-$ subset, but they also found a 50% downregulation in the NKG2D receptor through day 21 on the $CD56^{BRIGHT}$ subset. Furthermore, they found a downregulation of FasL on both $CD56^{BRIGHT}$ and $CD56^{DIM}$ subsets. On the $CD56^{DIM}$ subset, expression of the NK activating receptor NKp46 increased 2-fold through day 56 following HSCT, yet there was no change in NKp30 and NKp44. These results indicate the emergence of an immature phenotype [119, 122].

Several studies indicate that the immature NK cells have depressed function, but may even have an inhibitory effect on GvT effects [119, 123]. Shilling et al. followed NK reconstitution in 18 leukemia patients who were given HLA-matched HSC transplants [119]. The patients were analyzed for KIR genotype. From 6 to 9 months following HSCT, eight patients reconstituted a NK repertoire similar to their donor, but five other patients exhibited a depressed frequency of KIR expressing NK cells. Similarly, Cooley et al. have reported decreased surface KIR expression by NK cells following HSCT [124]. In unrelated T cell replete HSCT, recipients had increased NK cells, but KIR expression was diminished. NKG2A expression and IFN-γ production increased as seen in other studies, along with increased acute GvHD. Their results indicated that T cells in the graft had a detrimental effect on NK cell recovery. They also showed that increased KIR expression is an accurate predictor of survival. In a later study, Cooley et al. showed decreased KIR expression and increased NKG2A on $CD56^{BRIGHT}$ NK cells, yet showed a smaller $CD56^{DIM}$ subset that lacked KIR and NKG2A expression [125]. This unique subset displayed decreased cytotoxicity and IFN-γ production. They postulated that this $CD56^{DIM}$ subset is, in fact, an NK precursor subset, and not the mature cytotoxic NK subset.

The NK subsets can change rapidly, however, when exposed to increasing levels of cytokines, particularly IL-15 and IL-2 [116, 121]. Dulphy et al. showed similar results, but 3 months following HSCT, they found an increased intracellular IFN-γ production following stimulation with IL-2 in $CD56^{BRIGHT}$ cells, together with an increase in intracellular perforin and an ability to degranulate [115]. They also found that the $CD56^{BRIGHT}$ cells were undergoing a maturing process through the upregulation of CD117, TRAIL and CD25. It was postulated that the reconstitution following HSCT was driven by IL-15 due to the increase in serum levels of IL-15.

Donor NK cells also have an effect on reconstitution of other cell types, including T cells [7, 61, 62, 126]. Production of growth stimulating (GM-CSF, IL-1, IL-6) and growth inhibiting (TGF-β, IFN-γ, TNF-α) cytokines by NK cells can have an effect on reconstitution of other cell types. Adoptive transfer of IL-2 activated NK has been shown to promote reconstitution of granulocytes following syngeneic HSCT in mice [61, 62, 127]. In a 2008 study of nonmyeloablative T cell-replete HSCT, Sobecks et al. used recipient inhibitory KIR genotype and donor KIR/HLA ligand matches to generate an inhibitory KIR score for 31 patients undergoing HSCT. They found that patients with a higher inhibitory KIR score were more likely to have NK and T cells that were less active, and as a result, have a greater ability to achieve complete donor T cell engraftment [76]. Patients with low inhibitory KIR scores were less likely to achieve full engraftment, and were more likely to develop graft rejection. Interestingly, of those patients who achieved full donor engraftment, 48% lacked the donor KIR ligand HLA-Cw group for the recipient inhibitory KIR. These results may suggest that a balance must be maintained to achieve optimal engraftment and antitumor effects with minimal rejection.

Preclinical studies in mice have shown that reconstitution of NK cells and T cells can be increased by the administration of recombinant human (rh) IL-15. Alpdogan et al. demonstrated that exogenous administration of rhIL-15 following allogeneic HSCT significantly increased donor NK and T cell numbers in mice [128]. Furthermore, NK cytotoxicity significantly increased with the rhIL-15 administration as demonstrated by increased percentage of lysis of YAC-1 cells in a standard chromium release assay. Administration of rhIL-15 also significantly increased the proliferation of NK and T cells over a 28-day period, yet did not induce GvHD. Assessment of NK cell subsets was not performed in these studies and questions remain of the dependence of NK cells for continued cytokine administration.

The administration of drugs used for conditioning or to control GvHD can have a negative effect on NK reconstitution. In the late 1980s, cyclosporin A, along with several other drugs, was administered to patients as post-transplantation immunosuppression. Since this time, cyclosporin or other calcineurin inhibitor use has become widespread. Recently, however, researchers have found that in vitro administration of cyclosporin A to IL-2 or IL-15-activated PBMC reduced the NK cell population as a whole, particularly $CD56^{DIM}$ subsets, and specifically $CD56^+$ $CD16^+$ KIR^+ NK cells [129]. $CD56^{BRIGHT}$ NK cells not expressing CD16 and KIR were resistant. Surprisingly, the exposed NK cells retained their cytotoxicity. When stimulated with IL-12 and IL-18, more IFN-γ producing cells were observed. Another drug that has been used to treat GvHD is antithymocyte globulin or ATG. Unfortunately, while the drug is efficacious in immunosuppression, most patients suffer significant side effects [130]. In a recent study of 69 Swedish patients, subcutaneous alemtuzumab was compared with ATG in nonmyeloablative conditioning regimens. More of the ATG treated patients experienced full donor engraftment than the alemtuzumab patients. No differences were seen in NK cell engraftment levels between ATG or alemtuzumab, however. Yet, in a 2008 study, Penack et al. showed that both alemtuzumab and ATG were very potent inducers of

NK cell death [131]. Furthermore, they found a significant reduction in NK cytotoxic degranulation in the alemtuzumab group of HSCT patients at day 30. They hypothesize that while alemtuzumab and ATG are both toxic to NK cells, alemtuzumab has a longer half-life and may explain the significant reduction in NK cytotoxicity. Thus, immunosuppressive agents must be carefully investigated and selected in order to optimize the NK engraftment for GvT effects to be realized. In addition, this may explain differences on GvT effects obtained by NK cells dependent on the regimens used in allogeneic HSCT.

7 The Role of NK Cells in GvHD

GvHD is a significant problem for patients receiving allogeneic HSCT. Unfortunately, it is also associated with the positive effects of GvT. As discussed previously, the GvT effects significantly reduce the incidence of relapse. Acute GvHD can occur within days following HSCT. Conversely, initial presentation of chronic GvHD may occur soon after HSCT, or it may not occur for months following transplantation. It is hypothesized that T cells in the donor graft start the inflammation associated with GvHD. Donor T cells respond to antigen presented in recipient MHC as foreign, lysing the recipient cells, and producing inflammatory cytokines and chemokines that recruit other immune cells to the site. Donor NK cells respond, are activated by cytokines produced by T cells. Increased tissue damage, organ damage and/or failure, susceptibility to opportunistic infections, as well as partial or delayed engraftment are some of the effects of GvHD. Interestingly, however, donor-type NK cells can enhance engraftment and increase GvT effects, as previously discussed, yet without increasing GvHD. Furthermore, adoptive transfer of NK cells as an immunotherapy can maximize these effects and even be a preventative therapy for GvHD. Asai et al. found that activated NK cells had a protective effect against GvHD and increased GvT effects. In 11 experiments, NK cells activated by rhIL-2 either in vivo or ex vivo and transferred, significantly delayed the onset of GvHD. Activated NK cells and not rhIL-2 were found to be protective against GvHD, since administration of rhIL-2 after the third day following BMT exacerbated GvHD very likely by encouraging the expansion of T cell populations [56].

8 NK Cell Immunotherapy

Adoptive transfer of activated NK cells has been shown in preclinical studies with mice to exhibit GvT effects after HSCT and not increase GvHD [7, 16, 56, 104, 132–135]. Asai et al. found that in lethally irradiated mice given bone marrow and T cell-replete spleen cells, an adoptive transfer of IL-2 activated NK cells prevented GvHD, and also displayed antitumor effects [56]. Coadministration of an

NK-activating cytokine such as IL-2 or IL-15 leads to increased survival of adoptively transferred NK cells and increased antitumor effects. Alici et al. combined IL-2 administration with adoptive transfer of NK cells in a murine multiple myeloma model [132]. They noted that IL-2 activated NK cells killed the myeloma tumor in vitro and in vivo by localizing to the tumor sites to elicit their cytotoxic action. They further noted that the IL-2 activated NK cells prolonged the survival of the tumor bearing mice. Castriconi et al. found similar results with IL-2 or IL-15 activated NK cells against metastatic neuroblastoma in NOD/scid mice [104].

These studies, together with numerous others, have lead to clinical trials utilizing adoptive transfer of activated NK cells. Miller et al. conducted a trial of 43 patients presenting with metastatic melanoma, metastatic renal cell carcinoma, refractory Hodgkins disease and AML. They received haplo-identical NK cell infusions with either a low intensity conditioning (low dose cyclophosphamide/ methylprednisolone) or high intensity conditioning (high dose cyclophosphamide/ fludarabine). The NK cells were activated overnight with IL-2 prior to infusion into the patients. They found in vivo persistence and expansion of the NK cells in patients who had received the high intensity conditioning, and that these NK cells were functional at day 14 following infusion. The NK persistence and expansion was caused by an increase in IL-15 production brought about by the high intensity conditioning [36, 136, 137].

As stated previously, NK cell-mediated cytotoxicity is dependent upon activating receptors prevailing against inhibitory receptors. Koh et al. hypothesized that blocking NK inhibitory receptors augment cytotoxicity against tumor [58–60]. Using $F(ab')_2$ fragments, Ly49 inhibitory receptors were blocked, and the NK cells were transferred into tumor bearing mice. They found that NK inhibitory receptor blockade inhibited tumor growth in vitro, and increased the survival of tumor bearing mice. Furthermore, they found that the inhibitory receptor blockade did not inhibit reconstitution following HSCT, and augmented in vivo cytotoxicity against tumor cells. Use of similar inhibitory receptor blockades may provide greater NK cytotoxicity without adverse effects.

IL-2 is a potent regulatory cytokine for both T and NK cells. IL-2, along with other cytokines that increase proliferation and survival such as IL-15, IL-12 or IL-18, has been used to achieve greater antitumor effects [138–141]. Unfortunately, administered cytokines are rapidly cleared, thus requiring repeated injections of large amounts. The potential for toxicity dramatically increases with the increasing amount and frequency. In addition, high doses given frequently can increase unwanted cell types, such as T regulatory cells which can suppress NK cells, and/ or also cause activation induced cell death (AICD) of NK cells. Thus the challenge is to find a way to administer the cytokine and avoid the toxicity. Low dose subcutaneous IL-2 administration has been used successfully in lymphoma and breast cancer patients [142]. These patients saw increases in NK cell numbers and function. Coupled with ex vivo IL-2 activated NK cells, these two immunotherapies have shown great potential.

Gene therapy has been proposed as another way to circumvent toxicity issues related to cytokine administration. Ortaldo et al. hydrodynamically delivered IL-2 cDNA to mice to determine the in vivo effects on NK and T cells [134]. They found significant increases in NK and NKT numbers in the liver and the spleen, and significant increases of NK in the blood 72 h following hydrodynamic delivery, as well as substantial enhancement of NK cytotoxicity in both the spleen and liver. Furthermore, they found expansion and repopulation of bone marrow progenitors in the IL-2 mice following bone marrow transplantation, as well as inhibited progression of established metastases in the liver. Currently, our laboratory has utilized similar technology with IL-15 cDNA. We have found significant increases in murine NK cell numbers and function, particularly in the liver. Furthermore, when coupled with adoptive transfer of IL-2 activated NK cells, the IL-15 cDNA increases NK proliferation and survival.

Another approach to enhance the proliferation and survival of transferred NK cells using cytokines is to create complexes of cytokines with their receptors or antibodies. Prlic et al. complexed IL-2 with an anti-IL-2 antibody [143]. The IL-2/anti-IL-2 complex mediates the response through the IL-2Rβ and γc chains. They found that IL-2 complexes expand the donor NK population and not the host NK cells, and function effectively even bypassing the need for IL-15. Other studies have shown that complexing IL-15 with the IL-15Rα, or utilizing similar IL-15/IL-15Rα hyperagonists lead to increased NK cell numbers and function [51, 144, 145].

Gene therapy is also being used to enhance functionality of NK cells. Pegram et al. have developed a NK receptor gene specific for human erbB2 tumor-associated antigen [135]. They found that NK cells expressed the anti-erbB2 gene and elicited perforin-mediated cytotoxicity against murine leukemia tumor cells expressing the erbB2 antigen. Combining antigen recognition of the cytotoxic T cells along with MHC recognition of the cytotoxic NK cells may broaden the applicability for cancer patients.

Another innovative immunotherapy for the treatment of leukemia is the use of the human NK-92 cell line. NK-92 is unique in that it lacks inhibitory KIR [146, 147]. The main method of cytotoxicity is through perforin and granzymes, not unlike normal NK cells. Yet, because of the lack of inhibitory KIR, NK-92 cells create a simulated KIR mismatch situation, which may produce greater and more reliable GvL effects. In a murine model, Yan et al. noted that adoptive transfer of NK-92 cells produced antileukemia effects on T cell ALL [148]. AML and pre B ALL were also highly sensitive to the killing effects of NK-92 cells. Further studies have found that the NK-92 cell line did not induce cytotoxicity against bone marrow cells [148–150]. Currently, clinical trials are using NK-92 for other cancer immunotherapies [151].

9 Conclusion

Adoptive immunotherapy utilizing NK cells coupled with HSCT presents additional options for cancer patients, but several issues must be considered. Terme et al., in an excellent review on NK therapy, have suggested that NK inhibitory versus activating receptor manipulation, genotyping and phenotyping, subsets to be used, activation criteria, and delivery protocols are some of the issues that need consideration [152]. Additional issues related to the recipient need to be considered, including selection of conditioning regimen, type of graft to be used and post-transplant treatments. HSCT transplant centers and research centers are continuing to investigate the optimal methodology and use of NK immunotherapy in HSCT [36, 136]. Recently, Verheyden and Demanet published an excellent review focusing on NK cell receptors and their ligands in leukemia, and the paucity of studies focusing on specific escape pathways used of each type of leukemia [153]. Understanding of these pathways will lead us to similar escape mechanisms in lymphomas and other cancers. Further research into these escape pathways may help direct researchers to manipulating NK cells for optimal survival, proliferation and cytotoxicity in HSCT and related immunotherapies.

References

1. Lebkowski JS, Philip R et al (1997) Breast cancer: cell and gene therapy. Cancer Invest 15:568–576
2. Ringden O (1997) Allogeneic bone marrow transplantation for hematological malignancies–controversies and recent advances. Acta Oncol 36:549–564
3. Schetelig J, van Biezen A et al (2008) Allogeneic hematopoietic stem-cell transplantation for chronic lymphocytic leukemia with 17p deletion: a retrospective European group for blood and marrow transplantation analysis. J Clin Oncol 26:5094–5100
4. Storb R (1995) Bone marrow transplantation. Transplant Proc 27:2649–2652
5. Barao I, Murphy WJ (2003) The immunobiology of natural killer cells and bone marrow allograft rejection. Biol Blood Marrow Transplant 9:727–741
6. Handgretinger R, Lang P (2008) The history and future prospective of haplo-identical stem cell transplantation. Cytotherapy 10:443–451
7. Koh CY, Welniak LA et al (2000) Adoptive cellular immunotherapy: NK cells and bone marrow transplantation. Histol Histopathol 15:1201–1210
8. Murphy WJ, Longo DL (1997) The potential role of NK cells in the separation of graft-versus-tumor effects from graft-versus-host disease after allogeneic bone marrow transplantation. Immunol Rev 157:167–176
9. Welniak LA, Blazar BR et al (2007) Immunobiology of allogeneic hematopoietic stem cell transplantation. Annu Rev Immunol 25:139–170
10. Wagner JE, Thompson JS et al (2005) Effect of graft-versus-host disease prophylaxis on 3-year disease-free survival in recipients of unrelated donor bone marrow (T-cell Depletion Trial): a multi-centre, randomised phase II-III trial. Lancet 366:733–741
11. Schmitz N, Dreger P et al (1995) Primary transplantation of allogeneic peripheral blood progenitor cells mobilized by filgrastim (granulocyte colony-stimulating factor). Blood 85:1666–1672
12. Gluckman E, Rocha V et al (1997) Outcome of cord-blood transplantation from related and unrelated donors. Eurocord Transplant Group and the European Blood and Marrow Transplantation Group. N Engl J Med 337:373–381
13. Wagner JE, Kernan NA et al (1995) Allogeneic sibling umbilical-cord-blood transplantation in children with malignant and non-malignant disease. Lancet 346:214–219
14. Abdel-Rahman F, Hussein A et al (2008) Bone marrow and stem cell transplantation at King Hussein cancer center. Bone Marrow Transplant 42(Suppl 1):S89–S91

15. Bautista G, Cabrera JR, et al (2008) Cord blood transplants supported by co-infusion of mobilized hematopoietic stem cells from a third-party donor. Bone Marrow Transplant
16. Chen PM, Hsiao LT et al (2008) Current status of hematopoietic stem cell transplantation in Taiwan. Bone Marrow Transplant 42(Suppl 1):S133–S136
17. Confer D, Robinett P (2008) The US National Marrow Donor Program role in unrelated donor hematopoietic cell transplantation. Bone Marrow Transplant 42(Suppl 1):S3–S5
18. Gan G, Teh A et al (2008) Bone marrow and stem cell transplantation: Malaysian experience. Bone Marrow Transplant 42(Suppl 1):S103–S105
19. Kurtzberg J, Prasad VK et al (2008) Results of the Cord Blood Transplantation Study (COBLT): Clinical outcomes of unrelated donor umbilical cord blood transplantation in pediatric patients with hematologic malignancies. Blood
20. Laughlin MJ, Eapen M et al (2004) Outcomes after transplantation of cord blood or bone marrow from unrelated donors in adults with leukemia. N Engl J Med 351:2265–2275
21. Moretta A, Maccario R et al (2001) Analysis of immune reconstitution in children undergoing cord blood transplantation. Exp Hematol 29:371–379
22. Ooi J, Takahashi S, et al (2008) Unrelated cord blood transplantation after myeloablative conditioning in adults with ALL. Bone Marrow Transplant
23. Sauter C, Barker JN (2008) Unrelated donor umbilical cord blood transplantation for the treatment of hematologic malignancies. Curr Opin Hematol 15:568–575
24. van Be T, van Binh T et al (2008) Current status of hematopoietic stem cell transplantations in Vietnam. Bone Marrow Transplant 42(Suppl 1):S146–S148
25. Hallett WH, Murphy WJ (2004) Natural killer cells: biology and clinical use in cancer therapy. Cell Mol Immunol 1:12–21
26. Cudkowicz G, Bennett M (1971) Peculiar immunobiology of bone marrow allografts. I. Graft rejection by irradiated responder mice. J Exp Med 134:83–102
27. Cudkowicz G, Bennett M (1971) Peculiar immunobiology of bone marrow allografts. II. Rejection of parental grafts by resistant F 1 hybrid mice. J Exp Med 134:1513–1528
28. Herberman RB, Ortaldo JR (1981) Natural killer cells: their roles in defenses against disease. Science 214:24–30
29. Oldham RK, Herberman RB (1973) Evaluation of cell-mediated cytotoxic reactivity against tumor associated antigens with 125I-iododeoxyuridine labeled target cells. J Immunol 111:862–871
30. Ortaldo JR, Herberman RB (1984) Heterogeneity of natural killer cells. Annu Rev Immunol 2:359–394
31. Rosenberg EB, Herberman RB et al (1972) Lymphocyte cytotoxicity reactions to leukemia-associated antigens in identical twins. Int J Cancer 9:648–658
32. Takasugi M, Mickey MR et al (1973) Reactivity of lymphocytes from normal persons on cultured tumor cells. Cancer Res 33:2898–2902
33. Blum KS, Pabst R (2007) Lymphocyte numbers and subsets in the human blood. Do they mirror the situation in all organs? Immunol Lett 108:45–51
34. Caligiuri MA (2008) Human natural killer cells. Blood 112:461–469
35. Chiorean EG, Miller JS (2001) The biology of natural killer cells and implications for therapy of human disease. J Hematother Stem Cell Res 10:451–463
36. McKenna DH Jr, Sumstad D et al (2007) Good manufacturing practices production of natural killer cells for immunotherapy: a six-year single-institution experience. Transfusion 47:520–528
37. Miller JS (2002) Biology of natural killer cells in cancer and infection. Cancer Invest 20:405–419
38. Ferlazzo G, Tsang ML et al (2002) Human dendritic cells activate resting natural killer (NK) cells and are recognized via the NKp30 receptor by activated NK cells. J Exp Med 195:343–351
39. Herberman RB (1981) Natural killer (NK) cells and their possible roles in resistance against disease. Clin Immunol Rev 1:1–65

40. Raulet DH (2004) Interplay of natural killer cells and their receptors with the adaptive immune response. Nat Immunol 5:996–1002
41. Kim S, Iizuka K et al (2000) In vivo natural killer cell activities revealed by natural killer cell-deficient mice. Proc Natl Acad Sci USA 97:2731–2736
42. Smyth MJ, Cretney E et al (2005) Activation of NK cell cytotoxicity. Mol Immunol 42:501–510
43. Di Santo JP (2006) Natural killer cell developmental pathways: a question of balance. Annu Rev Immunol 24:257–286
44. Fan YY, Yang BY et al (2008) Phenotypically and functionally distinct subsets of natural killer cells in human PBMCs. Cell Biol Int 32:188–197
45. Fehniger TA, Cooper MA et al (2003) CD56bright natural killer cells are present in human lymph nodes and are activated by T cell-derived IL-2: a potential new link between adaptive and innate immunity. Blood 101:3052–3057
46. Colucci F, Di Santo JP et al (2002) Natural killer cell activation in mice and men: different triggers for similar weapons? Nat Immunol 3:807–813
47. Lanier LL (2008) Up on the tightrope: natural killer cell activation and inhibition. Nat Immunol 9:495–502
48. Hallett WH, Murphy WJ (2006) Positive and negative regulation of Natural Killer cells: therapeutic implications. Semin Cancer Biol 16:367–382
49. Colucci F, Caligiuri MA et al (2003) What does it take to make a natural killer? Nat Rev Immunol 3:413–425
50. Vosshenrich CA, Samson-Villeger SI et al (2005) Distinguishing features of developing natural killer cells. Curr Opin Immunol 17:151–158
51. Budagian V, Bulanova E et al (2006) IL-15/IL-15 receptor biology: a guided tour through an expanding universe. Cytokine Growth Factor Rev 17:259–280
52. Cooper MA, Bush JE et al (2002) In vivo evidence for a dependence on interleukin 15 for survival of natural killer cells. Blood 100:3633–3638
53. Kennedy MK, Glaccum M et al (2000) Reversible defects in natural killer and memory CD8 T cell lineages in interleukin 15-deficient mice. J Exp Med 191:771–780
54. Mrozek E, Anderson P et al (1996) Role of interleukin-15 in the development of human CD56+ natural killer cells from CD34+ hematopoietic progenitor cells. Blood 87:2632–2640
55. Puzanov IJ, Bennett M et al (1996) IL-15 can substitute for the marrow microenvironment in the differentiation of natural killer cells. J Immunol 157:4282–4285
56. Asai O, Longo DL et al (1998) Suppression of graft-versus-host disease and amplification of graft-versus-tumor effects by activated natural killer cells after allogeneic bone marrow transplantation. J Clin Invest 101:1835–1842
57. Farag SS, Fehniger TA et al (2002) Natural killer cell receptors: new biology and insights into the graft-versus-leukemia effect. Blood 100:1935–1947
58. Koh CY, Blazar BR et al (2001) Augmentation of antitumor effects by NK cell inhibitory receptor blockade in vitro and in vivo. Blood 97:3132–3137
59. Koh CY, Ortaldo JR et al (2003) NK-cell purging of leukemia: superior antitumor effects of NK cells H2 allogeneic to the tumor and augmentation with inhibitory receptor blockade. Blood 102:4067–4075
60. Koh CY, Raziuddin A et al (2002) NK inhibitory-receptor blockade for purging of leukemia: effects on hematopoietic reconstitution. Biol Blood Marrow Transplant 8:17–25
61. Murphy WJ, Bennett M et al (1992) Donor-type activated natural killer cells promote marrow engraftment and B cell development during allogeneic bone marrow transplantation. J Immunol 148:2953–2960
62. Murphy WJ, Keller JR et al (1992) Interleukin-2-activated natural killer cells can support hematopoiesis in vitro and promote marrow engraftment in vivo. Blood 80:670–677
63. Ruggeri L, Capanni M et al (2002) Effectiveness of donor natural killer cell alloreactivity in mismatched hematopoietic transplants. Science 295:2097–2100
64. Yokoyama WM (1995) Natural killer cell receptors. Curr Opin Immunol 7:110–120

65. Anderson SK, Ortaldo JR et al (2001) The ever-expanding Ly49 gene family: repertoire and signaling. Immunol Rev 181:79–89
66. Dorfman JR, Raulet DH (1998) Acquisition of Ly49 receptor expression by developing natural killer cells. J Exp Med 187:609–618
67. Long EO, Colonna M et al (1996) Inhibitory MHC class I receptors on NK and T cells: a standard nomenclature. Immunol Today 17:100
68. Raulet DH, Held W et al (1997) Specificity, tolerance and developmental regulation of natural killer cells defined by expression of class I-specific Ly49 receptors. Immunol Rev 155:41–52
69. Ryan JC, Seaman WE (1997) Divergent functions of lectin-like receptors on NK cells. Immunol Rev 155:79–89
70. Murphy WJ, Kumar V et al (1987) Rejection of bone marrow allografts by mice with severe combined immune deficiency (SCID). Evidence that natural killer cells can mediate the specificity of marrow graft rejection. J Exp Med 165:1212–1217
71. Murphy WJ, Kumar V et al (1990) Natural killer cells activated with interleukin 2 in vitro can be adoptively transferred and mediate hematopoietic histocompatibility-1 antigen-specific bone marrow rejection in vivo. Eur J Immunol 20:1729–1734
72. Murphy WJ, Kumar V et al (1990) An absence of T cells in murine bone marrow allografts leads to an increased susceptibility to rejection by natural killer cells and T cells. J Immunol 144:3305–3311
73. Yu YY, Kumar V et al (1992) Murine natural killer cells and marrow graft rejection. Annu Rev Immunol 10:189–213
74. O'Reilly RJ, Brochstein J et al (1986) Evaluation of HLA-haplotype disparate parental marrow grafts depleted of T lymphocytes by differential agglutination with a soybean lectin and E-rosette depletion for the treatment of severe combined immunodeficiency. Vox Sang 51(Suppl 2):81–86
75. Scott I, O'Shea J et al (1998) Molecular typing shows a high level of HLA class I incompatibility in serologically well matched donor/patient pairs: implications for unrelated bone marrow donor selection. Blood 92:4864–4871
76. Sobecks RM, Ball EJ et al (2008) Influence of killer immunoglobulin-like receptor/HLA ligand matching on achievement of T-cell complete donor chimerism in related donor nonmyeloablative allogeneic hematopoietic stem cell transplantation. Bone Marrow Transplant 41:709–714
77. Bennett M (1987) Biology and genetics of hybrid resistance. Adv Immunol 41:333–445
78. Bennett M, Taylor PA et al (1998) Cytokine and cytotoxic pathways of NK cell rejection of class I-deficient bone marrow grafts: influence of mouse colony environment. Int Immunol 10:785–790
79. Raziuddin A, Longo DL et al (2002) Increased bone marrow allograft rejection by depletion of NK cells expressing inhibitory Ly49 NK receptors for donor class I antigens. Blood 100:3026–3033
80. Raziuddin A, Longo DL et al (1998) Differential effects of the rejection of bone marrow allografts by the depletion of activating versus inhibiting Ly-49 natural killer cell subsets. J Immunol 160:87–94
81. Raziuddin A, Longo DL et al (1996) Ly-49 G2+ NK cells are responsible for mediating the rejection of H-2b bone marrow allografts in mice. Int Immunol 8:1833–1839
82. Ruggeri L, Mancusi A et al (2008) NK cell alloreactivity and allogeneic hematopoietic stem cell transplantation. Blood Cells Mol Dis 40:84–90
83. Kulkarni S, Martin MP, et al (2008) The Yin and Yang of HLA and KIR in human disease. Semin Immunol
84. Parham P (2005) MHC class I molecules and KIRs in human history, health and survival. Nat Rev Immunol 5:201–214
85. Lanier LL (2005) Missing self, NK cells, and The White Album. J Immunol 174:6565
86. Lanier LL (2005) NK cell recognition. Annu Rev Immunol 23:225–274

87. Diefenbach A, Raulet DH (2001) Strategies for target cell recognition by natural killer cells. Immunol Rev 181:170–184
88. Saether PC, Westgaard IH et al (2008) KLRE/I1 and KLRE/I2: a novel pair of heterodimeric receptors that inversely regulate NK cell cytotoxicity. J Immunol 181:3177–3182
89. Westgaard IH, Dissen E et al (2003) The lectin-like receptor KLRE1 inhibits natural killer cell cytotoxicity. J Exp Med 197:1551–1561
90. Hsu KC, Dupont B (2005) Natural killer cell receptors: regulating innate immune responses to hematologic malignancy. Semin Hematol 42:91–103
91. Kim S, Poursine-Laurent J et al (2005) Licensing of natural killer cells by host major histocompatibility complex class I molecules. Nature 436:709–713
92. Kim S, Sunwoo JB et al (2008) HLA alleles determine differences in human natural killer cell responsiveness and potency. Proc Natl Acad Sci USA 105:3053–3058
93. Garrido F, Ruiz-Cabello F et al (1997) Implications for immunosurveillance of altered HLA class I phenotypes in human tumours. Immunol Today 18:89–95
94. Karre K (2002) NK cells, MHC class I molecules and the missing self. Scand J Immunol 55:221–228
95. Karre K, Ljunggren HG et al (1986) Selective rejection of H-2-deficient lymphoma variants suggests alternative immune defence strategy. Nature 319:675–678
96. Kiessling R, Hochman PS et al (1977) Evidence for a similar or common mechanism for natural killer cell activity and resistance to hemopoietic grafts. Eur J Immunol 7:655–663
97. Ljunggren HG, Karre K (1990) In search of the "missing self": MHC molecules and NK cell recognition. Immunol Today 11:237–244
98. Parham P (2000) NK cell receptors: of missing sugar and missing self. Curr Biol 10:R195–R197
99. Raulet DH (2006) Missing self recognition and self tolerance of natural killer (NK) cells. Semin Immunol 18:145–150
100. Ruggeri L, Mancusi A et al (2007) Donor natural killer cell allorecognition of missing self in haploidentical hematopoietic transplantation for acute myeloid leukemia: challenging its predictive value. Blood 110:433–440
101. Giebel S, Locatelli F et al (2003) Survival advantage with KIR ligand incompatibility in hematopoietic stem cell transplantation from unrelated donors. Blood 102:814–819
102. Farag SS, Bacigalupo A et al (2006) The effect of KIR ligand incompatibility on the outcome of unrelated donor transplantation: a report from the center for international blood and marrow transplant research, the European blood and marrow transplant registry, and the Dutch registry. Biol Blood Marrow Transplant 12:876–884
103. Biron CA, Nguyen KB et al (1999) Natural killer cells in antiviral defense: function and regulation by innate cytokines. Annu Rev Immunol 17:189–220
104. Castriconi R, Dondero A et al (2004) Natural killer cell-mediated killing of freshly isolated neuroblastoma cells: critical role of DNAX accessory molecule-1-poliovirus receptor interaction. Cancer Res 64:9180–9184
105. Fernandez NC, Treiner E et al (2005) A subset of natural killer cells achieves self-tolerance without expressing inhibitory receptors specific for self-MHC molecules. Blood 105:4416–4423
106. Leung W, Handgretinger R et al (2007) Inhibitory KIR-HLA receptor-ligand mismatch in autologous haematopoietic stem cell transplantation for solid tumour and lymphoma. Br J Cancer 97:539–542
107. Leung W, Iyengar R et al (2004) Determinants of antileukemia effects of allogeneic NK cells. J Immunol 172:644–650
108. Leung W, Iyengar R et al (2005) Comparison of killer Ig-like receptor genotyping and phenotyping for selection of allogeneic blood stem cell donors. J Immunol 174:6540–6545
109. Hsu KC, Keever-Taylor CA et al (2005) Improved outcome in HLA-identical sibling hematopoietic stem-cell transplantation for acute myelogenous leukemia predicted by KIR and HLA genotypes. Blood 105:4878–4884

110. Yu J, Heller G et al (2007) Hierarchy of the human natural killer cell response is determined by class and quantity of inhibitory receptors for self-HLA-B and HLA-C ligands. J Immunol 179:5977–5989
111. Hsu KC, Gooley T et al (2006) KIR ligands and prediction of relapse after unrelated donor hematopoietic cell transplantation for hematologic malignancy. Biol Blood Marrow Transplant 12:828–836
112. Storek J (2008) Immunological reconstitution after hematopoietic cell transplantation – its relation to the contents of the graft. Expert Opin Biol Ther 8:583–597
113. Storek J, Zhao Z et al (2004) Recovery from and consequences of severe iatrogenic lymphopenia (induced to treat autoimmune diseases). Clin Immunol 113:285–298
114. Chklovskaia E, Nowbakht P et al (2004) Reconstitution of dendritic and natural killer-cell subsets after allogeneic stem cell transplantation: effects of endogenous flt3 ligand. Blood 103:3860–3868
115. Dulphy N, Haas P et al (2008) An unusual CD56(bright) CD16(low) NK cell subset dominates the early posttransplant period following HLA-matched hematopoietic stem cell transplantation. J Immunol 181:2227–2237
116. Jacobs R, Stoll M et al (1992) CD16- CD56+ natural killer cells after bone marrow transplantation. Blood 79:3239–3244
117. Koehl U, Bochennek K et al (2007) Immune recovery in children undergoing allogeneic stem cell transplantation: absolute CD8+ CD3+ count reconstitution is associated with survival. Bone Marrow Transplant 39:269–278
118. Schwinger W, Weber-Mzell D et al (2006) Immune reconstitution after purified autologous and allogeneic blood stem cell transplantation compared with unmanipulated bone marrow transplantation in children. Br J Haematol 135:76–84
119. Shilling HG, McQueen KL et al (2003) Reconstitution of NK cell receptor repertoire following HLA-matched hematopoietic cell transplantation. Blood 101:3730–3740
120. Hokland M, Jacobsen N et al (1988) Natural killer function following allogeneic bone marrow transplantation. Very early reemergence but strong dependence of cytomegalovirus infection. Transplantation 45:1080–1084
121. Schulze A, Schirutschke H et al (2008) Altered phenotype of natural killer cell subsets after haploidentical stem cell transplantation. Exp Hematol 36:378–389
122. Vitale C, Pitto A et al (2000) Phenotypic and functional analysis of the HLA-class I-specific inhibitory receptors of natural killer cells isolated from peripheral blood of patients undergoing bone marrow transplantation from matched unrelated donors. Hematol J 1:136–144
123. Nguyen S, Dhedin N et al (2005) NK-cell reconstitution after haploidentical hematopoietic stem-cell transplantations: immaturity of NK cells and inhibitory effect of NKG2A override GvL effect. Blood 105:4135–4142
124. Cooley S, McCullar V et al (2005) KIR reconstitution is altered by T cells in the graft and correlates with clinical outcomes after unrelated donor transplantation. Blood 106:4370–4376
125. Cooley S, Xiao F et al (2007) A subpopulation of human peripheral blood NK cells that lacks inhibitory receptors for self-MHC is developmentally immature. Blood 110:578–586
126. Murphy WJ, Koh CY et al (2001) Immunobiology of natural killer cells and bone marrow transplantation: merging of basic and preclinical studies. Immunol Rev 181:279–289
127. Siefer AK, Longo DL et al (1993) Activated natural killer cells and interleukin-2 promote granulocytic and megakaryocytic reconstitution after syngeneic bone marrow transplantation in mice. Blood 82:2577–2584
128. Alpdogan O, Eng JM et al (2005) Interleukin-15 enhances immune reconstitution after allogeneic bone marrow transplantation. Blood 105:865–873
129. Wang H, Grzywacz B et al (2007) The unexpected effect of cyclosporin A on CD56+CD16- and CD56+CD16+ natural killer cell subpopulations. Blood 110:1530–1539
130. Pihusch R, Holler E et al (2002) The impact of antithymocyte globulin on short-term toxicity after allogeneic stem cell transplantation. Bone Marrow Transplant 30:347–354

131. Penack O, Fischer L et al (2008) Serotherapy with thymoglobulin and alemtuzumab differentially influences frequency and function of natural killer cells after allogeneic stem cell transplantation. Bone Marrow Transplant 41:377–383
132. Alici E, Konstantinidis KV et al (2007) Anti-myeloma activity of endogenous and adoptively transferred activated natural killer cells in experimental multiple myeloma model. Exp Hematol 35:1839–1846
133. Chen G, Wu D et al (2008) Expanded donor natural killer cell and IL-2, IL-15 treatment efficacy in allogeneic hematopoietic stem cell transplantation. Eur J Haematol 81:226–235
134. Ortaldo JR, Winkler-Pickett RT et al (2005) In vivo hydrodynamic delivery of cDNA encoding IL-2: rapid, sustained redistribution, activation of mouse NK cells, and therapeutic potential in the absence of NKT cells. J Immunol 175:693–699
135. Pegram HJ, Jackson JT et al (2008) Adoptive transfer of gene-modified primary NK cells can specifically inhibit tumor progression in vivo. J Immunol 181:3449–3455
136. McKenna DH, Kadidlo DM et al (2005) The Minnesota Molecular and Cellular Therapeutics Facility: a state-of-the-art biotherapeutics engineering laboratory. Transfus Med Rev 19:217–228
137. Miller JS, Soignier Y et al (2005) Successful adoptive transfer and in vivo expansion of human haploidentical NK cells in patients with cancer. Blood 105:3051–3057
138. Weiss JM, Subleski JJ et al (2007) Immunotherapy of cancer by IL-12-based cytokine combinations. Expert Opin Biol Ther 7:1705–1721
139. Wigginton JM, Gruys E et al (2001) IFN-gamma and Fas/FasL are required for the antitumor and antiangiogenic effects of IL-12/pulse IL-2 therapy. J Clin Invest 108:51–62
140. Wigginton JM, Lee JK et al (2002) Synergistic engagement of an ineffective endogenous anti-tumor immune response and induction of IFN-gamma and Fas-ligand-dependent tumor eradication by combined administration of IL-18 and IL-2. J Immunol 169:4467–4474
141. Wigginton JM, Wiltrout RH (2002) IL-12/IL-2 combination cytokine therapy for solid tumours: translation from bench to bedside. Expert Opin Biol Ther 2:513–524
142. Miller JS, Tessmer-Tuck J et al (1997) Low dose subcutaneous interleukin-2 after autologous transplantation generates sustained in vivo natural killer cell activity. Biol Blood Marrow Transplant 3:34–44
143. Prlic M, Kamimura D et al (2007) Rapid generation of a functional NK-cell compartment. Blood 110:2024–2026
144. Dubois S, Patel HJ et al (2008) Preassociation of IL-15 with IL-15R alpha-IgG1-Fc enhances its activity on proliferation of NK and CD8+/CD44high T cells and its antitumor action. J Immunol 180:2099–2106
145. Mortier E, Quemener A et al (2006) Soluble interleukin-15 receptor alpha (IL-15R alpha)-sushi as a selective and potent agonist of IL-15 action through IL-15R beta/gamma. Hyperagonist IL-15 x IL-15R alpha fusion proteins. J Biol Chem 281:1612–1619
146. Gong JH, Maki G et al (1994) Characterization of a human cell line (NK-92) with phenotypical and functional characteristics of activated natural killer cells. Leukemia 8:652–658
147. Maki G, Klingemann HG et al (2001) Factors regulating the cytotoxic activity of the human natural killer cell line, NK-92. J Hematother Stem Cell Res 10:369–383
148. Yan Y, Steinherz P et al (1998) Antileukemia activity of a natural killer cell line against human leukemias. Clin Cancer Res 4:2859–2868
149. Maki G, Tam YK et al (2003) Ex vivo purging with NK-92 prior to autografting for chronic myelogenous leukemia. Bone Marrow Transplant 31:1119–1125
150. Reid GS, Bharya S et al (2002) Differential killing of pre-B acute lymphoblastic leukaemia cells by activated NK cells and the NK-92 ci cell line. Clin Exp Immunol 129:265–271
151. Suck G (2006) Novel approaches using natural killer cells in cancer therapy. Semin Cancer Biol 16:412–418
152. Terme M, Ullrich E et al (2008) Natural killer cell-directed therapies: moving from unexpected results to successful strategies. Nat Immunol 9:486–494
153. Verheyden S, Demanet C (2008) NK cell receptors and their ligands in leukemia. Leukemia 22:249–257

NK Cells, NKT Cells, and KIR in Solid Organ Transplantation

Cam-Tien Le and Katja Kotsch

Abstract In solid organ transplantation, natural killer (NK) cells have emerged as a particular focus of interest because of their ability to distinguish allogeneic major histocompatibility complex (MHC) antigens and their potent cytolytic activity. On the basis of the potential relevance of this, NK cells have recently been shown to participate in the immune response in both acute and chronic rejection of solid organ allografts. Meanwhile, it has been demonstrated by several experimental and clinical studies that NK cells and natural killer T (NKT) cells can determine transplant survival by rejecting an allograft not directly but indirectly by influencing the alloreactivity of T cells or by killing antigen-presenting cells (APCs). Moreover, NK cells are influenced by immuno-suppressive regimens such as calcineurin inhibitors, steroids, or therapeutic antibodies. Recent findings suggest that NK cells also play a profound role in allograft tolerance induction, suggesting that the role of this lymphocyte subset in graft rejection and tolerance induction needs to be reconsidered.

1 Introduction

Advances in immunosuppressive protocols, organ preservation, and perioperative management have significantly reduced the risk of acute rejection of solid organs in the early posttransplantation phase. However, the development of chronic rejection and therefore, restricted graft survival in the long-term still remain a serious problem in transplant medicine. For instance, after renal transplantation, the clinical outcome is dependent on various antigen-independent risk factors including donor brain death, age, sex, ischemia reperfusion injury (IRI), and antigen-dependent risk

---

C.-T Le and K. Kotsch (✉)
Institut für Medizinische Immunologie, Charité-Universitätsmedizin Berlin, Schumannstrasse 20/21, 10117, Berlin, Germany
e-mail: katja.kotsch@charite.de

factors such as *human leukocyte antigen* (HLA) matching resulting in inflammation and tissue injury and therefore playing a critical role in the initiation of chronic graft failure [1–5]. Especially following IRI, oxidative stress induces cytokine and chemokine upregulation, which might enhance the recruitment of recipient-derived inflammatory cells capable of mediating tissue injury directly or indirectly. Meanwhile, it has been comprehensively documented that $CD4^+$ T cells are necessary and sufficient for the initiation of acute rejection but, in the absence of costimulation, the participation of other lymphocyte subsets is evident. An emerging concept in the field of transplantation research is therefore to reveal the linkage between innate and adaptive immunity, as there is increasing evidence of potential interplay between NK cells and the adaptive immune responses of the host. Whereas the immune function of NK cells is well described during viral infections or tumor diseases, where either the downmodulation or induction of NK cell receptor ligands results in NK cell alloreactivity, the exact role of NK cells in the rejection or acceptance of solid organs still remains an obstacle. Although a function for alloreactive NK cells has been described in preventing graft-versus-host disease (GvHD) in the setting of bone-marrow transplantation (BMT) [6], NK cells seem to be by themselves neither necessary nor sufficient for rejection of organ allografts. However, an increasing number of studies support the concept that NK cells are important players mediating allograft destruction and that NK cells may be required during tolerance induction.

1.1 Biology of Natural Killer and Natural Killer T Cells

NK cells are large granular cytotoxic lymphocytes that represent a fundamental component of the innate immune system. They are derived from $CD34^+$ hematopoietic progenitor cells (HPCs) [7] and, once released, they comprise roughly 5–20% of lymphocytes in the spleen, liver, and peripheral blood and are present at lower frequencies in the bone marrow, thymus, and lymph nodes [8]. They were originally identified by their ability to spontaneously kill certain tumor target cells in vivo and in vitro without sensitization [9, 10], and this killing was not restricted by the target cell's expression of major histocompatibility complex (MHC) molecules [10, 11]. NK cells are not only an important source of innate immunoregulatory cytokines, but they also possess direct or natural cytotoxic activity against virus-infected, leukemic, and other tumor cells. They also mediate the antibody-dependent cellular cytotoxicity (ADCC) of targets through $Fc\gamma RIII$ (CD16), a receptor that binds the Fc portion of antibodies [12, 13]. In general, the traditional phenotype defining human NK cells is characterized by the absence of the T cell receptor complex (TCR, CD3) and expression of CD56, the 140-kDa isoform of neural cell adhesion molecule (NCAM) which is also found on a minority of T cells [14, 15]. On the basis of their CD56 receptor expression density, human NK cells can be distinguished as $CD56^{dim}$ or $CD56^{bright}$ NK cells. In contrast, murine NK cells can be identified by using antibodies against the NK1.1 molecule [16, 17], the pan-NK marker DX5 [18], or the asialo GM1 (ASGM1) surface molecule [19]. These markers can also be expressed on subsets of T lymphocytes and granulocytes and therefore are not lineage-specific.

Also, another important subset of nonclassical MHC class I-restricted cells are NKT cells, which comprise a very heterogeneous group of T cells but share properties with NK cells. They were originally characterized in mice as cells that express both a TCR and NK1.1 (CD161c in humans) [20]. The most studied and best-characterized NKT cell population in mice and humans is referred to as type I NKT cells, or iNKT cells. These NKT cells express a TCR formed by the rearrangement of the Vα24 gene segment (Vα14 in mice) to the Jα18 gene segment. iNKT cells recognize glycolipid antigens presented by the nonpolymorphic MHC class I-like molecule CD1d [21]. However, several studies have identified subsets of CD1d-dependent T cells that either express or do not express the invariant Vα24–Jα18 (Vα14–Jα18 in mice) TCR and/or CD161 (NK1.1 in mice). Although all subsets have been referred to as NKT cells, they probably represent functionally distinct cell types [22]. iNKT cells are found with the highest frequency in the liver and the bone marrow of mice, with significant numbers also in the thymus, spleen, and peripheral blood. In humans, the frequency of iNKT cells is usually much lower and a high degree of variability between individuals has been reported [23].

1.2 Inhibitory and Activating Receptors of NK and NKT Cells

Meanwhile it has become clear that NK cells possess a variety of inhibitory and activating receptors that engage MHC class I molecules, MHC class I-like molecules, and molecules unrelated to MHC, and that their cytolytic activity is controlled by the balance between their inhibitory and activating receptors. Thus, NK cells are restricted in engaging potential target cells depending on the expression of their ligands, but this occurs in a very complex fashion that is still not completely understood. According to the "missing-self" hypothesis, NK cells kill target cells that display reduced levels of MHC class I antigens such as virally transformed or tumor cells [24]. In humans, there are three types of major MHC class I specific inhibitory receptors expressed by NK cells. The human killer-cell immunoglobulin-like receptors (KIRs; Ly49 receptor family in mice) and the immunoglobulin-like transcripts (ILTs) bind classical and nonclassical HLA class I molecules, whereas the C-type lectin heterodimer CD94/NKG2A binds to the nonclassical MHC class I molecule HLA-E (Qa-1^b in mice) [25–27]. The binding of MHC class I complexes to KIRs or to the heterodimeric CD94/NKG2A receptor initiates inhibitory pathways that can override activating signals [28].

Among activating receptors, NK cells express the C-type lectin homodimer NKG2D, CD16, and natural cytotoxicity receptors (NCRs) such as NKp30, NKp44, and NKp46, whereas the latter is selectively expressed on both human and murine NK cells. Additionally NK cells express a variety of activating coreceptors, including 2B4 [CD244], NKp80, NTB-A, and DNAM-1 (CD226), which may also contribute to NK cell activation [29]. NKG2D binds to cellular stress-induced molecules including nonclassical MHC class I molecules (MHC class I polypeptide-related sequence A or B, MICA and MICB in humans), and the MHC class I-related UL-16 binding proteins (ULBP) 1–4 [30]. While the ligands for NCRs on tumor cells are still unknown, viral hemagglutinins have been suggested as ligands for NKp44 and NKp46 [31, 32]. In addition, poliovirus receptor (PVR; CD155) and nectin-2 (CD112) have been identified as DNAM-1 ligands [33], whereas CD48 engages 2B4 [34, 35].

2 The Role of NK Cells in Solid Allograft Rejection

As infiltration of NK cells in renal and cardiac allografts has been observed shortly after transplantation in both clinical and experimental models, activation of NK cells seems to be critical posttransplantation [36–40]. This infiltration often occurs before evidence of T cell infiltration and is consistent with the role of NK cells as early innate effector cells in response to inflammatory stimuli. Early studies have shown that IFNγ produced by NK cells during interaction with allogeneic endothelial cells stimulates MHC class I and class II expression, rendering them more susceptible to attack by alloantigen-specific T cells following priming and recruitment to the graft site [41–44] (Fig. 1). Moreover, NK cells are not only present but are activated to effector function following infiltration of solid organ allografts as it has been reported that rat NK cell-mediated cytolysis of donor target cells in vitro is increased in NK cell populations isolated from recipients of allogeneic heart grafts [39, 40].

The original assumption that NK cells do not participate in the rejection of solid allografts was supported by experiments where depletion of NK cells did not result in graft acceptance of skin, heart, or liver allografts [37, 45]. In contrast, the final evidence that NK cells play an important role in the rejection of solid transplants is the finding that depletion of NK cells in the absence of CD28 costimulation results in markedly prolonged graft survival in a complete MHC-mismatched mouse model of heterotopic heart transplantation (Balb/c, H-2^d→C57BL/6, H2^b). In this setting, NK and/or NKT cells are proposed to infiltrate the allograft tissue, become activated because of the absence of self MHC expression by the graft and provide help to CD28$^{-/-}$ T cells, thereby overcoming costimulation deficiency [46]. Another proof of the functional role of NK cells in allograft rejection has been provided by McNerney et al., who showed in the same model that NK but not NKT cells were required for cardiac rejection. NK cells which bear the inhibitory receptor Ly49G suppressed rejection, whereas a subset of NK cells lacking inhibitory Ly49 receptors for donor MHC class I molecules was sufficient to promote rejection. The same authors also illustrated that NK cells promote the expansion and effector function of alloreactive T cells in vitro and in vivo, and proposed that NK cells help the priming and differentiation of alloreactive T cells during a transplant response [47]. Although these studies suggested that rejection was

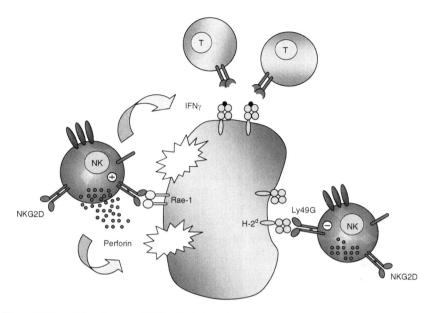

Fig. 1 Differential pathways of NK cells influencing allograft outcome. During the interaction with allogeneic endothelial cells NK cells become activated via receptor–ligand interactions (e.g., NKG2D-Rae-1) and kill allogeneic cells in a perforin-dependent pathway. Moreover, IFNγ produced by NK cells stimulates MHC class I and class II expression, rendering the allograft more susceptible to attack by antigen-specific T cells. On the contrary, NK cells positive for the inhibitory receptor Ly49G recognizing MHC H-2^d antigens seem to suppress rejection

independent of the activating receptors Ly49D and NKG2D, treatment with a neutralizing antibody against NKG2D was highly effective in preventing CD28-independent rejection of cardiac allografts in another study [48]. The same authors demonstrated the upregulation of NKG2D ligands during the rejection response, including Rae-1, Mult-1, and H60 in tissue grafts, thereby enhancing effector function of graft-infiltrating $CD8^+$ T cells and NK cells. These experimental data were supported by initial observations illustrating that stress-induced MIC proteins on human renal and pancreatic allografts are induced during acute rejection [49, 50]. In this context renal tubular epithelial cells express the Rae-1 protein as a consequence of IRI, thereby leading to NKG2D-mediated and perforin-dependent NK cell killing (Fig. 1). Consequently, depletion of NK cells attenuated kidney IRI whereas adoptive transfer of NK cells worsened kidney injury in NK, T, and B cell deficient mice [51]. NK cells have also been implicated in promoting chronic allograft destruction as early NK cell infiltration into transplants occurs as a consequence of IRI [52]. Applying a model of semiallogeneic cardiac transplants, NK cells were activated by the absence of self MHC class I molecules on donor endothelium and thus participated in the pathogenesis of cardiac allograft vasculopathy (CAV). However, this process requires functional interactions with T cells and recipient-derived IFNγ [53].

2.1 Role of Cytokines, Chemokines, and Chemokine Receptors on NK Cells

In alloantigen-induced T cell activation, additional antigen-independent factors including cytokines, chemokines, and their receptors were discovered to play a critical role in the activation of NK cells, thus contributing to the development of transplant rejection. In this context a recent study suggested a functional role of IL-15 in graft rejection. Kroemer et al. demonstrated that NK cells in $Rag^{-/-}$ mice in a resting state readily reject allogeneic cells, but not skin allografts. However, treatment with anIL-15/IL-15Rα complex resulted in activation of NK cells in vivo, expressing an activated phenotype and potent in mediating acute skin allograft rejection in the absence of adaptive immune cells [54]. These data suggest that differences in the activation status of NK cells may significantly affect their effector functions in vivo. Furthermore, NK cell activity stimulates adhesion molecule expression and production of chemokines (CXC and CC chemokines) during rejection, including MCP-1 or CX3CL1 (fractalkine) [55, 56]. In a model of acute allograft rejection of rat orthotopic liver transplantation the chemokines CCL3, CXCL10, and CX3CL1 were significantly increased in allografts posttransplantation, suggesting a role for these chemokines in the recruitment of alloantigen-primed T cells and other recipient leukocyte populations to the allograft. Additionally IFNγ levels were markedly increased in the serum of recipients, indicating that graft-infiltrating NK cells were the major source of this immunoregulatory cytokine [57]. The importance of chemokine receptor expression on NK cells was further examined by investigating the role of CXCR3, the binding receptor for chemokines including CCL9, CCL10, and CCL11, which is expressed on a small percentage of NK but to a greater extent on NKT cells [58] (Fig. 2). As previously shown in $CXCR3^{-/-}$ mice, the administration of an anti-CXCR3 antibody resulted in prolonged graft survival of heart and islet allografts. In combination with the administration of rapamycin this therapy even led to long-term graft acceptance [58, 59].

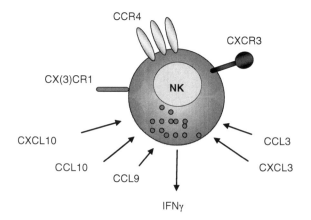

Fig. 2 Chemokine and chemokine receptor expression. Relevant chemokines influencing the activation status of NK cells and chemokine receptors expressed by NK cells themselves which have been demonstrated to be involved in solid graft rejection

Likewise, CCR1-deficient mice show a significantly prolonged survival of MHC-mismatched heart allografts [60] and this observation corresponds with a prolongation of transplant survival and a reduction of graft-infiltrating NK cells in the absence of the chemokine receptor CX(3)CR1 under low-dose cyclosporin A therapy [61]. Similar results illustrated that the percentage of NKT cells is markedly reduced in hearts grafted into CCR4-deficient recipients, suggesting that CCR4 expressed on host lymphocytes plays an important role in the recruitment of NKT cells into cardiac allografts [62].

3 NK Cells, NKT Cells, and Tolerance

Despite observed NK cell alloreactivity during rejection processes, it has become apparent that NK cells cannot be regarded only as "killers" but also have important regulatory properties. Beside T cells, antigen-presenting cells (APCs) are potential targets of NK cell regulation. NK cells are capable of killing autologous immature DCs though they can also drive DC maturation. However, it remains uncertain as to what extent activated NK cells are involved in regulating host APCs. In solid organ transplantation, the graft itself contains donor APCs which migrate to the recipient's secondary lymphoid organs where they directly prime alloreactive T cells [63]. Consequently, survival and distribution of donor-derived APCs in the host might affect either the initiation of rejection or the induction of tolerance.

3.1 The Role of NK Cells Contributing to Allograft Tolerance

There are several studies published indicating that the alloreactivity of NK cells may not only play an important role in influencing allograft rejection but may also affect the induction of tolerance by costimulatory blockade. A crucial role for host MHC class I-dependent NK cell reactivity in allograft tolerance could be demonstrated in mice by either inducing costimulation blockade using CD154-specific antibody therapy or by targeting cellular adhesion by blocking LFA-1 [64]. In this model of islet transplantation, tolerance induction was shown to require host expression of both MHC class I$^+$ and NK1.1$^+$ cells but was independent of CD8$^+$ T cell-dependent immunity. Additionally, CD154-specific antibody-induced allograft tolerance was demonstrated to be perforin-dependent, as perforin-competent NK cells were sufficient to restore allograft tolerance in perforin-deficient recipients. The authors therefore concluded that NK cells might promote allograft tolerance by eliminating activated alloreactive recipient-derived T cells [64]. In addition, Coudert et al. provided evidence that NK cells, through their interaction with allogeneic APCs, can quantitatively and qualitatively control allospecific CD4$^+$ T responses in vivo. Alloreactivity of host NK cells mediated by missing inhibitory MHC class I ligands expressed by donor APCs resulted in diminished allospecific Th cell responses associated with the development of effector Th cells producing IFNγ rather than type 2 cytokines. In contrast, alloreactive CD4$^+$ T cell priming and Th2 cell development were restored by neutralizing NK cells. Similar results were obtained by analyzing the effect of NK cell activation on CD4$^+$ T cell responses to skin allografts. Despite the dramatic effect of NK cells on alloreactive Th1/Th2 cell development, the kinetics of skin graft rejection were not affected [65].

The functional aspects of NK cell reactivity have been further demonstrated in a skin transplantation model. Skin allografts contain a subset of APCs which are usually destroyed by host NK cells. But in the absence of NK cells, surviving donor APCs migrate to the host lymphoid and extralymphoid sites. They directly stimulate the activation of alloreactive T cells which are more resistant to costimulatory blockade treatment, thereby preventing stable skin allograft survival [66]. These observations were further supported by a recent study in a model of CD4$^+$ T cell-mediated allogeneic skin graft rejection, where the absence of host NK-cell alloreactivity was characterized by enhanced expansion of alloreactive effector T lymphocytes, including Th2 cells in the rejected tissues. In addition it was demonstrated that blood-borne host NK cells (CD127$^-$) expressing the H-2^d-specific activating receptor Ly49D were recruited within draining lymph nodes in a L-Selectin-dependent manner and rapidly eliminated allogeneic H-2^d dendritic cells (DCs) through the perforin pathway, thus regulating alloreactive CD4$^+$ T cell responses in CD8$^+$ T cell-deficient C57BL/6 (H-2^b) recipients (Fig. 3). However, in wild-type mice, the authors showed that NK cells, by eliminating allogeneic DCs, strongly inhibited alloreactive CD8$^+$ T cell responses [67].

3.2 The Role of NKT Cells Contributing to Allograft Tolerance

The first data showing that Vα14 NKT cells are required for the induction of tolerance were provided by Ikehara et al. Administration of an anti-CD4 mAb allowed islet xenografts to be accepted by C57BL/6 mice, with no need for immunosuppressive drugs. This effect was associated with NKT cells, as rat islet xenografts were rejected in Vα14 NKT cell-deficient mice, despite the anti-CD4 mAb treatment [68]. Similarly, in models in which tolerance was induced against cardiac allografts by blockade of LFA-1/ICAM-1 or CD28/B7 interactions, long-term acceptance of the grafts was observed only in wild-type but not in Vα14 NKT cell-deficient mice. Adoptive transfer with Vα14 NKT cells restored long-term acceptance of allografts in Vα14 NKT cell-deficient mice. Experiments using IL-4- or IFNγ-deficient mice suggested a critical contribution of IFNγ to the Vα14 NKT cell-mediated allograft acceptance in vivo [69], although this could not be verified in a model of skin transplantation [70]. A possible mechanism in the maintenance of cardiac allograft tolerance mediated by NKT cells was suggested by a study demonstrating that blocking of the interaction between the chemokine receptor

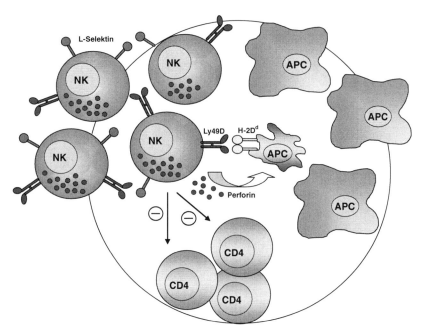

Fig. 3 NK cell-mediated killing of allogeneic APCs. NK cells positive for the $H\text{-}2^d$-specific activating receptor Ly49D were recruited within draining lymph nodes in an L-selectin-dependent manner and rapidly eliminate allogeneic $H\text{-}2^d$ DCs through the perforin pathway

CXCR6, highly expressed on Vα14 NKT cells, and its ligand, CXCL16, resulted in the failure to maintain graft tolerance and thus in the induction of acceleration of graft rejection. In this mouse transplant tolerance model, the expression of CXCL16 was upregulated in the tolerated allografts, and anti-CXCL16 mAb inhibited intragraft accumulation of NKT cells [71].

The functional role of CD1d-reactive NKT cells was further emphasized by the finding that anti-CD1d mAb abolished corneal graft survival in wild-type mice. In this model, the presence of CD1-restricted NKT cells was associated with the induction of allospecific regulatory T cells [72]. In contrast to activating receptors, the concept of Ly49 inhibitory receptors regulating immune reactivity to self by regulating the immune activity of individual cells has been addressed by Watte et al. in the same model. Blocking ligation of Ly49 C/I inhibitory receptor prevented NKT cell production of IL-10 and the subsequent development of tolerance and corneal graft survival. Furthermore, in the presence of TCR stimulation, cross-linking of Ly49 C/I on $CD4^+$ NKT cells stimulated an increase in IL-10 mRNA and a decrease in IFNγ [73]. Additionally it was demonstrated that NKT cells from transplant-tolerant recipients of cardiac allografts produced higher levels of IL-10, which is required for the maintenance of tolerance. DCs from wild-type tolerant recipients but not NKT cell-deficient recipients showed a higher IL-10-producing profile, a more immature phenotype, and tolerogenic capability. $CD4^+$ T cells from wild-type tolerant recipients but not NKT cell-deficient recipients also produced higher levels of IL-10 upon alloantigen stimulation and showed lower proliferative activity, which was reversed by blocking the IL-10 receptor [74].

4 The Functional Role of KIR on Solid Graft Outcome

In haploidentical BMT, reconstituting NK cells of donor origin develop alloreactivity when their inhibitory KIR receptors do not match with HLA-C ligands displayed by recipient cells. Especially in patients with acute myeloid leukemia (AML) alloreactive NK cells protect from the development of GvHD and leukemia relapse [75, 76]. As NK cell activity does not require prior sensitization, the allograft endothelium would appear to be an ideal target for the expression of NK cell effector function. In this context the absence of self MHC class I molecules on allogeneic cells would be expected to activate NK cells to express these functions early following solid organ transplantation. Activation of NK cells to mediate direct cytolysis of allogeneic target cells could indicate the ability of NK cells to directly mediate rejection of solid tissue allografts. Alternatively, NK cell activity may augment inflammation and facilitate the rejection process mediated by alloantigen-specific T cells following priming.

KIRs are named according to their structural and functional characteristics, i.e., by their number of extracellular Ig domains (2D or 3D) and the type of their intracellular tail mediating either an inhibitory (long [l]) or an activating (short [s]) intracellular signal. KIRs bind to different HLA class I antigens whose specificity is determined by amino acids in the C-terminal portion of the MHC class I $\alpha 1$ helix [77–79]. KIR2DL1 binds to HLA-C group 2 molecules (HLA-C2), which have amino acids Asn^{77} and Lys^{80}, KIR2DL2/2DL3 binds to HLA-C group 1 molecules (HLA-C1), which have amino acids Ser^{77} and Asn^{80}, and KIR3DL1 binds to HLA-B allele products, which contain the Bw4 epitope determined by amino acid positions 77–83 on the $\alpha 1$ helix of the heavy chain [80]. The Bw4 epitope is present on approximately one-third of all HLA-B alleles and to a smaller extent on HLA-A alleles, and the inhibitory receptor KIR3DL2 has been described to bind HLA-A*03, and *11 alleles [81]. In contrast, the ligands for activating KIRs (designated 2DS and 3DS) are less well defined, although it has been suggested that they bind to the same HLA-B or HLA-C molecules as recognized by their related inhibitory KIRs. On the basis of the presence or absence of multiple activating KIR receptors patients express either an A/A, A/B, or B/B genotype. Group A haplotypes consist of six genes encoding inhibitory receptors and one activating receptor (KIR2DS4). In contrast, the group B haplotypes contain a varying number of activating KIR genes (but no KIR2DS4) [82]. The genomic region that encodes KIRs exhibits extensive variability among individuals because of differences in gene content, gene copy number, and allelic polymorphism, thus creating a heterogenous NK cell population in each individual [83].

4.1 KIRs and Transplant Rejection

Several studies investigating the role of NK reactivity mediated by potential KIR–HLA interactions on solid graft outcome have been performed. For instance, by studying HLA ligand incompatibility in a large study cohort comprising more than 2,757 deceased-donor transplants, no correlation with kidney graft survival could be observed [84]. In contrast, Bishara et al. found that liver recipients transplanted from donors with matching HLA-C groups had significantly fewer rejection episodes in the first year after transplantation, whereas disparities for the HLA-Bw4 epitope did not affect the outcome [85]. However, neither investigation considered the KIR genotypes of transplant recipients for analysis although there are various possibilities that could affect NK cell alloreactivity against a solid allograft: (1) a mismatch is given when the recipient displays a certain KIR receptor but the donor graft does not have the corresponding HLA ligand; (2) a match is present, when a defined KIR receptor is expressed by the recipient and the corresponding HLA ligand is displayed by the allograft; or (3) when the allograft has a certain HLA allele but the recipient is lacking the corresponding receptor. Moreover, it has to be considered that recipients carrying an AA haplotype generate NK cells expressing only inhibitory KIRs whereas patients carrying an AA/AB haplotype can generate NK cells expressing both activating and inhibitory KIRs.

It is proposed that the inhibitory receptor KIR2DL1 and its corresponding ligand HLA-C group 2 show a stronger interaction in comparison to KIR2DL2/3 receptors and their corresponding HLA-C group 1 ligand, thus leading to enhanced inhibition of NK cells [86]. These results were supported by observations that NK cells from HLA-C group 1 homozygous subjects positive for the inhibitory receptor KIR2DL3 secreted more IFNγ at earlier time points after infection with influenza A virus than did NK cells from HLA-C group 2 homozygous subjects expressing the inhibitory KIR2DL1 receptor [87]. These results support the idea that the two subsets of HLA-C alleles exhibit significant differences in biological activity. Recently, the potential influence of the HLA-C donor type on allograft outcome was demonstrated suggesting a more potent inhibition of host NK cell function by the expression of an HLA-C allele by the donor allograft. Hanvesakul et al. showed that an HLA-C group 2 allele by the donor allograft was associated with less histological evidence of chronic rejection and graft cirrhosis, reduction in graft loss, and an improvement in patient survival at 10 years [88]. Although the authors did not observe a reduced risk of acute rejection in this transplant setting, Kunert et al. [89] documented that an HLA-C group 2 homozygous allograft resulted in a decreased risk of acute rejection in kidney transplantation. Furthermore, an association between a higher number of inhibitory receptors in the recipient´s genotype and stable renal function was found [89]. These results were supported by earlier studies in the same transplantation setting, showing in an ex vivo setting that NK recipient antidonor cytotoxicity was increased 3 days after transplantation even in the presence of immunosuppression. Interestingly, recipients exhibiting increased NK cytotoxicity were found to express more activating KIR genes for donor MHC class I antigens [90]. Similarly for lung transplantation, a recent study suggested that the lack of activating KIRs (A/A haplotype) resulted in a significant higher risk of the development of the bronchiolitis obliterans syndrome (BOS) [91]. Taken together, these findings support the concept that incompatibility for HLA–KIR ligand interactions and the presence of activating KIRs dominate NK cell activation in vitro and probably in vivo. Although it was demonstrated that neither the presence of self HLA epitopes nor particular KIR genes and haplotypes are associated with the occurrence of acute rejection after reduction of immunosuppression [92], donor HLA-C genotype might be a potential determinant of clinical outcome in the short- and long-term after solid organ transplantation.

4.2 KIRs and Viral Infections Posttransplantation

Viral infection is a common complication after solid organ transplantation. In the setting of BMT an association between the number of activating KIR receptors and cytomegalovirus (CMV) reactivation has been illustrated. Additional activating KIR genes in the donor compared to the recipient's genotype correlate with a lower incidence of CMV reactivation [93]. Similarly, kidney transplantation patients with a KIR A/A genotype showed a higher rate of CMV infection and reactivation as compared to transplant recipients with more than one activating KIR gene (B haplotype), whereas no association with the rate of other viral or nonviral infections, transplant function, and the number of clinical or subclinical rejection episodes was observed [94]. Importantly, this protection was dependent on the total number of KIR genes, i.e., the more KIR receptors an individual possessed the lower was the likelihood of CMV-related infections. In agreement with these data, it could be demonstrated in a large patient cohort that the absence of the HLA-C ligand for inhibitory KIR and the presence of activating KIR genes in renal recipients were both associated with a lower rate of CMV infection after transplantation [95]. In summary, these observations provide evidence for a protective effect of activating KIR on the rate of posttransplant CMV infections and reactivations.

5 Influence of Immunosuppressants on NK Cells

Initially, several in vivo and in vitro studies suggested that NK cells are not targeted by current clinical immunosuppressive drugs [39, 40, 96, 97]. However, with emerging studies investigating NK cells in the context of transplantation, recent data indicate that NK cell function is indeed targeted by various immunosuppressants.

5.1 NK Cells, Steroids, and Calcineurin Inhibitors

The influence of different immunosuppressive drugs on NK cell function has attracted particular interest, as it has recently been demonstrated that steroids and calcineurin inhibitors limit the function of IL-2-activated NK cells. For instance, cyclosporin A (CsA) induces a dose-dependent and selective inhibition in the IL-2- and IL-15-induced proliferation of human $CD56^{dim}$ NK in contrast to $CD56^{bright}$ NK cells. NK cells cultured in CsA retained cytotoxicity against the target cell line K562 and, following IL-12 and IL-18 stimulation, CsA-treated NK cells showed more IFNγ-producing cells [98]. In addition, Chiossone et al. showed that human NK cells cultured in either IL-2 or IL-15 display different susceptibility to methylprednisolone treatment in terms of cell survival, proliferation, and NK receptor-mediated cytotoxicity [99]. Steroid-induced inhibition of NK-cell cytotoxicity not only resulted in downregulation of surface expression or function of the activating receptors but also affected phosphorylation of the ERK1/2 signaling pathway, thus inhibiting granule exocytosis. Accordingly, methylprednisolone inhibited Tyr phosphorylation of STAT1, STAT3, and STAT5 in IL-2-cultured NK cells but only marginally in IL-15-cultured NK cells, whereas JAK3 was inhibited under both conditions. In contrast, rat NK cells demonstrated robust function in both the absence and presence of cyclosporin and FK506, whereas rapamycin significantly inhibited proliferation and cytotoxicity of NK cells in vitro. Experimental investigations in vivo illustrated that NK cell numbers remained stable in graft recipients treated with cyclosporin and FK506, whereas there was a significant decrease in NK cells in rapamycin-treated recipients [100].

5.2 NK Cells and Therapeutic Antibodies

An influence of NK cells after antibody therapy has been illustrated in patients with multiple sclerosis treated with a humanized monoclonal antibody directed against the anti-IL2Rα chain (daclizumab), which is also frequently used in solid organ transplantation. Application of daclizumab was associated with a significant expansion of $CD56^{bright}$ NK cells in vivo, thereby limiting the survival of activated T cells in a contact-dependent manner. Positive correlations between expansion of $CD56^{bright}$ NK cells and contraction of $CD4^+$ and $CD8^+$ T cell numbers in individual patients in vivo provide supporting evidence for NK cell-mediated negative immunoregulation of activated T cells during daclizumab therapy [101]. Moreover, rituximab, a therapeutic monoclonal antibody (anti-CD20) commonly applied for the treatment of B cell lymphoma, increases the killing frequency of both resting and IL-2-activated NK cells in vitro, therefore suggesting that strategies resulting in increased serial killing of NK cells can enhance the antitumor activities of NK cells [102]. In contrast, further observations illustrate that rabbit polyclonal antithymocyte globulin (rATG) and alemtuzumab (anti-CD52), both antibodies frequently used as induction agents in solid organ transplantation, induce rapid apoptosis in NK cells and a strong induction of inflammatory cytokines (e.g., TNFα, IFNγ, FasL). This effect is exclusively mediated via the binding of the IgG1 Fc part to the low-affinity receptor for IgG, CD16 (FcγRIII), and independently of antibody specificity [103]. As NK cells are functionally relevant for the effective clearance of opportunistic viral infections and antitumor activity posttransplantation, the latter observations should be considered in defining the optimal treatment dosage in clinical settings and for the generation of therapeutic antibodies in the future.

6 Conclusion

In order to achieve donor-specific tolerance to allogeneic organ transplants, it is essential to understand the various cell types involved in the rejection process. Although it was originally anticipated that NK cells are not involved in solid graft rejection, recent findings contribute to a new understanding of the functional role of NK and NKT cells in the context of solid organ transplantation. Both lymphocyte subsets cannot be regarded only as killers because of their cytolytic effector functions, but have been recognized as active participants in the development of acute and chronic rejection. More importantly, NK and NKT cells have been shown to be involved interactively with T cells and DCs for tolerance induction. Certainly, the exact mechanisms underlying NK cell tolerance in vivo remain to be clarified and there are still some questions left regarding whether mature NK cells can become tolerant with tolerance-inducing protocols. In conclusion, the potential duality of NK function influencing solid graft outcome requires a reconsideration of NK cell reactivity in the future.

References

1. Cecka JM (1994) Outcome statistics of renal transplants with an emphasis on long-term survival. Clin Transplant 8:324–327
2. Matas AJ, Gillingham KJ, Humar A, Dunn DL, Sutherland DE, Najarian JS (2000) Immunologic and nonimmunologic factors: different risks for cadaver and living donor transplantation. Transplantation 69:54–58
3. Prommool S, Jhangri GS, Cockfield SM, Halloran PF (2000) Time dependency of factors affecting renal allograft survival. J Am Soc Nephrol 11:565–573
4. Shoskes DA, Cecka JM (1998) Deleterious effects of delayed graft function in cadaveric renal transplant recipients independent of acute rejection. Transplantation 66:1697–1701
5. Terasaki PI, Cecka JM, Gjertson DW, Takemoto S (1995) High survival rates of kidney transplants from spousal and living unrelated donors. N Engl J Med 333:33–36
6. Lanier LL (1995) The role of natural killer cells in transplantation. Curr Opin Immunol 7:626–631
7. Miller JS, Verfaillie C, McGlave P (1999) The generation of human natural killer cells from CD34+/DR− primitive progenitors in long-term bone marrow culture. Blood 80:2182–2187

8. Lian RH, Kumar V (2002) Murine natural killer cell progenitors and their requirements for development. Semin Immunol 14:453–460
9. Herberman RB, Nunn ME, Lavrin DH (1975) Natural cytotoxic reactivity of mouse lymphoid cells against syngeneic and allogeneic tumors. I. Distribution of reactivity and specificity. Int J Cancer 16:216–229
10. Kiessling R, Klein E, Wigzell H (1975) "Natural" killer cells in the mouse. I. Cytotoxic cells with specificity for mouse Moloney leukemia cells. Specificity and distribution according to genotype. Eur J Immunol 5:112–117
11. Herberman RB, Nunn ME, Holden HT, Lavrin DH II (1975) Natural cytotoxic reactivity of mouse lymphoid cells against syngeneic and allogeneic tumors. II. Characterization of effector cells. Int J Cancer 16:230–239
12. Allavena P, Damia G, Colombo T, Maggioni D, D'Incalci M, Mantovani A (1989) Lymphokine-activated killer (LAK) and monocyte-mediated cytotoxicity on tumor cell lines resistant to antitumor agents. Cell Immunol 120:250–258
13. Landay AL, Zarcone D, Grossi CE, Bauer K (1987) Relationship between target cell cycle and susceptibility to natural killer lysis. Cancer Res 47:2767–2770
14. Lanier LL, Testi R, Bindl J, Phillips JH (1989) Identity of Leu-19 (CD56) leukocyte differentiation antigen and neural cell adhesion molecule. J Exp Med 169:2233–2238
15. Ritz J, Schmidt RE, Michon J, Hercend T, Schlossman SF (1988) Characterization of functional surface structures on human natural killer cells. Adv Immunol 42:181–211
16. Koo GC, Peppard JR (1984) Establishment of monoclonal anti-NK-1.1 antibody. Hybridoma 3:301–303
17. Sentman CL, Kumar V, Koo G, Bennett M (1989) Effector cell expression of NK1.1, a murine natural killer cell-specific molecule, and ability of mice to reject bone marrow allografts. J Immunol 142:1847–1853
18. Moore TA, von Freeden-Jeffry U, Murray R, Zlotnik A (1996) Inhibition of gamma delta T cell development and early thymocyte maturation in IL-7 2 /2 mice. J Immunol 157: 2366–2373
19. Charley MR, Mikhael A, Bennett M, Gilliam JN, Sontheimer RD (1983) Prevention of lethal, minor-determinate graft-host disease in mice by the in vivo administration of anti-asialo GM1. J Immunol 131:2101–2103
20. Bendelac A, Rivera MN, Park SH, Roark JH (1997) Mouse CD1-specific NK1 T cells: development, specificity, and function. Annu Rev Immunol 15:535–562
21. Matsuda JL, Naidenko OV, Gapin L, Nakayama T, Taniguchi M, Wang CR, Koezuka Y, Kronenberg M (2000) Tracking the response of natural killer T cells to a glycolipid antigen using CD1d tetramers. J Exp Med 192:741–754
22. Godfrey DI, MacDonald HR, Kronenberg M, Smyth MJ, Van Kaer L (2004) NKT cells: what's in a name? Nat Rev Immunol 4:231–237
23. Lee PT, Putnam A, Benlagha K, Teyton L, Gottlieb PA, Bendelac A (2002) Testing the NKT cell hypothesis of human IDDM pathogenesis. J Clin Invest 110:793–800
24. Ljunggren HG, Karre K (1990) In search of the 'missing self': MHC molecules and NK cell recognition. Immunol Today 11:237–244
25. Braud VM, Allan DS, O'Callaghan CA, Söderström K, D'Andrea A, Ogg GS, Lazetic S, Young NT, Bell JI, Phillips JH, Lanier LL, McMichael AJ (1998) HLA-E binds to natural killer cell receptors CD94/NKG2A, B and C. Nature 391:795–799
26. Gunturi A, Berg RE, Forman J (2004) The role of CD94/NKG2 in innate and adaptive immunity. Immunol Res 30:29–34
27. Karlhofer FM, Ribaudo RK, Yokoyama WM (1992) MHC Class I alloantigen specificity of Ly-49+ IL-2-activated natural killer cells. Nature 358:66–70
28. Natarajan K, Dimasi N, Wang J, Mariuzza RA, Margulies DH (2002) Structure and function of natural killer cell receptors: multiple molecular solutions to self, nonself discrimination. Annu Rev Immunol 20:853–885

29. Colucci F, Di Santo JP, Leibson PJ (2002) Natural killer cell activation in mice and men: different triggers for similar weapons? Nat Immunol 3:807–813
30. Raulet DH (2003) Roles of the NKG2D immunoreceptor and its ligands. Nat Rev Immunol 3:781–790
31. Arnon TI, Lev M, Katz G, Chernobrov Y, Porgador A, Mandelboim O (2001) Recognition of viral hemagglutinins by NKp44 but not by NKp30. Eur J Immunol 31:2680–2689
32. Mandelboim O, Lieberman N, Lev M, Paul L, Arnon TI, Bushkin Y, Davis DM, Strominger JL, Yewdell JW, Porgador A (2001) Recognition of haemagglutinins on virus-infected cells by NKp46 activates lysis by human NK cells. Nature 409:1055–1060
33. Bottino C, Castriconi R, Pende D, Rivera P, Nanni M, Carnemolla B, Cantoni C, Grassi J, Marcenaro S, Reymond N, Vitale M, Moretta L, Lopez M, Moretta A (2003) Identification of PVR (CD155) and Nectin-2 (CD112) as cell surface ligands for the human DNAM-1 (CD226) activating molecule. J Exp Med 198:557–567
34. Brown MH, Boles K, van der Merwe PA, Kumar V, Mathew PA, Barclay AN (1998) 2B4, the natural killer and T cell immunoglobulin superfamily surface protein, is a ligand for CD48. J Exp Med 188:2083–2090
35. Latchman Y, McKay PF, Reiser H (1998) Identification of the 2B4 molecule as a counter-receptor for CD48. J Immunol 161:5809–5812
36. Blancho G, Buzelin F, Dantal J, Hourmant M, Cantarovich D, Baatard R, Bonneville M, Vie H, Bugeon L, Soulillou JP (1992) Evidence that early acute renal failure may be mediated by CD3– CD16+ cells in a kidney graft recipient with large granular lymphocyte proliferation. Transplantation 53:1242–1247
37. Heidecke CD, Araujo JL, Kupiec-Weglinski JW, Abbud-Filho M, Araneda D, Stadler J, Siewert J, Strom TB, Tilney NL (1985) Lack of evidence for an active role for natural killer cells in acute rejection of organ allografts. Transplantation 4:441–444
38. Nemlander A, Saksela E, Häyry P (1983) Are "natural killer" cells involved in allograft rejection? Eur J Immunol 13:348–350
39. Petersson E, Qi Z, Ekberg H, Ostraat O, Dohlsten M, Hedlund G (1997) Activation of alloreactive natural killer cells is resistant to cyclosporine. Transplantation 63:1138–1144
40. Petersson E, Ostraat O, Ekber H, Hansson J, Simanaitis M, Brodin T, Dohlsten M, Hedlund G (1997) Allogeneic heart transplantation activates alloreactive NK cells. Cell Immunol 175:25–32
41. Ayalon O, Hughes EA, Cresswell P, Lee J, O'Donnell L, Pardi R, Bender JR (1998) Induction of transporter associated with antigen processing by interferon gamma confers endothelial cell cytoprotection against natural killer-mediated lysis. Proc Natl Acad Sci USA 95:2435–2440
42. McDouall RM, Batten P, McCormack A, Yacoub MH, Rose ML (1997) MHC class II expression on human heart microvascular endothelial cells: exquisite sensitivity to interferon-gamma and natural killer cells. Transplantation 64:1175–1180
43. Timonen T, Patarroyo M, Gahmberg CG (1988) CD11a-c/CD18 and GP84 (LB-2) adhesion molecules on human large granular lymphocytes and their participation in natural killing. J Immunol 141:1041–1046
44. Watson CA, Pezelbauer P, Zhou J, Pardi R, Bender JR (1995) Contact-dependent endothelial class II HLA gene activation induced by NK cells is mediated by IFN-gamma-dependent and -independent mechanisms. J Immunol 154:3222–3233
45. Markus PM, van den Brink M, Cai X, Harnaha J, Palomba L, Hiserodt JC, Cramer DV (1991) Effect of selective depletion of natural killer cells on allograft rejection. Transplant Proc 23:178–179
46. Maier S, Tertilt C, Chambron N, Gerauer K, Hüser N, Heidecke CD, Pfeffer K (2001) Inhibition of natural killer cells results in acceptance of cardiac allografts in CD28-/- mice. Nat Med 5:557–562
47. McNerney ME, Lee KM, Zhou P, Molinero L, Mashayekhi M, Guzior D, Sattar H, Kuppireddi S, Wang CR, Kumar V, Alegre ML (2006) Role of natural killer cell subsets in cardiac allograft rejection. Am J Transplant 6:505–513

48. Kim J, Chang CK, Hayden T, Liu FC, Benjamin J, Hamerman JA, Lanier LL, Kang SM (2007) The activating immunoreceptor NKG2D and its ligands are involved in allograft transplant rejection. J Immunol 179:6416–6420
49. Hankey KG, Drachenberg CB, Papadimitriou JC, Klassen DK, Philosophe B, Bartlett ST, Groh V, Spies T, Mann DL (2002) MIC expression in renal and pancreatic allografts. Transplantation 73:304–306
50. Sumitran-Holgersson S, Wilczek HE, Holgersson J, Söderström K (2002) Identification of the nonclassical HLA molecules, mica, as targets for humoral immunity associated with irreversible rejection of kidney allografts. Transplantation 74:268–277
51. Zhang ZX, Wang S, Huang X, Min WP, Sun H, Liu W, Garcia B, Jevnikar AM (2008) NK cells induce apoptosis in tubular epithelial cells and contribute to renal ischemia-reperfusion injury. J Immunol 181:7489–7498
52. Coulson MT, Jablonski P, Howden BO, Thomson NM, Stein AN (2005) Beyond operational tolerance: effect of ischemic injury on development of chronic damage in renal grafts. Transplantation 80:353–361
53. Uehara S, Chase CM, Kitchens WH, Rose HS, Colvin RB, Russell PS, Madsen JC (2005) NK cells can trigger allograft vasculopathy: the role of hybrid resistance in solid organ allografts. J Immunol 175:3424–3430
54. Kroemer A, Xiao X, Degauque N, Edtinger K, Wei H, Demirci G, Li XC (2008) The innate NK cells, allograft rejection, and a key role for IL-15. J Immunol 180:7818–7826
55. Kondo T, Morita D, Watarai Y, Auerbach MB, Taub DD, Novick AC, Toma H, Fairchild RL (2000) Early increased chemokine expression and production in murine allogeneic skin grafts is mediated by natural killer cells. Transplantation 69:969–977
56. Taub DD, Sayers TJ, Carter CR, Ortaldo JR (1995) Alpha and beta chemokines induce NK cell migration and enhance NK-mediated cytolysis. J Immunol 155:3877–3888
57. Obara H, Nagasaki K, Hsieh CL, Ogura Y, Esquivel CO, Martinez OM, Krams SM (2005) IFN-gamma, produced by NK cells that infiltrate liver allografts early after transplantation, links the innate and adaptive immune responses. Am J Transplant 5:2094–2103
58. Uppaluri R, Sheehan KC, Wang L, Bui JD, Brotman JJ, Lu B, Gerard C, Hancock WW, Schreiber RD (2008) Prolongation of cardiac and islet allograft survival by a blocking hamster anti-mouse CXCR3 monoclonal antibody. Transplantation 86:137–147
59. Hancock WW, Lu B, Gao W, Csizmadia V, Faia K, King JA, Smiley ST, Ling M, Gerard NP, Gerard C (2000) Requirement of the chemokine receptor CXCR3 for acute allograft rejection. J Exp Med 192:1515–1520
60. Gao W, Topham PS, King JA, Smiley ST, Csizmadia V, Lu B, Gerard CJ, Hancock WW (2000) Targeting of the chemokine receptor CCR1 suppresses development of acute and chronic cardiac allograft rejection. J Clin Invest 105:35–44
61. Hüser N, Tertilt C, Gerauer K, Maier S, Traeger T, Assfalg V, Reiter R, Heidecke CD, Pfeffer K (2005) CCR4-deficient mice show prolonged graft survival in a chronic cardiac transplant rejection model. Eur J Immunol 35:128–138
62. Haskell CA, Hancock WW, Salant DJ, Gao W, Csizmadia V, Peters W, Faia K, Fituri O, Rottman JB, Charo IF (2001) Targeted deletion of CX(3)CR1 reveals a role for fractalkine in cardiac allograft rejection. J Clin Invest 108:679–688
63. Austyn JM, Larsen CP (1990) Migration patterns of dendritic leukocytes. Implications for transplantation. Transplantation 49:1–7
64. Beilke JN, Kuhl NR, Van Kaer L, Gill RG (2005) NK cells promote islet allograft tolerance via a perforin-dependent mechanism. Nat Med 11:1059–1065
65. Coudert JD, Coureau C, Guéry JC (2002) Preventing NK cell activation by donor dendritic cells enhances allospecific CD4 T cell priming and promotes Th type 2 responses to transplantation antigens. J Immunol 169:2979–2987
66. Yu G, Xu X, Vu MD, Kilpatrick ED, Li XC (2006) NK cells promote transplant tolerance by killing donor antigen-presenting cells. J Exp Med 203:1851–1858

67. Laffont S, Seillet C, Ortaldo J, Coudert JD, Guéry JC (2008) Natural killer cells recruited into lymph nodes inhibit alloreactive T-cell activation through perforin-mediated killing of donor allogeneic dendritic cells. Blood 112:661–671
68. Ikehara Y, Yasunami Y, Kodama S, Maki T, Nakano M, Nakayama T, Taniguchi M, Ikeda S (2000) CD4(+) Valpha14 natural killer T cells are essential for acceptance of rat islet xenografts in mice. J Clin Invest 105:1761–1767
69. Seino KI, Fukao K, Muramoto K, Yanagisawa K, Takada Y, Kakuta S, Iwakura Y, Van Kaer L, Takeda K, Nakayama T, Taniguchi M, Bashuda H, Yagita H, Okumura K (2001) Requirement for natural killer T (NKT) cells in the induction of allograft tolerance. Proc Natl Acad Sci USA 98:2577–2581
70. Iwai T, Tomita Y, Shimizu I, Kajiwara T, Onzuka T, Okano S, Yasunami Y, Yoshikai Y, Nomoto K, Tominaga R (2007) The immunoregulatory roles of natural killer T cells in cyclophosphamide-induced tolerance. Transplantation 84:1686–1695
71. Jiang X, Shimaoka T, Kojo S, Harada M, Watarai H, Wakao H, Ohkohchi N, Yonehara S, Taniguchi M, Seino K (2005) Cutting edge: critical role of CXCL16/CXCR6 in NKT cell trafficking in allograft tolerance. J Immunol 175:2051–2055
72. Sonoda KH, Taniguchi M, Stein-Streilein J (2002) Long-term survival of corneal allografts is dependent on intact CD1d-reactive NKT cells. J Immunol 168:2028–2034
73. Watte CM, Nakamura T, Lau CH, Ortaldo JR, Stein-Streilein J (2008) Ly49 C/I-dependent NKT cell-derived IL-10 is required for corneal graft survival and peripheral tolerance. J Leukoc Biol 83:928–935
74. Jiang X, Kojo S, Harada M, Ohkohchi N, Taniguchi M, Seino KI (2007) Mechanism of NKT cell-mediated transplant tolerance. Am J Transplant 7:482–490
75. Ruggeri L, Capanni M, Urbani E, Perruccio K, Shlomchik WD, Tosti A, Posati S, Rogaia D, Frassoni F, Aversa F, Martelli MF, Velardi A (2002) Effectiveness of donor natural killer cell alloreactivity in mismatched hematopoietic transplants. Science 295:2097–2100
76. Ruggeri L, Mancusi A, Capanni M, Urbani E, Carotti A, Aloisi T, Stern M, Pende D, Perruccio K, Burchielli E, Topini F, Bianchi E, Aversa F, Martelli MF, Velardi A (2007) Donor natural killer cell allorecognition of missing self in haploidentical hematopoietic transplantation for acute myeloid leukemia: challenging its predictive value. Blood 110:433–440
77. Carrington M, Norman, P (2003) The KIR Gene Cluster. NAtional Library of Medicine (US), National Center for Biotechnology Information. Available from: http://www.ncbi.nih.gov/entrez/query.fcgi?db=Books
78. Lanier LL (2005) NK cell recognition. Annu Rev Immunol 23:225–274
79. Williams AP, Bateman AR, Khakoo SI (2005) Hanging in the balance. KIR and their role in disease. Mol Interv 5:226–240
80. Gumperz JE, Litwin V, Phillips JH, Lanier LL, Parham P (1995) The Bw4 public epitope of HLA-B molecules confers reactivity with natural killer cell clones that express NKB1, a putative HLA receptor. J Exp Med 181:1133–1144
81. Dohring C, Scheidegger D, Samaridis J, Cella M, Colonna M (1996) A human killer inhibitory receptor specific for HLA-A1, 2. J Immunol 156:3098–3101
82. Norman PJ, Carrington CV, Byng M, Maxwell LD, Curran MD, Stephens HA, Chandanayingyong D, Verity DH, Hameed K, Ramdath DD, Vaughan RW (2002) Natural killer cell immunoglobulin-like receptor (KIR) locus profiles in African and South Asian populations. Genes Immun 3:86–95
83. Hsu KC, Chida S, Geraghty DE, Dupont B (2002) The killer cell immunoglobulin-like receptor (KIR) genomic region: gene-order, haplotypes and allelic polymorphism. Immunol Rev 190:40–52
84. Tran TH, Mytilineos J, Scherer S, Laux G, Middleton D, Opelz G (2005) Analysis of KIR ligand incompatibility in human renal transplantation. Transplantation 80:1121–1123
85. Bishara A, Brautbar C, Zamir G, Eid A, Safadi R (2005) Impact of HLA-C and Bw epitopes disparity on liver transplantation outcome. Hum Immunol 66:1099–1105
86. Fan QR, Long EO, Wiley DC (2001) Crystal structure of the human natural killer cell inhibitory receptor KIR2DL1-HLA-Cw4 complex. Nat Immunol 2:452–460

87. Ahlenstiel G, Martin MP, Gao X, Carrington M, Rehermann B (2008) Distinct KIR/HLA compound genotypes affect the kinetics of human antiviral natural killer cell responses. J Clin Invest 118:1017–1026
88. Hanvesakul R, Spencer N, Cook M, Gunson B, Hathaway M, Brown R, Nightingale P, Cockwell P, Hubscher SG, Adams DH, Moss P, Briggs D (2008) Donor HLA-C genotype has a profound impact on the clinical outcome following liver transplantation. Am J Transplant 8:1931–1941
89. Kunert K, Seiler M, Mashreghi MF, Klippert K, Schönemann C, Neumann K, Pratschke J, Reinke P, Volk HD, Kotsch K (2007) KIR/HLA ligand incompatibility in kidney transplantation. Transplantation 84:1527–1533
90. Vampa ML, Norman PJ, Burnapp L, Vaughan RW, Sacks SH, Wong W (2003) Natural killer-cell activity after human renal transplantation in relation to killer immunoglobulin-like receptors and human leukocyte antigen mismatch. Transplantation 76:1220–8
91. Kwakkel-van Erp JM, van de Graaf EA, Paantjens AW, van Ginkel WG, Schellekens J, van Kessel DA, van den Bosch JM, Otten HG (2008) The killer immunoglobulin-like receptor (KIR) group A haplotype is associated with bronchiolitis obliterans syndrome after lung transplantation. J Heart Lung Transplant 27:995–1001
92. Kreijveld E, van der Meer A, Tijssen HJ, Hilbrands LB, Joosten I (2007) KIR gene and KIR ligand analysis to predict graft rejection after renal transplantation. Transplantation 84:1045–1051
93. Chen C, Busson M, Rocha V, Appert ML, Lepage V, Dulphy N, Haas P, Socié G, Toubert A, Charron D, Loiseau P (2006) Activating KIR genes are associated with CMV reactivation and survival after non-T-cell depleted HLA-identical sibling bone marrow transplantation for malignant disorders. Bone Marrow Transplant 38:437–444
94. Stern M, Elsässer H, Hönger G, Steiger J, Schaub S, Hess C (2008) The number of activating KIR genes inversely correlates with the rate of CMV infection/reactivation in kidney transplant recipients. Am J Transplant 8:1312–1317
95. Hadaya K, de Rham C, Bandelier C, Ferrari-Lacraz S, Jendly S, Berney T, Buhler L, Kaiser L, Seebach JD, Tiercy JM, Martin PY, Villard J (2008) Natural killer cell receptor repertoire and their ligands, and the risk of CMV infection after kidney transplantation. Am J Transplant 8:2674–2683
96. Lefkowitz M, Kornbluth J, Tomaszewski JE, Jorkasky DK (1988) Natural killer-cell activity in cyclosporine-treated renal allograft recipients. J Clin Immunol 8:121–127
97. Luo H, Chen H, Daloze P, Wu J (1992) Effects of rapamycin on human HLA-unrestricted cell killing. Clin Immunol Immunopathol 65:60–64
98. Wang H, Grzywacz B, Sukovich D, McCullar V, Cao Q, Lee AB, Blazar BR, Cornfield DN, Miller JS, Verneris MR (2007) The unexpected effect of cyclosporin A on CD56+CD16- and CD56+CD16+ natural killer cell subpopulations. Blood 110:1530–1539
99. Chiossone L, Vitale C, Cottalasso F, Moretti S, Azzarone B, Moretta L, Mingari MC (2007) Molecular analysis of the methylprednisolone-mediated inhibition of NK-cell function: evidence for different susceptibility of IL-2- versus IL-15-activated NK cells. Blood 109:3767–3775
100. Wai LE, Fujiki M, Takeda S, Martinez OM, Krams SM (2008) Rapamycin, but not cyclosporine or FK506, alters natural killer cell function. Transplantation 85:145–149
101. Bielekova B, Catalfamo M, Reichert-Scrivner S, Packer A, Cerna M, Waldmann TA, McFarland H, Henkart PA, Martin R (2006) Regulatory CD56(bright) natural killer cells mediate immunomodulatory effects of IL-2Ralpha-targeted therapy (daclizumab) in multiple sclerosis. Proc Natl Acad Sci USA 103:5941–5946
102. Bhat R, Watzl C (2007) Serial killing of tumor cells by human natural killer cells – enhancement by therapeutic antibodies. PLoS ONE 28:326
103. Stauch D, Dernier A, Sarmiento Marchese E, Kunert K, Volk HD, Pratschke J, Kotsch K (2009)Targeting of Natural Killer cells by rabbit antithymocyte globulin and Campath-1H: similar effects independent of specificity. PLoS ONE 4:e4709

NK Cells in Autoimmune and Inflammatory Diseases

Nicolas Schleinitz, Nassim Dali-Youcef, Jean-Robert Harle, Jacques Zimmer and Emmanuel Andres

Abstract The evidence of NK cell implication in human diseases has initially been shown for antiviral immunity and tumor surveillance. Nowadays, increased attention is being paid to the aspect of NK cells and innate immunity in studies on autoimmune diseases. However, despite a growing knowledge on NK cell function and regulation, their role in human autoimmune disease still remains controversial. In animal models, studies of NK cells have shown conflicting results toward a disease-promoting or disease-protective role. Similarly, in human diseases, available data suggest an unequivocal role of NK cells. Yet, the understanding of NK cell implication in human diseases is far from being achieved. We review here the current knowledge on NK cell biology in human autoimmune and inflammatory diseases and discuss their possible mechanisms of action in these complex pathologies.

1 Introduction

"Natural killer" (NK) cell implication in autoimmune disorders has been studied extensively in animal models of type 1 diabetes, experimental allergic encephalomyelitis, collagen-induced arthritis, experimental autoimmune myasthenia gravis

---

N. Schleinitz and J.-R Harle
Department of Internal Medicine, CHRU de Marseille, Marseille, France

N. Dali-Youcef
Laboratory of general and specialized Biochemistry, Hôpitaux Universitaires de Strasbourg and Department of Biochemistry, Faculty of Medicine, Strasbourg, France

J. Zimmer
Laboratoire d'Immunogénétique-Allergologie, Centre de Recherche Public de la Santé (CRP-Santé) de Luxembourg, Luxembourg

E. Andres (✉)
Service de Médecine Interne, Diabète et Maladies Métaboliques, Clinique Médicale B, Hôpital Civil – Hôpitaux Universitaires de Strasbourg, 1 porte de l'Hôpital, 67091, Strasbourg Cedex, France
e-mail: emmanuel.andres@chru-strasbourg.fr

and some lupus erythematosus-prone mice strains [1]. Controversial results reported in some models a disease-promoting role and in others a disease-protecting role of NK cells. These opposing roles of NK cells in autoimmunity underscore the difficult and incomplete understanding of NK cell biology in vivo. These conflicting studies included results obtained in human autoimmune diseases too. NK cell functions were either impaired or activated, unbalanced by inhibitory signals. Thus, their potential implication in autoimmune diseases was based on their known functions: cytotoxicity, cytokine and chemokine production and cell–cell interaction. Another important point is the complex mechanisms of their regulation based on the signaling balance between specific inhibitory and activating receptors, some of them recognizing MHC class I molecules. The balance between these signals is responsible for the induction of NK cell tolerance or NK cell activation [2]. Finally, it appears that NK cells are regulated and interact directly with dendritic cells (DC) and T regulatory cells (Treg), which are both implicated in the pathophysiology of autoimmune diseases. This complicates the picture of the potential implication of NK cells in autoimmune disease. An abnormal control at different steps, or "checkpoints," of NK cell regulation could result in (1) Excessive tissue inflammation/destruction by either cytotoxicity or by proinflammatory cytokine production; and (2) Orientation of the adaptive immune response toward autoimmunity. One must keep in mind that autoimmunity is a complex and dynamic process that implicates several immune cells. NK cells could act at the induction phase of the autoimmune reaction and/or maintain or control the autoimmune response during time. Most, if not all the data available on NK cell biology in humans were obtained during the chronic phase of the autoimmune disease (Table 1). Moreover, therapies such as high doses of corticosteroids and immunosuppressive agents have been shown to alter NK cell number and functions [3, 4].

The important role of NK cells in the control of viral infections has been indirectly linked to autoimmune diseases. As a matter of fact, NK cells act in the first line of defense against viruses; and viral infections were implicated in the induction of certain autoimmune diseases or autoantibodies [5, 6]. Naturally, an impaired NK cell response toward viruses could lead to the abnormal induction of an autoreactive adaptive immune response, mediated for example by interferon alpha (IFN-α) and DC as in systemic lupus erythematosus (SLE) [5].

We review here the current knowledge of the role of NK cells in autoimmune or inflammatory diseases in humans.

2 Disruption of NK Cell Inhibition Through MHC Class I Recognition

A few observations of human diseases have suggested a direct implication of NK cells or NK cell receptors in tissue destruction. A rare human disease associated with a reduced cell-surface expression of MHC class I expression has been related to excessive tissue inflammation/destruction by disrupting NK cell tolerance.

Table 1 NK cell characteristics and genotypic findings in some inflammatory and/or autoimmune diseases in humans

	NK cells changes		Genetic susceptibility
	Number	Function (cytotoxicity)	
Systemic lupus erythematosus	↓	↓	2DS1+/2DS2− [28]
Sjögren's disease	↓ [64]	Normal	Not analyzed
Systemic sclerosis	↓	↓ [65, 66]	2DS1/2DL1 [27] 2DS1+/2DS2− [28]
Antiphospholipid syndrome	↑[36]	Unknown	Not analyzed
Pemphigus vulgaris	↑ [57]	Unknown	Not analyzed
Rheumatoid arthritis (RA)	↑ (Synovial fluid) [48]	Unknown	Some association of KIR genes with clinical manifestations of RA [67] Association with KIR2DS4? [68]
Spondylarthropathies	↑ KIR3DL2 [41]	↑	Association to 3DS1 and protective role of 3DL1[69] [70]
Psoriasis/psoriatic arthritis	↓ blood [71] ↑ (Synovial fluid) [48]	Unknown	2DS1/2DS2,HLA-Cw homozygosity [31, 32]
Pregnancy Preeclampsia	Unknown	Unknown	Maternal KIR AA/ fetal HLA-C2 [33]
Recurrent fetal abortion	↑[36]	Unknown	Lack of inhibitory KIR for fetal HLA Cw alleles [34, 35]
Ulcerative colitis	Unknown	Unknown	Increase of the KIR 2DL2/2DS2 frequency Protective effect of KIR 2DL3 in the presence of its ligand [72]

Mutations in either subunits of transporter associated with antigen processing protein 1 or 2 (TAP1 or TAP2) were associated to damage of the skin and kidneys after a chronic infection. The ineffective control of NK cell activation through self MHC class I expression, recognized by inhibitory NK cell receptors, leads to tissue damage by NK cell lysis [7, 8]. These observations indicate that, in vivo, inappropriate or uncontrolled activation of innate immune effector cells (e.g., NK) could lead to autoimmune damage through autologous cell destruction. Hence, there is some evidence that NK cells are potentially highly autoreactive cells, as initially suggested by the missing self theory [9]. Moreover, these patients could develop vasculitis, which is considered as an autoimmune mediated disease [10]. This mechanism of disruption of autologous cell recognition through MHC class I recognition by NK cell receptors has been also implicated in a case of pure red cell aplasia, another autoimmune disease [11]. In this disease, a clonal T cell large granular lymphocytic expansion induced pure-red cell aplasia through cytotoxicity of erythroid progenitors expressing low levels of MHC class I antigens. This pathology is an additional argument showing that self tolerance disruption by the absence of recognition of MHC class I through NK cell inhibitory receptors, expressed either by NK or by T cell subsets, can lead to excessive inflammation and tissue destruction that is a hallmark of autoimmune diseases.

Another pathological situation where NK cells are directly associated with autoimmunity is NK large-type lymphoproliferative disease of granular lymphocytes (NK-LDGL). Most of large granular lymphocyte proliferations are of the T cell phenotype and are associated with autoimmune cytopenias and rheumatoid arthritis (RA) [12, 13]. NK-LDGL proliferation is a peculiar subgroup that represents about 10% of large granular lymphocytic leukemias [13]. NK-LDGL can be defined as a persistent state of natural killer cell excess in peripheral blood that is not associated with clinical lymphoma. NK cell proliferation in NK-LDGL is associated with an altered killer Ig-like receptor (KIR) expression that has been proposed as a marker of oligoclonality [14, 15]. Moreover in NK-LDGL, the KIR genotype and KIR expression at the cellular level is biased toward activating KIR receptors [16]. This could have a role in the proliferation of NK-LDGL and control of their cytotoxicity. In clinical studies, NK-LDGL have been associated with several autoimmune diseases: idiopathic thrombocytopenia purpura, autoimmune haemolytic anemia, autoimmune neutropenia, skin vasculitis, glomerulonephritis, idiopathic pulmonary fibrosis, Sjögren's disease and arthritis [14, 17–19]. In addition, in some cases, the clinical remission under treatment was associated with a significant decrease in NK cell number, suggesting thereby a direct causative role of NK cells in these autoimmune manifestations.

3 Association Between KIR Genotypes and Autoimmune and Inflammatory Diseases

The discovery of the NK cell inhibitory and activating receptors has led to the description of a family of highly homologous inhibitory and activating receptors recognizing MHC class I molecules, all located in the leukocyte region cluster on chromosome 19q13.4 [20, 21]. There is a high degree of variability at the genotypic level depending mostly on the set of activating NK cell receptors [22]. Several genotypes have been reported in humans and two major haplotypes have been individualized [23]. Moreover, an allelic variability in KIR genes has also been characterized [24]. Since NK cell regulation depends, at least in part, on MHC class I receptor interaction with MHC class I on the targeted cell, it has been suggested that the genetic variation of the MHC class I and NK cell-MHC class I receptor combination could confer susceptibility toward human diseases. This approach has been first evaluated in allogeneic bone marrow transplantation demonstrating that KIR/MHC class I mismatch can influence the rate of allograft rejection and the graft versus leukemia effect [25]. Knowing the role of viral infection in autoimmune diseases, some KIR/MHC class I genotypes were associated with the control of certain viral infections. Along another line, numerous reports unveiled the link between the susceptibility for inflammatory and autoimmune diseases and certain KIR or KIR/MHC class I genotypes association [26]. Likewise, susceptibility to systemic sclerosis was associated with the 2DS2$^+$/2DL2$^-$ genotype [27] and with the 2DS1$^+$/2DS2$^-$ genotypes [28]. Moreover, KIR 2DS1$^+$ patients with progressive systemic sclerosis displayed a significant decrease in the amount of the inhibitory KIR corresponding to the appropriate HLA-C ligand [28] (Table 1). In type 1 diabetes mellitus, an association with the 2DS2/HLA-C1 genotype [29] and a decrease in inhibitory KIR/HLA genotype combinations were reported [30]. In psoriatic arthritis, the 2DS1/2DS2 genotype and the HLA-Cw homozygosity were associated with a susceptibility to develop the disease [31, 32] (Table 1). Furthermore, in a Canadian cohort of SLE patients, an increase in the 2DS1$^+$/2DS2$^-$ genotype was documented [28].

In the uterus, NK cells are the most abundant component among the lymphocyte population and are implicated in the tolerance of the fetus during pregnancy [33]. Some autoimmune diseases can be responsible for severe complications during pregnancy such as preeclampsia and the antiphospholipid syndrome, which was associated with recurrent spontaneous abortion. In these conditions, KIR/MHC associations were studied and the cooperative connection between a maternal KIR AA genotype and a fetal HLA-C2 genotype was linked to the susceptibility to develop preeclampsia [33]. When the mother's genome is lacking the inhibitory KIR which normally recognizes certain fetal HLA-Cw alleles [34, 35], recurrent spontaneous abortion seems to be more frequent (Table 1). Taken together, given the abundance of NK cells in utero, these observations emphasize that NK cells are important for the tolerance of the fetus. It is likely that under some circumstances disruption of the immune fetal tolerance, controlled by uterine NK cells, can contribute to fetal loss. Along this line, it has been shown that women with recurrent spontaneous fetal abortions and antiphospholipid antibodies, exhibited an increased level of circulating NK cells [36].

Altogether, genetic data obtained from KIR/MHC class I association studies have identified genotypes associated with enhanced susceptibility to develop the disease or the severe form of it. Most of these situations seem to be associated with a either lack of NK cell inhibition or with an excessive NK cell activation through MHC class I receptor signaling. These interesting results need to be confirmed by gain or loss of function studies. Moreover, the genetic studies on KIR genes should be interpreted not only by focusing on NK cells but should also investigate T cell subsets expressing these receptors. This model has also to take into account the KIR variants and the variation in their clonal expression, and also KIR distribution on NK cells for a unique genotype.

4 Variation of NK Cell MHC Class I Receptor Expression in Autoimmune or Inflammatory Diseases

Aside the genetic studies on KIR/MHC class I association, another approach has evaluated the variation of KIR and other NK cell MHC class I receptors expression by NK and T cell subsets in human diseases. We have reported above the abnormal KIR expression profile of NK cells in NK-LDGL. Some T-LDGL also express KIR receptors and KIR expression has also been characterized in a subset of normal effector-memory T cells [37]. Moreover, NK cell receptors for MHC class I molecules include receptors of the NKG2 lectin-like family (NKG2A/B), expressed as heterodimers in association with CD94, and CD85j or ILT2, a member of the immunoglobulin-like transcript (ILT/LIR) family [2].

Some inflammatory human diseases are known to be linked to the genetic susceptibility to peculiar MHC class I alleles. In these situations, it was tempting to evaluate the variation of KIR expression at the cellular level (inhibitory versus activating receptors and/or level of expression by the median fluorescence intensity) and percentage of NK cells expressing a given NK cell receptor. In Behcet's disease, an inflammatory disorder with recurrent attacks of oral and genital aphtous ulcers, uveitis and skin lesions, the genetic susceptibility was associated with some HLA-B alleles. However, the expression of the KIR3DL1 inhibitory receptor that recognizes residues 77–83 of HLA-B alleles was not different between patients and controls [38, 39]. Consistent with these findings, genotypic studies in Behcet's disease found no associations [40].

In spondylarthritis, patients bearing the HLA-B27 displayed a significantly increased expression of the inhibitory receptor KIR3DL2 on NK and $CD4^+$ T lymphocytes, both in peripheral blood and in synovial fluid [41] (Table 1). Moreover, $KIR3DL2^+$ NK cells had an activated phenotype and showed an enhanced cytotoxicity suggesting an implication in the pathogenesis of the disease.

In RA, it has been shown that the clonally expanded auto reactive $CD4^+CD28^{null}$ T cells, a hallmark in patients with systemic forms, expressed preferentially functional activating KIR receptors. In vitro, it was demonstrated that the expression of activating receptors facilitated the expansion of autoreactive T cells by acting as a costimulatory receptor for TCR-mediated triggering [42]. More recently, studies have documented that $CD4^+CD28^{null}$ T cells expressing the natural cytotoxicity receptor NKG2D, the ligand for the stress inducible nonclassical MHC class I molecules MIC, were expanded in granulomatous lesions of Wegener's granulomatosis [43]. MIC molecules were expressed in stressed cells and were capable of activating T cells through NKG2D signaling. $CD4^+NKG2D^+$ T cells have also been reported to be increased in the lamina propria of patients with Crohn's disease and ulcerative colitis. They are one of the local inflammatory response mediators through the MICA–NKG2D interaction leading to the production of IFNγ [44].

We have shown that infiltrating CD8+ T cells in the muscle of polmyositis patients, an idiopathic inflammatory myopathy (IIM), express a unique NK cell MHC class I receptor: CD85j. The known function of CD85j suggested a protective role by downregulating the invading autoreactive $CD8^+$ T cells. Hence, it could be speculated that CD85j expression is induced locally by the inflammatory environment and contributes to the negative control of IIM pathogenesis [45].

In summary of this section, NK or T cell subsets in inflammatory diseases either alleviate the inhibitory signal by decreasing the expression of inhibitory class I receptors or, on the contrary, upregulate activating receptors and their ligands in inflamed tissues.

5 NK Cell Interactions with DC in Autoimmune Diseases

NK cells are a component of the innate immune system. As a primary effector of the immune response they can influence the adaptive immune response by cytokine and chemokine production at the site of tissue inflammation. Moreover, peculiar subsets of NK cells have been shown to be present in lymph nodes [46, 47] and in tissues where they can interact with other immune cells. Recently, several reports have analyzed the connection between NK cells and DC with some evidence that these interactions can impact the inflammation in human autoimmune diseases. In RA patients, NK cells from the synovial fluid can promote the differentiation of monocytes into DC_{NK} [48]. The NK cells are abundant in synovial fluid of RA patients and are KIR^-CD56^{bright} NK cells. Both synovial fluid and peripheral blood $CD56^{bright}$ NK cells of patients with RA and psoriatic arthritis can differentiate monocytes into DCs, initiated by local IL-15 production [48]. These DC_{NK} are functional antigen (Ag) presenting cells and can promote $CD4^+$ T cell activation and the polarization to TH1 cells. These results suggest that NK cells can sustain the joint inflammation in RA and psoriatic arthritis.

Although in the case of immune-mediated arthritis it has been suggested that NK cells promote the maturation of DC, and therefore, the TH1-mediated inflammatory response, the interaction of NK cells with DC is more complex. Interaction through cell–cell contact of NK cells and DC can lead to: (1) NK cell-mediated killing of immature monocyte-derived DC (iDC); (2) DC-induced NK cell proliferation; and (3) NK cell-dependent DC maturation. All these interactions are influenced by the local cytokine environment (Fig. 1). In tissues, NK cells can discriminate between myeloid iDC that typically underexpress MHC class I molecules, and mature DC that upregulate MHC class I expression after Ag uptake [49]. The killing of iDC by NK cells has been interpreted as a control of quality of DC, allowing only mature DC to migrate into the lymph node. Thus, NK cells can indirectly control the "quality" of the DC- induced adaptive immune response. Although there is no clear evidence that the crosstalk between NK and DC has a role in the pathophysiology of autoimmune diseases, it can be speculated that impaired NK cell functions, as observed in several autoimmune diseases, could be responsible for a decreased killing of immature DC, potentially presenting autoantigens with an appropriate costimulation to induce autoreactive T cells to proliferate.

Furthermore, DC are known to play a pivotal role in SLE by producing large amounts of IFNα [5, 50]. In this situation, it is well established that NK cell functions are altered. It could thus be hypothesized that the expansion of plasmacytoid DC is related to an impaired control of DC maturation by NK cells.

6 NK Cell Interactions with T Cells in Autoimmune Diseases

Treg cells are defined as $CD4^+CD25^{high}$ or $CD4^+FOXP3^+$ T cells and contribute to the immune tolerance through inhibition of the effector functions and proliferation of a large panel of leukocyte subsets. In humans, Treg inhibit the natural

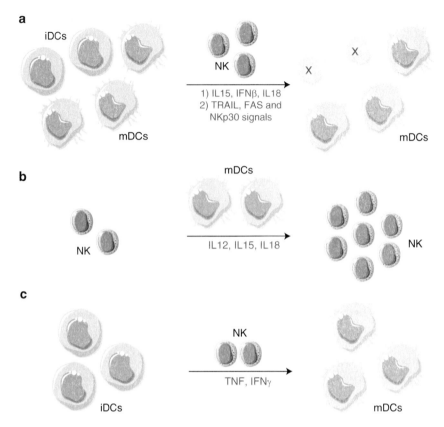

Fig. 1. Natural killer (NK) and dendritic cell (DC) interactions. a) NK cell-mediated destruction of immature DCs Cytokines IL 15 and IFNβ activate NK cell-mediated upregulation of death ligand TRAIL whereas IL18-activated NK cells induce FAS ligand responsible for immature DC's cytolysis. NKp30 signals are also involved in the killing of immature DCs. b) mature DC-produced cytokines IL12, IL15 and IL18 elicit NK cell proliferation. c) NK-mediated DC maturation requires soluble factors TNF and IFNγ. Abbreviations: TRAIL: tumor necrosis factorrelated apoptosis inducing ligand; IFN; interferon; IL: interleukin, FAS ligand also known as CD95 ligand. NKp30: natural killer protein 30

cytotoxicity of resting but not activated NK cells [51, 52]. This inhibition is dependent on membrane-bound transformed growth factor β (TGFβ). Treg also inhibit NK cell proliferation in vitro and can interfere with cytokine production in some conditions [53].

In autoimmune diseases, Treg have been shown to be deficient both numerically and functionally [54, 55]. Likewise, Treg would not control efficiently NK cells which could then be responsible for exaggerated cytokine production or tissue destruction. On the contrary, in several autoimmune diseases (i.e., SLE, Sjögren's syndrome and RA) peripheral NK cells are numerically and functionally impaired and paralleled the Treg deficiency. Perhaps the NK cell deficiency often observed could be related to the progression state of the autoimmune disease and that the interaction of Treg with NK might represent an important step in the initiation of the autoimmune response. Along this line, it has been suggested by observations in animal models that NK cell deficiency could be related to a reciprocal regulation with autoreactive T cells. In such a model, NK cells rapidly proliferate following induction of autoimmunity and are suppressed after autoreactive T cells have been activated. Hence, in this model NK cells are associated with a protective role [56].

In pemphigus vulgaris, an autoimmune blistering disease that affects the skin and multiple mucous membranes, NK cells are increased in number and overexpress MHC class II molecules in peripheral blood when compared to controls. The disease is mediated by autoantibodies to desmoglein. In this case it has been suggested that local infiltrating NK cells can induce autoreactive $CD4^+$ T cells. NK cells obtained from peripheral blood leukocytes (PBL) can induce the proliferation of autologous $CD4^+$ T cells isolated from PBL or skin lesions in the presence of desmoglein peptides [57, 58]. In some cases of possible interactions of NK and T cell there is a contrasting observation, which suggests that NK cells can act as a promoter of an autoimmune disease too, by acting as an efficient Ag presenting cell.

7 NK Cells in Systemic Lupus Erythematosus

SLE is a complex autoimmune disease associated with multiorgan damage and the presence of autoantibodies. In SLE, several authors have reported that NK cell number and functions are reduced [59–61]. NK cell cytotoxicity against normally sensitive targets is compromised. Thus, it has been suggested that the altered circulating NK cell compartment is implicated in the pathophysiology of SLE.

The reduction of the circulating NK cell count in SLE is variable and usually moderate. The decrease in NK cell cytotoxicity appears to be, only in part, correlated to the decrease in NK cell count. Studies on sorted NK cells in SLE patients have shown an intrinsic defect in NK cell cytotoxicity [3, 61]. Some studies have also correlated NK cell alteration to a flare-up of the disease [3]. However, it is important in such studies to take into account the ongoing treatments since high doses of steroids [4] and azathioprine [3] have been reported to affect both NK cell count and function. Very few studies have analyzed the expression of NK cell receptors in SLE patients in an attempt to explain the functional deficiency. In one report, the expression of the adaptor molecule DAP12 [62] (associated in NK cells to NKp44 and activating KIR) has been shown to be reduced by western blot analysis. In our experience, the expression of the DAP12 molecule on circulating NK cells of SLE patients analyzed by flow cytometry with a novel rat-antihuman DAP12 antibody revealed no difference with controls (manuscript in preparation, unpublished data). Recently, in a Swedish cohort of lupus patients, an increase in the proportion of $CD56^{bright}$ NK cells, without correlation with the severity of the disease, was observed. In this study, as in our hands, the expression of the different NK inhibitory and activating receptors was not statistically different from the control groups [63]. In another systemic disease, circulating NK cells have been characterized in patients with Sjögren's syndrome. Like in SLE patients, NK cells were reduced and had impaired cytotoxicity [64]. The increase in apoptotic annexin V-positive NK cells in Sjögren's patients explained the decreased NK cell number [64]. The percentage of NK cells expressing the activating NK receptors CD2 and NKG2D was decreased and the authors noted an increase in the median fluorescence intensity of NKp46 compared with healthy controls. However, these phenotypic changes were moderate and were not associated with a decrease in NK cell function in an analysis of the lytic activity of NK cells in a per-cell basis. The significance of decreased NK cell counts and altered NK cell functions in SLE has not been established so far. Two main questions remain unanswered: (1) Is NK cell quantitative and functional deficiency a "genetic" determinant of SLE, as proposed by some authors [3], or the consequence of the disease and/or associated treatments?; (2) What are the characteristics of NK cell function deficiency in SLE?

The hypothetical, but not exclusive, disease-promoting role of NK cell "deficiency" could be either due to the abnormal control of DC differentiation or to an enhanced susceptibility to viral infections, both associated with SLE.

8 Concluding Remarks

In this review, we have tried to present in a comprehensive manner the different aspects and the results of NK cells studies in inflammatory and autoimmune disease in humans. Different approaches to explain the role of NK cells in these diseases have been based not only on functional and phenotypical studies but also on genetic analyses. NK cell receptors have also been shown to be implicated in the regulation of some T cell subsets. Surprisingly, particular T cells subsets expressing NK cell receptors are associated with inflammatory disease (Table 2).

All these results strongly suggest that NK cells are implicated in the pathophysiology of some autoimmune diseases. However, their exact role remains largely unknown. Moreover, they are thought to play a dual role, either disease-promoting or disease-protecting, depending on the disease. Two important points should be addressed in the further for a better understanding: the dynamic changes of the NK cell compartment during the disease and the role of tissular NK cells. This analysis together with a finer comprehension of NK cell interactions with DC in autoimmune and inflammatory diseases would greatly improve our knowledge of their role. Depending on the disease it can be speculated that the modulation of the NK

Table 2 Expression of activating NK cell receptors by lymphocyte subsets in inflammatory diseases

	NK cell receptors	Lymphocyte subset
Wegener's Granulomatosis	NKG2D	$CD3^+CD4^+$
Crohn's disease	NKG2D	$CD3^+CD4^+CD28^{null}$
Rheumatoid arthritis	Activating KIR	$CD3^+CD4^+CD28^{null}$
Spondylarthropathies	KIR3DL2	NK cells

cell compartment could be a therapeutic approach in some autoimmune diseases. Currently, biotherapies modifying NK cell functions are already in phase I trials in hematologic malignancies.

Thus it is necessary to clarify the changes and the role of the NK cell compartment in autoimmune and inflammatory diseases. Routine phenotypical and functional analysis of the NK cell compartment can now be achieved by flow cytometrical studies and specific NK cell markers, as NKp46, can be used for specific staining on tissue sections. These tools together with genetic studies will give us a more precise understanding of NK cell involvement in autoimmune and inflammatory diseases, allowing the development of new therapeutic approaches.

References

1. Johansson S, Berg L, Hall H, Hoglund P (2005) NK cells: elusive players in autoimmunity. Trends Immunol 26(11):613–618
2. Biassoni R, Cantoni C, Pende D, Sivori S, Parolini S, Vitale M et al (2001) Human natural killer cell receptors and co-receptors. Immunol Rev 181:203–214
3. Green MR, Kennell AS, Larche MJ, Seifert MH, Isenberg DA, Salaman MR (2005) Natural killer cell activity in families of patients with systemic lupus erythematosus: demonstration of a killing defect in patients. Clin Exp Immunol 141(1):165–173
4. Vitale C, Chiossone L, Cantoni C, Morreale G, Cottalasso F, Moretti S et al (2004) The corticosteroid-induced inhibitory effect on NK cell function reflects down-regulation and/or dysfunction of triggering receptors involved in natural cytotoxicity. Eur J Immunol 34 (11):3028–3038
5. Pascual V, Banchereau J, Palucka AK (2003) The central role of dendritic cells and interferon-alpha in SLE. Curr Opin Rheumatol 15(5):548–556
6. James JA, Harley JB, Scofield RH (2001) Role of viruses in systemic lupus erythematosus and Sjogren syndrome. Curr Opin Rheumatol 13(5):370–376
7. Moins-Teisserenc HT, Gadola SD, Cella M, Dunbar PR, Exley A, Blake N et al (1999) Association of a syndrome resembling Wegener's granulomatosis with low surface expression of HLA class-I molecules. Lancet 354(9190):1598–1603
8. Markel G, Mussaffi H, Ling KL, Salio M, Gadola S, Steuer G et al (2004) The mechanisms controlling NK cell autoreactivity in TAP2-deficient patients. Blood 103(5):1770–1778
9. Ljunggren HG, Karre K (1990) In search of the 'missing self': MHC molecules and NK cell recognition. Immunol Today 11(7):237–244
10. de la Salle H, Zimmer J, Fricker D, Angenieux C, Cazenave JP, Okubo M et al (1999) HLA class I deficiencies due to mutations in subunit 1 of the peptide transporter TAP1. J Clin Invest 103(5):R9–R13
11. Handgretinger R, Geiselhart A, Moris A, Grau R, Teuffel O, Bethge W et al (1999) Pure red-cell aplasia associated with clonal expansion of granular lymphocytes expressing killer-cell inhibitory receptors. N Engl J Med 340(4):278–284
12. Lamy T, Loughran TP Jr (2003) Clinical features of large granular lymphocyte leukemia. Semin Hematol 40(3):185–195
13. Loughran TP Jr (1993) Clonal diseases of large granular lymphocytes. Blood 82(1):1–14
14. Pascal V, Schleinitz N, Brunet C, Ravet S, Bonnet E, Lafarge X et al (2004) Comparative analysis of NK cell subset distribution in normal and lymphoproliferative disease of granular lymphocyte conditions. Eur J Immunol 34(10):2930–2940

15. Zambello R, Falco M, Della Chiesa M, Trentin L, Carollo D, Castriconi R et al (2003) Expression and function of KIR and natural cytotoxicity receptors in NK-type lymphoproliferative diseases of granular lymphocytes. Blood 102(5):1797–1805
16. Scquizzato E, Teramo A, Miorin M, Facco M, Piazza F, Noventa F et al (2007) Genotypic evaluation of killer immunoglobulin-like receptors in NK-type lymphoproliferative disease of granular lymphocytes. Leukemia 21(5):1060–1069
17. Tefferi A (1996) Chronic natural killer cell lymphocytosis. Leuk Lymphoma 20(3–4): 245–248
18. Tefferi A, Li CY, Witzig TE, Dhodapkar MV, Okuno SH, Phyliky RL (1994) Chronic natural killer cell lymphocytosis: a descriptive clinical study. Blood 84(8):2721–2725
19. Lima M, Almeida J, Montero AG, Teixeira Mdos A, Queiros ML, Santos AH et al (2004) Clinicobiological, immunophenotypic, and molecular characteristics of monoclonal CD56-/+dim chronic natural killer cell large granular lymphocytosis. Am J Pathol 165(4):1117–1127
20. Trowsdale J, Barten R, Haude A, Stewart CA, Beck S, Wilson MJ (2001) The genomic context of natural killer receptor extended gene families. Immunol Rev 181:20–38
21. Barrow AD, Trowsdale J (2008) The extended human leukocyte receptor complex: diverse ways of modulating immune responses. Immunol Rev 224(1):98–123
22. Uhrberg M, Parham P, Wernet P (2002) Definition of gene content for nine common group B haplotypes of the Caucasoid population: KIR haplotypes contain between seven and eleven KIR genes. Immunogenetics 54(4):221–229
23. Uhrberg M, Valiante NM, Shum BP, Shilling HG, Lienert-Weidenbach K, Corliss B et al (1997) Human diversity in killer cell inhibitory receptor genes. Immunity 7(6):753–763
24. Middleton D, Meenagh A, Moscoso J, Arnaiz-Villena A (2008) Killer immunoglobulin receptor gene and allele frequencies in Caucasoid, Oriental and Black populations from different continents. Tissue Antigens 71(2):105–113
25. Ruggeri L, Aversa F, Martelli MF, Velardi A (2006) Allogeneic hematopoietic transplantation and natural killer cell recognition of missing self. Immunol Rev 214:202–218
26. Boyton RJ, Altmann DM (2007) Natural killer cells, killer immunoglobulin-like receptors and human leucocyte antigen class I in disease. Clin Exp Immunol 149(1):1–8
27. Momot T, Koch S, Hunzelmann N, Krieg T, Ulbricht K, Schmidt RE et al (2004) Association of killer cell immunoglobulin-like receptors with scleroderma. Arthritis Rheum 50(5):1561–1565
28. Pellett F, Siannis F, Vukin I, Lee P, Urowitz MB, Gladman DD (2007) KIRs and autoimmune disease: studies in systemic lupus erythematosus and scleroderma. Tissue Antigens 69 (Suppl 1):106–108
29. van der Slik AR, Alizadeh BZ, Koeleman BP, Roep BO, Giphart MJ (2007) Modelling KIR-HLA genotype disparities in type 1 diabetes. Tissue Antigens 69(Suppl 1):101–105
30. van der Slik AR, Koeleman BP, Verduijn W, Bruining GJ, Roep BO, Giphart MJ (2003) KIR in type 1 diabetes: disparate distribution of activating and inhibitory natural killer cell receptors in patients versus HLA-matched control subjects. Diabetes 52(10):2639–2642
31. Nelson GW, Martin MP, Gladman D, Wade J, Trowsdale J, Carrington M (2004) Cutting edge: heterozygote advantage in autoimmune disease: hierarchy of protection/susceptibility conferred by HLA and killer Ig-like receptor combinations in psoriatic arthritis. J Immunol 173(7):4273–4276
32. Martin MP, Nelson G, Lee JH, Pellett F, Gao X, Wade J et al (2002) Cutting edge: susceptibility to psoriatic arthritis: influence of activating killer Ig-like receptor genes in the absence of specific HLA-C alleles. J Immunol 169(6):2818–2822
33. Parham P (2004) NK cells and trophoblasts: partners in pregnancy. J Exp Med 200(8): 951–955
34. Varla-Leftherioti M, Spyropoulou-Vlachou M, Niokou D, Keramitsoglou T, Darlamitsou A, Tsekoura C et al (2003) Natural killer (NK) cell receptors' repertoire in couples with recurrent spontaneous abortions. Am J Reprod Immunol 49(3):183–191
35. Varla-Leftherioti M (2005) The significance of the women's repertoire of natural killer cell receptors in the maintenance of pregnancy. Chem Immunol Allergy 89:84–95

36. Kwak JY, Beaman KD, Gilman-Sachs A, Ruiz JE, Schewitz D, Beer AE (1995) Up-regulated expression of CD56+, CD56+/CD16+, and CD19+ cells in peripheral blood lymphocytes in pregnant women with recurrent pregnancy losses. Am J Reprod Immunol 34(2):93–99
37. Anfossi N, Doisne JM, Peyrat MA, Ugolini S, Bonnaud O, Bossy D et al (2004) Coordinated expression of Ig-like inhibitory MHC class I receptors and acquisition of cytotoxic function in human CD8+ T cells. J Immunol 173(12):7223–7229
38. Saruhan-Direskeneli G, Uyar FA, Cefle A, Onder SC, Eksioglu-Demiralp E, Kamali S et al (2004) Expression of KIR and C-type lectin receptors in Behcet's disease. Rheumatology 43(4):423–427
39. Takeno M, Shimoyama Y, Kashiwakura J, Nagafuchi H, Sakane T, Suzuki N (2004) Abnormal killer inhibitory receptor expression on natural killer cells in patients with Behcet's disease. Rheumatol Int 24(4):212–216
40. Middleton D, Meenagh A, Sleator C, Gourraud PA, Ayna T, Tozkir H et al (2007) No association of KIR genes with Behcet's disease. Tissue Antigens 70(5):435–438
41. Chan AT, Kollnberger SD, Wedderburn LR, Bowness P (2005) Expansion and enhanced survival of natural killer cells expressing the killer immunoglobulin-like receptor KIR3DL2 in spondylarthritis. Arthritis Rheum 52(11):3586–3595
42. Namekawa T, Snyder MR, Yen JH, Goehring BE, Leibson PJ, Weyand CM et al (2000) Killer cell activating receptors function as costimulatory molecules on CD4+CD28null T cells clonally expanded in rheumatoid arthritis. J Immunol 165(2):1138–1145
43. Capraru D, Muller A, Csernok E, Gross WL, Holl-Ulrich K, Northfield J et al (2008) Expansion of circulating NKG2D+ effector memory T-cells and expression of NKG2D-ligand MIC in granulomaous lesions in Wegener's granulomatosis. Clin Immunol 127(2):144–150
44. Allez M, Tieng V, Nakazawa A, Treton X, Pacault V, Dulphy N et al (2007) CD4+NKG2D+ T cells in Crohn's disease mediate inflammatory and cytotoxic responses through MICA interactions. Gastroenterology 132(7):2346–2358
45. Schleinitz N, Cognet C, Guia S, Laugier-Anfossi F, Baratin M, Pouget J et al (2008) Expression of the CD85j (leukocyte Ig-like receptor 1, Ig-like transcript 2) receptor for class I major histocompatibility complex molecules in idiopathic inflammatory myopathies. Arthritis Rheum 58(10):3216–3223
46. Ferlazzo G (2008) Isolation and analysis of human natural killer cell subsets. Methods Mol Biol 415:197–213
47. Ferlazzo G, Thomas D, Lin SL, Goodman K, Morandi B, Muller WA et al (2004) The abundant NK cells in human secondary lymphoid tissues require activation to express killer cell Ig-like receptors and become cytolytic. J Immunol 172(3):1455–1462
48. Zhang AL, Colmenero P, Purath U, Teixeira de Matos C, Hueber W, Klareskog L, et al (2007) Natural killer cells trigger differentiation of monocytes into dendritic cells. Blood 110(7):2484–2493
49. Ferlazzo G, Morandi B, D'Agostino A, Meazza R, Melioli G, Moretta A et al (2003) The interaction between NK cells and dendritic cells in bacterial infections results in rapid induction of NK cell activation and in the lysis of uninfected dendritic cells. Eur J Immunol 33(2):306–313
50. Blanco P, Palucka AK, Gill M, Pascual V, Banchereau J (2001) Induction of dendritic cell differentiation by IFN-alpha in systemic lupus erythematosus. Science 294(5546):1540–1543
51. Ghiringhelli F, Menard C, Martin F, Zitvogel L (2006) The role of regulatory T cells in the control of natural killer cells: relevance during tumor progression. Immunol Rev 214:229–238
52. Trzonkowski P, Szmit E, Mysliwska J, Mysliwski A (2006) CD4+CD25+ T regulatory cells inhibit cytotoxic activity of CTL and NK cells in humans-impact of immunosenescence. Clin Immunol 119(3):307–316
53. Ghiringhelli F, Menard C, Terme M, Flament C, Taieb J, Chaput N et al (2005) CD4+CD25+ regulatory T cells inhibit natural killer cell functions in a transforming growth factor-beta-dependent manner. J Exp Med 202(8):1075–1085

54. Costantino CM, Baecher-Allan CM, Hafler DA (2008) Human regulatory T cells and autoimmunity. Eur J Immunol 38(4):921–924
55. Sakaguchi S, Ono M, Setoguchi R, Yagi H, Hori S, Fehervari Z et al (2006) Foxp3+ CD25+ CD4+ natural regulatory T cells in dominant self-tolerance and autoimmune disease. Immunol Rev 212:8–27
56. Liu R, Van Kaer L, La Cava A, Price M, Campagnolo DI, Collins M et al (2006) Autoreactive T cells mediate NK cell degeneration in autoimmune disease. J Immunol 176(9):5247–5254
57. Stern JN, Keskin DB, Barteneva N, Zuniga J, Yunis EJ, Ahmed AR (2008) Possible role of natural killer cells in pemphigus vulgaris – preliminary observations. Clin Exp Immunol 152(3):472–481
58. Takahashi H, Amagai M, Tanikawa A, Suzuki S, Ikeda Y, Nishikawa T et al (2007) T helper type 2-biased natural killer cell phenotype in patients with pemphigus vulgaris. J Invest Dermatol 127(2):324–330
59. Riccieri V, Spadaro A, Parisi G, Taccari E, Moretti T, Bernardini G, et al (2000) Downregulation of natural killer cells and of gamma/delta T cells in systemic lupus erythematosus. Does it correlate to autoimmunity and to laboratory indices of disease activity? Lupus 9(5):333–337
60. Erkeller-Yusel F, Hulstaart F, Hannet I, Isenberg D, Lydyard P (1993) Lymphocyte subsets in a large cohort of patients with systemic lupus erythematosus. Lupus 2(4):227–231
61. Yabuhara A, Yang FC, Nakazawa T, Iwasaki Y, Mori T, Koike K et al (1996) A killing defect of natural killer cells as an underlying immunologic abnormality in childhood systemic lupus erythematosus. J Rheumatol 23(1):171–177
62. Tomasello E, Vivier E (2005) KARAP/DAP12/TYROBP: three names and a multiplicity of biological functions. Eur J Immunol 35(6):1670–1677
63. Schepis D, Gunnarsson I, Eloranta ML, Lampa J, Jacobson SH, Karre K, et al (2009) Increased proportion of CD56(bright) natural killer cells in active and inactive systemic lupus erythematosus. Immunology 125(1): 140–146
64. Izumi Y, Ida H, Huang M, Iwanaga N, Tanaka F, Aratake K et al (2006) Characterization of peripheral natural killer cells in primary Sjogren's syndrome: impaired NK cell activity and low NK cell number. J Lab Clin Med 147(5):242–249
65. Ercole LP, Malvezzi M, Boaretti AC, Utiyama SR, Rachid A (2003) Analysis of lymphocyte subpopulations in systemic sclerosis. J Invest Allergol Clin Immunol 13(2):87–93
66. Horikawa M, Hasegawa M, Komura K, Hayakawa I, Yanaba K, Matsushita T et al (2005) Abnormal natural killer cell function in systemic sclerosis: altered cytokine production and defective killing activity. J Invest Dermatol 125(4):731–737
67. Majorczyk E, Pawlik A, Luszczek W, Nowak I, Wisniewski A, Jasek M et al (2007) Associations of killer cell immunoglobulin-like receptor genes with complications of rheumatoid arthritis. Genes Immun 8(8):678–683
68. Yen JH, Lin CH, Tsai WC, Wu CC, Ou TT, Hu CJ et al (2006) Killer cell immunoglobulin-like receptor gene's repertoire in rheumatoid arthritis. Scand J Rheumatol 35(2):124–127
69. Lopez-Larrea C, Blanco-Gelaz MA, Torre-Alonso JC, Bruges Armas J, Suarez-Alvarez B, Pruneda L et al (2006) Contribution of KIR3DL1/3DS1 to ankylosing spondylitis in human leukocyte antigen-B27 Caucasian populations. Arthritis Res Ther 8(4):R101
70. Jiao YL, Ma CY, Wang LC, Cui B, Zhang J, You L et al (2008) Polymorphisms of KIRs gene and HLA-C alleles in patients with ankylosing spondylitis: possible association with susceptibility to the disease. J Clin Immunol 28(4):343–349
71. Cameron AL, Kirby B, Griffiths CE (2003) Circulating natural killer cells in psoriasis. Br J Dermatol 149(1):160–164
72. Jones DC, Edgar RS, Ahmad T, Cummings JR, Jewell DP, Trowsdale J et al (2006) Killer Ig-like receptor (KIR) genotype and HLA ligand combinations in ulcerative colitis susceptibility. Genes Immun 7(7):576–582

Natural Killer Cells and the Skin

Dagmar von Bubnoff

1 Atopy and NK Cells

Atopic allergic individuals are prone to develop one or more atopic diseases such as allergic rhinitis (AR), allergic kerato-conjunctivitis, allergic asthma or atopic dermatitis (AD). Under normal circumstances, innocuous proteins delivered through the outer surfaces into regional lymph nodes do not evoke strong immune reactions but induce antigen-specific hypo responsiveness. The cellular mechanisms that develop after allergen encounter are understood only very incompletely, and the contributions of the various types of cells involved in T cell hypo responsiveness toward allergens are insufficiently defined. In patients with atopic disorders, periods of disease aggravation alternate with phases of remission, even in the presence of the offending allergen(s). A complex interrelation of genetic, environmental, psychological, and immunological factors may account for the manifestation of allergic inflammation [1, 2].

Atopic individuals show an inherited dominance of T_H2 responses. Blood T cells of these patients respond to allergens such as birch pollen allergen, with the production of Interleukin (IL)-4, IL-5, and IL-13 rather than Interferon (IFN)-γ (which is secreted by T_H1 cells) as in healthy individuals. IL-4 and IL-13 are the principal mediators of the B cell antibody class, switching towards IgE and therefore are key initiators of IgE-dependent reactions. Interleukin-5 acts mainly on eosinophils as an activating cytokine. These different mediators produced by the same T cell subset account for the frequently observed high serum levels of IgE and activated eosinophils seen in T_H2-dominated diseases. In AD, the initial T_H2-cytokine-dominated acute phase switches into a second phase, this is predominated by T_H1 cytokines such as IFN-γ, leading to chronic skin lesions [3].

D. von Bubnoff
Department of Dermatology and Allergy, University Hospitals, Bonn, Germany
e-mail: d.bubnoff@uni-bonn.de

The hallmark of the lesional skin in AD is the increased expression of the high-affinity receptor for IgE, FcεRI on Langerhans cells in the epidermis and on the only recently identified population of inflammatory dendritic epidermal cells (IDECs). In AD, it is assumed that the skin has a reduced barrier function which allows allergens to penetrate the epidermis more easily. In the epidermis, the allergens are taken up by these local antigen-presenting cells (APC) by means of FcεRI-bound IgE molecules [4, 5].

NK cells have long been neglected in the context of allergy. It is now clear that these cells are able to shape the overall inflammatory response in allergic reactions. In human skin, NK cells are mostly associated with skin tumors, but these cells are also present in lesional skin from patients with distinct inflammatory diseases such as AD. Skin biopsies from healthy individuals and nonlesional areas from patients with AD do show only a few scattered $CD56^+CD3^-$ NK cells in the dermis, in close proximity to the epidermis. In contrast, in lesional skin of patients with AD, $CD56^+$ NK cells are also found in the epidermis and are numerous in the dermal atopic infiltrate. In accordance with these findings, increased levels of the NK cell chemoattractants MCP-1/CCL2 and MDC/CCL22, the latter of which is derived from $CD1a^+$ dendritic cells (DC) in the skin, are present in lesional skin of AD [6].

Similar to naïve T cells, where culture conditions can favor the differentiation to T_H1 or T_H2 cells, NK cells from human cord blood can be conditioned to NK1 cells producing IFN-γ and NK2 cells secreting predominantly IL-5 and IL-13. In accordance with that, the in vivo existence of freshly isolated IFN-γ-secreting and IFN-γ-non-secreting human NK cell subsets could be established [7, 8]. The possible role of NK cells in AD has been studied with somewhat contradictory results. The main focus has been the number of NK cells in peripheral blood and their level of activation. In AD patients, both a numerical decrease of NK cells in the periphery and a lower NK cell activity compared to individuals without AD has been most consistently reported [9, 10]. NK cells leaving the circulation and accumulating in tissues could account for the lower number of NK cells reported in these studies. This shift of NK cells to the site of allergic inflammation was also evident in prior studies involving animal models of allergic asthma [11] or allergic peritonitis [12]. In accordance with these findings, in *Malassezia* atopy-patch test positive skin, where the yeast *Malassezia* acts as an allergen in the skin of patients with AD, numerous $CD56^+CD3^-$ NK cells have been found in the dermal lesions [6]. Interestingly, these NK cells were located in close contact to $CD1a^+$ DC. In vitro, *Malassezia* can be taken up by immature monocyte-derived DC leading to their maturation and production of cytokines with a potential to skew the immune outcome towards a T_H2-like response. In addition, these DC that were pretreated with *Malassezia* showed a reduced susceptibility to NK cell-induced cell death and may therefore have an increased capacity to sustain the atopic inflammation. In addition to the recruitment of NK cells to sites of inflammation, preferential apoptosis of NK cells in the periphery was thought to account for the decreased numbers of these cells in the blood of patients with AD [10]. The preferential apoptosis of NK cells in individuals with AD was dependent on cell contact with activated monocytes. Although many studies have shown that an interaction between NK cells and monocytes has profound effects on NK cell function, it remains to be determined why NK cells are susceptible to the inhibitory signal from monocytes [13, 14].

Much effort has been also devoted to the delineation of alterations in cytokine production by NK cells in AD. A significant reduction of TNF-α and IFN-γ from NK cells at a single-cell level was established by several groups [15, 16]. However, one group found that patients with AD harbored highly activated NK cells in vivo as indicated by a high spontaneous release of IL-4, IL-5, IL-13 and IFN-γ from isolated lesional NK cells [8]. Analogous to the predominance of T_H2 cells in atopy, the percentages of IL-5 and IL-13 producing NK cells were significantly higher in AD patients than in healthy individuals in the same study. Nevertheless, a reduction in IFN-γ production has been found consistently in childhood AD, an alteration that was not restricted to NK cells but extended to $CD4^+$, $CD8^+$ and $\gamma\delta^+$ T cells [9, 10]. In considering the crucial role of IFN-γ that is rapidly produced by NK cells early in the defense against allergens and pathogens, the conclusion carries some weight that this early impairment of NK cell function in AD is likely also to determine the nature of subsequent adaptive immune responses and to contribute to the overall response in a type 2-direction and to increased susceptibility to allergy and infection. The defective production of especially IFN-γ by these innate immune cells in vitro in AD could be interpreted in several ways: (1) there is an "intrinsic" and generalized defect in IFN-γ production in AD; (2) it may result from an "exhaustion" of innate and adaptive immune cells that have been continuously activated in vivo; or, (3) preferential apoptosis of type-1 cytokine-producing NK cells and T cells may occur upon contact with activated atopic APC. In vernal keratoconjunctivitis (VKC), a complex type I mediated hypersensitivity reaction of the conjunctiva involving also IgE-independent pathogenic mechanisms, NK cells constitute a significant proportion of the immune cells infiltrating the conjunctiva [17]. Here again, concomitant with the accumulation of NK cells at the site of allergic infiltration, a decrease of circulating NK cells in the blood was observed.

Overall, these studies appear to suggest the involvement of NK cells in the unbalanced cytokine network in allergic inflammation.

2 NK Cells and Malignant Skin Tumors

2.1 Lymphomas

Extra nodal lymphomas most often involve the gastrointestinal tract but the skin is the second most common site of manifestation [18, 19]. The current WHO/EORTC classification of cutaneous lymphomas (2006) makes the distinction between (1) mature T cell and NK cell lymphomas; (2) mature B cell lymphomas; and (3) immature hematopoietic malignancies, their variants and subgroups [20]. Cutaneous lymphomas can primarily start in the skin or arise as a secondary manifestation of systemic disease. Cutaneous lymphomas are most often of T cell origin (T cell lymphomas; TCL), representing about 75–80% of skin lymphomas. This is in contrast to other anatomic sites where B cell lymphomas (BCL) predominate [21].

Most of the cutaneous T cell and NK cell tumor types display a highly aggressive behavior, if not classified to the most common $CD4^+$ T cell lymphoma *mycosis fungoides* (MF) or *cutaneous anaplastic large cell lymphoma* (C-ALCL). It is very important precisely to sub-classify and phenotype these rarer tumors by immunohistochemistry and the local tissue appearance using criteria such as involvement of subcutaneous tissue, zonal necrosis, vascular destruction and others. Clonality is a distinctive feature of malignant lymphocytes. The cells of a malignant clone in TCL express all the same T cell receptor (TCR).

2.2 Extranodal NK/TCL, Nasal Type

True NK cell lymphomas are extremely rare. NK cells very often exhibit homing to extranodal sites. Nasal NK/T cell lymphoma is thought to be of true NK cell origin [22]. This type of lymphoma is mostly localized to the mucosal nasal cavity or nasopharynx, with the skin as the second most common site of involvement. Cutaneous lesions may arise as primary or secondary disease; primary cutaneous lesions show 25–50% involvement of the nasal cavity or other sites such as the gastrointestinal tract, lung or testis [23, 24]. NK/T cell lymphomas phenotypically express CD2, CD7, CD56, and cytoplasmic CD3; there is an absence of surface CD3 and CD5 expression and in most, but not all [23], of the cases there is no evidence for a T cell-lineage and therefore no *TCR* gene clonality. The expression of cytotoxic granule proteins (TIA-1, granzyme B, or perforin) is very common. In situ hybridization for Epstein–Barr virus (EBV)-derived RNA shows transcription of viral genes in 75% of the cases of the cutaneous nasal-type NK/TCL, especially in patients from the Far East or Central and South America.

In Europe and the United States, the incidence of NK/TCL, nasal type, is very low (<1% of non-Hodgkin lymphomas; NHL). Patients are frequently from Asia, Central and South America (where the relative incidence is 3–8% of NHL). Patients are of a median age of 52–66 years, ranging from 19 years to over 76 years. NK/TCL, nasal type, is clinically highly aggressive, particularly when there is extracutaneous involvement. The median survival is then less than 15 months [25, 26]. Reactive haemophagocytic syndrome has often been described in peripheral NK cell or T cell malignancies [27]. It is characterized by systemic activation of nonmalignant macrophages that phagocytose haematopoietic cells. Clinical features are fever, hepatosplenomegaly, and cytopenia. In the skin, lesions are typically multiple tumor nodules or plaques on the trunk and extremities. Histologically, tumor cells vary from the small to the medium and the large size. Typically, there is angioinvasion and angiodestruction with large zonal areas of necrosis. The number of recognizable malignant cells can be limited. Extension of the infiltrate into the subcutaneous tissue is common and may predominate, mimicking subcutaneous panniculitis-like TCL (SPTCL) [28, 29].

Aggressive NK cell leukemia can in rare cases secondarily involve the skin and therefore raises the question as to its relationship to NK/TCL, nasal type [30]. These two tumors have a similar phenotype, and differentiation requires correlation with clinical features and staging studies with bone marrow examination. Nasal-type NK/TCL has a much higher frequency of skin involvement than NK cell leukemia. NK/TCL, nasal-type, only rarely involves the bone marrow or lymph nodes, whereas general lymphadenopathy is common in aggressive NK cell leukemia. Clinically, nasal-type NK/TCL is a disease of the middle-aged to the old adults whereas NK cell leukemia typically occurs in younger patients.

2.3 NK Cells and Melanoma

NK cells are known to be important in the host defense against tumors, including melanoma [31–33]. However, NK cells are only infrequently found at the site of primary melanocytic tumors or precursor lesions. This is in accord with functional studies that specifically dealt with melanomas and found no NK cell activity in these lesions [34–36]. From these data one can assume that either NK cells act at an early stage before clinically detectable malignancy arises in primary tumor masses or NK cells may be more important in the peripheral blood, where they circulate in substantial numbers, or at sites of metastases.

In lymph node metastases from melanomas, NK cells can be found right within the tumor whereas the rest of the node is virtually devoid of NK cells [34]. While lymphocytes tend to be located in the periphery of these tumors, NK cells tend to be admixed with the tumor cells and can be quite numerous (ratio of NK to tumor cells: from 1:20 to 1:3) [34, 37] Given the presence of NK cells in these tumors, it is not clear whether the NK cells are attempting to kill the tumor cells or if not, why not. As discussed elsewhere in this book, NK-mediated killing of target cells occurs when a signal threshold is reached that is determined by the input from activating and inhibiting receptors. One of the most extensively studied activating NK cell receptors is NKG2D, a cell surface glycoprotein expressed in human and mouse NK cells and $CD8^+$ T lymphocytes [38]. Human NKG2D ligands (NKG2DL) are the MHC class I chain-related genes A and B (MICA and MICB) and the GPI-bound cell surface molecules UL16-binding protein (ULBP)-1,-2, and -3. It has been recently found that several human melanomas (cell lines and freshly isolated metastases) do not express MICA on the cell surface but have, immature deposits of this NKG2DL, which are retained in the endoplasmic reticulum [39]. These observations could represent a novel strategy developed by melanoma cells to evade NK cell-mediated immune surveillance. Many more studies are needed to exactly determine the role of NK cells for melanoma surveillance [40, 41].

3 NK Cells and Alopecia Areata

In the mammalian body, immunoprivileged sites are found in a few, well-defined tissue compartments. These sites include the anterior chamber of the eye, the testes, the placenta, and also hair follicles in the growth stage of the hair cycle (anagen phase) [42, 43]. A collapse of the immunoprivilege of the hair follicle results in the loss of hair as seen in patients with the autoimmune disease alopecia areata (AA). Histologically, this disease is characterized by intra- and perifollicular inflammatory cell infiltrates consisting of $CD4^+$ and $CD8^+$ T lymphocytes, macrophages and Langerhans cells that target hair follicle keratinocytes, melanocytes, and dermal papillar fibroblasts [44]. Clinically, round hairless patches are seen on the scalp while hair follicles remain intact. Like other sites of immunoprivilege, the hair follicle area is characterized by down regulation of MHC class-I expression, reduced capacity of APC and expression of local immunosuppressants such as TGF-β from regulatory T cells [45]. It might be expected that the absence or low expression of MHC class I expression in the hair follicle would allow for NK cell activation since NK cells attack cells with low or absent MHC class I. But this is clearly not the case since multiple samples from normal human scalp show that there is no NK cell attack on normal anagen hair follicles [46]. In AA lesions, $CD4^+$ and $CD8^+$ T cells are clustered at a high density around the anagen hair bulb. In addition, many perifollicular $CD56^+$ NK cells with high expression of NK cell-activating receptor NKG2D are present in atrophic AA lesions unlike normal scalp or scalp from patients with AD [47]. Concomitantly, the ligand for NKG2D, MICA, was found to be expressed at very high levels in the proximal outer root sheath, the dermal papilla and the connective tissue sheath of AA hair follicles [48]. This suggests that the up-regulation of MICA in lesional AA enhances the susceptibility of these hair follicles for an attack by $NKG2D^+$ NK cells, which may then promote anagen termination and AA progression. Also in the peripheral blood, NK cells of AA patients express increased levels of the NK cell-activating receptors NKG2D and NKG2C and reduced expression of the NK cell inhibitory receptors KIR-2DL2/2DL3 compared to healthy controls or patients with other inflammatory diseases [48, 49]. This suggests that NK cells in AA patients are considerably more susceptible to activating stimuli than in normal control subjects. In addition to NK cell receptors and their ligands, macrophage migration inhibitory factor (MIF), which is normally present in immunoprivileged areas and can potently suppress NK cell activity, is also aberrantly expressed in lesional AA [48]. While normal anagen hair follicles show widespread MIF expression throughout most of the epithelium of normal anagen scalp hair follicles, epithelium of lesional skin AA hair follicles displays reduced or absent MIF expression. Hair follicles in AA may therefore have a decreased capacity to suppress undesired NK cell activity.

Interferon-γ is now appreciated as a key cytokine in AA pathogenesis, presumably by mediating the collapse of the immunoprivilege of anagen hair follicles [50]. Interferon-γ is able to upregulate NKG2D on normal NK cells, which suggests that this is the reason for the elevated expression of the activating receptor NKG2D on NK cells from AA patients. In conclusion, the level of NK cell activity seems to be an important parameter to maintain sites of immunoprivilege such as hair follicles.

4 NK Cells and Psoriasis

Psoriasis has long been thought of as a disorder of predominantly epidermal hyperproliferation, which in conjunction with a faster maturation and terminal differentiation of basal epidermal cells results in an abnormal scaling of the epidermal barrier. This view of psoriasis as an epidermal disease has now been replaced with the recognition that this disorder is a genetically inherited, immune-mediated and organ-specific (skin and/or joints) disease. Innate and adaptive immunity, especially intralesional T lymphocytes trigger primed abnormal basal stem keratinocytes to proliferate and perpetuate the disease process [51, 52]. Most of the infiltrating T cells belong to the T_H1 and cytotoxic T subtypes and release IFN-γ and TNF-α, both of which are considered critical for disease initiation and persistence. This is underlined by the dramatic improvement of disease symptoms when T cells are targeted or TNF-α is blocked [53, 54]. However, it has been more recently suggested that the innate immune system is dysfunctional in psoriasis [55]. In psoriatic plaques, about 5%–8% $CD3^-CD56^+$ NK cells are present [56]. These cells locate to the mid and papillary dermis, are mostly (up to about 80%) $CD56^{bright}CD16^-$ and are recruited from the about 5–10% circulating NK cells that home to secondary lymphoid organs and modulate adaptive immune responses. Here, in the papillary dermis, they may interact with recently emigrated DC and may affect T helper cell polarization through the release of T_H1 cytokines [57]. Isolated NK cells from psoriatic lesions can release a remarkable amount of IFN-γ and also, albeit in smaller amounts, TNF-α. Chemokine receptors CXCR3 and CCR5, which are highly expressed on NK cells infiltrating lesional psoriatic skin, bind keratinocyte chemoattractants CXCL10 and CCL5, respectively, and presumably play a role in sequestering these NK cells in the inflamed skin. Interestingly, NK cells infiltrating psoriatic lesions do not express CLA (cutaneous lymphocyte antigen) or the chemokine receptor CCR4, both of which are associated with lymphocyte homing to the skin, suggesting a different mechanism for NK recruitment to and engagement in the skin. It has been speculated that immigrating DC activate $CD56^{high}CD16^-$NK cells in psoriatic skin, leading to the release of high levels of IFN-γ from the NK cells [56, 58]. This cytokine may affect bystander cells, keratinocytes in particular. Psoriatic NK cell supernatants are potent in inducing ICAM-1 and MHC class II expression on cultured psoriatic keratinocytes and can promote the release of CXCL10, CCL5 and CCL20 by keratinocytes [56]. In conclusion, NK cells seem to participate in psoriatic skin inflammation thus shaping the overall inflammatory response.

5 NK Cells and Herpes Simplex

Herpes simplex virus (HSV) only infects humans, there is no animal reservoir known. HSV has two serotypes: HSV-1 and HSV-2. These types are identical morphologically but differ on a genetic level, in epidemiology and with regard to clinical manifestation. HSV-2 is typically found in genital lesions whereas HSV-1 is found primarily in the mouth of children (herpetic gingivostomatitis) and in the eye (herpetic keratoconjunctivitis) but also on the skin of healthy individuals (herpetic infection of the skin) [59]. HSV can readily infect and spread on the skin of an individual with atopic eczema (eczema herpeticum) [5]. NK cells have been shown to be crucial to containment of ocular infections [60]. A genetically inherited functional impairment of NK cells and their innate IFN-γ production is often causally linked with more severe HSV infections [61]. Normally, the primary HSV infection is asymptomatic. However, HSV persists latently in nerve root ganglia from where the infection can be reactivated during periods of immune suppression. The crucial element in the early restriction of virus replication is the concerted action of different arms of the innate immune response [62].

Macrophages and DC orchestrate a multitude of antiherpetic mechanisms during the first hours of the attack. Cytokines, especially type I interferons (IFN type I) and TNF are produced and exert a direct antiviral effect while at the same time activating macrophages and DC [63, 64]. In the next wave, IL-12 is produced from these APC, which induce the production of IFN-γ in NK cells and T cells, and IL-12 activates the cytotoxic potential of these cells [65]. Unlike T cells, NK cells readily and rapidly reply to IL-12 with the secretion of IFN-γ, since the IFN-γ locus in NK cells is constitutively demethylated and is thus ready for transcription of the gene [66]. Macrophages, DC and NK cells are then stimulated by IFN-γ, resulting in activation for enhanced antimicrobial capacity.

Many positive feedback mechanisms and synergistic interactions now intensify the struggle against the virus. The innate and acquired immune responses are ultimately linked together during the antigen-presentation process, where internalized and processed viral antigens are presented to naïve T cells. In particular, DC and NK cells and their cytokine milieu generated during the early innate immune response to HSV determine the activation of T cells in the draining lymph nodes. Here, IL-12 and IFN-γ are the main factors to activate and drive the adaptive immunity towards a T_H1-driven response, important for the long-term control of intracellular pathogens [67, 68] IL-12 stimulates proliferation of naïve T cells, and in conjunction with IFN-γ, inhibits T_H2 cell differentiation and the production of T_H2 cytokines (e.g., IL-4, IL-5, and IL-13). The optimal activation of NK cells seems to play a pivotal role in the defense against HSV infection. In a murine study, it has been shown that ablation of DC led to enhanced susceptibility to HSV-1 infection in the highly resistant C57BL/6 mouse strain [67]. The ablation of DC impaired the activation of NK cells and T cells in response to HSV-1, resulting in increased spread of HSV-1 into the nervous system and increased mortality. These data show that NK cells are crucial not only for the acquired immune response toward HSV-1 but also for the innate resistance to HSV. Interestingly, Langerhans cells, a subset of DC that are typically localized in the basal and suprabasal layers of the epidermis and along mucosal surfaces, do not seem to play a leading role in activating T cell mediated defenses against viral infections of the skin or mucosa [69]. Together, the early combined activation of macrophages, DC and NK cells and their antiviral activity, IL-12 and IFN-γ production are important mediators of innate resistance to HSV.

References

1. von Bubnoff D, Geiger E, Bieber T (2001) Antigen-presenting cells in allergy. J Allergy Clin Immunol. 108:329–339
2. Bieber T (2008) Atopic dermatitis. N Engl J Med. 358:1483–1494
3. Grewe M, Bruijnzeel-Koomen CA, Schopf E et al (1998) A role for Th1 and Th2 cells in the immunopathogenesis of atopic dermatitis. Immunol Today 19:359–361
4. von Bubnoff D, Novak N, Kraft S, Bieber T (2003) The central role of FcepsilonRI in allergy. Clin Exp Dermatol 28:184–187
5. Wollenberg A, Klein E (2007) Current aspects of innate and adaptive immunity in atopic dermatitis. Clin Rev Allergy Immunol. 33:35–44
6. Buentke E, Heffler LC, Wilson JL et al (2002) Natural killer and dendritic cell contact in lesional atopic dermatitis skin–Malassezia-influenced cell interaction. J Invest Dermatol 119:850–857
7. Erten G, Aktas E, Deniz G (2008) Natural killer cells in allergic inflammation. Chem Immunol Allergy 94:48–57
8. Aktas E, Akdis M, Bilgic S et al (2005) Different natural killer (NK) receptor expression and immunoglobulin E (IgE) regulation by NK1 and NK2 cells. Clin Exp Immunol 140: 301–309
9. Campbell DE, Fryga AS, Bol S, Kemp AS (1999) Intracellular interferon-gamma (IFN-gamma) production in normal children and children with atopic dermatitis. Clin Exp Immunol 115:377–382
10. Katsuta M, Takigawa Y, Kimishima M, Inaoka M, Takahashi R, Shiohara T (2006) NK cells and gamma delta+ T cells are phenotypically and functionally defective due to preferential apoptosis in patients with atopic dermatitis. J Immunol 176:7736–7744
11. Schuster M, Tschernig T, Krug N, Pabst R (2000) Lymphocytes migrate from the blood into the bronchoalveolar lavage and lung parenchyma in the asthma model of the brown Norway rat. Am J Respir Crit Care Med 161:558–566
12. Walker C, Checkel J, Cammisuli S, Leibson PJ, Gleich GJ (1998) IL-5 production by NK cells contributes to eosinophil infiltration in a mouse model of allergic inflammation. J Immunol 161:1962–1969
13. Kang SJ, Liang HE, Reizis B, Locksley RM (2008) Regulation of hierarchical clustering and activation of innate immune cells by dendritic cells. Immunity 29:819–833
14. Kloss M, Decker P, Baltz KM et al (2008) Interaction of monocytes with NK cells upon Toll-like receptor-induced expression of the NKG2D ligand MICA. J Immunol 181:6711–6719
15. Koning H, Baert MR, Oranje AP, Savelkoul HF, Neijens HJ (1996) Development of immune functions related to allergic mechanisms in young children. Pediatr Res 40:363–375
16. Liao SY, Liao TN, Chiang BL et al (1996) Decreased production of IFN gamma and increased production of IL-6 by cord blood mononuclear cells of newborns with a high risk of allergy. Clin Exp Allergy 26:397–405
17. Lambiase A, Normando EM, Vitiello L et al (2007) Natural killer cells in vernal keratoconjunctivitis. Mol Vis 13:1562–1567
18. Bouaziz JD, Bastuji-Garin S, Poszepczynska-Guigne E, Wechsler J, Bagot M (2006) Relative frequency and survival of patients with primary cutaneous lymphomas: data from a single-centre study of 203 patients. Br J Dermatol 154:1206–1207
19. Willemze R, Jaffe ES, Burg G et al (2005) WHO-EORTC classification for cutaneous lymphomas. Blood 105:3768–3785
20. Kinney MC, Jones D (2007) Cutaneous T-cell and NK-cell lymphomas: the WHO-EORTC classification and the increasing recognition of specialized tumor types. Am J Clin Pathol 127:670–686

21. Sander CA, Flaig MJ, Jaffe ES (2001) Cutaneous manifestations of lymphoma: a clinical guide based on the WHO classification. World Health Organization Clin Lymphoma 2:86–100; discussion 101–102
22. Mraz-Gernhard S, Natkunam Y, Hoppe RT, LeBoit P, Kohler S, Kim YH (2001) Natural killer/natural killer-like T-cell lymphoma, CD56+, presenting in the skin: an increasingly recognized entity with an aggressive course. J Clin Oncol 19:2179–2188
23. Natkunam Y, Smoller BR, Zehnder JL, Dorfman RF, Warnke RA (1999) Aggressive cutaneous NK and NK-like T-cell lymphomas: clinicopathologic, immunohistochemical, and molecular analyses of 12 cases. Am J Surg Pathol 23:571–581
24. Kobashi Y, Nakamura S, Sasajima Y et al (1996) Inconsistent association of Epstein-Barr virus with CD56 (NCAM)-positive angiocentric lymphoma occuring in sites other than the upper and lower respiratory tract. Histopathology 28:111–120
25. Abouyabis AN, Shenoy PJ, Lechowicz MJ, Flowers CR (2008) Incidence and outcomes of the peripheral T-cell lymphoma subtypes in the United States. Leuk Lymphoma 49:2099–2107
26. Chattopadhyay A, Slater DN, Hancock BW (2005) Cutaneous CD56 positive natural killer and cytotoxic T-cell lymphomas. Int J Oncol 26:1559–1562
27. Sandner A, Helmbold P, Winkler M, Gattenlohner S, Muller-Hermelink HK, Holzhausen HJ (2008) Cutaneous dissemination of nasal NK/T-cell lymphoma in a young girl. Clin Exp Dermatol 33:615–618
28. Lee J, Park YH, Kim WS et al (2005) Extranodal nasal type NK/T-cell lymphoma: elucidating clinical prognostic factors for risk-based stratification of therapy. Eur J Cancer 41:1402–1408
29. Kim TM, Park YH, Lee SY et al (2005) Local tumor invasiveness is more predictive of survival than International Prognostic Index in stage I(E)/II(E) extranodal NK/T-cell lymphoma, nasal type. Blood 106:3785–3790
30. Nava VE, Jaffe ES (2005) The pathology of NK-cell lymphomas and leukemias. Adv Anat Pathol 12:27–34
31. Poeck H, Besch R, Maihoefer C et al (2008) 5'-Triphosphate-siRNA: turning gene silencing and Rig-I activation against melanoma. Nat Med 14:1256–1263
32. Skak K, Frederiksen KS, Lundsgaard D (2008) Interleukin-21 activates human natural killer cells and modulates their surface receptor expression. Immunology 123:575–583
33. Hussein MR (2005) Tumour-infiltrating lymphocytes and melanoma tumorigenesis: an insight. Br J Dermatol 153:18–21
34. Kornstein MJ, Stewart R, Elder DE (1987) Natural killer cells in the host response to melanoma. Cancer Res 47:1411–1412
35. Klein E, Mantovani A (1993) Action of natural killer cells and macrophages in cancer. Curr Opin Immunol 5:714–718
36. Klein E, Vanky F, Galili U, Vose BM, Fopp M (1980) Separation and characteristics of tumor-infiltrating lymphocytes in man. Contemp Top Immunobiol 10:79–107
37. Watt SV, Andrews DM, Takeda K, Smyth MJ, Hayakawa Y (2008) IFN-gamma-dependent recruitment of mature CD27(high) NK cells to lymph nodes primed by dendritic cells. J Immunol 181:5323–5330
38. Zimmer J, Michel T, Andres E, Hentges F (2008) Up-regulation of NKG2D ligands by AML cells to increase sensitivity to NK cells: the tumour might strike back. Comment on "Differentiation-promoting drugs up-regulate NKG2D ligand expression and enhance the susceptibility of acute myeloid leukemia cells to natural killer cell-mediated lysis" by Rohner et al. Leuk Res 2007;31:1393–402. Leukocyt Res 32:676–677
39. Fuertes MB, Girart MV, Molinero LL et al (2008) Intracellular retention of the NKG2D ligand MHC class I chain-related gene A in human melanomas confers immune privilege and prevents NK cell-mediated cytotoxicity. J Immunol 180:4606–4614
40. Strid J, Roberts SJ, Filler RB et al (2008) Acute upregulation of an NKG2D ligand promotes rapid reorganization of a local immune compartment with pleiotropic effects on carcinogenesis. Nat Immunol 9:146–154

41. Cagnano E, Hershkovitz O, Zilka A et al (2008) Expression of ligands to NKp46 in benign and malignant melanocytes. J Invest Dermatol 128:972–979
42. Ito T, Meyer KC, Ito N, Paus R (2008) Immune privilege and the skin. Curr Dir Autoimmun 10:27–52
43. Paus R, Nickoloff BJ, Ito T (2005) A 'hairy' privilege. Trends Immunol 26:32–40
44. Christoph T, Muller-Rover S, Audring H et al (2000) The human hair follicle immune system: cellular composition and immune privilege. Br J Dermatol 142:862–873
45. Ito T, Ito N, Bettermann A, Tokura Y, Takigawa M, Paus R (2004) Collapse and restoration of MHC class-I-dependent immune privilege: exploiting the human hair follicle as a model. Am J Pathol 164:623–634
46. Ranki A, Kianto U, Kanerva L, Tolvanen E, Johansson E (1984) Immunohistochemical and electron microscopic characterization of the cellular infiltrate in alopecia (areata, totalis, and universalis). J Invest Dermatol 83:7–11
47. Chiarini C, Torchia D, Bianchi B, Volpi W, Caproni M, Fabbri P (2008) Immunopathogenesis of folliculitis decalvans: clues in early lesions. Am J Clin Pathol 130:526–534
48. Ito T, Ito N, Saatoff M et al (2008) Maintenance of hair follicle immune privilege is linked to prevention of NK cell attack. J Invest Dermatol 128:1196–1206
49. Baadsgaard O, Lindskov R (1986) Circulating lymphocyte subsets in patients with alopecia areata. Acta Derm Venereol 66:266–268
50. Freyschmidt-Paul P, McElwee KJ, Hoffmann R et al (2006) Interferon-gamma-deficient mice are resistant to the development of alopecia areata. Br J Dermatol 155:515–521
51. Chamian F, Krueger JG (2004) Psoriasis vulgaris: an interplay of T lymphocytes, dendritic cells, and inflammatory cytokines in pathogenesis. Curr Opin Rheumatol 16:331–337
52. Gaspari AA (2006) Innate and adaptive immunity and the pathophysiology of psoriasis. J Am Acad Dermatol 54:S67–S80
53. Griffiths CE, Iaccarino L, Naldi L et al (2006) Psoriasis and psoriatic arthritis: immunological aspects and therapeutic guidelines. Clin Exp Rheumatol 24:S72–S78
54. Paller AS, Siegfried EC, Langley RG et al (2008) Etanercept treatment for children and adolescents with plaque psoriasis. N Engl J Med 358:241–251
55. Bos JD, de Rie MA, Teunissen MB, Piskin G (2005) Psoriasis: dysregulation of innate immunity. Br J Dermatol 152:1098–1107
56. Ottaviani C, Nasorri F, Bedini C, de Pita O, Girolomoni G, Cavani A (2006) CD56brightCD16 (-) NK cells accumulate in psoriatic skin in response to CXCL10 and CCL5 and exacerbate skin inflammation. Eur J Immunol 36:118–128
57. Cameron AL, Kirby B, Fei W, Griffiths CE (2002) Natural killer and natural killer-T cells in psoriasis. Arch Dermatol Res 294:363–369
58. Giustizieri ML, Mascia F, Frezzolini A et al (2001) Keratinocytes from patients with atopic dermatitis and psoriasis show a distinct chemokine production profile in response to T cell-derived cytokines. J Allergy Clin Immunol 107:871–877
59. Koelle DM, Corey L (2008) Herpes simplex: insights on pathogenesis and possible vaccines. Annu Rev Med 59:381–395
60. Inoue Y (2008) Immunological aspects of herpetic stromal keratitis. Semin Ophthalmol 23:221–227
61. Reading PC, Whitney PG, Barr DP, Smyth MJ, Brooks AG (2006) NK cells contribute to the early clearance of HSV-1 from the lung but cannot control replication in the central nervous system following intranasal infection. Eur J Immunol 36:897–905
62. Ellermann-Eriksen S (2005) Macrophages and cytokines in the early defence against herpes simplex virus. Virol J 2:59
63. Halford WP, Balliet JW, Gebhardt BM (2004) Re-evaluating natural resistance to herpes simplex virus type 1. J Virol 78:10086–10095
64. Murphy EA, Davis JM, Brown AS, Carmichael MD, Ghaffar A, Mayer EP (2008) Effect of IL-6 deficiency on susceptibility to HSV-1 respiratory infection and intrinsic macrophage antiviral resistance. J Interferon Cytokine Res 28:589–595

65. Trinchieri G (2003) Interleukin-12 and the regulation of innate resistance and adaptive immunity. Nat Rev Immunol 3:133–146
66. Tato CM, Martins GA, High FA, DiCioccio CB, Reiner SL, Hunter CA (2004) Cutting edge: innate production of IFN-gamma by NK cells is independent of epigenetic modification of the IFN-gamma promoter. J Immunol 173:1514–1517
67. Kassim SH, Rajasagi NK, Zhao X, Chervenak R, Jennings SR (2006) In vivo ablation of CD11c-positive dendritic cells increases susceptibility to herpes simplex virus type 1 infection and diminishes NK and T-cell responses. J Virol 80:3985–3993
68. Malmgaard L, Paludan SR (2003) IFN-alpha/beta, IL-12 and IL-18 coordinately induce production of IFN-gamma during infection with herpes simplex virus type 2. J Gen Virol 84:2497–2500
69. Allan RS, Smith CM, Belz GT et al (2003) Epidermal viral immunity induced by CD8alpha+ dendritic cells but not by Langerhans cells. Science 301:1925–1928

NK Cells in Oncology

Sigrid De Wilde and Guy Berchem

Abstract NK cells have a dual role in human cancer. First they are responsible for various haematological malignancies like the aggressive NK cell leukemia, the extra nodal NK/T cell lymphoma of the nasal type or the blastic NK cell lymphoma. Secondly NK cells have shown great promise in the therapeutic area since the 1970s. Indeed NK cell therapeutic strategies are being tested in vitro and in vivo in human cancer. As the understanding of NK cell immunology increases we will probably be able to use these cells in adoptive immunotherapy strategies to tackle human cancer in the future.

1 Introduction

Since the discovery of NK cells in the early seventies by Thornthwaite et al. [1], clinicians have dreamed to use these unique killer cells to eradicate human cancer. Unfortunately all these years NK cells, have appeared in medical textbooks more often as the "culprit" than as the "hero" capable of eradicating human cancer.

If some heroic attempts at using NK cells in cancer treatment have unfortunately failed this is very probably due to insufficient knowledge of their implication in the immune system and more precisely in tumor cytotoxicity. Over the past few years this knowledge gap has been closing in and some encouraging advances have been made, at least in xenograft animal models.

As for the implication of NK cells in haematological malignancies, a few NK cell lymphomas and leukemias have been described, like the aggressive NK cell leukemia, the extranodal NK/T cell lymphoma of the nasal type or the blastic NK cell lymphoma and even today many of these diseases have a very dismal prognosis.

S. De Wilde and G. Berchem (✉)
Service d'Hémato-Cancérologie, Centre Hospitalier de Luxembourg, 4 rue Barblé, L-1210, Luxembourg
e-mail: berchem.guy@chl.lu

In the first part of this chapter we will try to summarize the multiple attempts at using NK cells to cure cancer and in the second part we will focus on the malignant transformation of NK cells.

2 Therapeutic Implications in Human Cancer

For the therapeutic use of NK cells, after years of research and some clinical studies, and after the pioneering work of Rosenberg et al. in the late seventies, still no NK-based therapies are available for any sort of cancer today.

If NK cells are well-known by releasing perforin and other molecules which damage the cell membrane and induce apoptosis, it is also appreciated that they have a significant role in host defense against invading pathogens. After IL2 was discovered in the 1980s, it rapidly appeared that this new cytokine could activate NK cells and transform them into what was called by Rosenberg et al. [2] LAK cells (lymphokine activated killer). These cells, although triggering initially enormous enthusiasm, proved rapidly to be disappointing. Indeed it was shown that even if IL2 significantly increased the number of circulating NK cells in vivo, the cells were not maximally cytotoxic in vitro [3].

NK cells are both controlled by positive and negative cytolytic signals. Some of these inhibiting molecules have been cloned over the past 10 years and their ligands are almost always class I MHC molecules. Some of these receptors like those of the killer cell immunoglobulin receptor family (KIR) are specific for determinants shared by certain class I alleles and are expressed by a subset of NK cells. It has been suggested by Ruggeri et al. in 2002 [4] that the amount of tumor killing is directly linked to the degree of KIR mismatch with the tumor. As autologous NK cell therapy is thus insufficient, the use of allogeneic NK cell infusions was suggested by Miller's group [5]. IL2 activated allogeneic haplo-identical NK cells were administered to patients with metastatic melanoma, metastatic renal cell carcinoma, refractory Hodgkin's disease or poor prognosis AML. They showed that NK cells can persist and expand in vivo and induce complete haematological remissions in 5 of 19 poor prognosis AML patients but unfortunately no activity on other tumors was observed. Moreover the results showed that patients which were KIR ligand mismatched were the only ones having remissions. One can thus say that nonspecifically activated NK cells may be useful against a subset of tumors but the donor cells must be allogeneic and are much more likely to be effective if they are HLA mismatched.

More recently it has been shown that NK cells activated in vitro, with a combination of IL12 and IL18 injected into syngeneic animals, were able to collaborate with the host's own NK cells. Some tumor effect could be shown for melanoma xenografts [6, 7]. These and other groups have shown in animal models that NK cell strategies promise future immunotherapy for human cancer. As of today no reports of human use are available.

3 NK Cell Neoplasms

NK cells share immunophenotypic and functional properties with T cells. Therefore the NK cell and T cell neoplasms are often considered together.

For the WHO classification, we can retain the following ICD-codes for the NK cell neoplasms [8].

- 9948/3: Aggressive NK cell leukemia
- 9719/3: Extranodal NK/T cell lymphoma, nasal type
- 9727/3: Blastic NK cell lymphoma

The following paragraphs give an overview of the clinical hematological situations in which NK cells are implicated.

Firstly, a transient reactive population of NK cell large granular lymphocytes (LGL) can occur after viral infections, auto-immune disease, malignancies and solid organ transplantation. In atomic-bomb survivors, NK cell proliferations may be seen in association with neutropenia.

The absolute number of LGL in normal peripheral blood is between 200 and 400/μl. They can have a phenotype of NK cells (CD3−, CD56+) or a phenotype of activated cytotoxic effector T cells (CD3+, CD56−, CD57+).

NK cell neoplasms are prevalent in Asia and South America but rare in Western countries. For example, nasal NK/T cell lymphoma accounts for 8% of all lymphomas in Hong Kong whereas for less than 1% in Europe and North America.

The lymphoma cells have large granular lymphocyte morphology with the following immunophenotype: CD16+, CD2+, CD56+, CD57+, cytoplasmic CD3 epsilon and sometimes CD7+ (T lineage associated). They have a T cell receptor gene germline and often an EBV DNA load. If this EBV DNA load is $>6 \times 10^7$ copies/ml, this means a negative impact on disease free survival. Actually it stays difficult to determine clonality for NK cells.

T cell and NK cell malignancies express cytotoxic proteins like perforin, granzyme B and T cell intracellular Antigen (TIA)-1. They show prominent apoptosis, necrosis and angio- invasion.

Soluble FAS-ligand is often increased in aggressive NK cell lymphoma/leukemia but seems to be undetectable in patients in remission.

A consistent cytogenetic aberration is the deletion of 6q− but also complex caryotypes as deletions in 11q, 13q and 17p are seen.

The choice for treatment depends on the risk for systemic relapse but the problem is identifying good risk factors for this relapse [9]. Treatment is mostly based on a symptomatic basis. For example, radiotherapy has high initial response rates but frequent relapses occur locally within the first year. In refractory disease, L-asparaginase seems effective. Nowadays, lots of different treatment modalities have been proposed:

- Low dose methotrexate + cyclofosfamide + cyclosporine
- Nucleoside analogs
- Growth factors

- Anti-CD52
- ATG
- Splenectomy
- Tipifarnib

3.1 Aggressive NK Cell Lymphoma/Leukemia

Aggressive NK cell lymphoma or leukemia presents itself at the median age of 30 years with acute symptoms as: weight loss, high fever, jaundice, lymphadenopathy or hepatosplenomegaly.

The peripheral blood shows lymphocytosis, severe anemia and thrombocytopenia with often a hemophagocytic syndrome. Infiltration of cerebrospinal fluids, gastro-intestinal tract and peritoneal fluids are possible.

The median survival time is 2 months because the illness is refractory to various chemotherapy regimens.

A multidrug resistance gene-encoded protein on the cell membrane extrudes various cytotoxic agents. The final cause of the death of the patient with aggressive NK cell leukemia is mostly multiorgan failure with coagulopathy. The treatment should be based on specific trials.

The lymphocytes look slightly immature with broad pale cytoplasm and azurophilic granules, occasional nucleoli and somewhat fine nuclear chromatin. They are CD2+, surface CD3−, cytoplasmic CD3 epsilon+, CD56+, CD16−/+, CD57+, CD94+ (mature cell). Germline configuration of T cell receptor genes (TRB), IgH and clonal EBV in the tumor cells are standard.

Sometimes the number of neoplastic cells in peripheral blood and bone marrow is very low. Serum soluble FAS ligand level can be very high, contributing to the associated multiorgan failure.

Mostly a deletion of chromosome 6q has been found. However other clonal cytogenetic abnormalities have been detected as: duplication of 1q, rearrangement at 3q, loss of chromosomes Y, 13 or 10 and trisomy 8 [10].

3.2 Chronic NK Cell Lymphocytosis or NK Cell LGL Lymphocytosis

This is an indolent affection that only exceptionally transforms into an aggressive phase. There is an association with vasculitis or nephrotic syndrome but no cytopenias nor organomegaly. It occurs mainly in men in their sixties.

The LGL has the immunophenotype CD2+, surface CD3−, CD56+, CD16+ and CD57+. Although viral infections are thought to be implicated, there is no evidence for EBV or HTLV infection.

These NK cells must be in peripheral blood for at least 6 months at 0.6×10^9/l or higher.

This chronic NK cell lymphocytosis regresses sometimes spontaneously, the median duration is 5 years. Usually patients with this indolent affection do not need treatment. There are cases treated with cyclophosphamide or methotrexate [11].

3.3 Extranodal NK/T Cell Lymphoma, Nasal Type

This was formerly called lethal midline granuloma and occurs mainly in males of the fifth decade. It is most common in Asia (Hong Kong) and in native populations in Peru. These lymphomas are also encountered in immunosuppressed or post transplant patients.

The lesions are found in the nose and the upper aero digestive tract causing facial swelling, nasal obstruction and bleeding, prooptosis, destruction of hard palate, impairment of ocular movement etc.

Lesions in testis, skin, soft tissue and gastro intestinal tract are not excluded. B symptoms can accompany the local symptoms. Overlap with aggressive NK cell leukemia is seen in case of marrow and blood involvement.

The tumor cells vary in size and are mixed with a lot of inflammatory cells (small lymphocytes, plasma cells, histiocytes and eosinophils). They can cause ischemic necrosis and apoptotic bodies. Cytoplasm is often pale to clear with azurophilic granules. Often, mitotic figures are present.

The immunophenotype of the tumor cells is CD2+, surface CD3−, cytoplasmic CD3 epsilon+, CD7−/+, CD30−/+, CD56+, CD16−, CD57−. EBV is almost always associated. Some patients show rather an EBV+, CD56 − cytotoxic T cell phenotype.

Del (6)(q21q25) or i(6)(p10) is a common finding.

Prognosis of nasal NK/T cell lymphoma is variable and the prognostic factors are not known. Some features have been proposed as poor risk factors but not always confirmed by other studies:

- Local tumor invasiveness in bone, skin etc
- B symptoms
- Increased serum LDH
- Stage III or IV
- Age above 60 years
- EBV DNA copy load
- Cox 2 expression
- Positivity for cutaneous lymphocyte antigen (homing receptor)
- Loss of expression of granzyme B protease inhibitor 9 (PI9)
- Low apoptotic index

The ratio stages I + II over III + IV is about 7/3. For stage I–II, a radiotherapy of 30–60 Gy is commonly used but with frequent relapses and a 5 year overall survival of about 50%.

Chemotherapy can be initially discussed if the patient is medically fit. For the more advanced stages; immediate chemotherapy is advised but in 30–40% of cases salvage radiotherapy is necessary. Concomitant chemoradiotherapy has also been used (dexamethasone + etoposide + ifosfamide + carboplatinum or dexamethasone + etoposide + methotrexate + cyclofosfamide).

When the nasal type NK/T cell lymphoma occurs without the nasal cavity affected, there is a very aggressive pattern with poor response to therapy [12].

3.4 Blastic NK Cell Lymphoma

This lymphoma is very rare and occurs rather in the elderly. It may involve skin, lymph nodes, soft tissue, peripheral blood, bone marrow, nasal cavity etc. Lymphadenopathy and disseminated disease at presentation are usual [8].

The precise lineage of blastic NK cell lymphoma is still not clear but some think it originates rather from plasmocytoid dendritic cells. The tumor cells have a lymphoblast like morphology and even the differential diagnosis with acute myeloblastic leukemia can be difficult.

The immunophenotype is CD56+, surface CD3−, CD4−/+, CD43−/+, CD19−, CD20−, CD13− and CD33−. Expression of CD2, CD7, cytoplasmic CD3 epsilon and cytotoxic molecules is usually negative. There is no association with EBV and the T cell receptor beta and IgH are in germline configuration.

A distinction with myeloid/NK cell precursor acute leukemia has to be made by immunophenotyping with the latter showing CD7+, CD33+, CD34+ and CD56+.

The disease is aggressive and chemotherapy of acute leukemia like regimen, is necessary but rarely successful. Median survival is 12 months but long-term survival is very rare and has only been described with allogeneic hematopoietic cell transplantation.

3.5 Nonnasal NK Cell Lymphoma

This disease is comparable with the Extranodal NK/T cell lymphoma, nasal type without nasal involvement. A nasal panendoscopy is therefore essential to exclude an occult manifestation [12].

The early stages disseminate rapidly and the ratio stages I + II over stage III + IV is rather 4–6 compared to 7–3 for the nasal type.

Radiotherapy is only given in palliative situations.

3.6 Tokura–Ishihara Disease

This affection occurs mostly in Japan with a median onset of disease at 6 years.

It is not known who may be at risk for the disease and genetic and environmental factors will be investigated in future studies.

It is the triad of:

- Hypersensitivity to mosquito bites (with ulcerative lesions)
- Chronic EBV infection
- NK cell leukemia or lymphoma

The NK cells are in an activated state and express a high level of surface CD94 molecules and over express surface FAS ligand. They produce IFN-gamma without expression of IL-4, IL-10 and IL-13. Furthermore, the immunophenotype is CD2+, CD56+, CD4 and CD8 negative, CD16− and CD20−.

The disease presents with a local intense skin reaction after a mosquito bite accompanied by fever, liver damage, hemophagocytosis, thrombopenia, hepatosplenomegaly, lymphadenopathy, hematuria and proteinuria.

The children are mostly treated with prednisolone.

After recovering from the severe systemic manifestations, the children go well till the next mosquito bite. However, this affection may be fatal.

The CD4 positive T cells are thought to be stimulated by mosquito bites and so inducing EBV reactivation and EBV oncogene expression. This could develop not only Tokura–Ishihara disease but also NK cell oncogenesis (by the expression of viral oncogene LMP1) [12, 13, 14].

3.7 NK Cell Intravascular Lymphomatosis

Only about six cases of NK cell intravascular lymphomatosis have been described.

It occurs at all ages but females are more affected than males. Skin manifestations are mostly present but the central nervous system or bone marrow may also be involved.

Large lymphoid cells proliferate exclusively within the vascular lumen and show a high proliferation activity. The immunophenotype of the tumor cells is CD3 epsilon +, CD56+, TIA-1+, CD20−, CD79q− and CD45RO−. T cell receptor genes are in germline configuration and EBV is present.

It seems essential to treat the patients with intensive chemotherapy and stem cell transplantation. However, the outcome is dismal, with a 5 year overall survival of 20% [15].

3.8 KIR Receptors in Allotransplantation

Several hematological malignancies need bone marrow transplantation to cure the patient and to overcome relapse after chemotherapy alone.

In the beginning, the major problem with transplantation of donor bone marrow or donor peripheral blood stem cells was the high rate of rejection or graft versus host disease (GVHD). This is mainly due to the number of T cells of the host (after conditioning) and donor (in the graft). We now know that donor derived NK cells may improve the outcome of hematopoietic stem cell transplantation by mediating the antileukemic effects and by attacking residual host lymphohematopoietic progenitors, including T lymphocytes, which are responsible for graft rejection.

These donor derived NK cells do not attack the other tissues and thus do not favor the presence of T cell mediated GVHD. The NK cells function by the mean of a balance between incoming activating and inhibitory signals. The most studied NK cell receptors are the KIR receptors (inhibitory Killer Immunoglobulin-like receptor). HLA and KIR genes are inherited independently.

The hypothesis is that an alloreaction between donor and host by a mismatch between KIR ligand of donor and host, leads to less relapse, less GVHD, less rejection and a better disease free survival.

However, multiple studies have been done and actually there are rather conflicting data. The contradictory findings may be related to differences in transplant protocol (including T cell depletion), differences in myeloid and lymphoid malignancies, differences in genotype and phenotype of KIR repertoire assessment.

References

1. Thornthwaite JT, Leif RC (1974) J Immunol 113:1897
2. Rosenberg SA, Lotze MT, Muul LM et al (1987) N Engl J Med 316:889
3. Miller JS, Tessmer-Tuck J, Pierson BA et al (1997) Biol Blood Marrow Transplant 3:34
4. Ruggeri L, Capanni M, Urbani E et al (2002) Science 295:2097
5. Miller JS, Soignier Y, Panoskaltsis-Mortari A, McNearney SA et al (2005) Blood 105:3051
6. Baxevanis CN, Gritzapis AD, Papamichail M (2003) J Immunol 171:2953
7. Subleski JJ, Hall VL, Back TC, Ortaldo JR, Wiltrout RH (2006) Cancer Res 66:11005
8. Jaffe ES, Harris HL, Stein H, Vardiman JW (2001) Tumours of haematopoietic and lymphoid tissues, WHO classification of tumours. WHO, Geneva
9. Greer JP (2006) Educational Book American Society of Hemetology, ASH Education Program Book 331. ASH, Orlando, USA
10. Murdock J, Jaffe ES, Wilson WH et al (2004) Leukemia Lymphoma 45:1269
11. Sokol L, Loughran TP (2006) The Oncologist 11:263
12. Sandner A, Helmbold P, Winkler M (2008) Clin Exp Dermatol 33:615
13. Asada H (2007) J Derm Sci 45:153
14. Cho JH, Kim H-S, Ko YH, Park C-S (2006) J Infect 52:e173
15. Nakamichi N, Fukuhara S, Aozasa K, Morii E (2008) Eur J Haematol 81:1

The Role of KIR in Disease

Salim I Khakoo

Abstract Following the cloning of the killer cell immunoglobulin-like (KIR) genes in 1995 (Colonna M, Samaridis J. Science 268(5209):405–408, 1995) their population diversity has become increasingly apparent. This has spawned a plethora of disease association studies. As the KIR genes need to be considered in combination with their MHC class I ligands, this has added complexity to the analysis of these studies. KIR, and KIR:MHC class I gene combinations have been associated with viral infections, autoimmunity, transplantation and pregnancy-associated disorders. Simple rules, with which to interpret these datasets, are often difficult to find and, as our understanding of the interaction between KIR and MHC class I increases, the analysis of these datasets will become even more complex. This review attempts to summarize our current knowledge whilst indicating areas of potential further complexity.

1 Introduction

Since the original description of natural killer cells in 1975 by Kiessling [1], our understanding of their function and role in immunology has gradually expanded. This has been facilitated by the identification of key cell surface markers, which allowed their distinction from T cells, and the generation of antibodies to novel receptors which were able to stimulate NK cells. Coupled with this was the key functional observation that they were under a constitutive inhibitory control, as described in the missing-self hypothesis [2]. This has led to a model of NK cell function in which their effector functions, cytotoxicity and cytokine secretion, are determined by the integration of activating and inhibitory signals derived from cell

---

S.I. Khakoo
Department of Hepatology, Division of Medicine, Imperial College, 10th floor QEQM Building, St Mary's Hospital Campus, South Wharf Road, London W2 1PG, London, UK
e-mail: skhakoo@imperial.ac.uk

surface receptors. Complexity and diversity of NK cell function is generated at many levels ranging from variation in the genes of different individuals to the multiple activating and inhibitory receptor:ligand interactions and diversity of signaling molecules that transduce signals from these interactions. Ultimately, this array of factors determines whether an NK cell becomes activated or remains quiescent following an interaction with a target. These cells recognize targets as diverse as virally infected epithelial cells, cancerous hematological cells and normal dendritic cells. Whilst NK cells express a plethora of activating and inhibitory receptors, the diversity in NK cell genetics and function is exemplified *par excellence* by the killer cell immunoglobulin-like receptors (KIR). Owing to their own diversity and that of their MHC class I ligands, there has been much investigation into their relationship to an array of diseases including viral infections, tumors, autoimmunity and pregnancy-related disorders.

2 The KIR Genes

These genes form a multigene family as part of the leukocyte receptor complex (LRC) on the short arm of chromosome 19q13.4. They encode type 1 transmembrane glycoproteins belonging to the immunoglobulin superfamily. There are 14 different expressed KIR genes encoding both inhibitory and activating receptors and three nonexpressed genes. KIR are named according to their structure, i.e., the number of extracellular immunoglobulin-like domains, either two or three ("2D" or "3D"); and the lengths of their cytoplasmic tails, short or long ("S" or "L"). In general, KIR with a long intracytoplasmic tail transduce inhibitory signals and those with a short intracytoplasmic tail transduce activating signals. Therefore, receptor nomenclature implies that a KIR3DL has three extracellular immunoglobulin domains (3D), with a long (L) intracytoplasmic tail and an inhibitory function.

The chromosomal region, encoding the KIR, also contains a number of other related genes including the leukocyte immunoglobulin-like receptor (LILR) family of genes and is located close to the natural cytotoxicity receptor gene NKp46 [3]. It is thought that the KIR gene family originated from KIRX, a single gene located within in the LILR locus, following duplication. This gene has subsequently undergone multiple further rounds of duplication to produce the extant KIR locus [4]. This locus consists of a series of tandemly arrayed KIR genes which share between 85 and 99% sequence similarity. Critically, this arrangement is conducive to evolution by nonhomologous recombination and so the locus has expanded and contracted over time. This has resulted in a remarkable heterogeneity in the gene content of individual KIR haplotypes, in which individuals can have from seven to fourteen expressed KIR genes.

The KIR haplotypes can be divided into two groupings dependent on their gene content. The A group of haplotypes is the most common and consists of five inhibitory KIR: *KIR2DL1, KIR2DL3, KIR3DL1, KIR3DL2* and *KIR3DL3* [5] and two activating KIR *KIR2L4* and *KIR2DS4*. Individuals with two group

A haplotypes are designated AA, and the frequency of AA haplotypes varies substantially across populations. For instance, it is represented at a frequency of 56% in the Japanese, 30% in Caucasians and 1.5% in the Australian Aborigines [6–9]. Although the gene content remains similar between the haplotypes, allelic variation generates genetic diversity within this grouping [10]. There may also be functional diversity within this group as in some haplotypes the *KIR2DS4* gene encodes for a nonfunctional gene. This is due to a 21 base pair deletion in the transmembrane domain [11, 12], and this null allele is found in approximately 64% of Caucasians [13]. The alternative haplotypes are designated the B grouping. These are much more heterogeneous in their gene content, especially in terms of activating receptors. Individuals can have from two to seven activating KIR (including KIR2DL4), whilst retaining similar numbers of inhibitory receptors as the A group of haplotypes. Thus, the overall pattern of the KIR locus is one of retention of the inhibitory receptors, whilst there is diversity of the activating receptors. This diversity is remarkably well tolerated and although KIR are implicated in many diseases, loss of specific KIR does not result in an immunodeficiency state. Indeed, recent analysis of the CEPH panel of families has demonstrated a KIR locus with only three genes KIR3DL2, KIR3DL3 and KIR2DS1 [14].

The ligands for KIR are the classical MHC class I molecules HLA-A, -B and -C. These genes are on chromosome 6 and hence segregate independently from the KIR genes. The specificities of the inhibitory receptor:ligand pairings have been most readily defined. KIR2DL1 binds HLA-C allotypes with a lysine at position 80 of the MHC class I heavy chain (Group 2 HLA-C allotypes), and KIR2DL2 and KIR2DL3, which are alleles, bind Group 1 HLA-C allotypes (asparagine at position 80). KIR3DL1 binds HLA-B allotypes with the Bw4 serological motif. This represents about 40% of all HLA-B allotypes, with the remainder having the alternate Bw6 motif [15–18]. KIR3DL2 binds HLA-A3 and HLA-A11, but the functionality of this molecule has been harder to determine, and its contribution to the overall function of NK cells is less clear.

Although the specificity for inhibitory KIR was originally defined on the basis of simple structural motifs in the MHC class I heavy chain, there is further subtlety in the system. For instance, the inhibitory KIR have all been shown to have a degree of peptide selectivity such that peptides that stabilize their cognate HLA ligands, but do not productively engage KIR, can be defined [19, 20]. Furthermore, a peptide that stabilizes HLA-Cw7 (a group 1 HLA-C allotype) has been shown to confer binding to KIR2DL1, which recognizes predominantly group 2 HLA-C allotypes [21]. In depth analysis of the HLA-C specificity has demonstrated further promiscuity in KIR binding.

Whilst the specificities of the inhibitory receptors have been defined, those of the activating receptors have been more elusive. In general, it appears that these receptors bind HLA class I with much lower affinities and, as NK cells appear to be under the control of predominantly inhibitory KIR, their function has also been more difficult to define. KIR2DS1 shares >95% amino acid identity in its extracellular domain with KIR2DL1 and, consistent with this, it has been demonstrated to bind to group 2 HLA-C allotypes. Similarly, on the basis of its sequence homology to KIR2DL2/3,

KIR2DS2 would be predicted to interact with group 1 HLA-C allotypes. However, binding to these alleles is at best weak. Additionally, no binding to HLA-B allotypes has been demonstrated for KIR3DS1 which shares 98% amino acid identity with KIR3DL1 in its extracellular domains. KIR2DS4 may interact with MHC class I and non-MHC class I ligands [22, 23], but the other activating receptors, KIR2DS3 and KIR2DS5, and also the inhibitory KIR2DL5 have no putative ligands.

The allelic diversity of KIR can impact on its binding to MHC. For instance, although KIR2DL2 and KIR2DL3 are alleles at a single locus, KIR2DL2 binds HLA-C more avidly than KIR2DL3. This is thought to be related to a pair of residues at positions 16 and 148 which influence the hinge angle between the two extracellular Ig domains of these KIR [24]. In addition to changes in binding, allelism at the KIR locus can also affect the level of gene expression. This has been best demonstrated for KIR3DL1 which can be divided into high, low and null expressing allotypes [25]. In addition, the MHC ligands for these genes may differ in their ability to bind KIR, again demonstrated most elegantly for KIR3DL1 and its HLA-B^{Bw4}-positive ligands [26]. Changes in affinity of KIR for its ligand may therefore directly affect the function of NK cells. Similarly, the strength of the KIR:HLA interaction may also influence the education of NK cells. Recent work has demonstrated that MHC class I is important for the development of functionally mature NK cells, which has been controversially termed "licensing" [27]. Thus, NK cells which express a cognate receptor for MHC class I are reactive to class I negative target cell lines. However, there may also be differences in the levels of reactivity depending on the class I allele of the individual. For instance, the NK cells expressing KIR2DL3 from individuals that have HLA-Cw*07 make more IFNγ in response to the class I negative target cell line K562 than KIR2DL3-positive NK cells from individuals with other group 1 HLA-C ligands [28]. The extent to which this extends to other KIR:HLA receptor:ligand pairings is not clear; however, it is clear that this system has multiple subtleties which we are just beginning to unearth.

The allelic diversity and its functional consequences challenges the geneticist in the interpretation of KIR:HLA disease association studies. Thus, in order to make sense of current datasets a reductionist approach is necessary. This requires the KIR:HLA system to be reduced to a simple motif model. Further insights can then be gained by following the leads from genetic studies with carefully controlled functional experiments. Thus, there should be caution against overinterpretation of these simplified datasets.

3 KIR and Disease

The KIR have been studied in a wide variety of different diseases (Fig. 1). The key features of this gene family which lend them to these types of study are their diversity and their expression on NK cells. Thus, they have a potential role in diseases in which NK cells have been implicated in human and murine models, and also those diseases in which there is a combination of diverse outcomes and/or

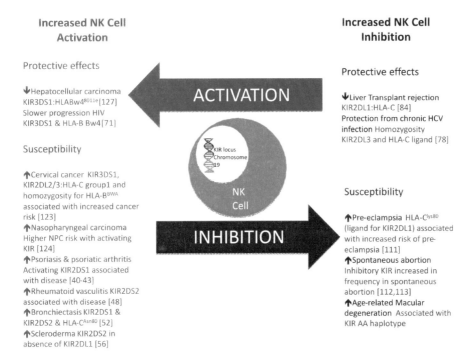

Fig. 1 Schematic diagram summarizing the diverse role of KIR in different disease states. The diseases are categorized as to their association with the KIR genes based on the presumed activating or inhibitory potential of that KIR or KIR:MHC genotype

heritability. Therefore, KIR diversity is predicted to be particularly pertinent for the outcome of viral infections and cancer. Their rapid evolution, in a manner suggestive of pathogen mediated selection, also hints that specific KIR:HLA combinations could be important for the resolution of infections. Furthermore, NK cells secrete Th1 type cytokines and can cross-talk with dendritic cells, implying that they may be involved in autoimmune and inflammatory disorders [29–35]. As KIR engage MHC class I molecules, it can be postulated that diseases which have an MHC association may also have a KIR association. Additionally, murine studies have demonstrated a critical role for MHC class I receptors in bone marrow transplantation in a "missing-self" model, inferring that KIR may have a role in human transplantation. Finally, NK cells are abundant in the pregnant uterus and so KIR may be involved in pregnancy-associated disorders.

3.1 KIR in Autoimmune and Inflammatory Disorders

Autoimmune disorders have long been associated with specific HLA class I and class II alleles. Thus, there is a possibility that KIR may confer additional disease susceptibility in combination with specific HLA class I ligands. In addition, the ability of NK cells to instruct downstream adaptive immune response implies that if NK cells are genetically predisposed towards activation then they may be a factor in breaking self-tolerance. Previous work has also shown that NK cell numbers and activity are reduced in a number of autoimmune diseases [36]. Furthermore, KIR may be expressed on the effector memory subset of T cells [37, 38] and so may have a direct effect on T cell-driven autoimmune responses. The overall theme of KIR genetics in inflammatory diseases is that there is a positive association with KIR haplotypes that have greater numbers of activating receptors. This is more relevant for a number of chronic inflammatory disorders rather than for the classical MHC class II-associated organ-specific autoimmune diseases.

Psoriasis is an inflammatory disorder with a number of systemic manifestations that has a strong association with the group 2 HLA-C allotypes HLA-Cw6 [39]. This HLA class I allele is a ligand for the inhibitory receptor KIR2DL1 and a putative ligand for the activating receptor KIR2DS1. Indeed, the KIR2DS1 gene has been associated with a predisposition to psoriasis in a Japanese study of 96 individuals [40]. KIR2DS1 is a component of the B group of haplotypes, which are relatively unusual in the Japanese who form a comparative genetically homogenous population that has an unusually high frequency of A haplotypes [8]. In the same study, there was a lower, but nonsignificant, frequency of the inhibitory receptor KIR2DL1, implying that in this disease the genetic predisposition is towards greater NK cell activation. Consistent with this observation, KIR2DS1 has been associated with psoriasis in studies from Sweden and Poland [41]. This gene was studied in association with its putative HLA–Cw6 ligand by Luszczek et al., but in isolation in that of Holm [42]. Further analysis of this data suggests that other KIR may also influence predisposition to this disease especially the activating KIR, KIR2DS3 and KIR2DS5.

One of the systemic manifestations of psoriasis is an inflammatory arthropathy. This shares the association with HLA–Cw6 and there may be additional genetic factors that can modulate the predisposition to this. In a cohort of individuals with psoriasis, in which 34.1% had the arthropathy, there was also a positive association with KIR2DS1 [43]. This susceptibility model has been expanded in a large study of individuals with psoriatic arthritis which demonstrated an association of KIR in combination with their HLA class I ligands with this disease phenotype [44]. Individuals that have either KIR2DS1 or KIR2DS2 and are homozygous for either group 1 or group 2 HLA-C allotypes are the most susceptible to psoriatic arthropathy. Conversely, individuals without activating KIR are relatively protected form this disease. A synthesis of this work is that it shows that individuals with more activating receptor:ligand interactions are most susceptible to psoriatic arthropathy, those with the most inhibitory receptor:ligand interactions are relatively protected, and those with an intermediate number of receptor:ligand interactions have a neutral risk. This study therefore establishes a novel genetic model that requires testing in other disease states.

The association of KIR and HLA with an inflammatory disorder was originally described in rheumatoid arthritis. Individuals with rheumatoid arthritis have an expansion of an unusual subpopulation of CD4+, CD28− T cells in the blood [45].

These T cells express the activating receptor KIR2DS2 in the absence of the corresponding inhibitory KIR, and a subset of these express the adaptor molecule KARAP/DAP-12. There is oligoclonal expansion of a subset of effector memory T cells that produce predominantly Th1-type cytokines [37]. Antibody ligation of KIR2DS2 on these cells results in both proliferation of, and cytokine release by, these cells [46, 47]. Thus, these T cells can be stimulated directly through KIR, thus bypassing signaling via the T cell receptor in an antigen independent fashion.

Although these T cells are found in individuals with rheumatoid arthritis, immunogenetic analysis of KIR:HLA in this condition did not demonstrate an association of KIR2DS2 with this inflammatory arthropathy [48]. Rather, it was associated with a systemic complication of this disorder, rheumatoid vasculitis, which affects a subgroup of these individuals. Indeed, there was a significant association with the vasculitis when compared to both individuals with rheumatoid arthritis but *without* vasculitis, and also to healthy controls. There was, however, no significant augmentation of the disease risk when KIR2DS2 was considered in combination with its putative group 1 HLA-C ligands. The study was however possibly underpowered to detect this association.

The association of KIR2DS2 has been described in acute coronary artery syndromes and similar to the findings in rheumatoid arthritis there is an expansion of CD4+, CD28−, KIR2DS2+ T cells in the peripheral blood of these individuals as compared to age-matched controls [49]. Additionally, KIR2DS2 expression has been found in the coronary arteries of individuals with coronary artery disease at autopsy. Thus there is some consistency in the association of this subpopulation with two different types of vascular disease.

A number of other inflammatory and autoimmune disorders have been associated with KIR:HLA interactions. In the inflammatory biliary disorder primary sclerosing cholangitis (PSC), the Bw4-positive HLA-B alleles and group 2 HLA-C allotypes are associated with protection from the disease. In this study, PSC was associated with the extended haplotype DRB1*0301 and DRB1*1501 in combination with the MIC5.1 allele and the absence of Bw4-positive HLA-B and group 2 HLA-C alleles [50]. The suggestion is that these alleles are protective in the context of inhibitory KIR, but no association was found with KIR genes. This may be related to the relatively high >95% frequencies of both KIR2DL1 and KIR3DS1, which reduces the study power. Conversely, age-related macular degeneration, which is the commonest cause for blindness in the Western World, is associated with the KIR AA haplotype which carries only one activating gene in combination with HLA-Cw*07 [51]. A much more rare eye disorder is that of birdshot chorioretinopathy. This unusual disease is associated with HLA-A29, and more recently has been shown to be associated with a tendency towards an activating KIR haplotype [52].

Thus, there is an overall trend towards activating KIR, especially *KIR2DS2* being susceptibility factors in chronic inflammatory conditions. A correlation with an "activating" KIR genotype was found in the chronic inflammatory condition idiopathic bronchiectasis [53]. Affected individuals in this study were more likely to have *KIR2DS1* or *KIR2DS2* in combination with homozygosity for

HLA-C^{Asn80} alleles, which is similar to the model that Martin et al. originally proposed for psoriatic arthropathy [54]. Furthermore, weak associations have been detected between diabetes mellitus and KIR2DS2:HLA-C^{Asn80} and also *KIR2DS2* and *KIR2DL2* in a Latvian sample [55, 56]. *KIR2DS2*, in the absence of its inhibitory counterpart *KIR2DL2*, has been found at high frequency in patients with scleroderma (12%) as compared to controls (2%) [57]. This is an unusual genotype as these KIR are in strong linkage disequilibrium. These data have not yet been correlated with their HLA-C ligands. Overall, autoimmune and inflammatory conditions appear associated with a surplus of activating KIR genotypes, and where tested these activating receptors appear to be involved in costimulation or activation of T cells rather than NK cells.

3.2 KIR in the Response Against Pathogens

NK cells are important for a successful immune response to viruses, bacteria and protozoa and thus there is potential for KIR to have a role in the immune response to a number of different infections. Rapid evolution of the KIR gene locus, which implies a pathogen-mediated selection process provides further evidence for this role [58–62]. Similar to the situation in inflammatory disease, this may be related to KIR expression on both NK cells and T cells. In understanding the association of these receptors with disease, it is important to bring together both genetic and functional data. Combined use of these two strategies can circumvent the main limitations associated with each of these two approaches. Genetic studies are restricted to being correlative in nature and functional studies are difficult to perform in the large numbers of samples necessary to cope with extensive KIR and MHC allelic diversity. In addition, reagents are often cross-reactive, for instance, monoclonal antibodies that recognize inhibitory receptors may also bind the activating forms with a similar specificity. To date, most insight into the role of KIR and HLA has been derived from a combined genetic and functional approach in HIV infection.

HIV affects more than 40 million people worldwide and is a significant problem in both the developing and developed world. This has permitted comparison of data sets from distinct populations, and given unique insights into the role of KIR in viral infections. Following exposure to HIV, the majority of individuals become chronically infected, but the clinical course may be variable. In particular, the rate at which the CD4 T cell count of an individual declines can vary amongst infected people, and thus the time at which susceptibility to opportunistic pathogens and AIDS supervenes can differ dramatically between individuals. Endpoints in HIV studies can therefore vary and it is important to take this into consideration when analyzing different studies. Furthermore, KIR protection might operate at the level of an AIDS defining opportunistic infection rather than at the level of the virus itself. With the successful treatment of HIV with highly active antiretroviral chemotherapy, longitudinal studies in the developed countries will rely more on stored samples and historical data, and whilst contemporaneous studies may be more readily performed in the developing world it is an important goal of global medicine to urgently bring effective treatments to these nations. Nevertheless, cross-sectional studies have and will continue to be extremely valuable in understanding the relationship between KIR and HIV infection. These have shown that there are substantial alterations in NK cell subsets, and their functions in chronic HIV infection [63, 64]. In essence, NK cells are hypofunctional in chronic HIV infection. This is, in part, related to an unusual subset of NK cells that express CD16, but not CD56 [65, 66]. These cells express inhibitory KIR and have lower than normal levels of natural cytotoxicity receptors. Thus, they do not lyse autologous targets. Interestingly, a similar subpopulation of NK cells have also been described in human umbilical cord blood and can differentiate to CD56+CD16+ NK cells raising the possibility that they are immature NK cells [67, 68]. It is likely that the presence of these cells is a consequence rather than a cause of chronic HIV infection and in acute HIV infection there is an expansion of $CD56^{dim}$, $CD16^+$ NK cells [66]. These express KIR and are cytolytic and hence it is in the acute phase of disease that KIR genetics is most likely to be important.

Insights into the potential for an influence of KIR on the outcome of HIV infection have also come from genetic studies. In particular, association of progression of HIV has been linked to specific HLA-B allotypes that express the Bw4 serological motif [69]. Furthermore, homozygosity for HLA-B allotypes with the Bw4 serological motif ($HLA-B^{Bw4}$) is associated with a slower decline in CD4 T lymphocyte counts in HIV-infected individuals [70]. These HLA-B allotypes interact with KIR3DL1, and, on the basis of sequence homology are also likely to bind KIR3DS1, although this interaction has yet to be formally shown.

The first description of a protective KIR:HLA interaction in a viral infection came from a study of over 1,000 HIV-infected individuals. This showed that progression to AIDS was slower in individuals that had KIR3DS1 in combination with its putative $HLA-B^{Bw4}$ ligands. However, only a subgroup of these allotypes, those with an isoleucine at position 80 of the MHC class I heavy chain, were protective [71]. This protective effect required the presence of both KIR and HLA, as one in the absence of the other was not protective. KIR3DS1 homozygosity is present at a significantly higher frequency in individuals exposed to, but not infected with HIV, as compared with their HIV-seropositive partners [72]. The observed genetic associations of KIR3DS1 have some functional correlates. Suppression of HIV infection in vitro has been shown to be associated with KIR3DS1-positive NK cells and Bw4-seropositive targets [73]. Furthermore NK cells from KIR3DS1-positive individuals tend to secrete more IFNγ than those from individuals who lack this receptor [74]. This activity appears to be further augmented in subjects that express either HLA-B57 or HLA-B58.

Interestingly, protection by $HLA-B^{Bw4}$ may not be conferred solely by KIR3DS1. KIR3DS1 segregates as an allele of the inhibitory receptor KIR3DL1 and there are currently 18 inhibitory and five activating alleles described. Within the inhibitory alleles, a hierarchy of expression (high, low and null) has been determined by flow cytometry [25, 75]. This may either be due to promoter polymorphisms or differences in trafficking amongst the alleles [76, 77]. The high-expressing alleles show a higher affinity for HLA-B^{Bw4} allotypes [26]. Interestingly, high expressing alleles of an inhibitory type are protective against AIDS progression in combination with the HLA-B^{Bw4} allotypes [78]. This seeming paradox may to some extent be explained by a model for NK cell education in which higher-affinity MHC class I interactions lead to greater activation of NK cells in the context of targets which do not express the cognate ligand for that inhibitory receptor. Thus, if HLA-B is downregulated these NK cells will liberate more IFNγ than NK cells from an individual with a low-expressing allele. However, this model is challenged by the observation that the null allele was most protective in combination with its ligand in this study. Further work is required in order to determine if there is a role for this allele in the function of NK or T cells.

Hepatitis C virus (HCV) is a common chronic viral infection that leads to cirrhosis of the liver and hepatocellular carcinoma. HCV was the first viral infection in which the pairing of an inhibitory KIR with its MHC ligand was found to be protective [79]. In this study, homozygosity for KIR2DL3 and its group 1 HLA-C ligand was associated with spontaneous resolution of infection in individuals that acquired HCV through mechanisms other than transfusion of blood products (mainly intravenous drug usage). KIR2DL2, which is an allele of KIR2DL3 and has the same MHC class I ligands, was not protective. These data suggest a quantitative model for NK cells in HCV. In this model, NK cells which express the KIR2DL3 receptor are protective, but this protective influence can be negated in the presence of either a KIR2DL2:HLA-C interaction or a KIR2DL1:HLA-C interaction. As KIR expression is "hard-wired" and coexpression of KIR stochastic, individuals that are homozygous for KIR2DL3 and also for group 1 HLA-C will have more NK cells expressing the protective KIR2DL3 gene in isolation than individuals who are heterozygous for KIR2DL3/KIR2DL2. Furthermore, KIR2DL1 is present in >95% of the population which means that nearly all individuals who have a group 2 HLA-C allotypes will have NK cells inhibited by KIR2DL1, and expression of this receptor will subtract from the pool of "protective" NK cells that are inhibited by KIR2DL3:HLA-C1. Finally, this dataset suggests that large inocula of HCV may overcome this protective effect. In murine CMV infection, NK cells are protective. This protection has been defined as relating to the interaction of Ly49H with the viral protein m152. In this infection, this protection can be overcome by a large infecting inoculum. Thus, this is a tenable model for KIR2DL3:HLA-C1 in HCV infection. [80–82]. The protective effect of this combination has now been confirmed in a second independent study of resolving HCV infection [83]. The data beg the question as to why KIR2DL3 is protective, but KIR2DL2 is not. One hypothesis is that KIR2DL2 binds to group 1 HLA-C allotypes more avidly than KIR2DL3. This appears to be the case in studies using KIR-Fc proteins to stain MHC class I transfectants [84]. More recent work has suggested that specific amino acid residues that affect the hinge angle between the D1 and D2 domains of KIR2DL2 and KIR2DL3 play a critical role in modulating this avidity [24]. Furthermore, this work demonstrates that KIR2DL2 may also have a greater avidity for some group 2 HLA-C allotypes.

The liver is a unique immunological environment with a strong bias towards the innate immune system. Liver transplantation is performed without MHC class I matching and therefore there is a strong potential to generate alloreactive NK cell clones. In 416 liver transplants performed for a variety of different disorders, the presence of group 2 HLA-C allotypes in the allograft was associated with less chronic rejection. This suggests that the KIR2DL1:HLA-C inhibitory interaction which is considered a strong inhibitory interaction is associated with less NK cell alloreactivity, and hence less rejection [85]. A similar, although weaker, association has also been found in renal transplantation [86].

Another pathogen that has long been associated with natural killer cells is cytomegalovirus. This herpes virus infects the majority of individuals and in most cases remains latent unless the individual becomes immunocompromised, in which case it can cause a severe systemic disease. NK cell deficient individuals are susceptible to herpes virus infection [87], and CMV has evolved multiple mechanisms for interfering with MHC class I and MHC class I-like molecules. These include MHC class I downregulation, upregulation of ligands for inhibitory receptors and blockade of an activating receptor [88–91]. Although the MHC class I receptor Ly49H has a dominant role in the clearance of CMV infection, the role of KIR in human CMV infection is not well established. Activating KIR may be protective against CMV following bone marrow transplantation, and there is a case report of an individual with an NK repertoire that was dominated by KIR2DL1, who suffered severe CMV-related disease [92–94]. Unfortunately, functional experiments have been unable to confirm or refute the role of KIR in the lysis of autologous infected fibroblasts [95].

Natural killer cells may also be important for a successful response to protozoan infections. [96]. In particular, NK cells may liberate the proinflammatory cytokine interferon-γ in response to stimulation by malaria infected red blood cells. This process is indirect and requires the presence of macrophages in a so-called "ménage a trois" [97, 98]. The level of this in vitro effect appears to be associated with specific KIR3DL2 alleles. In particular, those with KIR3DL2*002 secreted higher levels of this cytokine in response to Plasmodium falciparum-infected red blood cells [98]. Thus, in this situation KIR appear not to be responsible directly for pathogen recognition, but for the indirect amplification of the immune response via NK:macrophage cross-talk.

3.3 KIR in Pregnancy

Whilst natural killer cells are critical for immune responses to pathogens, a twist in the evolutionary tail comes from their involvement in pregnancy. NK cells are plentiful in the pregnant uterus, being most marked in the decidua early in gestation at the time of implantation of the fetal trophoblast [99]. These uterine natural killer cells (uNK) have a phenotype that distinguished them for those of peripheral blood and may represent a specialized population or even a distinct lineage in that they are

CD56^{bright} and have a distinct receptor profile including that of KIR [100–103]. These changes may relate to the unusual expression of MHC class I alleles in the extravillous trophoblast (EVT). The EVT cells express the HLA class I molecules HLA-C, -E and –G, each of which serve as ligands for distinct NK cell receptors, but they do not express HLA-A and –B [104, 105], and HLA-C has been shown to be expressed on the trophoblast in a stable β2-microglobulin-associated form [106]. Therefore, interactions between EVT and uNK cells are controlled by the HLA-C-specific KIR (KIR2DL2/3 and KIR2DL1), the HLA-E-specific receptors NKG2A, NKG2C and NKG2E, and the HLA-G-specific receptor KIR2DL4 [107–109]. Additionally, NK cells from the pregnant uterus are more likely to express KIR specific for HLA-C, but not HLA-B [110] and this expression seems to decline during the first trimester of pregnancy.

The role of NK cells in pregnancy is related to the invasion of the uterine spiral arteries and decidua by the trophoblast. Formation of an adequate blood supply to the placenta is a complex process involving the EVT cells of the fetus and the uNK of the mother. If this process does not occur effectively, then the blood supply to the placenta develops inadequately and this leads to preeclampsia, a condition that can progress to eclampsia which is associated with both fetal and maternal mortality [111]. Thus, pregnancy can exert a strong selective effect on NK cells and particularly on HLA-C and HLA-C specific KIR.

Seminal work by Hiby et al. in which both maternal *KIR* and fetal *HLA* types were analyzed, suggests that these genes may be risk factors for the development of preeclampsia [112]. KIR haplotype A contains relatively few genes, and is the most common haplotype worldwide. It includes the expressed *KIR* genes *KIR2DL1*, *KIR2DL3*, *KIR3DL1*, *KIR3DL2* and the two activating receptor genes *KIR2DL4* and *KIR2DS4*, each of which have nonfunctional alleles. KIR B haplotypes essentially consist of all other *KIR* combinations, and nearly always contains more activating receptors than the A haplotypes. Hiby et al. studied individuals with preeclampsia and showed that the maternal A haplotype was weakly associated with development of their disease. However, this association was strengthened if the fetus had two HLA-C^{Lys80} allotypes, the ligands for KIR2DL1. Furthermore, there was an inverse correlation between the number of activating *KIR* genes present in the mother and the prevalence of preeclampsia. As the KIR2DL1: HLA-C^{Lys80} interaction is considered to be one of stronger inhibition, it appears that a more activating KIR genotype (more activating receptors and weaker inhibitory interactions) was associated with protection from preeclampsia. This implies a model in which excessive inhibition of the maternal uNK cells by the fetal EVT prevents adequate invasion of fetal cells into the spiral arteries. This results in a limitation of the blood flow to the fetus, inadequate formation of the placenta and thence high risk for preeclampsia. Additionally, previous reports have suggested that an excess of inhibitory *KIR* may also lead to an increased frequency of spontaneous abortion [113, 114].

KIR2DL4 is the most evolutionarily conserved KIR, having an orthologous representation in apes and Old World monkeys [3]. An attractive hypothesis is that this conservation may be related to a role in for this receptor in pregnancy.

KIR2DL4 is expressed on all KIR haplotypes and is ubiquitously expressed at the RNA level by all NK cells. It may have a differential expression in uNKs, as it appears to be on the cell surface in uNKs but not those in peripheral blood [102]. In binding assays, KIR2DL4 can bind cell surface HLA-G [107, 108]. However, recent work has shown that the KIR2DL4:HLA-G interaction is most relevant for soluble than for surface-expressed HLA-G as KIR2DL4 resides in the Rab5-positive endocytic compartment [109]. NK cells can thus be activated by both cell-surface and soluble HLA-G. This leads to upregulation of a number of cytokines including TNF-α, IL-1β and IFN-γ, which may be important for the development of the uterine spiral arteries via the secretion of vascular-derived endothelial growth factor [115]. Interestingly, KIR2DL4 and HLA-G do not appear to be critical for a successful pregnancy as there are well-documented cases of individuals who are homozygous for null alleles of HLA-G alleles and a case report of a woman, without a KIR2DL4 gene, that have successfully given birth to several children [116–119]. Furthermore, there is one allele of KIR2DL4 that has a single nucleotide deletion, which results in a splicing change that causes loss of the cytoplasmic signaling tail. This allele was not, however, overrepresented in 45 preeclamptic women as compared to 48 normotensive pregnant controls [120]. However, it is in linkage disequilibrium with the A *KIR* haplotype. Therefore, although KIR2DL4:HLA-G is not critical to a successful pregnancy, it may contribute to a successful pregnancy by augmenting NK cell activation, and this may be alone or in combination with the favorable fetal HLA-C allotypes described by Hiby et al. [112].

3.4 KIR and Cancer

In models of tumorigenesis, NK cells have been shown to both increase survival and reduce tumor burden [121]. However, work in humans has been less clear. The rationale for their involvement in human tumor control includes MHC class I downregulation, such as the generation of alloreactive NK cells following haematopoietic stem cell transplantation (HSCT), a viral etiology to the tumor, the expression of ligands for activating receptors, and the upregulation of NKG2D ligands during cell cycling [122]. KIR have therefore been associated with both solid and hematological malignancies.

In terms of viral-driven tumors, carcinoma of the cervix is strongly associated with human papilloma virus. Cervical intraepithelial neoplasia progresses over the years to carcinoma [123]. There is also chronic inflammatory response to the tumor. In a large study involving three independent cohorts of individuals with cervical neoplasia, those with cervical neoplasia (CIN3 or squamous cell carcinoma) were found to be more likely to have *KIR* and *HLA* genes associated with a predisposition towards activation [124]. Thus, the strong inhibitory genotypes KIR2DL1:HLA-C group 2 and KIR3DL1:HLA-B^{Bw4} were relatively associated with protection from disease, whilst weaker inhibitory genotypes such as KIR2DL2/3:HLA-C group 1 and homozygosity for HLA-B^{Bw6} were associated with susceptibility to neoplasia.

Critically, this model held true for all three populations studied. Additionally, the activating gene KIR3DS1 was associated with susceptibility. This may be directly attributable to this gene or act as a marker of the group B KIR haplotypes which are associated with more activating KIR. The original model to explain this effect suggests that it is the chronic inflammatory response to the virus that induces ongoing cell turnover and hence the predisposition to cancer, rather than a direct failure of the antineoplastic immune response. However, in the light of recent work on the education of NK cells [28] it may be that NK cells educated on a strong inhibitory MHC class I ligand may generate a stronger IFNγ response to a class I-negative tumor, and hence group 2 allotypes HLA-C are protective rather than group 1 HLA-C allotypes for which the most frequent inhibitory receptors are KIR2DL3, rather than KIR2DL2.

The model of susceptibility caused by a more activating genotype is supported by data from nasopharyngeal carcinoma (NPC) [125]. This disease is associated with Epstein–Barr Virus (EBV) infection. In this study of 295 individuals with this disease and 252 controls matched for age, sex and geographic residence, the number of activating *KIR* was positively associated with having NPC. The strongest association was in individuals seropositive for EBV than for the whole study population, implying an effect of *KIR* on the anti-EBV response. Although there were no significant effects of individual *KIR* in combination with their cognate HLA-C ligands, the unusual *HLA-B* allele, *HLA-B*46*01*, was weakly associated with NPC in combination with *KIR2DS2* amongst the EBV-seropositive individuals. This allele has been previously associated with NPC [126], and is an interlocus recombinant between HLA-C and HLA-B in which HLA-C is thought to have donated the KIR-binding motif from HLA-C^{Asn80} allotypes. The presence of this motif in the HLA-B allotypes allows it to act as a ligand for KIR2DL2/3 and KIR2DS2 [127].

Hepatocellular carcinoma (HCC) is a complication of chronic liver disease including viral hepatitis due to hepatitis B and hepatitis C (HCV). In chronic HCV infection, the activating receptor *KIR3DS1* in combination with $Bw4^{80Ile}$ is protective against the development of HCC. KIR3DS1:HLA $Bw4^{80Ile}$ was found more commonly in asymptomatic HCV carriers compared with those who develop the complication of HCC with an odds ratio of 24.2 [128]. Interestingly, *KIR3DS1* and *HLA-B^{Bw4}* are also weakly protective against chronic HCV infection [79].

Tumor-infiltrating NK cells in HCC express lower levels of both KIR and NKG2A than the NK cells of healthy livers [129], suggesting a selective recruitment against those with inhibitory receptor expression. Colorectal tumors expressing low levels of HLA class I have an infiltrate with a bias towards T cells rather than NK cells [130]. In this context, NK cells can be inhibited by the tumor marker, carcinoembryonic antigen, through its interaction with CEA-related cell adhesion molecule 1 [131], and this inhibitory signal may be related to the absence of NK cells in the tumor infiltrate. T cells also express inhibitory KIR on tumor-infiltrating T lymphocytes which may modulate T cell function, preventing them from killing cancerous cells [132, 133]. T cell function may also be modulated by the presence of soluble MIC-A and MIC-B. These are ligands for the activating receptor NKG2D and may block their interaction with cancerous cells. Thus, modulation of the balance between activation and inhibition of both T cells and NK cells towards inhibition may impair the antitumor immune response [134, 135].

Tumor cells may also modulate their susceptibility to NK cell-mediated cytotoxicity by expressing factors that prevent apoptosis. The antiapoptotic factor livin-β, expressed by the Mel-B1 cell line may overcome the NK cell cytotoxicity that would be anticipated because of the MHC class I downregulation observed in this line [136]. In malignant melanoma, some KIR associations have been observed but these are weak and do not support a role for the activating receptor KIR2DS4 which has been reported to recognize a non-MHC class I ligand expressed on melanoma cells [22, 137].

KIR may also be important for the control and outcome of hematological malignancy. There is a wealth of data on survival following haematopoietic stem cell transplantation (HSCT) and KIR. This has been driven in part by the observations, from the Perugia group, of alloreactive NK cells in the unusual setting of haploidentical HSCT [138]. Subsequent work has demonstrated that this protection is not as straightforward as originally thought, and the conditioning regimen used may be important in determining the relative benefit of such NK cells. In particular, the graft vs. leukemia effect may also be determined by the underlying disease that necessitated the transplant. This is a complex area which is out of the scope of this review and hence the reader is directed towards reviews which examine this in depth [139, 140]. Whilst KIR may not be relevant in the pathogenesis of the more common leukemias, proliferations of large granular lymphocytes (LGL) may be clonal and consist of either cytotoxic T cells (T-LGL) or NK cells (NK-LGL) [141]. In T cell LGL, the clone has a terminally differentiated effector CD45RA+ CD27− CD28− CCR7-phenotype [142, 143]. The individuals may be asymptomatic of suffering from cytopenias, neuropathies, recurrent infections, splenomegaly or rheumatoid arthritis [144]. In this context, KIR may act as a marker of clonality in LGL or be involved in disease pathogenesis. They have been shown to be clonally expressed on up to 48% of T-LGL [145], and expression of a KIR without a cognate MHC class I ligand (five out of seven cases) may be associated with a more severe disease phenotype [146]. Consistent with an autoreactive model, expression of LILRB1 (ILT2/LIR-1), an inhibitory receptor with a broad MHC class I specificity, is associated with a lack of symptoms, and so may be preventing autoreactivity [147]. KIR is also expressed in approximately half of the NK-LGL cases [145, 148–150]. However, this may not be clonal [151] and these populations are associated with predominantly activating KIR and a group B KIR haplotype [149]. The activating nature of these KIR has been confirmed in functional experiments [150].

4 Conclusion

KIR have a multifunctional role in disease and disease pathogenesis. A number of genetic studies have suggested associations of KIR and their ligands with human diseases. In broad terms, the KIR genes present in an individual determine a specific functional consequence by affecting the degree of NK cell activation or inhibition. Some KIR haplotypes have a tendency towards more activation, or lower levels of inhibition, and this in combination with the appropriate ligands is associated with increased risk of inflammatory diseases, but a stronger antiviral response. Stronger inhibitory haplotypes may lead to disorders in pregnancy. Thus, the KIR locus appears to be subject to balancing selection pressures that maintain these diverse haplotypes [152]. Adequately powered population studies have the potential to reveal associations that confer additional benefit or susceptibility to diseases with multifactorial aetiologies. However, the observed correlations of KIR in disease are diluted by the stochastic expression of the KIR genes on NK cells and T cells, leading to a reduction in statistical power. At present, we have only a limited understanding of the control of these genes, which further constrains our ability to interpret the data. Receptor:ligand associations, where known, are predicted on the basis of simple structural motifs which likely underestimate the complexity of the KIR:HLA interactions, and by inference modulate our ability to interpret the functional role of a given genetic association. Also, a number of KIR receptors do not have well-defined ligands, which could significantly affect the data analysis if they were found to exhibit either polymorphism or substantial population diversity. Using our knowledge, gained from functional experiments to interpret the genetic data, can also be problematic, as is exemplified by the current controversy over the role of inhibitory receptors for MHC class I in the "licensing" of NK cells [27] versus mechanisms for NK cell self-tolerance [153]. For now, the interpretation of disease association studies remains constrained by these simple algorithms, and interpretation of the minutiae of these datasets is open to debate. Using large population-based studies, the geneticist has the opportunity to provide instructive hypotheses for cell biologists to test. Working together will be the key to unravelling these problems, and ultimately to translating them into clinical benefit.

Acknowledgments I would like to acknowledge the assistance of Dr Lucia Possami in the construction of the figure and valuable comments on the manuscript.

References

1. Kiessling R, Klein E, Wigzell H (1975) "Natural" killer cells in the mouse. I. Cytotoxic cells with specificity for mouse Moloney leukemia cells. Specificity and distribution according to genotype. Eur J Immunol 5(2):112–117
2. Ljunggren HG, Karre K (1990) In search of the 'missing self': MHC molecules and NK cell recognition. Immunol Today 11(7):237–244
3. Kelley J, Walter L, Trowsdale J (2005) Comparative genomics of natural killer cell receptor gene clusters. PLoS Genet 1(2):129–139
4. Sambrook JG, Bashirova A, Andersen H, Piatak M, Vernikos GS, Coggill P et al (2006) Identification of the ancestral killer immunoglobulin-like receptor gene in primates. BMC Genomics 7:209
5. Uhrberg M, Valiante NM, Shum BP, Shilling HG, Lienert-Weidenbach K, Corliss B et al (1997) Human diversity in killer cell inhibitory receptor genes. Immunity 7(6):753–763

6. Witt CS, Dewing C, Sayer DC, Uhrberg M, Parham P, Christiansen FT (1999) Population frequencies and putative haplotypes of the killer cell immunoglobulin-like receptor sequences and evidence for recombination. Transplantation 68(11):1784–1789
7. Norman PJ, Stephens HA, Verity DH, Chandanayingyong D, Vaughan RW (2001) Distribution of natural killer cell immunoglobulin-like receptor sequences in three ethnic groups. Immunogenetics 52(3–4):195–205
8. Yawata M, Yawata N, McQueen KL, Cheng NW, Guethlein LA, Rajalingam R et al (2002) Predominance of group A KIR haplotypes in Japanese associated with diverse NK cell repertoires of KIR expression. Immunogenetics 54(8):543–550
9. Toneva M, Lepage V, Lafay G, Dulphy N, Busson M, Lester S et al (2001) Genomic diversity of natural killer cell receptor genes in three populations. Tissue Antigens 57 (4):358–362
10. Shilling HG, Guethlein LA, Cheng NW, Gardiner CM, Rodriguez R, Tyan D et al (2002) Allelic polymorphism synergizes with variable gene content to individualize human KIR genotype. J Immunol 168(5):2307–2315
11. Maxwell LD, Wallace A, Middleton D, Curran MD (2002) A common KIR2DS4 deletion variant in the human that predicts a soluble KIR molecule analogous to the KIR1D molecule observed in the rhesus monkey. Tissue Antigens 60(3):254–258
12. Hsu KC, Liu XR, Selvakumar A, Mickelson E, O'Reilly RJ, Dupont B (2002) Killer Ig-like receptor haplotype analysis by gene content: evidence for genomic diversity with a minimum of six basic framework haplotypes, each with multiple subsets. J Immunol 169(9):5118–5129
13. Maxwell LD, Williams F, Gilmore P, Meenagh A, Middleton D (2004) Investigation of killer cell immunoglobulin-like receptor gene diversity: II. KIR2DS4. Hum Immunol 65 (6):613–621
14. Martin MP, Single RM, Wilson MJ, Trowsdale J, Carrington M (2008) KIR haplotypes defined by segregation analysis in 59 Centre d'Etude Polymorphisme Humain (CEPH) families. Immunogenetics 60(12):767–774
15. Colonna M, Borsellino G, Falco M, Ferrara GB, Strominger JL (1993) HLA-C is the inhibitory ligand that determines dominant resistance to lysis by NK1- and NK2-specific natural killer cells. Proc Natl Acad Sci USA 90(24):12000–12004
16. Wagtmann N, Rajagopalan S, Winter CC, Peruzzi M, Long EO (1995) Killer cell inhibitory receptors specific for HLA-C and HLA-B identified by direct binding and by functional transfer. Immunity 3(6):801–809
17. Cella M, Longo A, Ferrara GB, Strominger JL, Colonna M (1994) NK3-specific natural killer cells are selectively inhibited by Bw4-positive HLA alleles with isoleucine 80. J Exp Med 180(4):1235–1242
18. Gumperz JE, Litwin V, Phillips JH, Lanier LL, Parham P (1995) The Bw4 public epitope of HLA-B molecules confers reactivity with natural killer cell clones that express NKB1, a putative HLA receptor. J Exp Med 181(3):1133–1144
19. Malnati MS, Peruzzi M, Parker KC, Biddison WE, Ciccone E, Moretta A et al (1995) Peptide specificity in the recognition of MHC class I by natural killer cell clones. Science 267 (5200):1016–1018
20. Rajagopalan S, Long EO (1997) The direct binding of a p58 killer cell inhibitory receptor to human histocompatibility leukocyte antigen (HLA)-Cw4 exhibits peptide selectivity. J Exp Med 185(8):1523–1528
21. Maenaka K, Juji T, Nakayama T, Wyer JR, Gao GF, Maenaka T et al (1999) Killer cell immunoglobulin receptors and T cell receptors bind peptide-major histocompatibility complex class I with distinct thermodynamic and kinetic properties. J Biol Chem 274 (40):28329–28334
22. Katz G, Gazit R, Arnon TI, Gonen-Gross T, Tarcic G, Markel G et al (2004) MHC class I-independent recognition of NK-activating receptor KIR2DS4. J Immunol 173 (3):1819–1825

23. Katz G, Markel G, Mizrahi S, Arnon TI, Mandelboim O (2001) Recognition of HLA-Cw4 but not HLA-Cw6 by the NK cell receptor killer cell Ig-like receptor two-domain short tail number 4. J Immunol 166(12):7260–7267
24. Moesta AK, Norman PJ, Yawata M, Yawata N, Gleimer M, Parham P (2008) Synergistic polymorphism at two positions distal to the ligand-binding site makes KIR2DL2 a stronger receptor for HLA-C than KIR2DL3. J Immunol 180(6):3969–3979
25. Gardiner CM, Guethlein LA, Shilling HG, Pando M, Carr WH, Rajalingam R et al (2001) Different NK cell surface phenotypes defined by the DX9 antibody are due to KIR3DL1 gene polymorphism. J Immunol 166(5):2992–3001
26. Yawata M, Yawata N, Draghi M, Little AM, Partheniou F, Parham P (2006) Roles for HLA and KIR polymorphisms in natural killer cell repertoire selection and modulation of effector function. J Exp Med 203(3):633–645
27. Kim S, Poursine-Laurent J, Truscott SM, Lybarger L, Song YJ, Yang L et al (2005) Licensing of natural killer cells by host major histocompatibility complex class I molecules. Nature 436(7051):709–713
28. Yawata M, Yawata N, Draghi M, Partheniou F, Little AM, Parham P (2008) MHC class I-specific inhibitory receptors and their ligands structure diverse human NK-cell repertoires toward a balance of missing self-response. Blood 112(6):2369–2380
29. Zingoni A, Sornasse T, Cocks BG, Tanaka Y, Santoni A, Lanier LL (2004) Cross-Talk between activated human NK cells and CD4+ T cells via OX40-OX40 ligand interactions. J Immunol 173(6):3716–3724
30. Ferlazzo G, Pack M, Thomas D, Paludan C, Schmid D, Strowig T, et al (2004) Distinct roles of IL-12 and IL-15 in human natural killer cell activation by dendritic cells from secondary lymphoid organs. Proc Natl Acad Sci USA 101(47):16606–16611
31. Chiesa MD, Vitale M, Carlomagno S, Ferlazzo G, Moretta L, Moretta A (2003) The natural killer cell-mediated killing of autologous dendritic cells is confined to a cell subset expressing CD94/NKG2A, but lacking inhibitory killer Ig-like receptors. Eur J Immunol 33(6):1657–1666
32. Ferlazzo G, Tsang ML, Moretta L, Melioli G, Steinman RM, Munz C (2002) Human dendritic cells activate resting natural killer (NK) cells and are recognized via the NKp30 receptor by activated NK cells. J Exp Med 195(3):343–351
33. Gerosa F, Baldani-Guerra B, Nisii C, Marchesini V, Carra G, Trinchieri G (2002) Reciprocal activating interaction between natural killer cells and dendritic cells. J Exp Med 195(3):327–333
34. Piccioli D, Sbrana S, Melandri E, Valiante NM (2002) Contact-dependent stimulation and inhibition of dendritic cells by natural killer cells. J Exp Med 195(3):335–341
35. Fernandez NC, Lozier A, Flament C, Ricciardi-Castagnoli P, Bellet D, Suter M et al (1999) Dendritic cells directly trigger NK cell functions: cross-talk relevant in innate anti-tumor immune responses in vivo. Nat Med 5(4):405–411
36. Baxter AG, Smyth MJ (2002) The role of NK cells in autoimmune disease. Autoimmunity 35(1):1–14
37. van Bergen J, Thompson A, van der Slik A, Ottenhoff TH, Gussekloo J, Koning F (2004) Phenotypic and functional characterization of CD4 T cells expressing killer Ig-like receptors. J Immunol 173(11):6719–6726
38. Arlettaz L, Degermann S, De Rham C, Roosnek E, Huard B (2004) Expression of inhibitory KIR is confined to CD8+ effector T cells and limits their proliferative capacity. Eur J Immunol 34(12):3413–3422
39. Henseler T (1997) The genetics of psoriasis. J Am Acad Dermatol 37(2 Pt 3):S1–S11
40. Suzuki Y, Hamamoto Y, Ogasawara Y, Ishikawa K, Yoshikawa Y, Sasazuki T et al (2004) Genetic polymorphisms of killer cell immunoglobulin-like receptors are associated with susceptibility to psoriasis vulgaris. J Invest Dermatol 122(5):1133–1136
41. Holm SJ, Sakuraba K, Mallbris L, Wolk K, Stahle M, Sanchez FO (2005) Distinct HLA-C/KIR genotype profile associates with guttate psoriasis. J Invest Dermatol 125(4):721–730

42. Luszczek W, Manczak M, Cislo M, Nockowski P, Wisniewski A, Jasek M et al (2004) Gene for the activating natural killer cell receptor, KIR2DS1, is associated with susceptibility to psoriasis vulgaris. Hum Immunol 65(7):758–766
43. Williams F, Meenagh A, Sleator C, Cook D, Fernandez-Vina M, Bowcock AM et al (2005) Activating killer cell immunoglobulin-like receptor gene KIR2DS1 is associated with psoriatic arthritis. Hum Immunol 66(7):836–841
44. Nelson GW, Martin MP, Gladman D, Wade J, Trowsdale J, Carrington M (2004) Cutting edge: heterozygote advantage in autoimmune disease: hierarchy of protection/susceptibility conferred by hla and killer ig-like receptor combinations in psoriatic arthritis. J Immunol 173 (7):4273–4276
45. Snyder MR, Muegge LO, Offord C, O'Fallon WM, Bajzer Z, Weyand CM et al (2002) Formation of the killer Ig-like receptor repertoire on CD4+CD28null T cells. J Immunol 168 (8):3839–3846
46. Namekawa T, Snyder MR, Yen JH, Goehring BE, Leibson PJ, Weyand CM et al (2000) Killer cell activating receptors function as costimulatory molecules on CD4+CD28null T cells clonally expanded in rheumatoid arthritis. J Immunol 165(2):1138–1145
47. Snyder MR, Nakajima T, Leibson PJ, Weyand CM, Goronzy JJ (2004) Stimulatory killer Ig-like receptors modulate T cell activation through DAP12-dependent and DAP12-independent mechanisms. J Immunol 173(6):3725–3731
48. Yen JH, Moore BE, Nakajima T, Scholl D, Schaid DJ, Weyand CM et al (2001) Major histocompatibility complex class I-recognizing receptors are disease risk genes in rheumatoid arthritis. J Exp Med 193(10):1159–1167
49. Nakajima T, Goek O, Zhang X, Kopecky SL, Frye RL, Goronzy JJ et al (2003) De novo expression of killer immunoglobulin-like receptors and signaling proteins regulates the cytotoxic function of CD4 T cells in acute coronary syndromes. Circ Res 93(2):106–113
50. Karlsen TH, Boberg KM, Olsson M, Sun JY, Senitzer D, Bergquist A et al (2007) Particular genetic variants of ligands for natural killer cell receptors may contribute to the HLA associated risk of primary sclerosing cholangitis. J Hepatol 46(5):899–906
51. Goverdhan SV, Khakoo SI, Gaston H, Chen X, Lotery A (2008) Age related macular degeneration is associated with the HLA Cw*0701 genotype and the Natural Killer cell receptor AA haplotype. Invest Ophthalmol Vis Sci 49(11):5077–5082
52. Levinson RD, Du Z, Luo L, Monnet D, Tabary T, Brezin AP et al (2008) Combination of KIR and HLA gene variants augments the risk of developing birdshot chorioretinopathy in HLA-A*29-positive individuals. Genes Immun 9(3):249–258
53. Boyton RJ, Smith J, Ward R, Jones M, Ozerovitch L, Wilson R et al (2006) HLA-C and killer cell immunoglobulin-like receptor genes in idiopathic bronchiectasis. Am J Respir Crit Care Med 173(3):327–333
54. Martin MP, Nelson G, Lee JH, Pellett F, Gao X, Wade J et al (2002) Cutting edge: susceptibility to psoriatic arthritis: influence of activating killer Ig-like receptor genes in the absence of specific HLA-C alleles. J Immunol 169(6):2818–2822
55. van der Slik AR, Koeleman BP, Verduijn W, Bruining GJ, Roep BO, Giphart MJ (2003) KIR in type 1 diabetes: disparate distribution of activating and inhibitory natural killer cell receptors in patients versus HLA-matched control subjects. Diabetes 52(10):2639–2642
56. Nikitina-Zake L, Rajalingham R, Rumba I, Sanjeevi CB (2004) Killer cell immunoglobulin-like receptor genes in Latvian patients with type 1 diabetes mellitus and healthy controls. Ann N Y Acad Sci 1037:161–169
57. Momot T, Koch S, Hunzelmann N, Krieg T, Ulbricht K, Schmidt RE et al (2004) Association of killer cell immunoglobulin-like receptors with scleroderma. Arthritis Rheum 50(5):1561–1565
58. Biron CA, Byron KS, Sullivan JL (1989) Severe herpesvirus infections in an adolescent without natural killer cells. N Engl J Med 320(26):1731–1735
59. French AR, Yokoyama WM (2003) Natural killer cells and viral infections. Curr Opin Immunol 15(1):45–51

60. Orange JS, Ballas ZK (2006) Natural killer cells in human health and disease. Clin Immunol 118(1):1–10
61. Liese J, Schleicher U, Bogdan C (2008) The innate immune response against Leishmania parasites. Immunobiology 213(3–4):377–387
62. Denkers EY, Sher A (1997) Role of natural killer and NK1+ T-cells in regulating cell-mediated immunity during Toxoplasma gondii infection. Biochem Soc Trans 25(2):699–703
63. Meier UC, Owen RE, Taylor E, Worth A, Naoumov N, Willberg C et al (2005) Shared alterations in NK cell frequency, phenotype, and function in chronic human immunodeficiency virus and hepatitis C virus infections. J Virol 79(19):12365–12374
64. Mavilio D, Benjamin J, Daucher M, Lombardo G, Kottilil S, Planta MA et al (2003) Natural killer cells in HIV-1 infection: dichotomous effects of viremia on inhibitory and activating receptors and their functional correlates. Proc Natl Acad Sci USA 100(25):15011–15016
65. Mavilio D, Lombardo G, Benjamin J, Kim D, Follman D, Marcenaro E et al (2005) Characterization of CD56-/CD16+ natural killer (NK) cells: a highly dysfunctional NK subset expanded in HIV-infected viremic individuals. Proc Natl Acad Sci USA 102(8):2886–2891
66. Alter G, Teigen N, Davis BT, Addo MM, Suscovich TJ, Waring MT et al (2005) Sequential deregulation of NK cell subset distribution and function starting in acute HIV-1 infection. Blood 106(10):3366–3369
67. Fan YY, Yang BY, Wu CY (2008) Phenotypic and functional heterogeneity of natural killer cells from umbilical cord blood mononuclear cells. Immunol Invest 37(1):79–96
68. Gaddy J, Broxmeyer HE (1997) Cord blood CD16+56- cells with low lytic activity are possible precursors of mature natural killer cells. Cell Immunol 180(2):132–142
69. Carrington M, O'Brien SJ (2003) The influence of HLA genotype on AIDS. Annu Rev Med 54:535–551
70. Flores-Villanueva PO, Yunis EJ, Delgado JC, Vittinghoff E, Buchbinder S, Leung JY et al (2001) Control of HIV-1 viremia and protection from AIDS are associated with HLA-Bw4 homozygosity. Proc Natl Acad Sci USA 98(9):5140–5145
71. Martin MP, Gao X, Lee JH, Nelson GW, Detels R, Goedert JJ et al (2002) Epistatic interaction between KIR3DS1 and HLA-B delays the progression to AIDS. Nat Genet 31(4):429–434
72. Boulet S, Sharafi S, Simic N, Bruneau J, Routy JP, Tsoukas CM et al (2008) Increased proportion of KIR3DS1 homozygotes in HIV-exposed uninfected individuals. Aids 22(5):595–599
73. Alter G, Martin MP, Teigen N, Carr WH, Suscovich TJ, Schneidewind A et al (2007) Differential natural killer cell-mediated inhibition of HIV-1 replication based on distinct KIR/HLA subtypes. J Exp Med 204(12):3027–3036
74. Long BR, Ndhlovu LC, Oksenberg JR, Lanier LL, Hecht FM, Nixon DF et al (2008) Conferral of enhanced natural killer cell function by KIR3DS1 in early human immunodeficiency virus type 1 infection. J Virol 82(10):4785–4792
75. Gumperz JE, Valiante NM, Parham P, Lanier LL, Tyan D (1996) Heterogeneous phenotypes of expression of the NKB1 natural killer cell class I receptor among individuals of different human histocompatibility leukocyte antigens types appear genetically regulated, but not linked to major histocompatibililty complex haplotype. J Exp Med 183(4):1817–1827
76. Li H, Pascal V, Martin MP, Carrington M, Anderson SK (2008) Genetic control of variegated KIR gene expression: polymorphisms of the bi-directional KIR3DL1 promoter are associated with distinct frequencies of gene expression. PLoS Genet 4(11):e1000254
77. Pando MJ, Gardiner CM, Gleimer M, McQueen KL, Parham P (2003) The protein made from a common allele of KIR3DL1 (3DL1*004) is poorly expressed at cell surfaces due to substitution at positions 86 in Ig domain 0 and 182 in Ig domain 1. J Immunol 171(12):6640–6649
78. Martin MP, Qi Y, Gao X, Yamada E, Martin JN, Pereyra F et al (2007) Innate partnership of HLA-B and KIR3DL1 subtypes against HIV-1. Nat Genet 39(6):733–740

79. Khakoo SI, Thio CL, Martin MP, Brooks CR, Gao X, Astemborski J et al (2004) HLA and NK cell inhibitory receptor genes in resolving hepatitis C virus infection. Science 305 (5685):872–874
80. Smith HR, Heusel JW, Mehta IK, Kim S, Dorner BG, Naidenko OV et al (2002) Recognition of a virus-encoded ligand by a natural killer cell activation receptor. Proc Natl Acad Sci USA 99(13):8826–8831
81. Arase H, Mocarski ES, Campbell AE, Hill AB, Lanier LL (2002) Direct recognition of cytomegalovirus by activating and inhibitory NK cell receptors. Science 296 (5571):1323–1326
82. Scalzo AA, Fitzgerald NA, Simmons A, La Vista AB, Shellam GR (1990) Cmv-1, a genetic locus that controls murine cytomegalovirus replication in the spleen. J Exp Med 171(5): 1469–1483
83. Romero V, Azocar J, Zuniga J, Clavijo OP, Terreros D, Gu X et al (2008) Interaction of NK inhibitory receptor genes with HLA-C and MHC class II alleles in Hepatitis C virus infection outcome. Mol Immunol 45(9):2429–2436
84. Winter CC, Gumperz JE, Parham P, Long EO, Wagtmann N (1998) Direct binding and functional transfer of NK cell inhibitory receptors reveal novel patterns of HLA-C allotype recognition. J Immunol 161(2):571–577
85. Hanvesakul R, Spencer N, Cook M, Gunson B, Hathaway M, Brown R et al (2008) Donor HLA-C genotype has a profound impact on the clinical outcome following liver transplantation. Am J Transplant 8(9):1931–1941
86. Kunert K, Seiler M, Mashreghi MF, Klippert K, Schonemann C, Neumann K et al (2007) KIR/HLA ligand incompatibility in kidney transplantation. Transplantation 84(11):1527–1533
87. Orange JS (2002) Human natural killer cell deficiencies and susceptibility to infection. Microbes Infect 4(15):1545–1558
88. Lopez-Botet M, Llano M, Ortega M (2001) Human cytomegalovirus and natural killer-mediated surveillance of HLA class I expression: a paradigm of host-pathogen adaptation. Immunol Rev 181:193–202
89. Tomasec P, Wang EC, Davison AJ, Vojtesek B, Armstrong M, Griffin C et al (2005) Downregulation of natural killer cell-activating ligand CD155 by human cytomegalovirus UL141. Nat Immunol 6(2):181–188
90. Tomasec P, Braud VM, Rickards C, Powell MB, McSharry BP, Gadola S et al (2000) Surface expression of HLA-E, an inhibitor of natural killer cells, enhanced by human cytomegalovirus gpUL40. Science 287(5455):1031
91. Arnon TI, Achdout H, Levi O, Markel G, Saleh N, Katz G et al (2005) Inhibition of the NKp30 activating receptor by pp 65 of human cytomegalovirus. Nat Immunol 6(5):515–523
92. Cook M, Briggs D, Craddock C, Mahendra P, Milligan D, Fegan C et al (2006) Donor KIR genotype has a major influence on the rate of cytomegalovirus reactivation following T-cell replete stem cell transplantation. Blood 107(3):1230–1232
93. Gazit R, Garty BZ, Monselise Y, Hoffer V, Finkelstein Y, Markel G et al (2004) Expression of KIR2DL1 on the entire NK cell population: a possible novel immunodeficiency syndrome. Blood 103(5):1965–1966
94. Guma M, Angulo A, Vilches C, Gomez-Lozano N, Malats N, Lopez-Botet M (2004) Imprint of human cytomegalovirus infection on the NK cell receptor repertoire. Blood 104 (12):3664–3671
95. Carr WH, Little AM, Mocarski E, Parham P (2002) NK cell-mediated lysis of autologous HCMV-infected skin fibroblasts is highly variable among NK cell clones and polyclonal NK cell lines. Clin Immunol 105(2):126–140
96. Korbel DS, Finney OC, Riley EM (2004) Natural killer cells and innate immunity to protozoan pathogens. Int J Parasitol 34(13–14):1517–1528

97. Baratin M, Roetynck S, Lepolard C, Falk C, Sawadogo S, Uematsu S et al (2005) Natural killer cell and macrophage cooperation in MyD88-dependent innate responses to Plasmodium falciparum. Proc Natl Acad Sci USA 102(41):14747–14752
98. Artavanis-Tsakonas K, Eleme K, McQueen KL, Cheng NW, Parham P, Davis DM et al (2003) Activation of a subset of human NK cells upon contact with Plasmodium falciparum-infected erythrocytes. J Immunol 171(10):5396–5405
99. Moffett-King A, Entrican G, Ellis S, Hutchinson J, Bainbridge D (2002) Natural killer cells and reproduction. Trends Immunol 23(7):332–333
100. Kopcow HD, Allan DS, Chen X, Rybalov B, Andzelm MM, Ge B et al (2005) Human decidual NK cells form immature activating synapses and are not cytotoxic. Proc Natl Acad Sci USA 102(43):15563–15568
101. Verma S, King A, Loke YW (1997) Expression of killer cell inhibitory receptors on human uterine natural killer cells. Eur J Immunol 27(4):979–983
102. Ponte M, Cantoni C, Biassoni R, Tradori-Cappai A, Bentivoglio G, Vitale C et al (1999) Inhibitory receptors sensing HLA-G1 molecules in pregnancy: decidua-associated natural killer cells express LIR-1 and CD94/NKG2A and acquire p49, an HLA-G1-specific receptor. Proc Natl Acad Sci USA 96(10):5674–5679
103. Koopman LA, Kopcow HD, Rybalov B, Boyson JE, Orange JS, Schatz F et al (2003) Human decidual natural killer cells are a unique NK cell subset with immunomodulatory potential. J Exp Med 198(8):1201–1212
104. King A, Burrows TD, Hiby SE, Bowen JM, Joseph S, Verma S et al (2000) Surface expression of HLA-C antigen by human extravillous trophoblast. Placenta 21(4):376–387
105. King A, Allan DS, Bowen M, Powis SJ, Joseph S, Verma S et al (2000) HLA-E is expressed on trophoblast and interacts with CD94/NKG2 receptors on decidual NK cells. Eur J Immunol 30(6):1623–1631
106. Apps R, Gardner L, Hiby SE, Sharkey AM, Moffett A (2008) Conformation of human leucocyte antigen-C molecules at the surface of human trophoblast cells. Immunology 124 (3):322–328
107. Cantoni C, Falco M, Pessino A, Moretta A, Moretta L, Biassoni R (1999) P49, a putative HLA-G1 specific inhibitory NK receptor belonging to the immunoglobulin Superfamily. J Reprod Immunol 43(2):157–165
108. Rajagopalan S, Long EO (1999) A human histocompatibility leukocyte antigen (HLA)-G-specific receptor expressed on all natural killer cells. J Exp Med 189(7):1093–1100
109. Rajagopalan S, Bryceson YT, Kuppusamy SP, Geraghty DE, van der Meer A, Joosten I et al (2006) Activation of NK cells by an endocytosed receptor for soluble HLA-G. PLoS Biol 4 (1):e9
110. Sharkey AM, Gardner L, Hiby S, Farrell L, Apps R, Masters L et al (2008) Killer Ig-like receptor expression in uterine NK cells is biased toward recognition of HLA-C and alters with gestational age. J Immunol 181(1):39–46
111. Redman CW, Sargent IL (2005) Latest advances in understanding preeclampsia. Science 308 (5728):1592–1594
112. Hiby SE, Walker JJ, O'Shaughnessy KM, Redman CW, Carrington M, Trowsdale J, Hiby SE, Walker JJ, O'Shaughnessy KM, Redman CW, Carrington M, Trowsdale J et al (2004) Combinations of maternal KIR and fetal HLA-C genes influence the risk of preeclampsia and reproductive success. J Exp Med 200(8):957–965
113. Varla-Leftherioti M (2004) Role of a KIR/HLA-C allorecognition system in pregnancy. J Reprod Immunol 62(1–2):19–27
114. Varla-Leftherioti M, Spyropoulou-Vlachou M, Niokou D, Keramitsoglou T, Darlamitsou A, Tsekoura C et al (2003) Natural killer (NK) cell receptors' repertoire in couples with recurrent spontaneous abortions. Am J Reprod Immunol 49(3):183–191
115. Choi SJ, Park JY, Lee YK, Choi HI, Lee YS, Koh CM et al (2002) Effects of cytokines on VEGF expression and secretion by human first trimester trophoblast cell line. Am J Reprod Immunol 48(2):70–76

116. Gomez-Lozano N, de Pablo R, Puente S, Vilches C (2003) Recognition of HLA-G by the NK cell receptor KIR2DL4 is not essential for human reproduction. Eur J Immunol 33(3):639–644
117. Yan WH, Fan LA, Yang JQ, Xu LD, Ge Y, Yao FJ (2006) HLA-G polymorphism in a Chinese Han population with recurrent spontaneous abortion. Int J Immunogenet 33(1):55–58
118. Aldrich C, Verp MS, Walker MA, Ober C (2000) A null mutation in HLA-G is not associated with preeclampsia or intrauterine growth retardation. J Reprod Immunol 47(1):41–48
119. Ober C, Billstrand C, Kuldanek S, Tan Z. The miscarriage-associated HLA-G -725G allele influences transcription rates in JEG-3 cells. Hum Reprod 21:1743–1748
120. Witt CS, Whiteway JM, Warren HS, Barden A, Rogers M, Martin A et al (2002) Alleles of the KIR2DL4 receptor and their lack of association with pre-eclampsia. Eur J Immunol 32(1):18–29
121. Smyth MJ, Hayakawa Y, Takeda K, Yagita H (2002) New aspects of natural-killer-cell surveillance and therapy of cancer. Nat Rev Cancer 2(11):850–861
122. Garrido F, Ruiz-Cabello F, Cabrera T, Perez-Villar JJ, Lopez-Botet M, Duggan-Keen M et al (1997) Implications for immunosurveillance of altered HLA class I phenotypes in human tumours. Immunol Today 18(2):89–95
123. Snijders PJ, Steenbergen RD, Heideman DA, Meijer CJ (2006) HPV-mediated cervical carcinogenesis: concepts and clinical implications. J Pathol 208(2):152–164
124. Carrington M, Wang S, Martin MP, Gao X, Schiffman M, Cheng J et al (2005) Hierarchy of resistance to cervical neoplasia mediated by combinations of killer immunoglobulin-like receptor and human leukocyte antigen loci. J Exp Med 201(7):1069–1075
125. Butsch Kovacic M, Martin M, Gao X, Fuksenko T, Chen CJ, Cheng YJ et al (2005) Variation of the killer cell immunoglobulin-like receptors and HLA-C genes in nasopharyngeal carcinoma. Cancer Epidemiol Biomarkers Prev 14(11 Pt 1):2673–2677
126. Ren EC, Chan SH (1996) Human leucocyte antigens and nasopharyngeal carcinoma. Clin Sci 91(3):256–258
127. Barber LD, Percival L, Valiante NM, Chen L, Lee C, Gumperz JE et al (1996) The interlocus recombinant HLA-B*4601 has high selectivity in peptide binding and functions characteristic of HLA-C. J Exp Med 184(2):735–740
128. Lopez-Vazquez A, Rodrigo L, Martinez-Borra J, Perez R, Rodriguez M, Fdez-Morera JL et al (2005) Protective Effect of the HLA-Bw4I80 Epitope and the Killer Cell Immunoglobulin-Like Receptor 3DS1 Gene against the Development of Hepatocellular Carcinoma in Patients with Hepatitis C Virus Infection. J Infect Dis 192(1):162–165
129. Norris S, Doherty DG, Curry M, McEntee G, Traynor O, Hegarty JE et al (2003) Selective reduction of natural killer cells and T cells expressing inhibitory receptors for MHC class I in the livers of patients with hepatic malignancy. Cancer Immunol Immunother 52(1):53–58
130. Sandel MH, Speetjens FM, Menon AG, Albertsson PA, Basse PH, Hokland M et al (2005) Natural killer cells infiltrating colorectal cancer and MHC class I expression. Mol Immunol 42(4):541–546
131. Stern N, Markel G, Arnon TI, Gruda R, Wong H, Gray-Owen SD et al (2005) Carcinoembryonic antigen (CEA) inhibits NK killing via interaction with CEA-related cell adhesion molecule 1. J Immunol 174(11):6692–6701
132. Ikeda H, Lethe B, Lehmann F, van Baren N, Baurain JF, de Smet C et al (1997) Characterization of an antigen that is recognized on a melanoma showing partial HLA loss by CTL expressing an NK inhibitory receptor. Immunity 6(2):199–208
133. Bakker AB, Phillips JH, Figdor CG, Lanier LL (1998) Killer cell inhibitory receptors for MHC class I molecules regulate lysis of melanoma cells mediated by NK cells, gamma delta T cells, and antigen-specific CTL. J Immunol 160(11):5239–5245
134. Wu JD, Higgins LM, Steinle A, Cosman D, Haugk K, Plymate SR (2004) Prevalent expression of the immunostimulatory MHC class I chain-related molecule is counteracted by shedding in prostate cancer. J Clin Invest 114(4):560–568
135. Groh V, Wu J, Yee C, Spies T (2002) Tumour-derived soluble MIC ligands impair expression of NKG2D and T-cell activation. Nature 419(6908):734–738

136. Nachmias B, Mizrahi S, Elmalech M, Lazar I, Ashhab Y, Gazit R et al (2007) Manipulation of NK cytotoxicity by the IAP family member Livin. Eur J Immunol 37(12):3467–3476
137. Naumova E, Mihaylova A, Stoitchkov K, Ivanova M, Quin L, Toneva M (2005) Genetic polymorphism of NK receptors and their ligands in melanoma patients: prevalence of inhibitory over activating signals. Cancer Immunol Immunother 54(2):172–178
138. Ruggeri L, Capanni M, Urbani E, Perruccio K, Shlomchik WD, Tosti A et al (2002) Effectiveness of donor natural killer cell alloreactivity in mismatched hematopoietic transplants. Science 295(5562):2097–2100
139. Dupont B, Hsu KC (2004) Inhibitory killer Ig-like receptor genes and human leukocyte antigen class I ligands in haematopoietic stem cell transplantation. Curr Opin Immunol 16 (5):634–643
140. Ruggeri L, Aversa F, Martelli MF, Velardi A (2006) Allogeneic hematopoietic transplantation and natural killer cell recognition of missing self. Immunol Rev 214:202–218
141. Loughran TP Jr (1993) Clonal diseases of large granular lymphocytes. Blood 82(1):1–14
142. Mitsui T, Maekawa I, Yamane A, Ishikawa T, Koiso H, Yokohama A et al (2004) Characteristic expansion of CD45RA CD27 CD28 CCR7 lymphocytes with stable natural killer (NK) receptor expression in NK- and T-cell type lymphoproliferative disease of granular lymphocytes. Br J Haematol 126(1):55–62
143. Young NT, Uhrberg M, Phillips JH, Lanier LL, Parham P (2001) Differential expression of leukocyte receptor complex-encoded Ig-like receptors correlates with the transition from effector to memory CTL. J Immunol 166(6):3933–3941
144. Rabbani GR, Phyliky RL, Tefferi A (1999) A long-term study of patients with chronic natural killer cell lymphocytosis. Br J Haematol 106(4):960–966
145. Morice WG, Kurtin PJ, Leibson PJ, Tefferi A, Hanson CA (2003) Demonstration of aberrant T-cell and natural killer-cell antigen expression in all cases of granular lymphocytic leukaemia. Br J Haematol 120(6):1026–1036
146. Nowakowski GS, Morice WG, Phyliky RL, Li CY, Tefferi A (2005) Human leucocyte antigen class I and killer immunoglobulin-like receptor expression patterns in T-cell large granular lymphocyte leukaemia. Br J Haematol 128(4):490–492
147. Casado LF, Granados E, Algara P, Navarro F, Martinez-Frejo MC, Lopez-Botet M (2001) High expression of the ILT2 (LIR-1) inhibitory receptor for major histocompatibility complex class I molecules on clonal expansions of T large granular lymphocytes in asymptomatic patients. Haematologica 86(5):457–463
148. Pascal V, Schleinitz N, Brunet C, Ravet S, Bonnet E, Lafarge X et al (2004) Comparative analysis of NK cell subset distribution in normal and lymphoproliferative disease of granular lymphocyte conditions. Eur J Immunol 34(10):2930–2940
149. Zambello R, Falco M, Della Chiesa M, Trentin L, Carollo D, Castriconi R et al (2003) Expression and function of KIR and natural cytotoxicity receptors in NK-type lymphoproliferative diseases of granular lymphocytes. Blood 102(5):1797–1805
150. Epling-Burnette PK, Painter JS, Chaurasia P, Bai F, Wei S, Djeu JY et al (2004) Dysregulated NK receptor expression in patients with lymphoproliferative disease of granular lymphocytes. Blood 103(9):3431–3439
151. Fischer L, Hummel M, Burmeister T, Schwartz S, Thiel E (2006) Skewed expression of natural-killer (NK)-associated antigens on lymphoproliferations of large granular lymphocytes (LGL). Hematol Oncol 24(2):78–85
152. Gendzekhadze K, Norman PJ, Abi-Rached L, Layrisse Z, Parham P (2006) High KIR diversity in Amerindians is maintained using few gene-content haplotypes. Immunogenetics 58(5–6):474–80
153. Raulet DH (1999) Development and tolerance of natural killer cells. Curr Opin Immunol 11 (2):129–134
154. Colonna M, Samaridis J (1995) Cloning of immunoglobulin-superfamily members associated with HLA-C and HLA-B recognition by human natural killer cells. Science 268 (5209):405–408

Interactions Between NK Cells and Dendritic Cells

Guido Ferlazzo

Abstract Dendritic cells (DC) represent the most powerful antigen-presenting cells (APC). Located in peripheral tissues in an immature form, DC are efficient in capturing antigens such as pathogens and dying cells. In general, microbial stimulation causes DC to convert from the immature state, in which they induce abortive or tolerogenic T cell responses, to a mature state, in which they elicit a productive response.

The functional links between natural killer (NK) cells and DC have been widely investigated in the last years and different studies have demonstrated that reciprocal activations ensue upon NK/DC interactions. More recently, the anatomical sites where these interactions take place have been identified together with the related cell subsets involved. Remarkably, there is now "in vivo" evidence that this cellular cross-talk occurring during the innate phase of the immune response can deeply affect the magnitude and the quality of the subsequent adaptive response. Thus, NK cells are not merely cytotoxic lymphocytes competent in containing the spreading of viruses and tumors but can now rather be considered as crucial fine-tuning effector cells.

1 Introduction

Dendritic cells (DC) represent the most powerful antigen-presenting cells (APC) and are found in various tissues where they play a major role in antigen capture. Subsequent stimuli will induce further differentiation into mature DC and their migration into secondary lymphoid organs, primarily lymph nodes. These stimuli are represented either by microbial products or by cytokines of the innate immunity including tumor necrosis factor (TNF)-α, IL-1 and IL-6. Mature DCs acquire a

G. Ferlazzo
Laboratory of Immunology and Biotherapy, University of Messina, Italy
e-mail: guido.ferlazzo@unime.it

potent APC activity because of the expression of costimulatory molecules and of high levels of surface HLA molecules [1–3].

Until recently, no information existed on the possible interactions between natural killer (NK) and DC. However, recent reports revealed that these cells may interact with each other, thus providing evidence for novel, unexpected mechanisms of regulation at the interface between innate and adaptive immunity.

Thus, NK cells are not merely cytotoxic lymphocytes competent in containing the spread of viruses and tumors, but can now rather be considered as crucial fine-tuning effector cells widely involved in different phases of the immune response.

Indeed, while NK cells have long been defined as "primitive" and "nonspecific" effector cells, we have now a different perception of these cells and it is now clear that NK cells evolved to cooperate with the adaptive immunity.

First, it is evident that NK cells evolved to adapt to and cooperate with mechanisms of the specific immunity: they have evolved receptors for the Fc portion of IgG that allow killing of antibody-coated target cells or certain pathogens; they also release a number of cytokines that regulate T cell activation and function. Importantly, an early activation of NK cells during immune responses may influence the quality of the subsequent T cell response by inducing a Th1 polarization. Second, NK cells have evolved a mechanism allowing the rapid detection and killing of potentially dangerous cells characterized by an altered expression of MHC class I antigens due to infections. Human NK cells have been shown to express different human leukocyte antigen (HLA) class I-specific inhibitory receptors. A family of these receptors (termed killer Ig-like receptors (KIR)) detect shared allelic determinants of HLA class I molecules while others display a more "promiscuous" pattern of recognition and are characterized by a broad specificity for different HLA class I molecules (LIR1/ILT2) or recognize the HLA class Ib HLA-E molecules (CD94/NKG2A) [4–8]. This mechanism is sophisticated (KIR detect allelic determinants of HLA class I molecules) and of recent evolution since murine NK cells lack KIR (a similar function is mediated by structurally different receptors) [4] and major differences in the type and specificity of expressed KIR exist in chimpanzees, a species that diverged from humans only 5 million years ago [9]. This clearly means that KIR have evolved recently, paralleling the rapid evolution of HLA class I molecules.

2 Cytotoxic and Cytokine Secreting NK Cells

Although NK cells were initially described as cytotoxic effectors in peripheral blood and were found to lyse tumor cells without prior activation, recent studies suggest that a possibly immunoregulatory subset of NK cells responds to activation mainly with cytokine secretion. For instance, IFN-γ, a major cytokine released by activated NK cells, represents the principal phagocyte-activating factor, indicating the crucial function of NK cell activation during infections. Human peripheral blood mononuclear cells contain around 10% of NK cells [10]. The majority (around 95%) belongs to the $CD56^{dim}CD16^+$ cytolytic NK subset [11–13]. These cells carry homing markers for inflamed peripheral sites and express perforin to rapidly mediate cytotoxicity [11, 12]. The minor NK subset in blood (usually less than 5%) is $CD56^{bright}CD16^-$ cells [11–13]. These NK cells lack perforin (or have low level of it), but secrete large amounts of IFN-γ and TNF-β upon activation and are superior to $CD56^{dim}$ NK cells in these functions. In addition, they display homing markers for secondary lymphoid organs, namely CCR7 and CD62L [12, 13].

Another difference between these subsets can be found in their receptors mediating target recognition. Human NK cell recognition of target cells is guided by the balance of activating and inhibitory signals given by different groups of surface receptors. The main activating receptors constitutively found on all NK cells in peripheral blood are NKG2D and the natural cytotoxicity receptors (NCR) NKp30 and NKp46 [14]. All of them probably recognize molecules that are upregulated upon cellular stress [15, 16].

However, only the stress-induced NKG2D ligands MICA/B and ULBP have so far been identified [17, 18]. While cytotoxic peripheral blood $CD56^{dim}$ NK cells are able to target antibody-opsonized cells via their low-affinity FcγRIII/CD16 molecule, immunoregulatory $CD56^{bright}$ NK cells lack this receptor nearly entirely [12, 13].

Most inhibitory NK cell receptors engage MHC class I molecules on target cells. They can be distinguished into two groups, detecting either common allelic determinants of MHC class I, or MHC class I expression in general. The KIR receptors constituting the first group distinguish polymorphic HLA-A, -B and -C molecules. The inhibitory receptor surveying MHC class I expression in general are more heterogeneous. They include the LIR1/ILT2 molecule with a broad specificity for different HLA class I molecules and the CD94/NKG2A heterodimer, specific for HLA-E whose surface expression is dictated by the availability of HLA class I heavy chain signal peptides. The MHC class I allele-specific KIR receptors are expressed on subsets of $CD56^{dim}CD16^-$ cytolytic NK cells, while the immunoregulatory $CD56^{bright}CD16^+$ NK subset expresses uniformly CD94/NKG2A and lacks KIR [12].

These phenotypes are consistent with the hypothesis that $CD56^{dim}CD16^+$ NK cells are terminally differentiated effectors that carry the whole panel of sophisticated activating and inhibitory receptors to detect allelic HLA class I loss and can readily lyse aberrant cells at peripheral inflammation sites. On the contrary, $CD56^{bright}CD16^-$ NK cells might perform an immunoregulatory function in secondary lymphoid tissues and release large amount of effector cytokines able to control pathogen spreading.

3 NK Cells in Secondary Lymphoid Organs

Peripheral blood is the most accessible source of human NK cells. Therefore, most studies have been performed with NK cell populations from this organ and it was assumed that most NK cells circulate in human blood after their emigration from the bone marrow. Recently, however, it has been shown that a substantial amount of human NK cells homes to secondary lymphoid organs. These amount to around 5% of mononuclear cells in uninflamed lymph nodes and 0.4–1% in inflamed tonsils and lymph nodes [19, 20]. These NK cells constitute a remarkable pool of innate effector cells, since lymph nodes harbor 40% of all lymphocytes, while peripheral blood contains only 2% of all lymphocytes [21, 22]. Therefore, lymph node NK cells are in the absence of infection and inflammation ten times more abundant than blood NK cells.

As expected from CCR7 and CD62L expression on $CD56^{bright}$ NK cells in blood, the $CD56^{bright}$ NK subset is enriched in all secondary lymphoid organs analyzed so far (lymph nodes, tonsils and spleen) [20]. In spleen, around 15%, and in tonsils and lymph nodes around 75% of NK cells, belong to the $CD56^{bright}$ subset. Lymph node and tonsil NK cells, however, uniformly lack FcγRIII/CD16 and KIR, while the $CD56^{dim}$ NK cells in spleen express these molecules [20]. Surprisingly and in contrast to $CD56^{bright}$ blood NK cells, lymph node and tonsil NK cells show no or very low expression of the constitutive NCR NKp30 and NKp46 [19, 20]. On the other hand, the inducible NCR NKp44, undetectable on NK cells directly isolated from peripheral blood, is upregulated on NK cells harbored in inflamed tonsils [20]. Therefore, target cell recognition by lymph node and tonsil NK cells seems to be mainly influenced by NKG2D as activating and CD94/NKG2A as inhibitory receptor [19, 20]. This NK receptor repertoire of lymph node and tonsil NK cells is similar to $CD56^{bright}$ blood NK cells, but even more restricted by the absence of constitutive NCR.

With respect to NK function, secondary lymphoid tissue NK cells show an impressive plasticity. Although lymph node and tonsil NK cells are initially perforin-negative and show no cytolytic activity against MHC class I low and ULBP high targets, perforin and cytotoxicity can be upregulated by IL-2 within 3–7 days [20]. Interestingly, NK cells from secondary lymphoid organs gain during the same time-period expression of CD16, NCR NKp30, NKp46 and NKp44 as well as KIR [20]. Therefore, activation converts lymph node and tonsil NK cells into effectors, similar to the terminally differentiated $CD56^{dim}CD16^{+}$ blood NK cells with cytolytic function and the sophisticated set of inhibitory and activating receptors.

4 Dendritic Cells as Early Activators of NK Cell Functions

The mentioned human NK cell compartments are probably all involved in early innate immune responses. Recent studies have demonstrated that during the innate phase of the immune response NK cells can also mediate DC maturation [23, 24]. This activation is not unidirectional, because the interaction between mature, but not immature, DC and NK cells results in NK cell proliferation, IFN-γ production and induction of cytolytic activity [25, 26]. Thus, DC have now emerged as the activators of NK response in the early phases of the immune response, i.e., before an adaptive immune response had been evoked and T cell derived cytokines, such as IL-2, could be produced. Remarkably, DC-induced NK cell cytolytic activity was directed not only towards tumor cells, but also against immature DC (iDC). The NK-mediated killing of DC was mostly dependent on the NKp30 NCR.

Other activating receptors or coreceptors played virtually no role [26]. This would imply that DC express the ligand for NKp30, but not for other major triggering NK receptors.

During NK activation by myeloid DC, both soluble factors as well as cell-to-cell contact seem to be important. In mice, induction of NK cell cytotoxicity was entirely blocked by transwell separation of DC and NK cells, indicative for a major contribution of DC surface receptors in NK activation [25]. In addition, DC derived IFN-α/β, IL-12 and IL-18 have been reported to be crucial in murine NK activation [27–29].

While IL-12 was mainly implicated in NK mediated IFN-γ secretion, IFN-α/β seems required for cytotoxicity of NK cells [28, 30]. In addition, secretion of IL-2 by DC stimulated with microbial stimuli might also contribute to NK activation [31, 32].

In human, NK activation by DC was not significantly disrupted by transwell separation of the two cell types, indicating a major contribution of soluble factors in NK activation [33, 34]. In one study, DC were unable to activate NK cells in the presence of neutralizing antibodies for IL-12 and IL-18 [34]. Other studies demonstrated that NK activation by DC subsets correlates with IL-12 secretion by these DC, while IL-15 and IL-18 secretion were not indicative for NK activation [35, 36]. Therefore, IL-12 might play an important role in human NK activation by DC.

Apart from myeloid DC that efficiently activate human and mouse NK cells, plasmacytoid DC (pDC) might also contribute to NK activation. This DC subset has been found to produce 200–1,000 times more type I IFN than other blood cells after viral challenge [37] and IFN-α/β are critical cytokines for inducing NK cell-mediated lysis of virus-infected targets [38–40]. IFN-α/β secretion seems, however, not only the signature of pDC, but also myeloid DC can secrete substantial amounts of these cytokines upon direct viral infection [41], as well as upon bacterial infection [42]. Therefore, the contributions of myeloid and pDC in type I IFN-mediated NK activation remains to be established.

While cytokines play a dominant role in human NK cell activation by DC, IFN-α treatment of DC in addition upregulates the NKG2D ligands MICA/B on monocyte-derived DC and these molecules seem to activate resting NK cells in a cell contact dependent manner [43]. MICA/B upregulation on DC upon IFN-α exposure seems to be mediated by DC derived autocrine/paracrine IL-15 [44]. Therefore, cell contact might contribute under certain inflammatory conditions to NK activation by human DC.

5 DC/NK Interactions During Infections

It is conceivable that DC/NK cell interactions may occur primarily during infections. Therefore, it was important to analyze the effect of live bacteria on the cross-talk between DC and NK cells. Two different models of bacterial infection have been analyzed [35]. One represented by the extracellular bacteria *Escherichia coli*, the other by the intracellular mycobacterium BCG, capable of efficiently infecting DC [45, 46]. In both systems, bacterial infection of DC led to a particularly rapid NK cell activation.

Since mature DC are capable of inducing proliferation of autologous NK cells [26], it was then further investigated whether the infection of DC with BCG had any effect on this capability. BCG was employed as infective agent because of its ability to efficiently infect DC without undergoing substantial proliferation (thus, not interfering with ^{3}H-thymidine incorporation assays).

In this study, it was confirmed that DC, derived from monocytes in the presence of GM-CSF and IL-4, were able to induce NK cell proliferation. It is of note that both the NK cell proliferation and the number of viable NK cells recovered after 5 days of culture were significantly increased in the presence of BCG. Control experiments with culture containing NK cells, and BCG alone, did not lead to NK cell proliferation.

The observed NK cell proliferation is likely to be sustained by lymphokines such as IL-2 and IL-15 [31, 47, 48]. Although these cytokines were detected at extremely low levels in the supernatants derived from DC cultured in the presence of bacteria, it is not possible to exclude their role in DC-mediated NK cell expansion. Indeed, a more recent study indicates a relevant role of the membrane-bound form of IL-15 on human DC stimulated by different inflammatory stimuli, including LPS, in DC-dependent NK cell proliferation. Notably, in this experimental model, $CD56^{bright}$ lymph node NK cells were preferentially expanded during DC/NK coculture.

Previous studies have shown that, following stimulation with LPS, DC induce the expression, in NK cells, of the early activation marker CD69 [23, 49]. Accordingly, DC that had been exposed to living bacteria could also induce activation markers on NK cells. Both CD69 and HLA-DR were also upregulated in the presence of bacteria. It is of note that whereas CD69 and HLA-DR can be detected after interaction of NK cells with noninfected immature DC, DC infection resulted in a greater increase of the expression of both molecules on NK cell surface. In addition, a de novo expression of CD25 could be detected. This de novo expression may be functionally relevant since the expression of CD25, a component of the IL-2 receptor complex, renders NK cells highly responsive to IL-2. Neither BCG nor *E. coli* could induce the expression of these activation markers in the absence of DC.

Since the NKp30 receptor is primarily involved in DC recognition and lysis by NK cells [26], it was also analyzed whether mAb-mediated masking of NKp30 could interfere with the NK cell activation induced by DC and bacteria. The addition of anti-NKp30 mAb did not modify the expression of CD69, HLA-DR and CD25. These data clearly indicate that the induction of an activated phenotype in NK cells is not mediated via NKp30. Therefore, this triggering receptor, which plays a major role in the recognition and lysis of DC, does not appear to play any substantial role in the activating signal delivered by DC to NK cells.

6 Dendritic Cell Editing by Activated NK Cells

Activated NK cells can lyse autologous iDC while they are less effective against mature DC [26, 50]. In addition, a short-term coculture of resting NK cells with DC that had been pulsed with LPS or heat-killed *Mycobacterium tuberculosis*, has been reported to result in increase of NK-mediated cytotoxicity against the Daudi target cell line [23]. It was further investigated whether NK cells cocultured with DC that had been infected with live bacteria could lyse autologous iDC. After 24 h of coculture, only NK cells cultured in the presence of infected DC could lyse autologous iDC. After this time interval, uninfected iDC failed to induce NK cell cytotoxicity.

Remarkably, infected DC were less susceptible to the lysis, as compared to iDC cultured in the absence of bacteria. These results were obtained after as few as 24 h of NK/DC coculture.

In the same set of experiments, it was also analyzed whether polyclonal NK cell lines cultured for over 1 week in IL-2 could discriminate between infected and noninfected autologous DC. IL-2-cultured NK cells lysed uninfected iDC very efficiently, but were less effective against infected DC. Therefore, the resistance of infected DC to NK-mediated lysis does not depend upon the degree of NK cell activation, but rather reflects an intrinsic property of infected DC themselves.

In this context, the arrival of NK cells to inflamed tissues and their encounter with iDC may appear paradoxical, as it would lead to depletion of APC. Nevertheless, DC exposed to bacteria become highly resistant to NK cell-mediated lysis. Therefore, while they can positively influence NK cells, they are not damaged by NK cells themselves and are allowed to migrate to secondary lymphoid organs.

In turn, NK cells undergo both activation and proliferation, display a rapid increase in their cytolytic activity and may release large amounts of cytokines, including TNF-α, GM-CSF and IFN-γ [47], which may further amplify the inflammatory response.

The different susceptibility of iDC versus mature DC is primarily related to differences in surface expression of HLA class I molecules [51]. In order to determine whether the early resistance of infected DC to NK-mediated lysis reflected a rapid upregulation of HLA class I molecules, the surface density of HLA class I on DC, cultured alone or in the presence of either BCG or *E. coli*, was comparatively analyzed on a quantitative basis. Cell size and the expression of HLA class I molecules were evaluated by flow cytometry. By the simultaneous evaluation of cell size and fluorescence intensity, the number of HLA class I molecules/m^2 of DC cell surface could be calculated. Two days (48 h) after infection, DC increased the number of HLA class I molecules/m^2 approximately tenfold with both BCG and *E. coli*. In view of the high expression of HLA class I in infected DC, it appears conceivable that resistance to NK-mediated lysis could be a distinct consequence of this phenomenon. On the other hand, another possible explanation could be that BCG and *E. coli* could downregulate the expression of the ligands for triggering receptors of NK cells. In order to discriminate between these two possibilities, cytolytic tests in the presence of mAbs (IgM isotype) specific for HLA class I molecules were performed. Upon mAb-mediated masking of HLA class I molecules, a sharp increase of cytolytic activity against infected DC could be detected. This clearly indicates that the resistance of infected DC to NK-mediated lysis is due to increased inhibitory interactions occurring between HLA class I and inhibitory receptors expressed on NK cells. The anti-HLA class I mAb-induced restoration of the cytolytic activity, and the inhibition of this activity by anti-NKp30 mAb, manifestly pointed out that infected DC also express levels of ligands for this triggering NK receptor sufficient to induce NK cell activation.

A relevant question related to the above-presented data is why bacterial infection should lead to a rapid induction of NK cell activation and cytotoxicity. Thus, while NK cell recruitment and activation result in production of cytokines and chemokines which may contribute to the defense against bacterial spreading, the cytolytic activity of NK cells does not exert any direct effect against bacteria. In this regard, a physiopathologic mechanism, in which activated NK cells could play a role in the homeostasis of the immune response during bacterial infections, has been proposed [26, 52]. This model is based on the evidence that activated NK cells are inefficient in killing infected DC (as discussed above) but they can efficiently lyse uninfected DC. Accordingly, the presence of activated NK cells in tissues and lymph nodes may limit an overwhelming recruitment of iDC at a stage in which pathogens have already been eliminated and infection has been controlled. Therefore, the ability of NK cells to discriminate between infected and uninfected DC may suggests that NK cells play an important regulatory role by selectively editing APC during bacterial infection and thus switching off an excessive immune response. This mechanism may be particularly useful in preventing tissue damage.

Exposure of DC to bacteria rapidly induces DC maturation and expression of functionally important surface molecules, including CD80 and CD86 coreceptors, HLA molecules and CCR7 [35]. Thus, DC acquire rapidly the ability to efficiently function as professional APC and to migrate to secondary lymphoid organs, where they can interact with T cells and evoke a prompt adaptive immune response against infecting bacteria. In addition, DC that had encountered bacteria can also rapidly and markedly potentiate an important effector arm of the innate immunity by inducing a rapid activation of NK cells.

A conceivable question may now be how and where DC and NK cells can meet each other during the immune response against pathogens.

7 Sites of NK/DC Interaction During the Immune Response

The recent data on NK cell activation by DC suggest that these APC initiate the early and innate NK activation during the immune response. The question remains as to where this interaction takes place.

No direct evidence exists that DC and NK cells encounter at sites of infection. Nevertheless, microbial invasion causes tissue inflammation and one site of DC/NK interaction is inflamed tissue. NK cells have been found in close contact to DC in lesions of allergen-induced atopic eczema/dermatitis syndrome [53]. Moreover, the chemokine receptor repertoire and the chemokine responsiveness of $CD56^{dim}$ $CD16^+$ blood NK cells suggest that they can home to sites of inflammation efficiently [11]. The cytotoxic blood NK subset migrates efficiently in response to IL-8 and soluble fractalkine and expresses the respective receptors for these chemokines, CXCR1 and CX3CR1. Both chemokines are induced by proinflammatory cytokines like IL-1 and TNF-α [54, 55]. Fractalkine mediates adhesion to endothelia and emigration of NK cells from the blood stream, while IL-8 mediates further migration to the site of inflammation [56]. Therefore, cytotoxic $CD56^{dim}CD16^+$ blood NK cells are able to home to inflamed tissues where they can encounter DC, which are resident in peripheral tissue sites.

The DC/NK encounter at sites of inflammation can either result in DC maturation by modest NK infiltration or in immature DC lysis due to large NK cell infiltrates [49]. The maturation of DC upon NK encounter has been largely attributed to TNF-α secretion by NK cells [23, 49]. Both NK effector mechanisms will deplete the inflamed tissue of DC either by maturation-induced migration or by killing. This will also deprive DC-trophic bacteria of their host cells at the site of infection.

8 Dendritic Cells Induce Maturation of NK Cells

Until recently, only limited information had been available on NK cells located in lymphoid tissues, and therefore NK cells have mainly been considered as effector cells harbored in the blood stream and able to promptly extravasate to inflamed tissues. The evidence that a large amount of NK cells is located in uninflamed lymph nodes [20] suggests secondary lymphoid organs as important sites of NK cell activation.

Indeed, $CD56^{bright}CD16^-$ NK cells isolated from uninflamed human lymph nodes become strongly cytolytic upon stimulation with IL-2.

The de novo acquired cytotoxic properties were accompanied by the expression of both activating and inhibitory receptors. In addition, perforin-negative NK cells located in secondary lymphoid organs might play different roles prior to maturation into cytolytic effectors, such as secretion of critical immunoregulatory cytokines upon activation. In this regard, it has been demonstrated that peripheral blood $CD56^{bright}CD16^-$ NK cells produce significantly higher levels of cytokines than their $CD56^{dim}CD16^+$ counterpart [47, 49]. Similarly, the $CD56^{bright}CD16^-$ NK subset located in secondary lymphoid organs produces relevant cytokines prior to maturation into cytolytic effector cells [20].

Interestingly, several reports have recently shown that DC elicit IFN-γ secretion by autologous NK cells [23–26, 49]. Specifically, NK cells from secondary lymphoid organs are particularly effective in this function and when they are cocultured with autologous DC stimulated by bacterial products, they can produce IFN-γ

within 6 h. This is of interest since IFN-γ represents a main cytokine for phagocyte activation. Indeed, CD56^{bright}CD16$^-$ NK cells produce significantly higher levels of IFN-γ, TNF-β, GM-CSF, IL-10 and IL-13 protein in response to monokines produced by DC, such IL-12, IL-15, IL-18 and IL-1β [13]. On the basis of these findings, NK cells in secondary lymphoid organs should not be referred to as merely "immature" NK cells but rather as effector cells whose functional plasticity enables them to accomplish different sequential tasks during immune responses.

Since DC mature and migrate to secondary lymphoid tissues following an encounter with bacteria, they might encounter CD56^{bright}CD16$^-$ NK cells there in the very early phase of an immune response prior to T cell activation. As discussed in detail below, this could result in local cytokine release by NK cells, which might be able to shape the following adaptive immune response and probably also APC functions [57].

9 Role of NK Cells in DC-Mediated T Cell Polarization

We discussed above that a recent "in vitro" model proposed that NK cells might play an important regulatory role by selectively editing APC during the course of immune responses. NK-mediated lysis of immature, but not mature DC might select an immunogenic DC population during the initiation of immune responses.

In addition to removal of nonimmunogenic DC, NK cells also secrete IFN-γ upon encounter with DC [26]. As a consequence, subsequent T cell polarization may be influenced. Indeed, in vitro studies in both mouse and human systems have demonstrated the importance of IFN-γ in the polarization of type 1 immune response. Interestingly, in a murine model of skin graft rejection, the recognition of donor DC by NK cells led to modulation of Th1/Th2 cell development. Namely, turning host NK cells off was sufficient to skew the alloresponse to Th2 [58].

Thus, as already mentioned, the encounter of mature DC with perforin-negative NK cells located in secondary lymphoid organs should lead to a critical immunoregulatory role [59], as DC can induce NK cells of human secondary lymphoid organs to secrete IFN-γ. Moreover, DC selectively stimulate the CD56^{bright}CD16$^-$ cell subset to produce IFN-γ and the production is extremely rapid and fully dependent on IL-12 released by mature DC. In vivo, murine lymph node NK cells, activated by LPS-stimulated autologous DC migrating into the lymph node, secrete IFN-γ with a peak of cytokine release after 48 h. NK cell depletion and reconstitution experiments show that NK cells provide an early source of IFN-γ that is absolutely required for Th1 polarization [60]. Therefore, in a model of DC stimulation by bacterial products, NK cells play a crucial regulatory role during DC-dependent T cell priming and subsequent polarization in the T cell areas of secondary lymphoid organs.

In conclusion, early activation of NK cells by DC activated by bacteria may play an immunoregulatory role in shaping the emerging adaptive immune responses.

The ability to edit APC as well as secrete immunomodulatory cytokines might result in increased and predominantly Th1-polarized immune responses.

10   DC Activation by NK Cells

Recent studies have demonstrated that during the innate phase of the immune response DC maturation can be also mediated by NK cells [23, 24, 49]. This might be particularly relevant for evoking an adaptive immune response against cancer cells, since the absence of pathogen related molecules and of inflammation, at least in the early phases, does not lead to DC maturation and, as a result, to an effective tumor antigen presentation. Similarly, NK cell-mediated DC maturation should be crucial in infections caused by viruses unable to trigger DC maturation [61–63].

Recent studies shed light on the molecular mechanisms that regulate this specific part of the NK cell/DC cross-talk. It has been found that at low NK cell:DC ratio (1:5) NK cell/DC interaction induces cytokine production (especially TNF-α and IL-12) by DC, and this stimulating effect may depend on cell-to-cell contact as well as TNF-α released by NK cells [49]. The physical interaction between these cells might allow the engagement of NCR that recognize ligands on the DC surface. It has been recently reported that NK-mediated DC maturation depends on the triggering of NKp30 on NK cells which in turn secrete TNF-α and IFN-γ [64]. When they used neutralizing monoclonal antibodies against TNF-α, IFN-γ and NKp30, they could not detect DC maturation, demonstrating that these factors may represent different players of the same event.

11   Conclusions

In summary, immunomodulatory as well as cytotoxic NK cells can be activated via DC early during the immune responses. Possible interaction sites for this encounter are sites of inflammation and secondary lymphoid tissues. DC-activated NK cells might then exert effector functions against infected cells, but mainly influence the emerging adaptive immune response via DC editing and/or activation as well as releasing immunomodulatory cytokines.

The complex cross-talk occurring between these two major players of the innate immunity provides a novel mechanism by which NK cells could cooperate in the defense against pathogens and cancer cells.

References

1. Banchereau J, Steinman RM (1998) Dendritic cells and the control of immunity. Nature 392:245–252
2. Thery C, Amigorena S (2001) The cell biology of antigen presentation in dendritic cells. Curr Opin Immunol 13:45–51

3. Banchereau J, Briere F, Caux C et al (2000) Immunobiology of dendritic cells. Annu Rev Immunol 18:767–811
4. Yokoyama WM, Seaman WE (1993) The Ly49 and NKR-P1 gene families encoding lectin-like receptors on natural killer cells: the NK gene complex. Annu Rev Immunol 11:613–635
5. Moretta A, Bottino C, Vitale M et al (1996) Receptors for HLA-class I-molecules in human natural killer cells. Annu Rev Immunol 14:619–648
6. Lanier LL (1998) NK cell receptors. Annu Rev Immunol 16:359–393
7. Long EO (1999) Regulation of immune response through inhibitory receptors. Annu Rev Immunol 17:875–904
8. Lopez-Botet M, Pe´rez Villar M, Carretero M (1997) Structure and function of the CD94 C-type lectin receptor complex involved in the recognition of HLA class I molecules. Immunol Rev 155:165–174
9. Khakoo SI, Rajalingam R, Shum BP et al (2000) Rapid evolution of NK cell receptor systems demonstrated by comparison of chimpanzees and humans. Immunity 12:687–698
10. Robertson MJ, Ritz J (1990) Biology and clinical relevance of human natural killer cells. Blood 76:2421–2438
11. Campbell JJ, Qin S, Unutmaz D, Soler D, Murphy KE, Hodge MR, Wu L, Butcher EC (2001) Unique subpopulations of CD56+ NK and NK-T peripheral blood lymphocytes identified by chemokine receptor expression repertoire. J Immunol 166:6477–6482
12. Jacobs R, Hintzen G, Kemper A, Beul K, Kempf S, Behrens G, Sykora KW, Schmidt RE (2001) CD56bright cells differ in their KIR repertoire and cytotoxic features from CD56dim NK cells. Eur J Immunol 31:3121–3127
13. Cooper MA, Fehniger TA, Turner SC, Chen KS, Ghaheri BA, Ghayur T, Carson WE, Caligiuri MA (2001) Human natural killer cells: a unique innate immunoregulatory role for the CD56bright subset. Blood 97:3146–3151
14. Moretta A, Bottino C, Vitale M, Pende D, Cantoni C, Mingari MC, Biassoni R, Moretta L (2001) Activating receptors and coreceptors involved in human natural killer cell-mediated cytolysis. Annu Rev Immunol 19:197–223
15. Moretta L, Bottino C, Pende D, Mingari MC, Biassoni R, Moretta A (2002) Human natural killer cells: their origin, receptors and function. Eur J Immunol 32:1205–1211
16. Moretta L, Ferlazzo G, Mingari MC, Melioli G, Moretta A (2003) Human natural killer cell function and their interactions with dendritic cells. Vaccine 21(Suppl 2):S38
17. Bauer S, Groh V, Wu J, Steinle A, Phillips JH, Lanier LL, Spies T (1999) Activation of NK cells and T cells by NKG2D, a receptor for stress- inducible MICA. Science 285:727–729
18. Cosman D, Mullberg J, Sutherland CL, Chin W, Armitage R, Fanslow W, Kubin M, Chalupny NJ (2001) ULBPs, novel MHC class I-related molecules, bind to CMV glycoprotein UL16 and stimulate NK cytotoxicity through the NKG2D receptor. Immunity 14:123–133
19. Fehniger TA, Cooper MA, Nuovo GJ, Cella M, Facchetti F, Colonna M, Caligiuri MA (2003) CD56bright natural killer cells are present in human lymph nodes and are activated by T cell-derived IL-2: a potential new link between adaptive and innate immunity. Blood 101: 3052–3057
20. Ferlazzo G, Lin SL, Goodman K, Thomas D, Morandi B, Muller WA, Moretta A, Mu¨nz C (2004) The abundant NK cells in human lymphoid tissues require activation to become cytolytic. J Immunol 172:1455–1462
21. Westermann J, Pabst R (1992) Distribution of lymphocyte subsets and natural killer cells in the human body. Clin Investig 70:539–544
22. Trepel F (1974) Number and distribution of lymphocytes in man. A critical analysis. Klin Wochenschr 52:511–515
23. Gerosa F, Baldani-Guerra B, Nisii C, Marchesini V, Carra G, Trinchieri G (2002) Reciprocal activating interaction between natural killer cells and dendritic cells. J Exp Med 195:327–333
24. Gerosa F, Gobbi A, Zorzi P, Burg S, Briere F, Carra G, Trinchieri G (2005) The reciprocal interaction of NK cells with plasmacytoid or myeloid dendritic cells profoundly affects innate resistance functions. J Immunol 174:727–734

25. Fernandez NC, Lozier A, Flament C, Ricciardi-Castagnoli P, Bellet D, Suter M, Perricaudet M, Tursz T, Maraskovsky E, Zitvogel L (1999) Dendritic cells directly trigger NK cell functions: cross-talk relevant in innate anti-tumor immune responses in vivo. Nat Med 5:405–411
26. Ferlazzo G, Tsang ML, Moretta L, Melioli G, Steinman RM, Munz C (2002) Human dendritic cells activate resting natural killer (NK) cells and are recognized via the NKp30 receptor by activated NK cells. J Exp Med 195:343–351
27. Andrews DM, Scalzo AA, Yokoyama WM, Smyth MJ, Degli-Esposti MA (2003) Functional interactions between dendritic cells and NK cells during viral infection. Nat Immunol 4: 175–181
28. Dalod M, Hamilton T, Salomon R, Salazar-Mather TP, Henry SC, Hamilton JD, Biron CA (2003) Dendritic cell responses to early murine cytomegalovirus infection: subset functional specialization and differential regulation by interferon alpha/beta. J Exp Med 197:885–898
29. Orange JS, Biron CA (1996) An absolute and restricted requirement for IL-12 in natural killer cell IFN-g production and antiviral defense. Studies of natural killer and T cell responses in contrasting viral infections. J Immunol 156:1138–1142
30. Nguyen KB, Salazar-Mather TP, Dalod MY, Van Deusen JB, Wei XQ, Liew FY, Caligiuri MA, Durbin JE, Biron CA (2002) Coordinated and distinct roles for IFN-alpha beta, IL-12, and IL-15 regulation of NK cell responses to viral infection. J Immunol 169:4279–4287
31. Granucci F, Vizzardelli C, Pavelka N, Feau S, Persico M, Virzi E, Rescigno M, Moro G, Ricciardi-Castagnoli P (2001) Inducible IL-2 production by dendritic cells revealed by global gene expression analysis. Nat Immunol 2:882–888
32. Granucci F, Feau S, Angeli V, Trottein F, Ricciardi-Castagnoli P (2003) Early IL-2 production by mouse dendritic cells is the result of microbial-induced priming. J Immunol 170:5075–5081
33. Nishioka Y, Nishimura N, Suzuki Y, Sone S (2001) Human monocyte-derived and CD83+ blood dendritic cells enhance NK cell-mediated cytotoxicity. Eur J Immunol 31:2633–2641
34. Yu Y, Hagihara M, Ando K, Gansuvd B, Matsuzawa H, Tsuchiya T, Ueda Y, Inoue H, Hotta T, Kato S (2001) Enhancement of human cord blood CD34þ cell-derived NK cell cytotoxicity by dendritic cells. J Immunol 166:1590–1600
35. Ferlazzo G, Morandi B, D'Agostino A, Meazza R, Melioli G, Moretta A, Moretta L (2003) The interaction between NK cells and dendritic cells in bacterial infections results in rapid induction of NK cell activation and in the lysis of uninfected dendritic cells. Eur J Immunol 33:306–313
36. Munz C, Dao T, Ferlazzo G, de Cos MA, Goodman K, Young JW (2005) Mature myeloid dendritic cell subsets have distinct roles for activation and viability of circulating human natural killer cells. Blood 105:266–273
37. Siegal FP, Kadowaki N, Shodell M, Fitzgerald-Bocarsly PA, Shah K, Ho S, Antonenko S, Liu YJ (1999) The nature of the principal type 1 interferon-producing cells in human blood. Science 284:1835–1837
38. Bandyopadhyay S, Perussia B, Trinchieri G, Miller DS, Starr SE (1986) Requirement for HLA-DR+ accessory cells in natural killing of cytomegalovirus-infected fibroblasts. J Exp Med 164:180–195
39. Dalod M, Salazar-Mather TP, Malmgaard L, Lewis C, Asselin-Paturel C, Briere F, Trinchieri G, Biron CA (2002) Interferon α/β and interleukin 12 responses to viral infections: pathways regulating dendritic cell cytokine expression in vivo. J Exp Med 195:517–528
40. Feldman M, Howell D, Fitzgerald-Bocarsly P (1992) Interferon α- dependent and -independent participation of accessory cells in natural killer cell-mediated lysis of HSV-1-infected fibroblasts. J Leukoc Biol 52:473–482
41. Diebold SS, Montoya M, Unger H, Alexopoulou L, Roy P, Haswell LE, Al-Shamkhani A, Flavell R, Borrow P, Reis eSousa C (2003) Viral infection switches non-plasmacytoid dendritic cells into high interferon producers. Nature 424:324–8
42. Granucci F, Zanoni I, Pavelka N, Van Dommelen SL, Andoniou CE, Belardelli F, Degli Esposti MA, Ricciardi-Castagnoli P (2004) A contribution of mouse dendritic cell-derived IL-2 for NK cell activation. J Exp Med 200:287–295

43. Jinushi M, Takehara T, Kanto T, Tatsumi T, Groh V, Spies T, Miyagi T, Suzuki T, Sasaki Y, Hayashi N (2003) Critical role of MHC class I-related chain A and B expression on IFN-a-stimulated dendritic cells in NK cell activation: impairment in chronic hepatitis C virus infection. J Immunol 170:1249–1256
44. Jinushi M, Takehara T, Tatsumi T, Kanto T, Groh V, Spies T, Suzuki T, Miyagi T, Hayashi N (2003) Autocrine/paracrine IL-15 that is required for type I IFN-mediated dendritic cell expression of MHC class I-related chain A and B is impaired in hepatitis C virus infection. J Immunol 171:5423–5429
45. Inaba K, Inaba M, Naito M, Steinman RM (1993) Dendritic cell progenitors phagocytose particulates, including bacillus Calmette-Guerin organisms, and sensitize mice to mycobacterial antigens in vivo. J Exp Med 178:479–488
46. Demangel C, Bean AG, Martin E, Feng CG, Kamath AT, Britton WJ (1999) Protection against aerosol Mycobacterium tuberculosis infection using Mycobacterium bovis Bacillus Calmette Guerin-infected dendritic cells. Eur J Immunol 29:1972–1979
47. Cooper MA, Fehniger TA, Caligiuri MA (2001) The biology of human natural killer-cell subsets. Trends Immunol 22:633–640
48. Mattei F, Schiavoni G, Belardelli F, Tough DF (2001) IL-15 is expressed by dendritic cells in response to type I IFN, double-stranded RNA, or lipopolysaccharide and promotes dendritic cell activation. J Immunol 167:1179–1187
49. Piccioli D, Sbrana S, Melandri E, Valiante NM (2002) Contact-dependent stimulation and inhibition of dendritic cells by natural killer cells. J Exp Med 195:335–341
50. Wilson JL, Heffler LC, Charo J, Scheynius A, Bejarano MT, Ljunggren HG (1999) Targeting of human dendritic cells by autologous NK cells. J Immunol 163:6365–6370
51. Ferlazzo G, Semino C, Melioli G (2001) HLA class I molecule expression is up-regulated during maturation of dendritic cells, protecting them from natural killer cell-mediated lysis. Immunol Lett 76:37–41
52. Moretta A (2002) Natural killer and dendritic cells: rendezvous in abused tissues. Nat Rev Immunol 2:1–8
53. Buentke E, Heffler LC, Wilson JL, Wallin RP, Lofman C, Chambers BJ, Ljunggren HG, Scheynius A (2002) Natural killer and dendritic cell contact in lesional atopic dermatitis skin-Malassezia-influenced cell interaction. J Invest Dermatol 119:850–857
54. Bazan JF, Bacon KB, Hardiman G, Wang W, Soo K, Rossi D, Greaves DR, Zlotnik A, Schall TJ (1997) A new class of membrane-bound chemokine with a CX3C motif. Nature 385:640–644
55. Hoffmann E, Dittrich-Breiholz O, Holtmann H, Kracht M (2002) Multiple control of interleukin-8 gene expression. J Leukoc Biol 72:847–855
56. Nishimura M, Umehara H, Nakayama T, Yoneda O, Hieshima K, Kakizaki M, Dohmae N, Yoshie O, Imai T (2002) Dual functions of fractalkine/CX3C ligand 1 in trafficking of perforin +/granzyme B+ cytotoxic effector lymphocytes that are defined by CX3CR1 expression. J Immunol 168:6173–6180
57. Dowdell KC, Cua DJ, Kirkman E, Stohlman SA (2003) NK cells regulate CD4 responses prior to antigen encounter. J Immunol 171:234–239
58. Coudert JD, Coureau C, Guery JC (2002) Preventing NK cell activation by donor dendritic cells enhances allospecific CD4 T cell priming and promotes Th type 2 responses to transplantation antigens. J Immunol 169:2979–2987
59. Ferlazzo G, Munz C (2004) NK cell compartments and their activation by dendritic cells. J Immunol 172:1333–1339
60. Martin-Fontecha A, Thomsen LL, Brett S, Gerard C, Lipp M, Lanzavecchia A, Sallusto F (2004) Induced recruitment of NK cells to lymph nodes provides IFN-gamma for T(H)1 priming. Nat Immunol 5:1260–1265
61. Redpath S, Angulo A, Gascoigne NR, Ghazal P (1999) Murine cytomegalovirus infection down-regulates MHC class II expression on macrophages by induction of IL-10. J Immunol 162:6701–6707

62. Basta S, Bennink J (2003) A survival game of hide and seek: cytomegaloviruses and MHC class I antigen presentation pathways. Viral Immunol 16:231–242
63. Granelli-Piperno A, Golebiowska A, Trumpfheller C, Siegal FP, Steinman RM (2004) HIV-1-infected monocyte-derived dendritic cells do not undergo maturation but can elicit IL-10 production and T cell regulation. Proc Natl Acad Sci USA 101:7669–7674
64. Vitale M, Della Chiesa M, Carlomagno S, Pende D, Arico M, Moretta L, Moretta A (2005) NK-dependent DC maturation is mediated by TNFalpha and IFNgamma released upon engagement of the NKp30 triggering receptor. Blood 106(2):566–571

Modulation of T Cell-Mediated Immune Responses by Natural Killer Cells

Alessandra Zingoni, Cristina Cerboni, Michele Ardolino, and Angela Santoni

Abstract Natural Killer (NK) cells are lymphoid cells that participate in innate immunity and in early defense against infections and tumors. NK cells express cell surface activating and inhibitory receptors that allow them to recognize and kill infected or tumor cells and rapidly produce cytokines and chemokines. Increasing evidence asserts an important immunomodulatory role of NK cells. We will discuss different mechanisms used by NK cells to regulate T cell-mediated immune responses.

1 Introduction

NK cells are an important component of innate immunity. They represent a highly specialized subpopulation of lymphocytes that mediate cellular cytotoxicity, and produce chemokines and cytokines such as interferon-γ (IFN-γ) and tumor necrosis factor-α (TNF-α) in response to target cell recognition or stimulation by proinflammatory cytokines or interferons. NK cells distinguish between normal and abnormal cells (such as virus-infected or tumor cells) by using a repertoire of cell surface receptors that control their activation, proliferation, and effector functions [1]. Different sophisticated recognition strategies are used by NK cells. These include direct recognition of pathogen-derived molecules mediated by activating or Toll-like receptors; recognition of self-proteins whose expression is upregulated on abnormal cells, and which is mediated by the interaction with activating receptors, such as NKG2D; recognition of MHC class I molecules whose levels are frequently downmodulated on transformed or virus-infected cells, mediated by inhibitory receptors. The amount of activating and inhibitory receptors on NK cells

A. Zingoni (✉), C. Cerboni, M. Ardolino and A. Santoni
Department of Experimental Medicine, University "Sapienza", Viale Regina Elena 324, 00161, Rome, Italy
e-mail: alessandra.zingoni@uniroma1.it

together with the amount of ligands on target cells determines the outcome of the NK cell response [1].

NK cells constitute a heterogeneous population with distinct functional subsets whose proportions differ according to their localization. NK cells are present in lymphoid organs (such as lymph nodes, bone marrow, spleen, and thymus), in peripheral blood, and in nonlymphoid organs including liver, lung, intestine, and uterus [2]. Therefore, there are substantial functional differences and distinct anatomical sites for NK cells. Moreover, NK cells are known to interact with various cellular components of the immune system, including dendritic cells (DC), macrophages, B cells, and $CD4^+$ and $CD8^+$ T lymphocytes. Thus, NK cells have the potential to function as regulatory cells. In fact, increasing evidence suggests that NK cells, in addition to their role as effectors in the innate immunity, can strongly contribute to the development and regulation of adaptive immune responses. NK cell modulation of adaptive immunity strongly depends on the specific subset involved as well as on the site where the interaction occurs. Importantly, both in vivo and in vitro studies suggest that NK cells can either promote or restrain T cell-mediated immune responses, depending on the physiological and/or pathological situation.

Here, we will discuss some of the current knowledge about the different mechanisms used by NK cells to modulate T cell-mediated immune responses.

2 NK Cell Regulation of T Cell-Mediated Responses Through the Production of Cytokines and Chemokines

NK cells can contribute to maintain the homeostasis of the immune system and to regulate adaptive immune responses through their production of cytokines and chemokines. NK cells represent the major early source of IFN-γ after viral [3], bacterial [4, 5], and protozoal [6, 7] infection in both conventional and T cell-deficient mice. Importantly, several in vivo studies have shown that NK cell-derived IFN-γ is critical for T helper 1 (Th1) polarization. In particular, NK cells are a crucial component in the resistance to the protozoan parasite *Leishmania major* by promoting an IFN-γ-dependent Th1 response [4]. A similar immunoregulatory role has been demonstrated in the response to *Chlamydia trachomatis* [5], *Listeria monocytogenes* [6], and *Toxoplasma gondii* [7]. In all these models, the pathogens are thought to activate NK cell cytokine production, especially IFN-γ, indirectly through their induction of IL-12 production by macrophages or DC. In addition, recent evidence has shown that migrating mature DC and some adjuvants can promote the recruitment to the draining lymph nodes of NK cells that provide an early source of IFN-γ necessary for Th1 polarization [8]. Moreover, using real-time in vivo imaging, NK cells have been found to be a functionally important component of the lymph node compartment [9, 10]. Indeed, the study of Bajenoff et al. [9] revealed that NK cells localize near the DC and interact with them in resting lymph nodes, while, after *Leishmania major* infection, abundant numbers of NK cells are recruited into the draining lymph nodes to produce cytokines. Importantly, a similar effect has been observed also in humans, where NK-cell-derived IFN-γ was found to enhance the expansion of Th1 cells [11].

These data denote that NK–DC–T cell interactions are likely to occur within the lymph node environment, thus affecting the nature of antigen (Ag)-specific immune responses.

However, since NK cells are not homogeneous, the outcome of an immune response can be profoundly influenced also by the type of the NK cell subset involved. Human NK cells are characterized phenotypically by the presence of CD56 and the lack of CD3. The NK cell subset primarily located in the parafollicular T cell- and Antigen Presenting Cell (APC)-rich region of the secondary lymphoid tissue is characterized by a $CD56^{high}CD16^-$ surface phenotype and can rapidly produce substantial amounts of cytokines and chemokines upon activation, but displays poor cytotoxic capacity [2]. By contrast, the majority of circulating human NK cells (about 90%) with low surface density of CD56 and high levels of CD16 ($CD56^{dim}CD16^{bright}$) have a lower ability to produce cytokines in response to activation but are highly cytotoxic [2]. Notably, the MHC class I allele-specific killer immunoglobulin (Ig)-like receptors (KIR) are expressed on a considerable fraction of $CD56^{dim}CD16^+$ NK cells, whereas the $CD56^{bright}CD16^-$ NK subset lacks KIR [12]. Interestingly, only $CD56^{high}$ NK cells express secondary lymphoid organ homing markers such as CCR7, CD62L, and CXCR3 [13]. Recent studies have shown that the two human NK-cell subsets represent different stages of sequential maturation, with $CD56^{dim}$ NK cells deriving from $CD56^{high}$ NK cells [14, 15].

Although CD56 is not expressed in rodents, subpopulations of NK cells similar to the two main human NK cell subsets have been also described in mice. Mouse $CD11b^{high}$ NK cells include $CD27^{high}$ and $CD27^{low}$ subsets that differ in terms of expression of NK cell inhibitory receptors, chemokine receptors and are functionally different: $CD27^{low}$ cells are mostly found in nonlymphoid organs (blood, liver, and lungs) and their cytotoxic and cytokine production activities are more tightly regulated [16, 17]. The $CD27^{high}$ cells have many features similar to that of $CD56^{high}$ cells. They predominate in the lymph nodes and are responsive to IL-12 and IL-18, but unlike $CD56^{high}$ cells, they are also highly cytotoxic.

Distinct subsets of human NK cells can also produce type 1 (IFN-γ and TNF-α) and type 2 (IL-5, IL-13, IL-10) cytokines [18, 19]. Expansion of type 2 cytokine-producing NK cells is also observed in IFN-γ knockout mice, showing that NK cells could functionally polarize in vivo [20]. Interestingly, it has been found that remission of multiple sclerosis is associated with the inhibition of autoimmune Th1 cells by NK-cell-derived IL-5 [21]. A role for NK-cell-produced IL-5 in the recruitment of eosinophils into the lungs has also been demonstrated in a mouse model of allergic inflammation [22]. In addition, NK cells have been described to regulate $CD4^+$ T cell responses before Ag presentation through their IL-10 production [23]. Beside IL-10, NK cells are also a source of transforming growth factor-β (TGF-β) [24]. Recently, both in humans and mice, an NK cell subset has been characterized that is located in the mucosa-associated lymphoid tissues which produces IL-22 providing an innate source of this cytokine that may protect mucosal sites during inflammation [25].

NK cells can affect not only the nature of T cell-mediated immune response but also the recruitment of specific T cell subsets, by modulating the release and/or releasing different chemokines. During murine cytomegalovirus (MCMV) infection, the production of IFN-γ by NK cells in the liver is essential for the local release of the chemokine MIG which attracts activated T cells to the site of infection [26]. More recently, it has been shown that during viral (MCMV) as well as bacterial (*Listeria monocytogenes*) infections, NK cells coordinately produce IFN-γ, and a number of chemokines (MIP-1α, MIP-1β, RANTES, lymphotactin) that are important to recruit and activate other inflammatory cells [27, 28].

3 NK Cell–DC Crosstalk

Increasing evidence shows that NK cells and DC can be reciprocally activated during an immune response [29] (Fig. 1). NK cells are directly involved in DC maturation, which is a crucial step for the induction of adaptive immune responses.

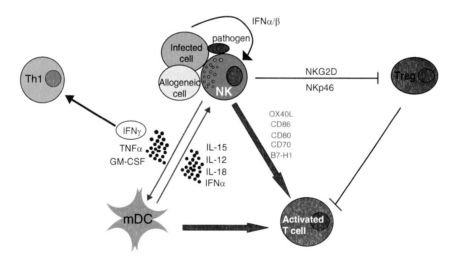

Fig. 1 NK cell promotion of T cell immune responses. NK cells are activated by direct recognition of abnormal cells, by pathogen-encoded molecules or by IFN-α/β produced from virally infected cells. Activated NK cells can promote T cell responses by inducing maturation of DC through the secretion of several cytokines (GM-CSF, TNF-α, and IFN-γ). Also, NK-cell-derived IFN-γ can contribute to Th1 polarization. In turn, mature DC produce different cytokines (IFNα/β, IL-12, IL-18, IL-15) capable of enhancing NK cell effector functions. In addition, activated NK cells expressing the ligands for several costimulatory receptors can directly costimulate antigen-induced T cell proliferation and effector functions. Moreover, NK cells can promote T cell immune responses through the direct killing of Treg cells which is mediated by NKG2D- and NKp46-activating receptors

Several in vitro experiments have demonstrated that IL-2-activated human NK cells can induce DC maturation by a process dependent on cell-to-cell contact and TNF-α [30–32]. Further, the interaction of immature DC with IL-2-activated NK cells can result in either maturation or cell death [30–32], depending on the NK-DC ratios used in the coculture system [32], suggesting that NK cells can eliminate immature DC from inflamed tissues. Studies aimed at identifying the receptors involved in the NK-cell-mediated cytotoxicity of both immature and mature DC indicate an important role of NKp30 and DNAM-1/nectin-2 receptor pairs [33]. Moreover, also the ligands for other NK cell receptors can be induced on DC such as MICA/B molecules [34, 35] and the lectin-like transcript-1 (LLT1) [36], which bind to NKG2D and NKRP1A, respectively, thus suggesting that several receptor/ligand interactions regulate the crosstalk between DC and NK cells.

The in vivo relevance of DC activation by NK cells has been described in several murine models. Interestingly, during MCMV infection, functional interactions between murine NK cells and $CD8\alpha^+$ DC occur [37]. In addition, NK cells activated in vivo by tumor cells expressing very low levels of MHC class I molecules, induce a strong and protective $CD8^+$ T cell response in an IFN-γ-dependent manner [38], by priming DC to produce IL-12. Interestingly, it has been shown that the interaction between DC and NK cells can completely replace $CD4^+$ T cell help in the induction of an antitumor $CD8^+$ T cell response, providing a novel alternative pathway for CTL induction [39]. In a murine model of skin graft rejection, the recognition of donor DC by host NK cells strongly affects alloreactive T cell polarization by inhibiting Th2 development [40, 41].

In turn, mature DC also stimulate NK cell effector functions and proliferation, through cell-to-cell contact and the production of several cytokines including IL-12, IL-15, IL-18, and IFN-α/β [37, 42–46]. In particular, it has been observed that IL-12 is required for DC to promptly stimulate IFN-γ production by $CD56^{high}CD16^-$ NK cells, whereas IL-15 is essential for the induction of NK cell proliferation [46].

Several studies show that the interaction of NK cells with the different DC populations might take place both in peripheral inflamed tissues [37, 47] and in secondary lymphoid organs [8–10, 41, 46, 48].

4 Promotion of T Cell-Mediated Immune Responses

Much of the findings available in the literature show that NK cells can shape T cell-mediated adaptive immune responses by indirect mechanisms, involving the secretion of cytokines or chemokines or the activation of DC. However, more recent studies suggest the possibility that activated NK cells might also communicate directly with T cells by a process involving cognate cell-to-cell interactions. Indeed, in vitro activated human NK cells express several ligands for T cell costimulatory receptors, such as CD86, CD80, CD70, and OX40L [49, 50]. Costimulatory receptors can be divided into two main groups: those belonging to the Ig superfamily (such as CD28 which binds to CD80 and CD86) and those belonging to the tumor necrosis factor receptor (TNFR) superfamily (including OX40 (CD134), 4-1BB (CD137), and CD27 which bind to OX40L, 4-1BBL, and CD70, respectively) [51]. Interestingly, it has been shown that the induction of costimulatory ligands on NK cells requires distinct conditions of stimulation. In particular, IL-2 alone, crosslinking of different activating NK cell receptors such as NKp30, NKp46, CD16, or NK cell coculture with susceptible targets are sufficient to induce the expression of CD86 on human NK cells [49, 50]. By contrast, the induction of OX40L requires a stronger stimulation involving both IL-2 or innate cytokines (i.e., IL-12, IL-15) and the ligation of the activating NK receptors (*i.e.*, CD16 or NKG2D) [49]. Upregulation of costimulatory molecules is observed not only in vitro but more importantly in vivo on human NK cells from inflamed tonsils and CMV-infected uterine decidual samples [50]. Our preliminary studies also show that NK cells isolated from the peripheral blood of kidney transplanted patients with a chronic graft rejection are activated and express CD80, CD86 and OX40L molecules (Zingoni A., Cerboni C., and Santoni A., unpublished observations). These molecules are functionally relevant as activated NK cells have been shown to costimulate TCR-induced proliferation and cytokine production of autologous $CD4^+$ T cells, and this process requires OX40-OX40L [49] and CD28-B7 interactions [49, 50].

Moreover, several studies indicate that direct interaction between NK and T cells might also involve other costimulatory receptors pairs. Murine NK cells can efficiently enhance $CD4^+$ as well as $CD8^+$ T cell proliferation in response to CD3 crosslinking and specific antigen through interactions between 2B4 on NK cells and CD48 on T cells [52]. In addition, it has been reported that the induction of B7-H1 molecule after exposure to inflammatory chemokines on both mouse and human NK cells allows them to stimulate T cell proliferation and secretion of IFN-γ and TNF-α [53].

Interestingly, NK cell depletion experiments in mice have demonstrated a role for NK cells in the generation of antigen-specific cytotoxic T cells (CTL) [54, 55]. Furthermore, studies performed in vitro suggest that NK cells are required for the differentiation of fully competent effector CTL in mixed lymphocyte cultures [56]. Although at present the molecules involved in this interaction have not been identified, recent reports demonstrate that the induction of costimulatory molecules on activated NK cells such as CD137 [57] and LIGHT [58] might be a critical signal to support the development of CTL in tumor models.

A functional interaction between NK and T cells able to promote T cell immune response has been described also in vivo. Indeed, an in vivo murine tumor model demonstrates that NK cells can promote the development of an antitumor immune response mediated by memory $CD4^+$ T cells in an NKG2D-dependent manner without the requirement of conventional type-1 cytokines [59]. In addition, two different studies performed in mice indicate that NK cells contribute to the alloresponse against solid organs through functional interactions with T cells [60, 61].

On the other hand, NK cells can also act as potential modulators of regulatory T cells (Treg); increased numbers and activity of Treg are often associated with reduced NK cell activity in several diseases [62, 63]. During the immune response to *Mycobacterium tuberculosis* infection, Treg cells can prevent efficient pathogen clearance in infected mice [64]. Interestingly, a recent study shows that upon *M. tuberculosis* infection, NK cells can inhibit the generation of Treg cells through direct lysis, with a mechanism dependent on NKG2D- and NKp46-activating receptors [65].

Similar to the NK cell–DC interaction, it is likely that NK and T cells might interact both in peripheral tissues and in secondary lymphoid organs depending on the T cell populations (naïve, effector, memory) they are contacting, as it has been shown in cardiac allograft vasculopathy lesions [60, 61] and in the liver of MCMV-infected mice [26].

In summary, during the course of an immune response NK cells can be activated by innate cytokines and/or by the interaction with abnormal cells (virus-infected, tumor, or allogeneic cells). NK cell activation results in the induction of different ligands for T cell costimulatory molecules and promotion of T cell responses. NK cells may costimulate activated T cells by enhancing the initial activation of naïve T cells, providing additional biochemically distinct signals to promote T cell division, survival, or effector functions (Fig. 1). In addition, NK cells can also contribute to the positive regulation of an immune response by direct killing of Treg cells.

5 Downregulation of T Cell-Mediated Immune Responses

The negative regulation of adaptive immunity is relevant to maintain lymphocyte homeostasis and to prevent inappropriate T cell activation that can ultimately result in autoimmune or lymphoproliferative diseases. Different reports suggest that NK cells can contribute to the negative regulation of T cell responses. In vivo depletion studies established that the presence of NK cells in MCMV-infected mice negatively affects $CD4^+$ and $CD8^+$ T cell-dependent IFN-γ production and proliferation [66]. Similarly, depletion of NK cells was shown to enhance the CTL response to MHC class I positive lymphomas [67]. Furthermore, expression of the class Ib MHC molecule Qa-1-Qdm by activated $CD4^+$ T cells is required to prevent lysis by NK cells expressing the inhibitory receptor CD94-NKG2A, and it was essential for T cell expansion and development of immunological memory [68].

Moreover, studies in animal models have suggested a regulatory role of NK cells in the initiation and progress of autoimmune disorders [69, 70]. In an animal model of multiple sclerosis, in vivo depletion of NK cells resulted in a more severe form of experimental autoimmune encephalomyelitis (EAE) [71]. The inhibitory role for NK cells in rodent EAE has been further strengthened by the finding that NK cells inhibit proliferation and cytokine production of T cells specific for the myelin basic protein in vitro [72] by a cell-to-cell contact-dependent mechanism [73]. Likewise, in an in vivo mouse model of colitis, it has been reported that NK cells inhibit $CD4^+$ T effector cells by a mechanism dependent on perforin, suggesting that NK cells can directly lyse T cells or some other intermediate immune cells such as DC [74]. In addition, NK cells have been shown to have a protective role in type I diabetes; treatment with complete Freund's adjuvant (CFA) prevents diabetes in nonobese diabetic (NOD) mice, through the downregulation of self-reactive CTL, with a mechanism dependent on NK cells [75]. A study from Ogasawara et al. further revealed that NK cell functions mediated by the NKG2D-activating receptor were impaired in NOD mice [76].

In accordance with the protective role of NK cells against autoimmune diseases shown in murine studies, decreased NK cell activity and numbers are found in the peripheral blood of patients with multiple sclerosis, rheumatoid arthritis, systemic lupus erythematosus, and type I diabetes [69, 70, 77].

NK cells can attenuate T cell adaptive immune responses by several mechanisms including killing of DC [30–33, 41] and/or of activated T cells [78–80] and secretion of inhibitory cytokines [23, 24].

With regard to the NK-cell-mediated killing of T cells, it has been shown that IL-2-activated mouse NK cells recognize and lyse syngeneic T cell blasts in a perforin-dependent manner through the NK-activating receptor NKG2D [79]. Interestingly, expression of NKG2D ligands (NKG2DLs) such as MICA, ULBP-1, ULBP-2, and ULBP-3 on T cells has been reported also in humans. MICA is induced on the surface of T cells following anti-CD3 stimulation [81, 82] and the presence of ULBP-1,2,3 transcripts on activated $CD8^+$ T cells cultured with IL-7 and IL-15 has been reported [82]. Interestingly, our recent studies indicate that activation of human T cells by alloantigens, superantigens, or a specific antigenic peptide is sufficient to induce surface expression of MICA, MICB, ULBP-1, -2 and -3 on $CD4^+$ and $CD8^+$ T lymphocytes [80, 83]. Similar to the mouse, we have also shown that activated T cells became susceptible to autologous NK lysis *via* NKG2D/NKG2DLs interaction [80].

Previously, expression of NKG2DLs was thought to be mostly restricted to transformed, infected, and/or stressed cells. Nowadays, however, this view is changing, as several studies report that MIC and ULBPs can be expressed also on distinct hematopoietic cells, including bone-marrow cells, mature DC [34, 35], activated macrophages and monocytes [84, 85], antigen-activated T cells [80, 83], expanded Treg cells [65] indicating a more general NKG2D-dependent immunoregulatory role of NK cells.

Induction of NKG2DLs on activated T lymphocytes during the course of an immune response might allow the establishment of a crosstalk with NK cells. This interaction may trigger a perforin-dependent lysis of T cells expressing NKG2DLs and act as a negative regulator of T cell responses. Interestingly, a role for perforin as immune regulator has been demonstrated. Perforin-deficient patients show lymphoproliferative disorders [86] and perforin-mediated killing is involved in downregulating T cell responses in vivo, since perforin-deficient mice show a huge expansion of activated T cells upon lymphocytic choriomeningitis virus (LCMV) infection [87]. Moreover, mice deficient in both Fas and perforin have a dramatic acceleration of the spontaneous lymphoproliferative disease seen in Fas-deficient mice [87, 88].

Thus, the receptor/ligand interactions that trigger a perforin-mediated cytotoxicity play a key role in controlling T cell responses during viral infections and autoimmunity.

Also, during allogeneic hematopoietic cell transplantations, donor T cells in the graft can mediate graft-versus-host disease (GVHD) which is initiated by host DC presenting alloantigens to donor T cells [89]. Strikingly, an in vivo study performed by Ruggeri et al. has shown that GVHD was the cause of death of mice engrafted with bone marrow cells containing nonalloreactive NK cells; inclusion of alloreactive NK cells in the grafts resulted in complete survival. GVHD has been suggested to be prevented through killing of host DC by alloreactive NK cells [90]. Conceivably, these NK cells could kill also activated donor T cells (expressing NKG2DLs), thus providing an additional protective mechanism from the deleterious effects of GVHD.

In summary, NK cells can contribute to the suppression of T cell responses and to the maintenance of T lymphocyte homeostasis through the direct elimination of activated T cells and antigen presenting cells and/or the release of inhibitory cytokines such as TGF-β and IL-10 which can inhibit DC maturation or T cell activation and functions (Fig. 2).

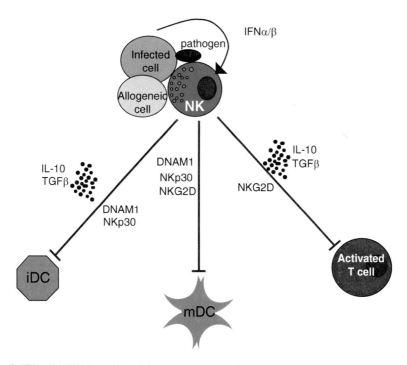

Fig. 2 NK cell inhibition of T cell immune responses. Activated NK cells can downmodulate T cell responses by secretion of TGF-β and IL-10 which can block DC maturation and/or directly inhibit T cell proliferation and effector functions. Activated NK cells kill both immature and mature DC as well as activated T cells through different NK-cell-activating receptors (NKG2D, NKp30, DNAM)

6 Concluding Remarks

Several in vivo and in vitro studies assert an important immunomodulatory role of NK cells. Thus, NK cells can shape adaptive immune responses through multiple mechanisms, but it remains unclear under which conditions NK cells promote or inhibit immune responses.

Promotion of T cell responses may occur by secretion of several cytokines and chemokines which can affect T cell polarization and DC maturation and T cell recruitment, respectively. The expression of ligands for T cell costimulatory molecules on activated NK cells reveals new and unexpected insights into the direct interaction between NK and T cells. In addition, the direct killing of Treg cells may contribute to promote the immune response in some pathological conditions. We envisage that NK-cell-mediated enhancement of T cell responses might have important implications in the context of tumor immune surveillance, infectious diseases, and transplantation.

On the other hand, NK cells can also limit T cell immune responses through the elimination of DC and of activated T cells, and/or the secretion of inhibitory cytokines. The negative regulation of adaptive immunity can be relevant in the maintenance of lymphocyte homeostasis subsequent to an immune response, as well as in the context of autoimmunity or GVHD.

Further studies are necessary to elucidate the mechanisms underlying these two opposite functions of NK cells and to validate the relevance of NK-T cell interactions both in physiological and pathological conditions.

References

1. Lanier LL (2005) Annu Rev Immunol 23:225
2. Caligiuri MA (2008) Blood 112:461
3. French A, Yokoyama W (2003) Curr Opinion Immunol 15:45
4. Unanue E (1997) Immunol Rev 158:11
5. Tseng CT, Rank RG (1998) Infect Immunol 66:5867
6. Scharton TM, Scott P (2003) J Exp Med 178:567
7. Gazzinelli RT, Hieny S, Wynn TA (1993) Proc Natl Acad Sci USA 90:6115
8. Martin-Fontecha A, Thomsen LL, Brett S, Gerard C, Lipp M, Lanzavecchia A, Sallusto F (2004) Nat Immunol 5:1260
9. Bajenoff M, Breart B, Huang AYC, Qi H, Cazareth J, Braud VM, Germain RN, Glaichenhaus N (2006) J Exp Med 203:619
10. Garrod KR, Sindy HW, Parker W, Cahalan MD (2007) Proc Natl Acad Sci USA 104:12081
11. Morandi B, Bougras G, Muller WA, Ferlazzo G, Munz C (2006) Eur J Immunol 36:2394
12. Jacobs R, Hintzen G, Kemper A, Beul K, Kempf S, Behrens G, Sykora KW, Schmidt RE (2001) Eur J Immunol 31:3121
13. Strowig T, Brilot F, Munz C (2008) J Immunol 180:7785
14. Romagnani C, Juelke K, Falco M, Morandi B, D'Agostino A, Costa R, Ratto G, Forte G, Carrega P, Lui G, Conte R, Strowig T, Moretta A, Münz C, Thiel A, Moretta L, Ferlazzo G (2007) J Immunol 178:4947

15. Chan A, Hong DL, Atzberger A, Kollnberger S, Filer AD, Buckley CD, McMichael A, Enver T, Bowness P (2007) J Immunol 179:89
16. Hayakawa Y, Huntington ND, Nutt SL, Smyth MJ (2006) Immunol Rev 214:47
17. Hayakawa Y, Smyth MJ (2006) J Immunol 176:517
18. Perussia B, Loza MJ (2003) Trend Immunol 24:235
19. Fehniger TA, Shah MH, Turner MJ, VanDeusen JB, Whitman SP, Cooper MA, Suzuki K, Wechser M, Goodsaid F, Caligiuri MA (1999) J Immunol. 162:4511
20. Hoshino T, Winkler-Pinkett RT, Mason AT, Ortaldo JR, Young HA (1999) J Immunol 162:51
21. Takahashi K, Miyake S, Kondo T, Terao K, Hatakenaka M, Hashimoto S, Yamamura T (2001) J Clin Invest 107:23
22. Walker C, Checkel J, Cammisuli S, Leibson PJ, Gleich GJ (1998) J Immunol 161:1962
23. Dowdell KC, Cua DJ, Kirkman E, Stohlman SA (2003) J Immunol 171:234
24. Li MO, Wan YY, Sanjabi S, Robertson AK, Flavell RA (2006) Annu Rev Immunol 24:4
25. Cella M, Fuchs A, Vermi W, Facchetti F, Otero K, Lennerz JK, Doherty JM, Mills JC, Colonna M (2009) Nature 457:722
26. Salazar-Mather TP, Hamilton TA, Biron CA (2000) J Clin Invest 105:985
27. Dorner BG, Schefflod A, Huser RMS, MB KSH, Radbruch A, Flesch IE, Kroczek RA (2002) Proc Natl Acad Sci USA 9:6181
28. Dorner BG, Smith HRC, French AR, Kim S, Poursine-Laurent J, Beckman DL, Pingel JT, Kroczek RA, Yokoyama WM (2004) J Immunol 172(119):3131
29. Degli-Esposti MA, Smyth M (2005) Nat Rev Immunol 5:112
30. Ferlazzo G, Tsang ML, Moretta L, Melioli G, Steinman RM, Münz C (2002) J Exp Med 195:343
31. Gerosa F, Baldani-Guerra B, Nisii B, Marchesini V, Carra G, Trinchieri G (2002) J Exp Med 195:327
32. Piccioli D, Sbrana S, Melandri E, Valiante NM (2002) J Exp Med 195:335
33. Pende D, Castriconi R, Romagnani P, Spaggiari GM, Marcenaro S, Dondero A, Lazzeri E, Lasagni L, Martini S, Rivera P, Capobianco A, Moretta L, Moretta A, Bottino C (2006) Blood 107:2030
34. Jinushi M, Takehara T, Tatsumi T, Kanto T, Groh V, Spies T, Suzuki T, Miyagi T, Hayashi N (2003) J Immunol 171:5423
35. Jinushi M, Takehara T, Kanto T, Tatsumi T, Groh V, Spies T, Miyagi T, Suzuki T, Sasaki Y, Hayashi N (2003) J Immunol 170:1249
36. Rosen DB, Cao W, Avery DT, Tangye SG, Liu YJ, Houchins JP, Lanier LL (2008) J Immunol 180:6508
37. Andrews DM, Scalzo AA, Yokoyama WM (2002) Nat Immunol 4:175
38. Mocikat R, Braumuller H, Gumy A, Egeter O, Ziegler H, Reusch U, Bubeck A, Louis J, Mailhammer R, Riethmüller G, Koszinowski U, Röcken M (2003) Immunity 19:561
39. Adam C, King S, Allgeier T, Braumüller H, Lüking C, Mysliwietz J, Kriegeskorte A, Busch DH, Röcken M, Mocikat R (2005) Blood 106:338
40. Coudert JD, Coureau C, Guery JC (2002) J Immunol 169:2979
41. Laffont S, Seillet C, Ortaldo J, Coudert JD, Guery JC (2008) Blood 112:661
42. Fernandez NC, Lozier A, Flament C, Ricciardi-Castagnoli P, Bellet D, Suter M, Perricaudet M, Tursz T, Maraskovsky E, Zitvogel L (1999) Nat Med 5:405
43. Yu Y, Hagihara M, Ando K, Gansuvd B, Matsuzawa H, Tsuchiya T, Ueda Y, Inoue H, Hotta T, Kato S (2001) J Immunol 166:1590
44. Dalod M, Hamilton T, Salomon R, Salazar-Mather TP, Henry SC, Hamilton JD, Biron CA (2003) J Exp Med 197:885
45. Mailliard RB, Son YI, Redlinger R, Coates PT, Giermasz A, Morel PA, Storkus WJ, Kalinski P (2003) J Immunol 171:2366
46. Ferlazzo G, Pack M, Thomas D, Paludan C, Schmid D, Strowig T, Bougras G, Muller WA, Moretta L, Münz C (2004) Proc Natl Acad Sci USA 101:16606

47. Buentke E, Heffler L, Wilson J, Wallin RP, Löfman C, Chambers BJ, Ljunggren HG, Scheynius A (2002) J Invest Dermatol 119:850
48. Ferlazzo G, Munz C (2004) J Immunol 172:1333
49. Zingoni A, Sornasse T, Cocks B, Tanaka Y, Santoni A, Lanier LL (2004) J Immunol 173:3716
50. Hanna J, Gonen-Gross T, Fitchett J, Rowe T, Daniels M, Arnon TI, Gazit R, Joseph A, Schjetne KW, Steinle A, Porgador A, Mevorach D, Goldman-Wohl D, Yagel S, LaBarre MJ, Buckner JH, Mandelboim O (2004) J Clin Invest 114:1612
51. So T, Lee SW, Croft M (2006) Int J Hematol 83:1
52. Assarsson E, Kambayashi T, Schatzle JD, Cramer SO, von Bonin A, Jensen PE, Ljunggren HG, Chambers BJ (2004) J Immunol 173:174
53. Saudemont A, Jouy N, Hetuin D, Quesnel B (2005) Blood 105:2418
54. Kos FJ, Engleman EG (1996) Cell Immunol 173:1
55. Kurosawa S, Harada M, Matsuzaki G, Shinomiya Y, Terao H, Kobayashi N, Nomoto K (1995) Immunology 85:338
56. Kos F, Engleman E (1995) J Immunol 155:578
57. Ryan A, Wilcox T, Tamada K, Scott E, Strome SE, Chen L (2002) J Immunol 169:4230
58. Fan Z, Yu P, Wang Y, Wang Y, Fu ML, Liu W, Sun Y, Fu YX (2006) Blood 107:1342
59. Westwood JA, Kelly JM, Tanner JE, Kershaw MH, Smyth MJ, Hayakawa Y (2003) J Immunol 171:757
60. Maier S, Tertilt C, Chambron N, Gerauer K, Hüser N, Heidecke CD, Pfeffer K (2001) Nat Med 7:557
61. Uehara S, Chase CM, Kitchens WH, Rose HS, Colvin RB, Russell PS, Madsen JC (2005) J Immunol 175:3424
62. Barao I, Hanash AM, Hallett W, Welniak LA, Sun K, Redelman D, Blazar BR, Levy RB, Murphy WJ (2006) Proc Natl Acad Sci USA 103:5460
63. Smyth MJ, Teng MW, Swann J, Kyparissoudis K, Godfrey DI, Hayakawa Y (2006) Clin Cancer Res 9:606
64. Kursar M, Koch M, Mittrucker HW, Nouailles G, Bonhagen K, Kamradt T, Kaufmann SH (2007) J Immunol 178:2661
65. Roy S, Barnes PF, Garg A, Wu S, Cosman D, Vankayalapati R (2008) J Immunol 180:1729
66. Su HC, Nguyen KB, Salazar-Mather TP, Ruzek MC, Dalod M, Biron CA (2001) Eur J Immunol 31:3048
67. Barber MA, Zhang T, Gagne BA, Sentman CL (2007) J Immunol 178:6140
68. Lu LL, Ikizawa K, Hu D, Werneck MBF, Wucherpfennig KW, Cantor H (2007) Immunity 26:593
69. French AR, Yokoyama WM (2004) Arthritis Res Ther 6:8
70. Lünemann JD, Münz C (2008) Brain 131:1681
71. Zhang B, Yamamura T, Kondo T, Fujiwara M, Tabira T (1997) J Exp Med 186:1677
72. Smeltz RB, Wolf NA, Swanborg RH (1999) J Immunol 163:1390
73. Trivedi PP, Roberts PC, Wolf NA, Swanborg RH (2005) J Immunol 174:4590
74. Madeline M, Leach MW, Rennick DM (1998) J Immunol 161:3256
75. Lee I, Qin H, Trudeau J, Dutz J, Tan R (2004) J Immunol 172:937
76. Ogasawara K, Hamerman J, Hsin H, Chikuma S, Bour-Jordan H, Chen T, Pertel T, Carnaud C, Bluestone JA, Lanier LL (2003) Immunity 18:41
77. Rodacki M, Svoren B, Butty V, Besse W, Laffel L, Benoist C, Mathis D (2007) Diabetes 56:177
78. Schott E, Bonasio R, Ploegh HL (2003) J Exp Med 19:1213
79. Rabinovich B, Li J, Shannon J, Hurren R, Chalupny J, Cosman D, Miller RG (2003) J Immunol 170:3572
80. Cerboni C, Zingoni A, Cippitelli M, Piccoli M, Frati L, Santoni A (2007) Blood 110:606
81. Molinero L, Fuertes M, Rabinovich G, Fainboim L, Zwirner N (2002) J Leukoc Biol 71:791
82. Maasho K, Opoku-Anane J, Marusina AI, Coligan JE, Borrego F (2005) J Immunol 174:4480
83. Cerboni C, Ardolino M, Santoni A, Zingoni A (2009) Blood 113:2955

84. Hamerman JA, Ogasawara K, Lanier LL (2004) J Immunol 172:2001
85. Kloss M, Decker P, Baltz KM, Baessler T, Jung G, Rammensee HG, Steinle A, Krusch M, Salih HR (2008) J Immunol 181:6711
86. Stepp S, Dufourcq-Lagelouse R, Deist FL, Bhawan S, Certain S, Mathew PA, Henter JI, Bennett M, Fischer A, de Saint Basile G, Kumar V (1999) Science 286:1957
87. Matloubian M, Suresh M, Glass A, Galvan M, Chow K, Whitmire JK, Walsh CM, Clark WR, Ahmed R (1999) J Virol 73:2527
88. Peng SL, Moslehi J, Robert ME, Craft J (1998) J Immunol 160:652
89. Shlomchik WD, Couzens MS, Tang CB, McNiff J, Robert ME, Liu J, Shlomchik MJ, Emerson SG (1999) Science 285:412
90. Ruggeri L, Capanni M, Urbani E, Perruccio K, Shlomchik WD, Tosti A, Posati S, Rogaia D, Frassoni F, Aversa F, Martelli MF, Velardi A (2002) Science 295:2097

Interactions Between NK Cells and Regulatory T Cells

Magali Terme, Nathalie Chaput, and Laurence Zitvogel

Abstract Regulatory T cells (Treg) maintain peripheral tolerance and prevent the development of autoimmune diseases. They control the function of different immune cells such as differentiated $CD4^+$ and $CD8^+$ T cells, B cells, and dendritic cells. We and others contributed to demonstrate that Treg also regulate NK cell functions in mice and humans. Treg hamper the homeostatic proliferation of NK cells, NKG2D-dependent NK cell cytotoxic activity, and IL-12-induced NK cell IFNγ secretion. In vivo, in the absence of Tregs, NK cell proliferation ensures in lymph nodes through an autoreactive crosstalk between resident dendritic cells (DC) and $CD4^+$ T lymphocytes leading to IL-15Rα exposure on DC. During tumor progression, Treg play a dominant role in suppressing the innate arm of antitumor immune responses. We will discuss strategies currently developed to impede Treg inhibitory effects in cancer.

Abbreviations

DC	Dendritic cells
DLN	Draining lymph nodes
IFN	Interferon

M. Terme, N. Chaput and L. Zitvoge (✉)
U805 and CICBT507 INSERM, Institut Gustave Roussy, 39 rue Camille Desmoulins, 94805, Villejuif, France
e-mail: zitvogel@igr.fr

Institut Gustave Roussy, Villejuif, France

N. Chaput and L. Zitvoge
Center of Clinical Investigations, CBT507, IGR, Villejuif, France

N. Chaput
Cellular Therapy Unit, Villejuif, France

NK   Natural killer cells
TGFβ   Transforming growth factor
Treg   Regulatory T cells
WT   Wild type

1 Introduction on Regulatory T cells

Regulatory T cells are a subset of T lymphocytes essential in maintaining peripheral tolerance [1–3]. They contribute to exert a dominant tolerance during infections [4, 5], tumor progression [6, 7], and allogeneic transplantations [8]. Traditionally, mouse and human Treg were defined as $CD4^+CD25^{hi}$ T cells. They also express Cytotoxic T Lymphocyte Antigen-4 (CTLA-4) [9] or the Glucocorticoid-induced TNF Receptor (GITR). The search for more specific markers, specifically in mice, led to the seminal identification of the Foxp3 transcription factor involved in Treg development [10, 11]. More recently, Sakaguchi's team identified the Folate Receptor 4 (FR4) as a specific marker of mouse Treg [12]. Indeed, mouse Treg constitutively express high amounts of the folate receptor 4 (FR4), a subtype of the vitamin folic acid receptor, in contrast to naive or activated T cells [12]. Since human conventional T cells ($CD4^+$ $CD25^-$) may also express Foxp3 after activation [13], other markers have been used to discriminate human Treg from effector T cells, such as IL-7 receptor alpha (IL-7Rα). Indeed, $CD4^+CD25^{hi}IL-7R\alpha^{low}$ T cells are highly enriched in naturally occurring Treg. About 85% of $CD4^+CD25^{hi}IL-7R\alpha^{low}$ T cells are $Foxp3^+$ [14].

    Treg suppress the proliferation of naive T cells and their differentiation into effector T cells in vivo. They also negatively control the functions of B lymphocytes, dendritic cells (DC), and macrophages [15–18]. Various controversial regulatory mechanisms have been described for Treg. Contact-dependent suppression and/or immunosuppressive cytokines have been proposed [19]. The role of inhibitory cytokines such as membrane-bound or secreted TGF-β and IL-10 has been largely studied [20]. Recently, a novel immunosuppressive cytokine, IL-35, belonging to the IL-12 subfamily members, has been described, which appears to play a key role in the suppressive function of Treg against conventional T cells [21]. Ectopic expression of IL-35 confers regulatory activity to naive T cells. Furthermore, IL-35 suppresses T cell proliferation in vitro [21]. The use of neutralizing antibodies and loss-of-function mice highlighted that CTLA-4-induced Treg-mediated suppression of the DC antigen-presenting functions [22]. Treg could condition DC to produce indoleamine-2,3 oxygenase (IDO), a tryptophan-degrading enzyme, via the interaction between CTLA-4 and B7 molecules on the DC surface. IDO expression suppresses effector T cell responses [23]. Treg can also hamper DC maturation via LAG-3 molecules [24] and prevent TLR-induced TRAIL expression on myeloid DC [25]. A recent study established that, through the generation of the immunosuppressive factor adenosine, the ectoenzymes CD39 and CD73 are important contributors to the regulatory activity of Treg cells [26].

Finally, Treg express granzyme B molecules during activation and may not only suppress but also delete peripheral T cell effectors in a perforin/granzyme-dependent manner [27–29]. Since Treg depletion aggravates diseases in some autoantibody-mediated autoimmune disease models, it has been proposed that Treg could also control B cells [30]. In vitro, Treg kill antigen-presenting B cells in a Fas-Fas ligand-dependent manner [31]. In vivo, Treg induce peripheral tolerance of B cells specific for autoantigen in spleen and lymph nodes by suppressing their proliferation and inducing killing in a perforin/granzyme-dependent manner [29, 32].

Therefore, it was well established that Treg could keep in check DC and/or T cells and B cells i.e., the adaptive arm of immune responses. Later, we and others unraveled that Treg also control the innate arm of immunity as detailed in the following section.

2 Regulatory T cells Control NK Cell Functions In Vitro

A pioneering study performed by Shimizu et al. in 1999 suggested the capacity of Treg to inhibit NK cell effector functions [33]. In this work, Balb/c Nude mice bearing the RLO♂1 leukemia were adoptively transferred by splenocytes depleted or not from Treg. Tumor regression only occurred in mice receiving Treg-depleted splenocytes. The cytotoxic activity of splenocytes depleted or not from Treg and restimulated with RLO♂1 was assessed. Treg-depleted splenocytes displayed significant lytic function attributable to $TCR^-CD4^-CD8^-$ cells, while total splenocytes failed to do so. The putative scenario was that these $TCR^-CD4^-CD8^-$ effector cells became activated by IL-2 secreted by the conventional $CD25^-CD4^+$ T cells in the absence of Treg [33].

Our group assessed the direct effects of Treg onto human $CD3^-CD56^+$ blood NK cells [34]. Treg were purified from normal volunteers' peripheral blood by cell sorting or magnetic purification. Autologous or allogeneic NK cells were cocultured with purified Treg, and analyzed for their cytolytic capacity against K562 and the GIST cell line GIST882. Freshly isolated Treg inhibited NK cell cytotoxicity at 1/1 to 1/5 T: NK ratios against cell lines expressing NKG2D ligands. Treg not only suppressed NK cell cytotoxic functions, but also cytokine secretion. Treg control IFNγ secretion, but it depends on the mode of NK cell activation. IFNγ secretion was inhibited by Treg when human NK cells were stimulated by IL-12 [34] but not by IL-2Rγ chain-dependent cytokines (such as IL-2, IL-4, IL-7, or IL-15). [34].

We next studied the mechanisms involved in Treg-mediated NK cell suppression. After formaldehyde fixation, Treg retained their suppressive activity toward NK cells suggesting a role for membrane-bound molecules [34]. TGF-β is an essential mediator used by Treg to suppress T cells [20, 35]. Moreover, resting human Treg express membrane-bound TGF-β but do not release this cytokine in soluble form [36]. Addition of anti-TGF-β neutralizing antibodies could restore the NK cell cytotoxicity and IFNγ secretion underlining the role of membrane-bound TGF-β in the Treg-mediated suppression of NK cells in vitro [34].

Surprisingly, in contrast to human data, only activated and not resting mouse Treg could inhibit NK cells [37]. Indeed, Smyth's group showed that Treg derived from C57Bl/6 mice required a prestimulation by anti-CD3- and anti-CD28-activating antibodies to suppress NK cells. Activated Treg could block NK cell-dependent killing of different tumor cells such as RMA-S-Rae-1β and B16-Rae1ε [37]. This inhibition was observed at a 1:1 and 1:3 T/NK ratio. As in humans, TGF-β was involved in this phenomenon since anti-TGFβ neutralizing antibody restored NK cell-mediated cytotoxicity. The discrepancy between mouse and human data concerning the requirement for an activation step of Treg could be explained by the expression of membrane-bound TGF-β. Resting human Treg constitutively express membrane-bound TGF-β, while mouse Treg require a TCR-driven signaling or IL-2 to do so [38]. Moreover, human Treg downregulate their membrane expression of TGF-β upon activation [39].

How did Treg-associated TGF-β impact on NK cell functions? Treg could specifically inhibit NKG2D expression on human and mouse NK cells that could be restored by anti-TGFβ neutralizing antibodies [34]. In humans, the NK cell-dependent cytotoxicity against K562, which is by large dependent on NKG2D, could be blocked by Treg and restored by anti-TGF-β neutralizing antibodies. This finding has been corroborated in the mouse system since Treg could only hamper the NKG2D-dependent cytotoxicity (that directed against specific targets overexpressing NKG2D ligands such as B16Rae1ε, or RMA-S-Rae1β but not the parental lines) [37]. In mice, Treg harvested from TGF-β loss-of-function mice (at 6 weeks of age) lost their inhibitory function on NK cell cytotoxicity after adoptive transfer. Moreover, Scurfy mice bearing a loss-of-function mutation in Foxp3 exhibited high proliferation index in the NK cell compartment in lymphoid organs [34]. Soluble TGF-β is also known to decrease the expression of NK-cell-activating receptors such as NKG2D and NKp30 [40]. In cancer patients, not only Treg but also soluble bioactive TGF-β account for the reduction of NKG2D expression on NK cells in glioma or colon carcinoma-bearing patients [41, 42].

These lines of evidence point out to the direct regulatory role of resting or activated Tregs onto NK cells in vitro and in vivo, in mice and humans.

3 Treg Control DC/NK Cell Crosstalk in Homeostatic and Inflammatory Conditions

Since NK cell homeostasis and functions are dictated by DC [43, 44], the indirect role of Treg during the DC/NK cell dialogue has been studied in homeostatic and inflammatory conditions.

Ablation of Treg in neonates or in adult mice induced the development of autoimmune disorders, supporting the notion that Treg are required to maintain peripheral tolerance and homeostasis of the immune system [45]. In Scurfy mice, which lack a functional Foxp3 transcription factor and thus have no Treg [46, 47], a deregulated proliferation of NK cells occurred [34]. This observation was confirmed in mice treated with metronomic cyclophosphamide or anti-CD25 (PC61) antibodies, both resulting in transient Treg depletion [34, 48, 49]. Using Foxp3-DTR mice, Kim et al. also reported a sevenfold and a fourfold increase in NK cell numbers in LN and spleen of Treg-depleted mice, respectively [1]. These observations suggested that Treg could control NK cell proliferation in vivo at the steady state. Two reports addressed the mechanisms involved in the Treg-mediated NK cell control. First, Giroux et al. brought up the evidence that Treg could block the NK cell differentiation pathway in LN. Using HY-TCR-Rag2$^{-/-}$ transgenic mice devoid of Treg, they detected an increase in the number of mature NK cells (i.e., CD122$^+$ CD11b^{hi} CD43$^+$) in LN. Moreover, the adoptive transfer of Treg repressed the development of mature NK cells in LN. In vitro, coculture of preactivated Treg with NK cell precursors suppressed NK cell development, while coculture with conventional CD4$^+$ CD25$^-$ T cells failed to do so. Thus, their data supported the notion whereby Treg hamper the generation of mature NK cells in LN through short interactions with NK cell precursors [50]. Nevertheless, this study did not rule out the possibility that Treg impair the entry of mature NK cells in the lymph nodes through the high endothelial veinules.

Secondly, we examined the effects of Treg on the DC/NK cell crosstalk in the LN using two different experimental approaches leading to either Treg inhibition (by low-dose CTX administration) or Treg depletion (by anti-CD25 Ab). We observed that NK cell increase in LN was mainly due to an active proliferation, as detected by BrdU incorporation [48, 49]. NK cells could interact with LN resident-DC which resulted in NK cell recruitment or proliferation and activation in LN [51]. In the absence of functional Treg, CD11c^{hi} I-Ab$^+$ DC were recruited in LN in a CCR5-dependent manner. NK cell proliferation involved IL-15Rα-expressing DC, since (1) no proliferation occurred in IL-15Rα$^{-/-}$ mice; (2) proliferation was abolished in DC-depleted mice (CD11c-DTR mice). Furthermore, in Treg-depleted mice, IL-15Rα$^+$ DC were recruited in LN. This phenomenon was not observed in mice depleted from both Treg and conventional T cells by anti-CD4 antibody. We reconstituted the system in in vitro studies. Coculture of DC with conventional T cells led to chemokine secretion, especially CCR5 ligands, and DC maturation characterized by the acquisition of IL-15Rα [48, 49]. The addition of Treg in the coculture inhibited chemokine release and IL-15Rα expression on DC. Thus, LN Treg control two major checkpoints (chemokine secretion and DC maturation) resulting in NK cell proliferation [48, 49].

In our work, we showed that IL-15Rα expression on DC was controlled by TGF-β, suggesting that NK cell homeostasis could also be controlled by TGF-β. In a CD11c-TGF-βRII-transgenic model, where TGF-β signaling is disrupted in DC and in NK cells, the NK cell number was greatly enhanced in spleen, bone marrow, and liver [52].

Not only DC but also monocytes can regulate NK cell functions [53]. In a setting mimicking infection where monocytes were activated by anti-influenza vaccine, Trzonkowski et al. showed that the presence of CD4$^+$CD25$^+$ Treg impaired the monocyte-mediated NK cell cytotoxicity against K562 [54] and to a lesser extent IFNγ or perforin levels in NK cells. Anti-IL-10 neutralizing antibodies did not influence Treg immunosuppressive activity on NK cells suggesting that IL-10 was not involved in this suppression [54]. In another report where human plasmacytoid DC could induce the selective proliferation of CD56^{bright} CD16$^-$ NK cells [55], this proliferation was strongly enhanced in the presence of conventional T cells in an IL-2-dependent manner. CD4$^+$ CD25$^+$ Treg abrogated the IL-2-dependent proliferation of NK cells. However, Treg could not directly inhibit pDC-induced NK cell proliferation [55].

Therefore, these lines of evidence establish the capacity of Treg in interfering in the DC/NK cell crosstalk in homeostatic and potentially pathological conditions.

4 Regulatory T cells Control NK Cell Functions in Pathology

4.1 In Cancer

During tumor development, Treg accumulate in tumor beds, tumor-draining lymph nodes (LN), and peripheral blood [7, 56, 57]. Numerous studies reported the association between high Treg numbers and poor prognosis in cancer patients. In mouse models, the impact of Treg depletion on tumor growth has been extensively studied. Different strategies have been utilized to deplete or inhibit Treg. First, the administration of anti-CD25 antibodies led to Treg depletion. Second, the administration of cyclophosphamide (CTX) at low dosages [48, 49, 58] allowed the transient inhibition of Treg functions. Third, the recent generation of the Foxp3-DTR knock-in-mice allowed conditional depletion of Treg [1]. Administration of anti-CD25 mAb or CTX could promote tumor-specific T-cell-mediated immune responses leading to tumor rejection [33, 59–61].

In the first study suggesting that Treg could impact on NK cells, the depletion of Treg using anti-CD25 antibodies prior to tumor inoculation eradicated the tumor. Tumor rejection depended upon CD8$^+$ T cells and CD4$^-$CD8$^-$ cells defined as NK cells [33]. In a Balb/c model of methylcholanthrene (MCA)-induced carcinoma, immunization with a series of SEREX (serological identification of antigens by recombinant expression cloning)-defined self-antigens resulted in the development of CD4$^+$CD25$^+$ Treg and tumor progression [62]. However, tumor growth could be delayed by administration of anti-CD25. Given that NK and NKT cells are essential in the immunosurveillance against MCA-induced carcinoma [63] and that SEREX Ag-induced immunization was associated with decreased innate functions, an inhibitory role of Treg onto NK and NKT cells was anticipated in this model. In immunized mice, the number of NK cells in spleen and lungs was unchanged compared with naive mice. However, NK cells displayed reduced cytolytic activity against YAC-1 target cells compared to NK cells obtained from control mice. Adoptive transfer of Treg obtained from immunized mice into naive Balb/c mice resulted in a decreased cytotoxic activity against YAC-1 [62], demonstrating that Treg could inhibit NK cell functions in vivo. In athymic Nude mice lacking conventional and regulatory T lymphocytes, we demonstrated that adoptive transfer of Treg abrogated NK-cell-dependent cytotoxic activity against YAC-1 cells unlike conventional CD4$^+$CD25$^-$ T cells [34]. MHC class I-deficient tumors such as RMA-S injected intraperitoneally induced an NK cell recruitment in the peritoneal cavity that could be blocked by adoptively transferred Tregs [34, 64].

In another study, depletion of Treg with anti-CD25 antibody before 3LL or B16Rae tumor inoculation enhanced the establishment of pulmonary metastases. Elimination of NK cells using anti-NK1.1 or asialo-GM1 antibodies reduced the antitumor effect induced by Treg depletion, demonstrating the capacity of naturally occurring Treg to keep in check NK-cell-controlled metastases. In vitro, different studies underscored the capacity of Treg to inhibit NKG2D-mediated NK cell cytotoxicity. In vivo, Treg depletion only impacted on the growth of NKG2D ligands-expressing tumors. For example, Treg elimination decreased metastases number in B16-Rae1ε model, which overexpress the Rae1ε NKG2D ligand, but had no effect on B16F10, the NKG2D ligand negative parental cell line [37]. The adoptive transfer of activated Treg cells in Nude mice, which lack T cells, enhanced B16-Rae1β tumor metastases, but failed to compromise lung metastasis establishment of B16F10 [34]. These results were corroborated in another mouse model lacking T cells (B6.RAG1$^{-/-}$ mice) [34, 37]. Therefore, Treg can suppress NK cell-mediated tumor immunosurveillance against NKG2D ligand-expressing tumor cells. Thus, strategies aimed at neutralizing Treg (by anti-CD25 Ab, or CTX) to increase NK-cell-dependent antitumor response were developed. Exogenous rIL-12 (a NK-cell-stimulating cytokine) could synergize with anti-CD25 antibodies to eradicate 3LL metastases [37].

The negative impact of Treg on human NK cells has also been demonstrated in tumor patients. We previously examined the effect of imatinib mesylate (IM) on NK cell effector functions. IM is a tyrosine kinase inhibitor blocking signaling through c-kit, BCR/ABL, c-abl, and PDGF-R receptors [65–67]; therefore, it is widely used in the treatment of chronic myeloid leukemia and gastrointestinal sarcoma (GIST). We previously showed that IM promoted NK cell activation by triggering the DC/NK cell crosstalk in vitro and in vivo in mouse models and in patients [68]. We performed a prospective analysis of NK cell IFNγ secretion before and after 2 months of IM in 77 GIST patients treated with IM and identified that those patients exhibiting stable or enhanced levels of NK cell IFNγ secretion at 2 months of IM were long-term survivors (in a Cox regression model) (Menard, Cancer Res in press). However, about 40% of GIST patients failed to exhibit NK cell activation despite IM therapy. The monitoring of the percentages and absolute numbers of CD4$^+$CD25^{high} Treg by flow cytometry revealed that Treg numbers were significantly augmented by two to threefold in patients without NK cell activation compared to the other GIST patients, independent of tumor burdens in the two cohorts [34]. In another Phase I clinical trial where melanoma patients where immunized with dendritic-cell-derived exosomes (Dex), we found that Dex were able to trigger NK cell activation in 7/14 pts (*Viaud et al., submitted to PlOs Biology*). There was an inverse correlation between Tregs numbers and Dex-mediated NK cell activation [69].

In cancer patients, administration of low-dose CTX referred to as "metronomic" therapy resulted in a profound and selective depletion of CD4$^+$ CD25^{high} Treg [70]. Peripheral blood mononuclear cells obtained from these advanced cancer patients displayed a reduced killing activity against the specific NK cell target K562 compared with normal volunteers [70]. After metronomic CTX treatment, Treg were depleted and NK cell cytolytic activity was restored [70]. Therefore, the management of NK-controlled cancers could take advantage of suppressing Treg functions using metronomic CTX.

4.2 In Bone Marrow Graft

The effect of Treg on NK cells has also been addressed in the "hybrid resistance" model. In this setting, parental bone marrow cells are rejected by lethally irradiated F1 recipients, since they only express half of the MHC Class I genes present in the recipient cells and are eliminated by NK cells according to the missing self-hypothesis [71]. Bone marrow cell engraftment is determined by measuring the colony-forming unit-granulocytes/monocytes (CFU-GM) in spleens. In this setting, administration of anti-CD25 antibodies to the recipient strongly enhanced the rejection of the parental bone marrow graft. Depletion of NK cells by anti-NK1.1 restored the engraftment, demonstrating the role of NK cells in the bone marrow rejection [72]. Once again, anti-TGFβ neutralizing antibodies enhanced bone marrow graft rejection suggesting that TGF-β was likely involved in Treg-mediated tolerance of allogeneic BMT. Finally, the adoptive transfer of Treg along with the graft suppressed the NK-cell-mediated rejection of the bone marrow [72].

4.3 In Infection

Lund et al. have investigated the role of Treg in a mouse model of herpes simplex virus (HSV)-2 mediated genital infection using Foxp3-DTR mice. After HSV-2 infection, regulatory and conventional T cells accumulated in the draining lymph nodes and at the site of infection. Treg played a key role in controlling the development of infection since Treg-ablated mice succumbed more rapidly than Treg-sufficient mice [73]. Type I IFN produced by pDC and IFNγ produced by CD4$^+$ T cells or NK cells were involved in the protection against virus infection [74, 75]. In the absence of Treg, local IFNγ secretion was impaired in the vaginal tract but not in draining lymph nodes. The proportion of immune effector cells, such as conventional T cells, NK cells, and subsets of DC (plasmacytoid DC and CD11b$^+$ DC), was greatly decreased in the vaginal tract but was enhanced in the DLN in Treg-depleted mice, compared to Treg-sufficient mice at 2 days after infection. In the absence of Treg, recruitment of effector cells, especially NK cells, was strikingly reduced at the site of infection early after infection. Treg depletion was associated with a sharp increase in proinflammatory chemokines (CXCL10, CXCL9, CCL2) in DLN. Thus, Lund et al. showed that Treg control chemokine production in DLN or in peripheral sites during mucosal infection. Treg facilitate the homing of immune effector cells such as NK cells at the site of infections. In the absence of Treg, chemokine production is greatly enhanced in the LN, leading to the recruitment of effector cells in the DLN, thereby compromising the mucosal immunity [73].

5 Concluding Remarks

It is now well accepted that Treg can control NK cell effector functions (recruitment, proliferation, cytotoxicity, and cytokine production). But depending on the stimuli-inducing NK cell activation, NK cells may not be repressable by Treg. When NK cells are activated by IL-2Rγ chain-signaling cytokines, Treg do not prevent IFNγ secretion. Treg abolish NK cell cytolytic function against tumors only when tumor cells overexpress NKG2D ligands. Therefore, it is likely that other checkpoints, yet to be defined, be active against NK cells in these circumstances.

In the context of cancer, NK cells have been shown to control metastases [76] and play a role at the effector phase in some circumstances (neuroblastoma [77], acute myeloid leukemia [78], GIST [68]). These NK cells can exhibit innate dysfunctions (bcr/abl expressing NK cells in CML patients [79], NK cells in myelodysplastic syndromes [80]), or acquired dysfunctions (STAT3 phosphorylation [81], soluble NKG2DL [82, 83], soluble TGF-β [84]). In this setting, Treg accumulation may add deleterious effects which may be solved, at least in part, by depleting or inhibiting their function. Metronomic CTX is currently used to selectively hamper Treg in cancer patients [70]. Alternatively, peptides blocking TGF-β could provide a new tool in enhancing antitumor responses [85]. In parallel, strategies harnessing NK cell effector functions could be developed [48, 49].

Another potential relevance of the role of Treg in controlling the DC/CD4[+] T cell crosstalk could be discussed in cases of autoimmune disorders (Fig. 1). Indeed, NK cell proliferation is controlled at the steady state by Treg [48, 49]. NK cells have been involved in the development of autoimmune disorders, but their precise role remains unclear. In two different models of diabetes (NOD and BDC2.5-TCR-Transgenic NOD), depletion of NK cells by anti-NK1.1 or anti-asialoGM1 antibody reduced the severity of the disease [86, 87]. Moreover, in distinct autoimmune disorders, Treg may be deficient [88] and may fail to control NK cells. Our data lead us to hypothesize that blocking IL-15Rα/IL-15 and CCR5 by pharmacological interventions could compromise an uncontrolled NK cell proliferation and activation and could ameliorate the exacerbation of NK-cell-dependent autoimmune disorders.

In the context of viral infections, the NK to Treg interaction needs further clarifications. It appears that Treg is indispensable for the appropriate recruitment of NK cells at infected sites [73]. Moreover, IFNα may render NK cells resistant to

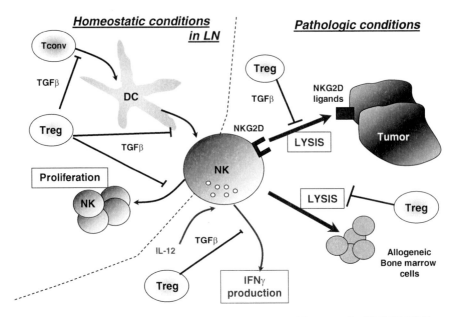

Fig. 1 Treg lymphocytes control NK cell effector functions and homeostasis. CD4$^+$CD25$^+$ Treg inhibit cytokine production, cytotoxic activity against tumor or allogeneic bone marrow cells, and thereby tumor elimination and bone marrow rejection. In homeostatic conditions, inadequate NK cell proliferation induced by DC is kept in check by Treg

the inhibitory effects of Treg. Conversely, human NK cells can also keep in check Treg. NK cells activated by monokines or *M. tuberculosis*-stimulated monocytes are able to lyse expanded Treg in response to *M. tuberculosis* infection, but not resting Treg [89]. Thus, the NK cells to Treg interaction is bidirectional and highly regulated.

In the near future, we anticipate that the comprehensive characterization of distinct NK cell subsets together with the increased knowledge on the regulatory pathways used by Treg will facilitate the pathophysiological relevance of the retrocontrol of Treg on innate immunity.

References

1. Kim JM, Rasmussen JP, Rudensky AY (2007) Regulatory T cells prevent catastrophic autoimmunity throughout the lifespan of mice. Nat Immunol 8:191–197
2. Sakaguchi S (2004) Naturally arising CD4+ regulatory t cells for immunologic self-tolerance and negative control of immune responses. Annu Rev Immunol 22:531–562
3. Shevach EM (2002) CD4+ CD25+ suppressor T cells: more questions than answers. Nat Rev Immunol 2:389–400
4. Belkaid Y, Piccirillo CA, Mendez S, Shevach EM, Sacks DL (2002) CD4+CD25+ regulatory T cells control Leishmania major persistence and immunity. Nature 420:502–507

5. Kullberg MC, Jankovic D, Gorelick PL, Caspar P, Letterio JJ, Cheever AW, Sher A (2002) Bacteria-triggered CD4(+) T regulatory cells suppress *Helicobacter hepaticus*-induced colitis. J Exp Med 196:505–515
6. Bates GJ, Fox SB, Han C, Leek RD, Garcia JF, Harris AL, Banham AH (2006) Quantification of regulatory T cells enables the identification of high-risk breast cancer patients and those at risk of late relapse. J Clin Oncol 24:5373–5380
7. Curiel TJ, Coukos G, Zou L, Alvarez X, Cheng P, Mottram P, Evdemon-Hogan M, Conejo-Garcia JR, Zhang L, Burow M, Zhu Y, Wei S, Kryczek I, Daniel B, Gordon A, Myers L, Lackner A, Disis ML, Knutson KL, Chen L, Zou W (2004) Specific recruitment of regulatory T cells in ovarian carcinoma fosters immune privilege and predicts reduced survival. Nat Med 10:942–949
8. Trenado A, Charlotte F, Fisson S, Yagello M, Klatzmann D, Salomon BL, Cohen JL (2003) Recipient-type specific CD4+CD25+ regulatory T cells favor immune reconstitution and control graft-versus-host disease while maintaining graft-versus-leukemia. J Clin Invest 112:1688–1696
9. Takahashi T, Kuniyasu Y, Toda M, Sakaguchi N, Itoh M, Iwata M, Shimizu J, Sakaguchi S (1998) Immunologic self-tolerance maintained by CD25+CD4+ naturally anergic and suppressive T cells: induction of autoimmune disease by breaking their anergic/suppressive state. Int Immunol 10:1969–1980
10. Fontenot JD, Gavin MA, Rudensky AY (2003) Foxp3 programs the development and function of CD4+CD25+ regulatory T cells. Nat Immunol 4:330–336
11. Hori S, Nomura T, Sakaguchi S (2003) Control of regulatory T cell development by the transcription factor Foxp3. Science 299:1057–1061
12. Yamaguchi T, Hirota K, Nagahama K, Ohkawa K, Takahashi T, Nomura T, Sakaguchi S (2007) Control of immune responses by antigen-specific regulatory T cells expressing the folate receptor. Immunity 27:145–159
13. Morgan ME, van Bilsen JH, Bakker AM, Heemskerk B, Schilham MW, Hartgers FC, Elferink BG, van der Zanden L, de Vries RR, Huizinga TW, Ottenhoff TH, Toes RE (2005) Expression of FOXP3 mRNA is not confined to CD4+CD25+ T regulatory cells in humans. Hum Immunol 66:13–20
14. Liu W, Putnam AL, Xu-Yu Z, Szot GL, Lee MR, Zhu S, Gottlieb PA, Kapranov P, Gingeras TR, Fazekas de St Groth B, Clayberger C, Soper DM, Ziegler SF, Bluestone JA (2006) CD127 expression inversely correlates with FoxP3 and suppressive function of human CD4+ T reg cells. J Exp Med 203:1701–1711
15. Azuma T, Takahashi T, Kunisato A, Kitamura T, Hirai H (2003) Human CD4+ CD25+ regulatory T cells suppress NKT cell functions. Cancer Res 63:4516–4520
16. Lim HW, Hillsamer P, Banham AH, Kim CH (2005) Cutting edge: direct suppression of B cells by CD4+ CD25+ regulatory T cells. J Immunol 175:4180–4183
17. Sakaguchi S, Yamaguchi T, Nomura T, Ono M (2008) Regulatory T cells and immune tolerance. Cell 133:775–787
18. von Boehmer H (2005) Mechanisms of suppression by suppressor T cells. Nat Immunol 6:338–344
19. Vignali D (2008) How many mechanisms do regulatory T cells need? Eur J Immunol 38:908–911
20. Gorelik L, Flavell RA (2002) Transforming growth factor-beta in T-cell biology. Nat Rev Immunol 2:46–53
21. Collison LW, Workman CJ, Kuo TT, Boyd K, Wang Y, Vignali KM, Cross R, Sehy D, Blumberg RS, Vignali DA (2007) The inhibitory cytokine IL-35 contributes to regulatory T-cell function. Nature 450:566–569
22. Oderup C, Cederbom L, Makowska A, Cilio CM, Ivars F (2006) Cytotoxic T lymphocyte antigen-4-dependent down-modulation of costimulatory molecules on dendritic cells in CD4+ CD25+ regulatory T-cell-mediated suppression. Immunology 118:240–249

23. Fallarino F, Grohmann U, Hwang KW, Orabona C, Vacca C, Bianchi R, Belladonna ML, Fioretti MC, Alegre ML, Puccetti P (2003) Modulation of tryptophan catabolism by regulatory T cells. Nat Immunol 4:1206–1212
24. Liang B, Workman C, Lee J, Chew C, Dale BM, Colonna L, Flores M, Li N, Schweighoffer E, Greenberg S, Tybulewicz V, Vignali D, Clynes R (2008) Regulatory T cells inhibit dendritic cells by lymphocyte activation gene-3 engagement of MHC class II. J Immunol 180: 5916–5926
25. Roux S, Apetoh L, Chalmin F, Ladoire S, Mignot G, Puig PE, Lauvau G, Zitvogel L, Martin F, Chauffert B, Yagita H, Solary E, Ghiringhelli F (2008) CD4+CD25+ Tregs control the TRAIL-dependent cytotoxicity of tumor-infiltrating DCs in rodent models of colon cancer. J Clin Invest 118:3751–3761
26. Deaglio S, Dwyer KM, Gao W, Friedman D, Usheva A, Erat A, Chen JF, Enjyoji K, Linden J, Oukka M, Kuchroo VK, Strom TB, Robson SC (2007) Adenosine generation catalyzed by CD39 and CD73 expressed on regulatory T cells mediates immune suppression. J Exp Med 204:1257–1265
27. Cao X, Cai SF, Fehniger TA, Song J, Collins LI, Piwnica-Worms DR, Ley TJ (2007) Granzyme B and perforin are important for regulatory T cell-mediated suppression of tumor clearance. Immunity 27:635–646
28. Gondek DC, Lu LF, Quezada SA, Sakaguchi S, Noelle RJ (2005) Cutting edge: contact-mediated suppression by CD4+CD25+ regulatory cells involves a granzyme B-dependent, perforin-independent mechanism. J Immunol 174:1783–1786
29. Zhao DM, Thornton AM, DiPaolo RJ, Shevach EM (2006) Activated CD4+CD25+ T cells selectively kill B lymphocytes. Blood 107:3925–3932
30. Curotto de Lafaille MA, Lafaille JJ (2002) CD4(+) regulatory T cells in autoimmunity and allergy. Curr Opin Immunol 14:771–778
31. Janssens W, Carlier V, Wu B, VanderElst L, Jacquemin MG, Saint-Remy JM (2003) CD4+CD25+ T cells lyse antigen-presenting B cells by Fas-Fas ligand interaction in an epitope-specific manner. J Immunol 171:4604–4612
32. Ludwig-Portugall I, Hamilton-Williams EE, Gottschalk C, Kurts C (2008) Cutting edge: CD25+ regulatory T cells prevent expansion and induce apoptosis of B cells specific for tissue autoantigens. J Immunol 181:4447–4451
33. Shimizu J, Yamazaki S, Sakaguchi S (1999) Induction of tumor immunity by removing CD25 +CD4+ T cells: a common basis between tumor immunity and autoimmunity. J Immunol 163:5211–5218
34. Ghiringhelli F, Menard C, Terme M, Flament C, Taieb J, Chaput N, Puig PE, Novault S, Escudier B, Vivier E, Lecesne A, Robert C, Blay JY, Bernard J, Caillat-Zucman S, Freitas A, Tursz T, Wagner-Ballon O, Capron C, Vainchencker W, Martin F, Zitvogel L (2005) CD4+CD25+ regulatory T cells inhibit natural killer cell functions in a transforming growth factor-beta-dependent manner. J Exp Med 202:1075–1085
35. Li MO, Wan YY, Flavell RA (2007) T cell-produced transforming growth factor-beta1 controls T cell tolerance and regulates Th1- and Th17-cell differentiation. Immunity 26:579–591
36. Nakamura K, Kitani A, Fuss I, Pedersen A, Harada N, Nawata H, Strober W (2004) TGF-beta 1 plays an important role in the mechanism of CD4+CD25+ regulatory T cell activity in both humans and mice. J Immunol 172:834–842
37. Smyth MJ, Teng MW, Swann J, Kyparissoudis K, Godfrey DI, Hayakawa Y (2006) CD4+ CD25+ T regulatory cells suppress NK cell-mediated immunotherapy of cancer. J Immunol 176:1582–1587
38. Nakamura K, Kitani A, Strober W (2001) Cell contact-dependent immunosuppression by CD4(+) CD25(+) regulatory T cells is mediated by cell surface-bound transforming growth factor beta. J Exp Med 194:629–644
39. Jonuleit H, Schmitt E (2003) The regulatory T cell family: distinct subsets and their interrelations. J Immunol 171:6323–6327

40. Castriconi R, Cantoni C, Della Chiesa M, Vitale M, Marcenaro E, Conte R, Biassoni R, Bottino C, Moretta L, Moretta A (2003) Transforming growth factor beta 1 inhibits expression of NKp30 and NKG2D receptors: consequences for the NK-mediated killing of dendritic cells. Proc Natl Acad Sci USA 100:4120–4125
41. Friese MA, Wischhusen J, Wick W, Weiler M, Eisele G, Steinle A, Weller M (2004) RNA interference targeting transforming growth factor-beta enhances NKG2D-mediated antiglioma immune response, inhibits glioma cell migration and invasiveness, and abrogates tumorigenicity in vivo. Cancer Res 64:7596–7603
42. Lee JC, Lee KM, Kim DW, Heo DS (2004) Elevated TGF-beta1 secretion and downmodulation of NKG2D underlies impaired NK cytotoxicity in cancer patients. J Immunol 172:7335–7340
43. Fernandez NC, Lozier A, Flament C, Ricciardi-Castagnoli P, Bellet D, Suter M, Perricaudet M, Tursz T, Maraskovsky E, Zitvogel L (1999) Dendritic cells directly trigger NK cell functions: cross-talk relevant in innate anti-tumor immune responses in vivo. Nat Med 5:405–411
44. Lucas M, Schachterle W, Oberle K, Aichele P, Diefenbach A (2007) Dendritic cells prime natural killer cells by trans-presenting interleukin 15. Immunity 26:503–517
45. Lahl K, Loddenkemper C, Drouin C, Freyer J, Arnason J, Eberl G, Hamann A, Wagner H, Huehn J, Sparwasser T (2007) Selective depletion of Foxp3+ regulatory T cells induces a scurfy-like disease. J Exp Med 204:57–63
46. Chang X, Gao JX, Jiang Q, Wen J, Seifers N, Su L, Godfrey VL, Zuo T, Zheng P, Liu Y (2005) The Scurfy mutation of FoxP3 in the thymus stroma leads to defective thymopoiesis. J Exp Med 202:1141–1151
47. Godfrey VL, Wilkinson JE, Rinchik EM, Russell LB (1991) Fatal lymphoreticular disease in the scurfy (sf) mouse requires T cells that mature in a sf thymic environment: potential model for thymic education. Proc Natl Acad Sci USA 88:5528–5532
48. Terme M, Ullrich E, Delahaye NF, Chaput N, Zitvogel L (2008) Natural killer cell-directed therapies: moving from unexpected results to successful strategies. Nat Immunol 9:486–494
49. Terme M, Chaput N, Combadiere B, Ma A, Ohteki T, Zitvogel L (2008) Regulatory T cells control dendritic cell/NK cell cross-talk in lymph nodes at the steady state by inhibiting CD4+ self-reactive T cells. J Immunol 180:4679–4686
50. Giroux M, Yurchenko E, St-Pierre J, Piccirillo CA, Perreault C (2007) T regulatory cells control numbers of NK cells and CD8alpha+ immature dendritic cells in the lymph node paracortex. J Immunol 179:4492–4502
51. Terme M, Tomasello E, Maruyama K, Crepineau F, Chaput N, Flament C, Marolleau JP, Angevin E, Wagner EF, Salomon B, Lemonnier FA, Wakasugi H, Colonna M, Vivier E, Zitvogel L (2004) IL-4 confers NK stimulatory capacity to murine dendritic cells: a signaling pathway involving KARAP/DAP12-triggering receptor expressed on myeloid cell 2 molecules. J Immunol 172:5957–5966
52. Laouar Y, Sutterwala FS, Gorelik L, Flavell RA (2005) Transforming growth factor-beta controls T helper type 1 cell development through regulation of natural killer cell interferon-gamma. Nat Immunol 6:600–607
53. Newman KC, Riley EM (2007) Whatever turns you on: accessory-cell-dependent activation of NK cells by pathogens. Nat Rev Immunol 7:279–291
54. Trzonkowski P, Szmit E, Mysliwska J, Dobyszuk A, Mysliwski A (2004) CD4+CD25+ T regulatory cells inhibit cytotoxic activity of T CD8+ and NK lymphocytes in the direct cell-to-cell interaction. Clin Immunol 112:258–267
55. Romagnani C, Della Chiesa M, Kohler S, Moewes B, Radbruch A, Moretta L, Moretta A, Thiel A (2005) Activation of human NK cells by plasmacytoid dendritic cells and its modulation by CD4+ T helper cells and CD4+ CD25hi T regulatory cells. Eur J Immunol 35:2452–2458

56. Sasada T, Kimura M, Yoshida Y, Kanai M, Takabayashi A (2003) CD4+CD25+ regulatory T cells in patients with gastrointestinal malignancies: possible involvement of regulatory T cells in disease progression. Cancer 98:1089–1099
57. Woo EY, Chu CS, Goletz TJ, Schlienger K, Yeh H, Coukos G, Rubin SC, Kaiser LR, June CH (2001) Regulatory CD4(+)CD25(+) T cells in tumors from patients with early-stage non-small cell lung cancer and late-stage ovarian cancer. Cancer Res 61:4766–4772
58. Lutsiak ME, Semnani RT, De Pascalis R, Kashmiri SV, Schlom J, Sabzevari H (2005) Inhibition of CD4(+)25+ T regulatory cell function implicated in enhanced immune response by low-dose cyclophosphamide. Blood 105:2862–2868
59. Ghiringhelli F, Larmonier N, Schmitt E, Parcellier A, Cathelin D, Garrido C, Chauffert B, Solary E, Bonnotte B, Martin F (2004) CD4+CD25+ regulatory T cells suppress tumor immunity but are sensitive to cyclophosphamide which allows immunotherapy of established tumors to be curative. Eur J Immunol 34:336–344
60. Liu JY, Wu Y, Zhang XS, Yang JL, Li HL, Mao YQ, Wang Y, Cheng X, Li YQ, Xia JC, Masucci M, Zeng YX (2007) Single administration of low dose cyclophosphamide augments the antitumor effect of dendritic cell vaccine. Cancer Immunol Immunother 56:1597–1604
61. Onizuka S, Tawara I, Shimizu J, Sakaguchi S, Fujita T, Nakayama E (1999) Tumor rejection by in vivo administration of anti-CD25 (interleukin-2 receptor alpha) monoclonal antibody. Cancer Res 59:3128–3133
62. Nishikawa H, Jager E, Ritter G, Old LJ, Gnjatic S (2005) CD4+ CD25+ regulatory T cells control the induction of antigen-specific CD4+ helper T cell responses in cancer patients. Blood 106:1008–1011
63. Smyth MJ, Crowe NY, Godfrey DI (2001) NK cells and NKT cells collaborate in host protection from methylcholanthrene-induced fibrosarcoma. Int Immunol 13:459–463
64. Glas R, Franksson L, Une C, Eloranta ML, Ohlen C, Orn A, Karre K (2000) Recruitment and activation of natural killer (NK) cells in vivo determined by the target cell phenotype. An adaptive component of NK cell-mediated responses. J Exp Med 191:129–138
65. Apperley JF, Gardembas M, Melo JV, Russell-Jones R, Bain BJ, Baxter EJ, Chase A, Chessells JM, Colombat M, Dearden CE, Dimitrijevic S, Mahon FX, Marin D, Nikolova Z, Olavarria E, Silberman S, Schultheis B, Cross NC, Goldman JM (2002) Response to imatinib mesylate in patients with chronic myeloproliferative diseases with rearrangements of the platelet-derived growth factor receptor beta. N Engl J Med 347:481–487
66. Buchdunger E, O'Reilly T, Wood J (2002) Pharmacology of imatinib (STI571). Eur J Cancer 38 (Suppl 5):S28–S36
67. Heinrich MC, Blanke CD, Druker BJ, Corless CL (2002) Inhibition of KIT tyrosine kinase activity: a novel molecular approach to the treatment of KIT-positive malignancies. J Clin Oncol 20:1692–1703
68. Borg C, Terme M, Taieb J, Menard C, Flament C, Robert C, Maruyama K, Wakasugi H, Angevin E, Thielemans K, Le Cesne A, Chung-Scott V, Lazar V, Tchou I, Crepineau F, Lemoine F, Bernard J, Fletcher JA, Turhan AG, Blay JY, Spatz A, Emile JF, Heinrich MC, Mécheri S, Tursz T, Zitvogel L (2004) Novel mode of action of c-kit tyrosine kinase inhibitors leading to NK cell-dependent antitumor effects. J Clin Invest 114:379–388
69. Ghiringhelli F, Menard C, Martin F, Zitvogel L (2006) The role of regulatory T cells in the control of natural killer cells: relevance during tumor progression. Immunol Rev 214:229–238
70. Ghiringhelli F, Menard C, Puig PE, Ladoire S, Roux S, Martin F, Solary E, Le Cesne A, Zitvogel L, Chauffert B (2007) Metronomic cyclophosphamide regimen selectively depletes CD4+CD25+ regulatory T cells and restores T and NK effector functions in end stage cancer patients. Cancer Immunol Immunother 56:641–648
71. Cudkowicz G, Bennett M (1971) Peculiar immunobiology of bone marrow allografts. II. Rejection of parental grafts by resistant F 1 hybrid mice. J Exp Med 134:1513–1528
72. Barao I, Hanash AM, Hallett W, Welniak LA, Sun K, Redelman D, Blazar BR, Levy RB, Murphy WJ (2006) Suppression of natural killer cell-mediated bone marrow cell rejection by CD4+CD25+ regulatory T cells. Proc Natl Acad Sci USA 103:5460–5465

73. Lund JM, Hsing L, Pham TT, Rudensky AY (2008) Coordination of early protective immunity to viral infection by regulatory T cells. Science 320:1220–1224
74. Lund JM, Linehan MM, Iijima N, Iwasaki A (2006) Cutting edge: plasmacytoid dendritic cells provide innate immune protection against mucosal viral infection in situ. J Immunol 177:7510–7514
75. Milligan GN, Bernstein DI (1997) Interferon-gamma enhances resolution of herpes simplex virus type 2 infection of the murine genital tract. Virology 229:259–268
76. Kim S, Iizuka K, Aguila HL, Weissman IL, Yokoyama WM (2000) In vivo natural killer cell activities revealed by natural killer cell-deficient mice. Proc Natl Acad Sci USA 97:2731–2736
77. Main EK, Lampson LA, Hart MK, Kornbluth J, Wilson DB (1985) Human neuroblastoma cell lines are susceptible to lysis by natural killer cells but not by cytotoxic T lymphocytes. J Immunol 135:242–246
78. Ruggeri L, Capanni M, Urbani E, Perruccio K, Shlomchik WD, Tosti A, Posati S, Rogaia D, Frassoni F, Aversa F, Martelli MF, Velardi A (2002) Effectiveness of donor natural killer cell alloreactivity in mismatched hematopoietic transplants. Science 295:2097–2100
79. Pierson BA, Miller JS (1996) CD56+bright and CD56+dim natural killer cells in patients with chronic myelogenous leukemia progressively decrease in number, respond less to stimuli that recruit clonogenic natural killer cells, and exhibit decreased proliferation on a per cell basis. Blood 88:2279–2287
80. Kiladjian JJ, Fenaux P, Caignard A (2007) Defects of immune surveillance offer new insights into the pathophysiology and therapy of myelodysplastic syndromes. Leukemia 21:2237–2239
81. Kortylewski M, Kujawski M, Wang T, Wei S, Zhang S, Pilon-Thomas S, Niu G, Kay H, Mule J, Kerr WG, Jove R, Pardoll D, Yu H (2005) Inhibiting Stat3 signaling in the hematopoietic system elicits multicomponent antitumor immunity. Nat Med 11:1314–1321
82. Groh V, Wu J, Yee C, Spies T (2002) Tumour-derived soluble MIC ligands impair expression of NKG2D and T-cell activation. Nature 419:734–738
83. Salih HR, Antropius H, Gieseke F, Lutz SZ, Kanz L, Rammensee HG, Steinle A (2003) Functional expression and release of ligands for the activating immunoreceptor NKG2D in leukemia. Blood 102:1389–1396
84. Kim R, Emi M, Tanabe K, Uchida Y, Toge T (2004) The role of Fas ligand and transforming growth factor beta in tumor progression: molecular mechanisms of immune privilege via Fas-mediated apoptosis and potential targets for cancer therapy. Cancer 100:2281–2291
85. Gil-Guerrero L, Dotor J, Huibregtse IL, Casares N, Lopez-Vazquez AB, Rudilla F, Riezu-Boj JI, Lopez-Sagaseta J, Hermida J, Van Deventer S, Bezunartea J, Llopiz D, Sarobe P, Prieto J, Borras-Cuesta F, Lasarte JJ (2008) In vitro and in vivo down-regulation of regulatory T cell activity with a peptide inhibitor of TGF-beta1. J Immunol 181:126–135
86. Maruyama T, Watanabe K, Takei I, Kasuga A, Shimada A, Yanagawa T, Kasatani T, Suzuki Y, Kataoka K, Saruta, et al. (1991) Anti-asialo GM1 antibody suppression of cyclophosphamide-induced diabetes in NOD mice. Diabetes Res 17:37–41
87. Poirot L, Benoist C, Mathis D (2004) Natural killer cells distinguish innocuous and destructive forms of pancreatic islet autoimmunity. Proc Natl Acad Sci USA 101:8102–8107
88. Dejaco C, Duftner C, Grubeck-Loebenstein B, Schirmer M (2006) Imbalance of regulatory T cells in human autoimmune diseases. Immunology 117:289–300
89. Roy S, Barnes PF, Garg A, Wu S, Cosman D, Vankayalapati R (2008) NK cells lyse T regulatory cells that expand in response to an intracellular pathogen. J Immunol 180:1729–1736

Interactions Between B Lymphocytes and NK Cells: An Update

Dorothy Yuan, Ning Gao, and Paula Jennings

Abstract The major role of NK cells has traditionally been assigned to their cytotoxic activity both against tumor cells and against virus-infected cells. Only recently has there been greater appreciation for their more diverse functions. The interaction between NK and B cells is unique in that NK cells have not been reported to kill B cells, allowing, therefore productive interactions between the two cell types. In this review we have focused on the ability of NK cells to influence the two major functions of B cells, their ability to secrete antibodies and to present antigen to T cells. This review provides the definitive evidence obtained from in vitro experiments that NK and B cells can interact productively in both directions. In addition, much progress has been made toward identification of the receptor–ligand pairs required for the interaction. This review also points out that whereas there is evidence for similar interactions to occur in vivo, unraveling the mechanisms by which this occurs is more challenging due to the many players involved. For this same reason we have focused the review in the mouse system which is more amenable to in vivo manipulation. We realize that much insights derived from work with human cells should be informative, but complete correlation with findings between the two system is beyond the scope of this review.

1 Introduction

There is ample documentation that the initiation of specific B cell antibody secretion requires, in addition to activation of the antigen receptor, interaction with either T cells or with pathogens that can directly trigger specific Toll-like receptors (TLRs) [1]. However, there is also accumulating evidence that the various functions of B cells can be also modulated by another cell type, natural killer (NK) cells.

D. Yuan (✉), N. Gao and P. Jennings
Department of Molecular Pathology, University of Texas Southwestern Medical Center, Texas, USA

Unlike B or T cells, NK cells do not possess antigen-specific receptors. On the other hand, they are equipped with receptors that allow them to distinguish self vs. nonself as well as specific pathogen motifs (reviewed in [2]). In addition, they express a number of cytokine receptors that can overcome the inhibitory receptors that keep them quiescent. These receptors are important for their expansion as well as further activation. In this review we will highlight recent findings that show how NK cells can direct the path of antigen-specific B cell responses. Among the most important of these functions is the ability of NK cells to selectively alter the effector function of B cells by modifying isotype switching, most commonly associated with an increase in the expression of the immunoglobulin (Ig) G2a/c subclass. IgG2a/c has been shown to be particularly adept at mediating antibody-dependent cytotoxicity (ADCC), binding to Fc receptors on phagocytes, as well as initiating complement fixation. Therefore a rapid induction by NK cells of this subclass prior to T cell participation can provide significant advantages in early responses to pathogens. In addition to the influence on Ig switch recombination more recent findings suggest that NK cells can also modulate yet another function of B cells, that of antigen presentation. This more subtle effect may thus also influence the adaptive response during the memory phase.

Conversely, we will also review evidence that B cells can in turn affect the differentiation of NK cells. Thus, in addition to activation by cytokines, NK cells can be activated by novel interactions with B cells resulting also in the initiation of cytokine production. These interactions are important points to consider in any dissection of the immunological response to pathogens as well as treatment regiments that implicate the immune system.

2 NK Modulation of B Cell Responses In Vivo

2.1 NK Cell Modulation of B Cell Responses to T-Independent Antigens

The most direct test of whether NK cells can modulate specific B cell responses independent of factors that can affect the quality of T cell help is to examine antibody responses to T-independent (TI) antigens. Moreover, the use of well-studied synthetic antigens representative of those expressed on pathogens offers the advantage of determining minimal effects that are independent of complications arising from the infective process. This approach further targets B cell responses to individual antigens that may be obscured by additional responses to other determinants expressed by the pathogen. Classical studies have shown that responses to TI-I antigens require activation of the B cell receptor together with TLRs on B cells, while responses to TI-II antigens require only extensive crosslinking of the B cell receptor as well as a source of cytokines which is not derived from direct MHCII-restricted interaction with T cells [3]. In addition, stimulation of the TACI, a receptor for the activators, BAFF or APRIL on B cells appears to be required for the TI-II response [4]. Importantly, the source of these TAC1-binding factors can be derived from a number of cell types [5]. Interestingly, a recent report indicates that CD11c(hi) dendritic cells (DCs) are not required for the TI-II response [6] further confirms that T cell participation via antigen presentation by DCs is not necessary. Our initial studies, using the simple approach of examining B cell responses after depletion of NK cells prior to challenge with either a TI-I antigen, TNP-LPS, or a TI-II antigen, NP-Ficoll, showed that neither the magnitude of serum Ig levels nor the extent of subsequent switching to various IgG subclasses was affected [7]. Significantly, the depletion of NK cells also does not result in increased responses to these antigens; therefore, at least in C57BL/6 mice, without specific stimulation, NK cells exert neither stimulatory nor inhibitory effects on B cell responses.

In contrast to these findings, activation of NK cells prior to challenge with these antigens showed that B cell responses, can be significantly modulated ([7] and Yuan, personal observations). These studies utilized Poly(I:C), a frequently used reagent for the stimulation of the cytolytic activity of NK cells. Poly(I:C) is a ligand for TLR3. Unlike human NK cells [8] TLR3 is not functionally expressed on mouse NK cells [9] and the activation of NK cells requires stimulation by cytokines secreted by either macrophages or DCs that express these receptors. Both IFN-α/β and IL-12 produced by these cells are potent stimulators of IFN-γ secretion by NK cells. IFN-γ in turn can stimulate cells expressing the receptor for this cytokine thus creating a cytokine circuit resulting in significant amplification of these cytokines. Because IFN-γ can enhance IgG2a/c synthesis by B cells activated by either LPS or by antigen, it is not surprising that activation of NK cells by this route can increase the production of this antibody subclass in response to both TI-I and TI-II antigens. The obligatory role of NK cells can be shown by depleting NK cells prior to the administration of Poly(I:C), resulting in the abrogation of the enhancement of antigen-specific IgG2c production [7]. In other studies we have utilized a B cell tumor that can directly stimulate NK-cell IFN-γ production in vitro. Injection of the tumor also enhances IgG2a/c production. Not only is this enhancement eliminated by depletion of NK cells, the enhancement is also dependent on the presence of IL-12, showing that despite the direct stimulation of NK-cell IFN-γ secretion amplification of a cytokine circuit is also required [10]. It should be noted that, for these early studies an anti-IgG2a antibody was used. We subsequently became aware that the gene for γ2a is replaced by γ2c in C57BL/6 mice [11]; therefore, the low levels of IgG2a reported are most likely representative of cross-reactivity for much higher levels of IgG2c.

In addition to Poly(I:C), other adjuvants such as RIBI or incomplete Freund's adjuvant (CFA) that can enhance immune responses can also stimulate NK cells sufficiently to result in their ability to significantly increase B cell responses to TI antigens [12]. In contrast to Poly(I:C) stimulation, how NK cells are activated in this case is not clear since the question of whether stimulation by these agents requires activation of TLRs is still controversial [13].

In conclusion, the relatively low level of B cell responses to TI antigens can be significantly enhanced by NK cells, if and only if, they are first activated.

2.2 NK Cell Modulation of B Cell Responses to Viral Infection

Classical studies have clearly documented a role for NK cells in the early control of some viruses [14, 15]. Modification of cell surface antigens by the infectious process can activate NK cells by either reducing the function of inhibitory receptors or by induction of specific activating ligands. A well-studied gene product expressed by the murine Cytomegalovirus (MCMV) is m157 ORF which can activate Ly49H expressed on NK cells of MCMV-resistant mice [16]. Thus, during MCMV infection, activation of NK cells can occur via two phases, the first involving the amplification of the cytokine circuit due to initial activation of macrophages or DCs via their TLRs, and the second occurring upon direct recognition of the virally expressed antigen expressed by the infected cells [17]. Direct cytopathic effects of the activated NK cells as well as IFN-γ produced by the cells play an important role in control of MCMV infection. Whereas the increased IFN-γ [15] may not always be required for the control of some viruses [18], it is notable, however, that IFN-γ produced upon MCMV infection is correlated with increased levels of IgG2a with specificity both against the virus and against a number of other polyclonal antigens. The polyclonal antibody production was found to be independent of T cell activation [19]. Whereas depletion of NK cells increases pathogenicity of MCMV, the effect on antibody levels has not been assessed. Thus it is possible that MCMV-activated NK cells can directly stimulate B cells via a BCR-independent pathway.

Antibody production by B cells plays an important role in the resolution of many, although clearly not, all viral infections. Antibodies are usually required for neutralization of preformed viruses, for prevention of adhesion, and for mediating ADCC of infected cells thus reducing virus spread. In fact there is evidence that in some cases, the function of antibodies may be necessary and sufficient for the control of infection. In B cell deficient or in SCID mice, injection of preformed antibodies into vesicular stomatitis virus (VSV)-infected mice can provide complete protection. Similarly in the absence of B or T cell function, SCID mice injected with various monoclonal antiviral antibodies can cure an influenza-virus-induced pneumonia [20]. Furthermore, immunization regiments in intact mice that allow the preferential generation of IgG2a antibodies have been shown to provide greater protection against viral influenza infection [21].

The induction of both T and B cell responses against most virus infections complicates the assessment of whether NK cells constitute a necessary or auxiliary component contributing to the resolution of the infection. However, the ability of viruses to activate TLR-3 on infected cells is comparable with our results showing NK-dependent enhancement of the IgG2c response by Poly(I:C) which functions by stimulating TLR-3 on accessory cells suggesting a role for NK cells in helping with viral clearance. Indeed many virus infections elicit a predominant IgG2a antibody response [22] which is presumably due to the development of a cytokine circuit resulting in the activation of NK cells and the production of IFN-γ. Thus, in the absence of IFN-γ, virus-specific IgG2a levels have been shown to be significantly reduced in influenza-infected mice [23, 24]. Nonetheless, in order to establish whether NK cell induction of B cell Ig production is sufficient for the control of some viruses it is necessary to dissociate the role of T cells. A number of viruses have been shown to display cell surface determinants that are sufficiently repetitive so that they can adequately crosslink the B cell antigen receptor resulting in the activation of B cells independent of MHCII-mediated T cell help. Analysis of the role of NK cells as an alternate source of help has utilized these viruses. For example, response to polyoma virus does not require T cell participation [25]. Significantly, in mice deficient in both T and NK cells, infection by this virus results in IgG2a levels that are more reduced than those lacking only T cells implicating a role for NK cell help in obtaining an optimal response [25]. Another virus shown to be able to induce a response similar to that of a TI-II antigen is VSV, which expresses sufficient repetitive determinants to result in extensive crosslinking of the BCR [26]. Therefore the response to VSV also does not require T cell help. Further studies showed that resolution of the infection was not greatly compromised in the absence of C3 or TNF [27]. The studies also concluded that NK cell help is not involved because enhanced NK cytotoxicity was not detected. However, it should be noted that since induction of NK cell cytokine production is often dissociated from cytotoxicity [28], the possibility remains that NK cell help play an important role in the generation of antibodies in response to this infection.

Despite the prominent role of IFN-γ in directing B cell IgG2a production it is worthy to note that there is convincing evidence that IFN-γ is not absolutely required for the induction of this isotype by virus infection. The level of IgG2a antibodies produced upon infection by lactate dehydrogenase-elevating virus (LDV) was not substantially decreased in IFN-γ knockout (KO) mice [29]. It is not known whether NK cells participate in the induction because it is only partially blocked by anti-CD40 antibodies. Furthermore, NK cells can also be involved in the control of this infection since IgG2a antibodies have been shown to be more effective for NK-cell-mediated ADCC [30].

Recent studies have also revealed that, in contrast to stimulation by double-stranded RNA, the ssRNA genome of influenza virus, is recognized by TLR7 [31]. Furthermore, virus infection can directly activate B cells via TLR7. It is interesting that this stimulation can only result in activation of IgG1 secretion. For switching to IgG2a activation of CD40 is required [32]. Whereas this finding implicates a role for T cells, the possibility that stimulation of other ligands on B cells can substitute for this costimulator requires further investigation.

In summary, a number of studies have suggested that upon infection by some viruses, NK cells, as well as IFN-γ produced by activated NK cells, can result in the enhancement of IgG2a production by B cells. Whereas in some cases antibodies and, in particular IgG2a, can be shown to be protective, the relative role of direct cytotoxicity against the infected cells vs. antibody-mediated effects can vary greatly. In addition, whether the isotype preference can be attributed to direct NK-B cell interaction also requires further elucidation.

2.3 NK Cell Modulation of B Cell Responses to Bacterial Infections

Classical studies using *Listeria monocytogenes* have shown that NK cells play an important role in the early response against infection by this organism mainly via amplification of the cytokine circuit initiated by activation of macrophages [33]. The increase in IFN-γ secretion is important for bactericidal activity of infected macrophages; however, antibodies produced during the normal course of the infection were shown not to be important for control of this mainly intracellular bacterium. Analyses of the requirement for NK cells in the resolution of the intracellular phase of a number of other bacterial infections have also not revealed a critical role [34–36].

On the other hand, it has long been known that antibodies are critically important in recovery from infections caused by most extracellular bacteria by virtue of their ability to activate the complement pathway, to enhance phagocytosis by macrophages, and to neutralize bacterial toxins. Since the relative efficacy of these pathways depends on the antibody subclass, the ability of NK cells to modulate antibody subclass distribution should be relevant in resolving some of these infections. As with viral infections, macrophages or neutrophils are activated to produce cytokines, including Types I IFNs, TNF-α, IL-12, and IL-15, that can activate NK cells. In this case activation by bacteria occurs mainly via TLR-4 and TLR-2. Thus a model antigen for this response can be TNP-LPS, where TLR-4 is a necessary component for the activation of TNP-specific B cells. However, depletion of NK cells prior to immunization by this antigen does not alter the magnitude or isotype distribution of the antibodies produced. Thus it appears that stimulation via TLR4 is insufficient for the activation of NK cells. On the other hand, if Poly(I:C) is injected prior to immunization with the same antigen, then IgG2c levels can be significantly enhanced and this enhancement is abolished by NK cell depletion [7]. Since mouse NK cells do not express TLR-3 [9], the role of Poly(I:C) cannot be attributed to direct stimulation of NK cells. The mechanism could be due to either alterations in the quantity or quality of cytokines induced by the combination of the two TLR ligands as well as the LPS receptor.

A class of bacteria including *Streptococcus pneumoniae*, *Haemophilus influenzae*, and *Klebsiella pneumoniae* are coated by large polysaccharides. The repetitive units of these polysaccharides are highly effective crosslinkers of BCR and therefore should activate B cells in a TI manner. However, resistance against these infections is compromised in the young, presumably due to the paucity of appropriate cytokines. The deficit has been effectively remedied by conjugating vaccines made with these carbohydrates with proteins that can stimulate MHC II mediated T cell help [37]. We have found that immunization in the presence of RIBI, an adjuvant that is an effective stimulator of NK cells, can augment the immune response against these carbohydrate antigens. Indeed the enhancement of the response is dependent on the presence of NK cells [12]. Thus, these results show that, as with the Poly(I:C) enhancement of the response against the representative

TI-II antigen, NP-Ficoll, activation of NK cells can serve to substitute the requirement for T-cell-derived cytokines.

In addition to the role of NK cells in propagating the cytokine circuit, the expression of TLRs on NK cells themselves may be altered by specific cytokines resulting in alterations of their activation status such that they may be able to activate B cells directly. We have recently examined this possibility by using infection by the microbe *Brucella abortis* (BA). The consequence of BA infection differs from many other bacterial infections in the induction of substantial polyclonal IgG2 production which is T cell independent [38]. Whereas early experiments have shown that IFN-γ plays a role in this induction, whether NK cells are required has not been examined. Heat-inactivated BA has been shown to be as effective as infective BA in its ability to induce the polyclonal B cell response [39]. Thus despite the inability of fixed BA to multiply, various cell types are necessary for this response. The polyclonal nature of the B cell response suggests that B cells themselves can be directly activated via receptors other than the BCR. However, since infection is accompanied by an increase in NK cell cytotoxicity [34], it is also possible that the induction requires direct interactions between NK and B cells. In order to determine the role of NK cell involvement in the induction of polyclonal antibodies, we assessed the effect of depletion of NK cells on the response to fixed BA. Figure 1 shows that polyclonal IgG2c production was dramatically reduced by the absence of NK cells. The extent of reduction (70%) was reproducibly found to be greater than the effect of IFN-γ (50%) or IL-12 (50%) depletion [Dang and Yuan, personal observations] suggesting that other than the effect of the cytokine

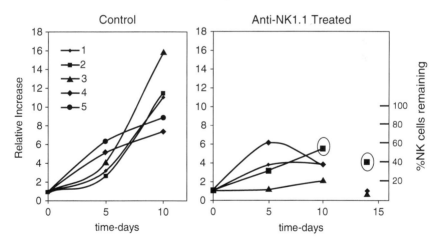

Fig. 1 Effect of NK depletion on increase in total IgG2c serum levels. Prior to immunization with fixed BA a set of animals was depleted of NK cells by anti-NK1.1 treatment. Induction of polyclonal IgG2c was determined by ELISA analysis of serum samples. The relative increase was obtained by dividing the IgG2c levels at each time point by the prebleed value of each animal. The *right-hand axis* indicates the percentage of NK cells remaining in each animal at d14 after treatment. Note significant recovery in the animal (*encircled*) with the highest titer

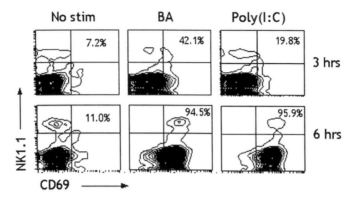

Fig. 2 Comparative kinetics of activation of NK cells by fixed BA vs. Poly(I:C). Total splenocytes stimulated with each reagent as indicated were stained at the times indicated and analyzed by FACS

circuit on the increase in IgG2c levels NK cells may play an additional role. Indeed in an experiment shown in Fig. 2, as early as 3 h after injection of BA, a large fraction of NK cells were activated. The stimulation of CD69 expression occurred prior to that on all the other splenic populations, including B cells.

It is not known how NK cells are directly activated by BA; however, their expression of TLR2 has been shown to be enhanced by IL-12. Thus, it is possible that for sufficient activation of NK cells resulting in production of IFN-γ they require, in addition, a second signal such as upregulation of their TLR2 [40]. Alternatively, BA-activated B cells could directly stimulate NK cells resulting eventually in cells that can further induce B cells directly. Inasmuch as the stimulation of polyclonal Ig secretion is more affected by NK depletion than by IFN-γ depletion it is possible that dual signals are responsible for the induction of B cell Ig secretion, one derived from cytokines and the other from direct interaction between B and NK cells. The role of putative ligands for this stimulation using in vitro studies is presently being pursued. Based on our in vivo findings we have developed a working hypothesis illustrated in Fig. 3.

In conclusion, activation of NK cells by bacterial pathogens either directly or indirectly has importance both for their bactericidal and deleterious effects due to excess activation of the inflammatory network. Whereas antibodies and especially those of the IgG2a subclass, induced during the immune response undeniably play an important role in recovery from and prevention of subsequent infections, the requirement for NK activation in this aspect requires further investigation. Our studies utilizing the ability of fixed BA to stimulate TI polyclonal Ig production may provide further insight.

It should be noted that this discussion of the role of NK cells in bacterial infections has not included the effect of NKT cells because we have ascertained that our regiment of NK depletion does not affect NKT cells [12]. More relevant is the fact that the majority of NKT cells express a defined TCR specificity that recognizes CD1d-restricted glycolipid antigens, the most prominent of which being

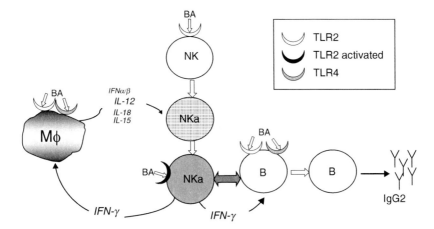

Fig. 3 Possible pathways of NK cell effects on induction of polyclonal Ig secretion. BA can activate both NK and B cells via TLR2 as well as other receptors but activation level is insufficient until accessory cells are also activated resulting in production of IL-12 which can upregulate TLR2. NK cells can now stimulate B cells directly and together with IFN-γ can program B cells to increase polyclonal IgG2c secretion

alphagalactosylceramide (alphaGC). These lipids are expressed on some Gram-negative, LPS-negative microbes [41, 42] and activation of appropriate T cells and associated cytokines can increase nonantigen specific help for B cells [43]. However the immunogenicity of these lipids in the context of most bacterial infections remains an open question [44].

2.4 Role of NK Cells on Antigen Presentation by B cells: Evidence of NK Cell Effect on T Cell Responses

While the antibody response to pathogens can occur in the absence of T cells, this response does not result in the generation of either memory B or T cells. For a robust secondary response T cell activation is required. Whereas the involvement of NK cells is usually considered to be part of the innate system that is rapidly activated prior to the initiation of T cell functions it is also possible that the early response of NK cells can affect B cell antibody production directly or through the quality of T cell help for B cells. To investigate this question we examined the primary antibody response to a classical T-dependent antigen, TNP-KLH and did not find detectable effects on the level of antigen-specific serum antibodies produced in NK-cell-depleted mice [7]. This finding was somewhat surprising in view of the fact that the antigen was injected in adjuvant which should have activated NK cells. In contrast, other studies using similar routes of antigenic challenge, but in a transgenic mouse model, concluded that the extent of T-cell-dependent switching to

IgG2a was dependent on NK cells [45]; however, the absence of NK cells in these transgenic mice was accompanied by other abnormalities, including cytokine dysregulation [46], that may affect the antibody response. Contradictory conclusions were obtained in two different laboratories in which the T-dependent antigen OVA was administered by aerosol [47–49]. These inconsistent results could be explained by subtle effects of NK cells on the different cell types involved in the generation of a T-cell-dependent response. One factor affecting these inconsistent results could be attributed to differences in direct NK cell effects on T cells which in turn influences the B cell antibody levels. For instance, NK cells can enhance Th1-mediated immunity via cytokine production in the lymph node [50]. NK cells may also directly enhance T cell proliferation through CD244 (2B4)–CD48 interactions [51]. Another possible pathway involves interactions between DCs and NK cells. NK cells can induce maturation of DCs [52] as well as eliminate mature DCs [53] thus potentially modulating the extent of T cell activation (reviewed in [54]). Recently it has been shown that NK cells can decrease antibody responses to Hepatitis B Surface antigen (HBsAg), which is correlated with decreased ability of DCs to present HBsAg to memory T cells [55]. In addition, human NK cells may acquire an antigen-presenting cell (APC) phenotype directly regulating T cell activation [56], although this ability has since been challenged [57].

In recent experiments we have confirmed earlier findings that depletion of NK cells did not affect the primary antibody response. Interestingly, however, we found that their absence during priming resulted in a decreased T cell memory response that is reflected in reduced antigen-specific serum IgG1 levels [58] upon secondary challenge. Significantly, the decrease was observed after complete recovery of the NK cell compartment. In order to determine more directly whether these reduced levels are attributable to NK effects on T cell responses we tested whether NK depletion can influence the extent of proliferation of CFSE labeled T cells from OT-II transgenic mouse transferred into C57BL/6 mice in response to ovalbumin administered in the adjuvant, Ribi. The greater representation of OVA peptide-specific T cells facilitated the assessment of antigen-specific T cell activation by either CD69 upregulation or proliferation. NK cell depletion was found to cause a small but reproducibly significant decrease in T cell activation and proliferation [58]. This apparent effect of NK cells could be a reflection of three independent but possibly related interactions. (1) NK cells could directly alter T cell responses, or (2) NK cells could influence the manner by which professional APCs present antigen to T cells or (3) NK cells could affect B cell presentation of antigen. We chose to examine the last possibility because a number of studies point toward a prominent role of B cells for T cell priming [59] which is correlated with more effective induction of memory T cells [60]. We generated radiation chimeric mice reconstituted with MHCII deficient B cells together with other APCs that were MHCII sufficient as described previously [61]. T cell priming was found to be greatly reduced in these mice confirming the role of B cells in antigen presentation. More importantly, the reduction was not further affected by the absence of NK cells. Furthermore, T cell priming can be restored by reconstitution of B cells only if NK cells were also present [58]. These results show, at least in this experimental system, that the effect of NK cells on T cell memory development can be mediated by modulation of antigen presentation by B cells.

In further in vitro experiments we have directly explored the effect of NK cells on B cell antigen presentation in an effort to determine the mechanism of enhancement (See Sect. 3.2). These experiments allowed us to conclude that direct cell–cell interaction between NK and B cells is required for the enhancement of antigen presentation by B cells and that soluble factors produced by NK cells are not sufficient for the induction. Moreover, it is important that NK cells directly isolated from the animal that has not been activated in vitro cannot enhance the B cell antigen presentation.

In these in vivo studies, the question that frequently arises is whether NKT cells were deleted along with NK cells when they were injected with anti-NK1.1 and therefore may be responsible for the effects. We have shown that NKT cells remain intact in the liver even with chronic depletion with this antibody [12]. Moreover, this question may not be critical for protein antigens such as KLH since greater than 80% of NKT cells express a restricted TCR specificity that recognizes CD1d-restricted glycolipid antigens. There is now clear evidence that upon recognition of these lipids, B cells can present antigen to NKT cell and can provide help for B cell Ig secretion even in the absence of adjuvant, CD4+ T cells [43], or MHC class II [62]. Interestingly, NKT cells from autoimmune mice may be particularly adept at providing CD1-restricted T cell help [63]. Clearly, however, this non-MHC class II-mediated help does not account for the role of conventional NK cells in the generation of antigen-specific CD4+ memory T cells.

The augmentation of B cell antigen presentation by NK cells resulting in enhanced T cell priming is an important factor to consider for blood-borne infections since the offending organism is likely to encounter and activate NK cells in the red pulp prior to localization in the marginal zone of the splenic follicles where large numbers of B cells can process the antigen. Further investigations should reveal whether migration of activated NK cells into the same area can result in encounter with the B cells that can process antigen. Of particular relevance in this regard is the finding by Linton et al. that costimulation via OX40L expressed by B cells is sufficient for the generation of memory T cells [64]. OX40 has been shown by gene expression profiling to be present on NK cells as well as T cells [65].

2.5 Possible Role of NK Cells in Autoantibody Production

Over the years a number of studies have explored the role of NK cells in the development of autoantibodies. These studies have been reviewed previously [66]. Due to the complexity of factors that could contribute to both autoantibody production and the further disease manifestations, the exact role of NK cells has not been easy to pinpoint. To examine the participation of NK cells in a more restricted context we have recently utilized a model system in which production of autoantibodies has been shown to be regulated by a gene segment defined as the

B6.Sle1b gene cluster. This gene cluster was derived from the NZM2410 mouse strain which spontaneously develop autoantibodies including antinuclear antibodies (ANA) and kidney disease. *B6.Sle1b* mice also develop autoantibodies late in life with high penetrance but do not progress to kidney disease. Further experiments also showed that for the development of ANA, B cells from this strain are the only necessary component. This locus contains the SLAM/CD2 gene cluster, which encodes numerous cellular interaction molecules expressed by B, T, as well as NK cells. One of these genes encodes CD244 which is expressed predominantly on NK cells. Interestingly, we have previously shown that activation of NK cells by B cells can occur via the interaction between CD48, another gene contained within the SLAM/CD2 cluster, and CD244 [67]. Therefore the interaction between NK cells and B cells may be a critical factor for the development of autoantibodies. Precursors of autoimmune B cells can arise by a number of mechanisms. First, they could be "ignorant B cells" which have arisen due to receptor editing of previously tolerized cells, or they could be anergic B cells that can nevertheless be triggered by strong T cell help ([68] and references within). B cells in the *B6.Sle1b* strain have been shown to undergo increased receptor editing possibly due to dysregulated expression of Ly108 [69]. Thus, these cells may fit the criteria of "ignorant" B cells that can be more easily triggered by antigens. In addition, although T cells, derived from the *B6.Sle1b* strain cannot on their own, initiate autoimmunity, they are nonetheless more easily activated. Therefore they may function as helper T cells that can overcome the activation barrier presented by anergic B cells. If autoantibodies in the *B6.Sle1b* mouse are produced by B cells that are more easily triggered, then it is possible that they would produce a higher response to a T-independent antigen. Indeed, we found that at an early age, before manifestation of any autoimmune symptoms, *B6.Sle1b* mice produce significantly higher levels of antibodies to NP-Ficoll than B6 littermates [70]. Moreover, the increased response can be eliminated by the depletion of NK cells. Importantly, the increased response is also eliminated by the inactivation of the SH2D1A/DSHP/*SAP* gene which encodes SLAM-associated protein (SAP), a signal transducer utilized by all members of the SLAM-CD2 family. Thus, even though NK cells can enhance responses to T-independent antigens if they are activated by the cytokine circuit; in this case, where there is no apparent overt stimulation of the circuit, it appears that interactions between NK and B cells, which may exist in a higher activation state due to developmental dysregulation, is sufficient to induce the higher response. Furthermore, this induction may require signaling mediated by SAP. Finally, we have also shown that when the *SAP* gene is inactivated in the *B6.Sle1b* strain by the creation of the *B6.Sle1b.SAP* −/− strain, the percentage of animals that produce ANA is much reduced but, the level of ANA produced in the animals that succumb is not reduced [70]. Since SAP is required for a T-dependent response, the ANA produced must be derived from responses to T-independent antigens. Significantly, the subclass of antibodies produced were predominantly Ig2b and IgG2c but not IgG1 which is most dependent on T cell help. Thus these experiments unequivocally show that ANA can be produced by the activation of B cells via T-independent antigens. Although these studies provide clear evidence that NK cells play a possible role in the induction of autoimmunity further analysis utilizing this system to explore the underlying mechanisms should enhance the understanding of the disease.

3 In Vitro Dissection of NK-B Cell Interactions

3.1 NK Cell Effects on B cells

Due to the multiplicity of cell types that can contribute to B cell activity in vivo we have expanded considerable effort toward dissecting possible effects of NK cells on B cell responses using in vitro systems. By this approach, various B cell populations can be isolated and their responses separately analyzed. Due to the heterogeneous nature of B cells that are directly isolated from the animal, analysis of the response is complicated by varying levels of activation signals that are required. Fractionation of B cells by density gradient sedimentation allows the preparation of large numbers of high-density B cells that contain a minimal proportion that had already been activated in vivo. Cell surface phenotype of this population showed that they do not express detectable levels of CD69, low levels of CD86, and does not contain B1 or marginal zone B cells as defined by B220+CD55+ or B220+, CD23lo and CD21hi markers, respectively. Greater than 85% of the cells have characteristics of follicular B cells (IgDhi and CD23hi). Since memory B cells can be also present in the high-density population, these cells were further depleted by means of anti-IgG columns resulting in cells that do not express mRNA for any IgG isotype that can be detected by semiquantitative RT-PCR. The response of these cells to NK cells was assessed by coculture of the two cell types. NK cells had to be propagated by IL-2 stimulation in order to obtain sufficient numbers. Initial studies first established that coculture of these NK cells at various ratios with B cells do not result in B cell lysis [28] or apoptosis (Dang and Yuan, unpublished observations) even when MHC disparate sources of cells were used. The NK cells also do not induce significant B cell proliferation [71]. On the other hand, coculture with NK cells increased B cell viability. Experiments are underway to determine if this effect could be due to NK cell induction of BAFF which can promote B cell viability.

Using these coculture conditions definitive induction of B cell differentiation by NK cells can be shown. First, B cell CD69 as well as CD86 expression was increased within as early as 6 h. Second, induction of germline transcripts for $\gamma 2a$ accompanied by upregulation of mRNA for AID [72] as well as T-bet [71] can be detected by semiquantitative RT-PCR analysis within 24 h. Thus NK cells can initiate the events required for both of the important functions mediated by B cells, antigen presentation, and Ig secretion. It should be noted that although it is possible that NKT cells can also function in this capacity, we have vigorously eliminated these cells from the propagated cells used to activate B cells. The NK cultures also do not contain MHCII positive cells; therefore, macrophages and DCs do not play a role. On the other hand, freshly isolated NK cells cannot effectively activate B cells;

therefore, partial activation of NK cells by IL-2 occurring during propagation is required for their activity.

3.2 NK Cell Enhancement of Antigen Presentation by B cells

The increase in CD86 on resting B cells upon interaction with NK cells suggested that the interaction can enhance their ability to present antigen to T cells. We assessed antigen presentation by determining the extent of proliferation of purified T cells from OT-II transgenic mice as measured by dilution of CFSE fluorescence intensity upon stimulation by OVA. As shown in Fig. 4a, highly purified T cells do not respond measurably to the addition of only antigen in vitro. Addition of B cells increased the proliferation, and as expected, activated low-density B cells can induce greater anti-OVA-dependent T cell proliferation than resting B cells (Fig. 4b, 4c). Most significantly, coincubation of resting B cells with IL-2 propagated NK cells enhanced proliferation mediated by the resting B cells (Fig. 4d).

The proliferation measured by CFSE dilution is mirrored by measurement of ^{3}HTdR incorporation after the cells have been cocultured in a similar manner. Table 1 shows a representative experiment.

Although the extent of enhancement of antigen presentation by resting B cells is not extensive due to the low responsiveness of resting B cells, the results have been reproduced in many independent experiments [58]; therefore, it is possible to further analyze the molecular basis for the enhancement. For example, by separating NK cells from B cells by a transwell membrane we showed further that the effect of NK cells on T cell proliferation requires direct cell contact and therefore is unlikely to be mediated by cytokines produced by NK cells (Table 1). Investigations are underway in attempts to identify the cell surface molecules involved in this interaction. One possible candidate is that between OX40L on B cells and OX40 [64, 73] on NK cells.

3.3 NK Cell Activation of Switch Recombination

The ability of NK cells to influence the extent of switch recombination to IgG2a in vivo has been usually attributed to their ability to produce IFN-γ. However, our results, showing that NK cells from IFN-γ KO mice can also induce germline transcription of the γ2 gene in a contact-dependent fashion [72], show that they have the ability to initiate this process in the absence of IFN-γ. Transcription of the Iγ2a exons is the first step required for B cells to initiate the process of DNA recombination from the variable region exons to the γ2a heavy chain exons resulting in appropriate coding sequences for the functional γ2a heavy chain. Interestingly, the induction by NK cells is limited to the first step of differentiation and productive switch recombination does not occur. In these in vitro cultures of highly

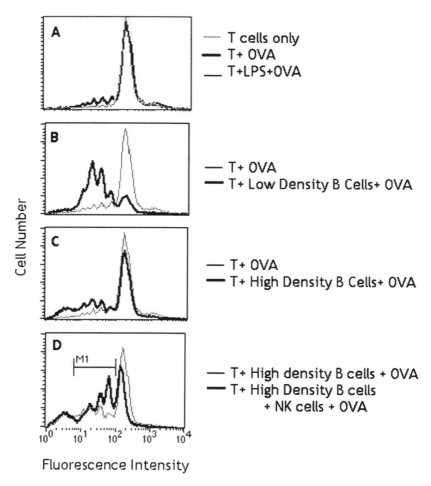

Fig. 4 NK cell enhancement of B cell antigen presentation. After 3 days of culture with the reagents indicated or with B cells (ratio of 1:1) CD4$^+$T cells in the cultures were identified by staining and assessed by FACS analysis to determine the fraction of T cells that had undergone cell division resulting in a reduction of CFSE fluorescence intensity. Profiles were normalized to equal areas to allow easier comparison. Coculture with low-density B cells in the presence of antigen enhanced T cell proliferation to a much greater extent than did resting, high-density B cells (compare B and C). Prior incubation of the resting B cells with NK cells (ratio of 2:1) resulted in a significant increase in the ability of the cells to induce T cell proliferation (D) The fraction of total cells present in the M1 gate was increased from 35.5 to 53.5%

purified B cells we also cannot detect germline or switched transcripts for γ1. The absence of these transcripts also indicates the absence of participation of inadvertent factors that can be a source of ligands that can directly stimulate TLRs expressed on B cells [32, 74].

Table 1 Enhancement of B-cell Antigen Presentation by NK cells

Coculture conditions	Day 3 ^{3}HTdR Incorporation (+/−SD)	
	No antigen	OVA (200 µg/ml)
T+B	2,605 cpm (+/−96)	4,297 cpm (+/−535)
T+B+NK	3,140 cpm (+/−476)	7,303 cpm (+/−1,196)
T+B +(NK in transwell)	1,687 cpm (+/−544)	2,655 cpm (+/−57)

Resting B cells were cultured alone or in the presence of IL-2-activated NK cells with or without 200 µg/ml ovalbumin. After 16 h cells were irradiated and purified primary OT-II cells were added. Proliferation was measured by ^{3}HTdR incorporation 3 days later

The observation that coculture with NK cells only results in induction of germline transcripts for γ2a is consistent with the unlikely probability that in vivo any chance encounter of activated NK cells would result in induction of B cell Ig secretion. The case may be different, however, for B cells that have encountered specific antigen. Indeed, if NK cells are incubated with B cells expressing a transgene encoding the specific antigen receptor for NP, together with NP-Ficoll to crosslink the BCR, IgG2a can be induced [71]. Again, the induction can occur in the absence of IFN-γ but requires direct cell contact. Thus, NK help is sufficient for switch recombination in response to a TI-II antigen, and activation of CD40 on B cells is not required [75].

In contrast to the induction of Iγ2a, it is interesting that incubation of B cells from mice expressing the transgene for NP with antigen alone is sufficient for the induction of germline transcripts for γ1 (Iγ1), although productive switch to γ1 mRNA also does not occur unless NK cells are added [71]. Thus, whereas the interaction with NK cells is required for both Iγ2a and mature γ2a mRNA expression, the signal is only needed for induction of the mature γ1 mRNA. Thus activation by NK cells appears to impinge on signals required for two different steps during B cell differentiation, the first one for induction of Iγ2a and the second one for induction of productive DNA rearrangement to either γ1 or γ2a.

These experiments were performed under stringent conditions in which high-density, IgG-negative, B cells with the phenotype of the follicular subset were used. Whereas the results allow us to determine the minimal conditions required for induction they do not negate the possibility, that in vivo, NK cells can stimulate other B cell subsets such as marginal zone B cells that are more accessible to NK cells that usually reside in the red pulp. Analysis of this population is precluded by their propensity to express switched transcripts without overt stimulation; therefore, unequivocal demonstration of NK induction is not easy.

The induction of both germline transcripts and switch recombination by NK cells requires direct contact with B cells [71]. In order to determine the nature of this interaction a number of antibodies for receptors known to be expressed on NK cells were used in attempts to inhibit the interaction. The results show that although NK cells can express T cell costimulators such as CD40L and CD28, they were neither sufficient nor necessary for activation of B cells. The interaction was also not inhibited by anti-TCR, further validating the fact that any residual NKT cells in the culture could not be responsible. Until now, the only antibody found to be able to inhibit the interaction was anti-CD48. Although CD48 is expressed by both NK and B cells, we showed that the expression of this molecule on B cells is essential for activation by NK cells. Whereas two different receptors, CD2 and CD244, can specifically interact with CD48, by the use of NK cells derived from either CD2 or 2B4 KO mice, we showed that CD2 is the more effective inducer for B cells [67]. Although CD48 is required for induction of both germline and switch recombination, ligation of this GPI-anchored receptor does not appear to be sufficient because crosslinking by anti-CD48 antibodies in either soluble or immobilized form cannot substitute for the effect of NK cells (Gao and Yuan, personal observations). Thus ligation of CD48 on B cells appears to be necessary but not sufficient for activation of B cells. Cellular activation by CD48 is unusual in that it is anchored to the membrane via glycosyl phosphatidylinositol. In the absence of cytoplasmic domains how the signal from CD2 is transmitted to B cells is not clear.

However, GPI-anchored receptors have been shown to coimmunoprecipitate with various members of the SRC family of tyrosine kinases in glycolipid-enriched microdomains [76] as well as G protein subunits [77]. Considering the weak charge–charge interactions of CD2 with its ligands one possibility is that any encounter between CD2 and CD48 only serves to initiate the interaction between the two cell types such that another receptor–ligand pair can provide the final stimulation. Efforts are underway to identify these additional receptors.

Additional actors to be considered are factors that can stimulate TACI on B cells since activation by TI-II antigens has been shown to be dependent on the expression of TACI on B cells [4]. The activation of TACI appears to be more important for downstream events of Ig secretion rather than for the initial activation of B cells [78]. Two different factors can stimulate TACI, BAFF, and APRIL. BAFF can be produced by a number of cells types although expression by NK cells has not been extensively analyzed [79]. Importantly, however, stimulation of B cells via TLRs can increase both BAFF secretion and TACI expression [80]. Thus it is possible that stimulation of B cells via CD48 also initiates an autocrine pathway in B cells and initiation of germline transcription and/or switch recombination in the presence of antigen is mediated by BAFF. A similar pathway may be responsible for the finding that human plasmacytoid DCs can activate switch recombination via the production of BAFF and APRIL [81].

It should be noted that if BAFF is the critical stimulator in the induction by NK cells it is most likely effective only as an autocrine factor and insufficient amounts are produced for the activation of B cells that have not been directly activated by NK cells. Thus, if NK and B cells are cocultured in the upper chamber of transwells only B cells within the same chamber can be stimulated to produce Iγ2a or VDJCγ2a while B cells cultured in the lower chamber remained inert [71]. Additionally, it is worthy to consider that whereas there is much evidence that both responder and regulatory B cells can produce a plethora of cytokines, including IL-4 and IFN-γ [82], that can potentially activate their own switch recombination, our transwell experiments show that if NK cell activation involves only the stimulation of B cells to produce cytokines then the amounts must be too low to be registered functionally in this system.

3.4 B Cell Activation of NK Cells

Early studies of NK cells have shown that they migrate from the red pulp to the marginal zone of B cell follicles upon activation by Poly(I:C) or by viruses [83]. This close association suggests that the large number of B cells in this zone may contribute to further activation of NK cells. Indeed, we, as well as others, have shown that low-density B cells, which contain the marginal zone subset, can enhance the secretion of IFN-γ by IL-2 propagated NK cells (reviewed in [66]). Various ligand–receptor interactions required for the induction have been implicated (reviewed in [66]). However, we found it difficult to delineate the role of these interactions by using the induction of IFN-γ mRNA as an indicator of further NK cell differentiation because IL-2 propagated NK cells produce variable amounts of IFN-γ even without overt stimulation. Furthermore, the mechanism of B cell enhancement of NK cell IFN-γ production occurs mainly via stabilization of the mRNA [84]; therefore, the extent of induction depends significantly on the initial level. For this reason we investigated the synthesis of other products by NK cells that may be a better indicator of further NK cell differentiation. Another cytokine produced by NK cells is IL-13 [85]. The induction of IL-13 appears to require more rigorous conditions in that IL-2 propagated NK cells produce virtually no detectable mRNA for IL-13. Thus it was possible to assess the inductive stimuli for NK cell differentiation under more stringent conditions. We showed that B cells can induce IL-13 mRNA expression as well as synthesis of low amounts of the protein by NK cells [86]. The level of mRNA expression can sometimes approximate that induced by the cytokines IL-12 and IL-18 [87]. Human NK cells have been divided into distinct subpopulations that produce different cytokines [88]; thus, it may be that induction by B cells only targets a fraction of the cells. At this point we cannot rule out this possibility despite the fact that signals that have been shown to enhance IFN-γ secretion by NK cells, including anti-NK1.1 as well as other activating ligands for NK receptors, can also induce IL-13 mRNA expression [86, 87]. Nonetheless, the more stringent conditions of induction of IL-13 mRNA allow the requisite cell interaction molecules to now be defined. Interestingly, we found that CD48 expression on B cells is also required for the activation of NK cells [86]. In this case, however, in which B cell CD48 is the ligand, the counter-receptor on NK cells is 2B4 rather than CD2. Since NK cells from 2B4 KO mice were completely inert to stimulation by B cells other possible ligand–receptor interactions, while they may contribute to the interaction, appear to be insufficient for the induction of IL-13 gene activation. Furthermore, our finding that activation by B cells requires the signaling pathways that depend upon expression of the X-linked lymphoproliferative disease gene product, SAP , rules out the participation of a number of possible candidates such as NKRP1, recently shown to be activated by lectin-like transcript -1 (LLT1) expressed on human DCs and B cells [89].

The interaction of 2B4 with CD48 has been investigated extensively ([reviewed in [90]) and the receptor has not been shown to interact with any other ligand. Although stimulation via 2B4 was originally shown to activate NK cell cytotoxicity in mouse cells [91] recently 2B4 was also found to serve as an inhibitory signal for stimulation by other ligands [92]. In contrast, studies of mature human NK cells have shown that 2B4 can only function as an activating receptor [93]. It is interesting that despite the expression of CD48 on virtually all hematopoetic cells including NK cells themselves, IL-2 propagated NK cell cultures, in which the cells can interact freely, do not produce IL-13. Therefore, the ligation of 2B4 by self CD48 in these cultures must provide an inhibitory signal. Alternatively, due to the early interaction of NK cells with CD48 during ontogeny, they may have become hyporesponsive to this ligand. How then does CD48 expressed on B cells overcome the inhibition? Our studies have shown that activation of 2B4 proceeds via the expected pathway of being transduced by SAP. However, although CD48 is absolutely required, other costimulators may be also involved. Since marginal zone B cells appear to be much more effective stimulators of NK cells than follicular B cells these costimulators may be differentially expressed by the subsets. It is also interestingly that, despite the similar level of expression of CD48 on T cells, they are not effective stimulators of IL-13 mRNA expression by NK cells (Gao and Yuan, personal observations). Until further information regarding the type of costimulator involved is obtained, alternative explanations for the stimulatory ability B cells can be entertained. For example, it is possible that the configuration of CD48 on NK cells and follicular B cells differs from marginal zone B cells such that the manner of engagement of 2B4 confers different signals. Studies are underway to dissect this problem.

Finally, it should be noted that IL-13 is a Th2 cytokine that functions in many similar ways as IL-4; however, IL-13 has not been shown to induce B cell switch recombination to IgG1. Correspondingly, IL-13 produced in cocultures of NK and B cells also does not induce switch to IgG1 [71]. However, depletion of NK cells can affect the IgE response to aerosolized antigen [47] as well as IgG1 responses to pneumococcal vaccines given in adjuvant [12]. In addition, although IL-13 has generally been associated with allergic inflammation [94] the cytokine may be also involved in other diseases as well [95]. Experimentally it will be difficult to sort out how the B cell induction of NK cell differentiation to this pathway impacts the activation conditions that result in T cell production of the cytokine in vivo. However, a plausible scenario can be envisioned in which the cytokine produced under certain conditions can amplify a cytokine burst involving mast cells and eosinophils in the same manner that NK production of IFN-γ amplifies the cytokine loop initiated by macrophages.

4 Conclusions

Investigations of the participation of NK cells as a major part of the innate immune system on antigen-specific responses mediated by B and T cells continue to generate important information regarding the effect of early responses to the eventual outcome of how organisms can counter infections by pathogens. In addition to the specific effects of NK cells on antibody production, our new findings showing that the initial interaction between the two cell types can impact memory responses have further implications yet to be explored. In addition, the recent identification of the role of CD48 on B lymphocytes as both receptor and a stimulatory ligand for the interaction with NK cells should provide an important tool for the dissection of the relative importance of each of the plethora of actors that participate in the immune response.

References

1. Medzhitov R, Janeway CA Jr (2002) Decoding the patterns of self and nonself by the innate immune system. Science 296:298–300
2. Tripathy SK, Keyel PA, Yang L, Pingel JT, Cheng TP, Schneeberger A, Yokoyama WM (2008) Continuous engagement of a self-specific activation receptor induces NK cell tolerance. J Exp Med 205:1829–1841
3. Vos Q, Lees A, Wu ZQ, Snapper CM, Mond JJ (2000) B-cell activation by T-cell-independent type 2 antigens as an integral part of the humoral immune response to pathogenic microorganisms. Immunol Rev 176:154–170
4. von Bulow GU, van Deursen JM, Bram RJ (2001) Regulation of the T-independent humoral response by TACI. Immunity 14:573–582
5. MacLennan I, Vinuesa C (2002) Dendritic cells, BAFF, and APRIL: innate players in adaptive antibody responses. Immunity 17:235–238
6. Hebel K, Griewank K, Inamine A, Chang HD, Muller-Hilke B, Fillatreau S, Manz RA, Radbruch A, Jung S (2006) Plasma cell differentiation in T-independent type 2 immune responses is independent of CD11c(high) dendritic cells. Eur J Immunol 36:2912–2919
7. Wilder JA, Koh CY, Yuan D (1996) The role of NK cells during in vivo antigen-specific antibody responses. J Immunol 156:146–152
8. Schmidt KN, Leung B, Kwong M, Zarember KA, Satyal S, Navas TA, Wang F, Godowski PJ (2004) APC-independent activation of NK cells by the Toll-like receptor 3 agonist double-stranded RNA. J Immunol 172:138–143
9. Lee CK, Rao DT, Gertner R, Gimeno R, Frey AB, Levy DE (2000) Distinct requirements for IFNs and STAT1 in NK cell function. J Immunol 165:3571–3577
10. Koh CY, Yuan D (1997) The effect of NK cell activation by tumor cells on antigen-specific antibody responses. J Immunol 159:4745–4752
11. Jouvin-Marche E, Morgado MG, Leguern C, Voegtle D, Bonhomme F, Cazenave PA (1989) The mouse Igh-1a and Igh-1b H chain constant regions are derived from two distinct isotypic genes. Immunogenetics 29:92–97
12. Yuan D, Bibi R, Dang T (2004) The role of adjuvant on the regulatory effects of NK cells on B cell responses as revealed by a new model of NK cell deficiency. Int Immunol 16:707–716
13. Gavin AL, Hoebe K, Duong B, Ota T, Martin C, Beutler B, Nemazee D (2006) Adjuvant-enhanced antibody responses in the absence of toll-like receptor signaling. Science 314: 1936–1938
14. Bukowski J, Woda B, Habu S, Okumura K, Welsh R (1983) Natural killer cell depletion enhances virus synthesis and virus-induced hepatitis in vivo. J. Immunol 131:1531–1538
15. Orange JS, Wang B, Terhorst C, Biron C (1995) Requirement for natural killer cell-produced interferon gamma in defense against murine cytomegalovirus infection and enhancement of this defense pathway by interleukin 12 administration. J Exp Med 182:1045–1056
16. Brown MG, Dokun AO, Heusel JW, Smith HR, Beckman DL, Blattenberger EA, Dubbelde CE, Stone LR, Scalzo AA, Yokoyama WM (2001) Vital involvement of a natural killer cell activation receptor in resistance to viral infection. Science 292:934–937

17. French AR, Yokoyama WM (2003) Natural killer cells and viral infections. Curr Opin Immunol 15:45–51
18. Orange JS, Biron CA (1996) An absolute and restricted requirement for IL-12 in natural killer cell IFN-gamma production and antiviral defense. Studies of natural killer and T cell responses in contrasting viral infections. J Immunol 156:1138–1142
19. Karupiah G, Sacks TE, Klinman DM, Fredrickson TN, Hartley JW, Chen JH, Morse HC 3rd (1998) Murine cytomegalovirus infection-induced polyclonal B cell activation is independent of CD4+ T cells and CD40. Virology 240:12–26
20. Palladino G, Mozdzanowska K, Washko G, Gerhard W (1995) Virus-neutralizing antibodies of immunoglobulin G (IgG) but not of IgM or IgA isotypes can cure influenza virus pneumonia in SCID mice. J Virol 69:2075–2081
21. Huber VC, McKeon RM, Brackin MN, Miller LA, Keating R, Brown SA, Makarova N, Perez DR, Macdonald GH, McCullers JA (2006) Distinct contributions of vaccine-induced immunoglobulin G1 (IgG1) and IgG2a antibodies to protective immunity against influenza. Clin Vaccine Immunol 13:981–990
22. Coutelier J-P, JTMvd L, Heessen FWA, Warnier G, Snick JV (1987) IgG2a restriction of murine antibodies elicited by viral infections. J Exp Med 165:64–69
23. Baumgarth N, Kelso A (1996) In vivo blockade of gamma interferon affects the influenza virus-induced humoral and the local cellular immune response in lung tissue. J Virol 70:4411–4418
24. Graham MB, Dalton DK, Giltinan D, Braciale VL, Stewart TA, Braciale TJ (1993) Response to influenza infection in mice with a targeted disruption in the interferon gamma gene. J Exp Med 178:1725–1732
25. Szomolanyi-Tsuda E, Welsh RM (1996) T cell-independent antibody-mediated clearance of polyoma virus in T cell-deficient mice. J Exp Med 183:403–411
26. Zinkernagel RM (1997) Protective antibody responses against viruses. Biol Chem 378: 725–729
27. Fehr T, Skrastina D, Pumpens P, Zinkernagel RM (1998) T cell-independent type I antibody response against B cell epitopes expressed repetitively on recombinant virus particles. Proc Natl Acad Sci USA 95:9477–9481
28. Yuan D, Koh CY, Wilder JA (1994) Interactions between B lymphocytes and NK cells. FASEB J 8:1012–1018
29. Markine-Goriaynoff D, van der Logt JT, Truyens C, Nguyen TD, Heessen FW, Bigaignon G, Carlier Y, Coutelier JP (2000) IFN-gamma-independent IgG2a production in mice infected with viruses and parasites. Int Immunol 12:223–230
30. Koh CY, Yuan D (2000) The functional relevance of NK-cell-mediated upregulation of antigen-specific IgG2a responses. Cell Immunol 204:135–142
31. Diebold SS, Kaisho T, Hemmi H, Akira S, Reis e Sousa C (2004) Innate antiviral responses by means of TLR7-mediated recognition of single-stranded RNA. Science 303:1529–1531
32. Heer AK, Shamshiev A, Donda A, Uematsu S, Akira S, Kopf M, Marsland BJ (2007) TLR signaling fine-tunes anti-influenza B cell responses without regulating effector T cell responses. J Immunol 178:2182–2191
33. Tripp CS, Wolf SF, Unanue ER (1993) Interleukin-12 and tumor necrosis factor are costimulators of interferon-_ production by natural killer cells in severe combined immunodeficiency mice with listeriosis, and interleukin 10 is a physiologic antagonist. Proc Natl Acad Sci USA 90:3725–3729
34. Fernandes DM, Benson R, Baldwin CL (1995) Lack of a role for natural killer cells in early control of Brucella abortus 2308 infections in mice. Infect Immun 63:4029–4033
35. Newton DW Jr, Runnels HA, Kearns RJ (1992) Enhanced splenic bacterial clearance and neutrophilia in anti-NK1.1-treated mice infected with Pseudomonas aeruginosa. Nat Immun 11:335–344
36. Saito M, Yamaguchi T, Kawata T, Ito H, Kanai T, Terada M, Yokosuka M, Saito TR (2006) Effects of methamphetamine on cortisone concentration, NK cell activity and mitogen response of T-lymphocytes in female cynomolgus monkeys. Exp Anim 55:477–481

37. Selman S, Hayes D, Perin LA, Hayes WS (2000) Pneumococcal conjugate vaccine for young children. Manag Care 9:49–52, 54, 56–57 passim
38. Betts M, Beining P, Brunswick M, Inman J, Angus RD, Hoffman T, Golding B (1993) Lipopolysaccharide from Brucella abortus behaves as a T-cell-independent type 1 carrier in murine antigen-specific antibody responses. Infect Immun 61:1722–1729
39. Finkelman FD, Katona IM, Mosmann TR, Coffman RL (1988) IFN-gamma regulates the isotypes of Ig secreted during in vivo humoral immune responses. J Immunol 140:1022–1027
40. Szomolanyi-Tsuda E, Liang X, Welsh RM, Kurt-Jones EA, Finberg RW (2006) Role for TLR2 in NK cell-mediated control of murine cytomegalovirus in vivo. J Virol 80: 4286–4291
41. Bendelac A, Savage PB, Teyton L (2007) The biology of NKT cells. Annu Rev Immunol 25:297–336
42. Tupin E, Kinjo Y, Kronenberg M (2007) The unique role of natural killer T cells in the response to microorganisms. Nat Rev Microbiol 5:405–417
43. Barral P, Eckl-Dorna J, Harwood NE, De Santo C, Salio M, Illarionov P, Besra GS, Cerundolo V, Batista FD (2008) B cell receptor-mediated uptake of CD1drestricted antigen augments antibody responses by recruiting invariant NKT cell help in vivo. Proc Natl Acad Sci USA 105:8345–8350
44. Kinjo Y, Pei B, Bufali S, Raju R, Richardson SK, Imamura M, Fujio M, Wu D, Khurana A, Kawahara K, Wong CH, Howell AR, Seeberger PH, Kronenberg M (2008) Natural sphingomonas glycolipids vary greatly in their ability to activate natural killer T cells. Chem Biol 15:654–664
45. Satoskar AR, Stamm LM, Zhang X, Okano M, David JR, Terhorst C, Wang B (1999) NK cell-deficient mice develop a Th1-like response but fail to mount an efficient antigen-specific IgG2a antibody response. J Immunol 163:5298–5302
46. Wang B, Hollander GA, Nichogiannopoulou A, Simpson SJ, Orange JS, Gutierrez-Ramos JC, Burakoff SJ, Biron CA, Terhorst C (1996) Natural killer cell development is blocked in the context of aberrant T lymphocyte ontogeny. Int Immunol 8:939–949
47. Korsgren M, Persson CGA, Sundler F, Bjerke T, Hansson T, Chambers BJ, Hong S, Van Kaer L, Ljunggren HG, Korsgren O (1999) Natural killer cells determine development of allergen-induced eosinophilic airway inflammation in mice. J Exp Med 189:553–562
48. Shi F-D, Wang H-B, Li H, Hong S, Taniguchi M, Link H, Kaer LV, Ljunggren H-G (2000) Natural killer cells determine the outcome of B cell-mediated autoimmunity. Nat Immunol 1:245–251
49. Wang M, Ellison CA, Gartner JG, HayGlass KT (1998) Natural killer cell depletion fails to influence initial CD4 T cell committment in vivo in exogenous antigen-stimulated cytokine and antibody responses. J Immunol 160:1098–1105
50. Martin-Fontecha A, Thomsen LL, Brett S, Gerard C, Lipp M, Lanzavecchia A, Sallusto F (2004) Induced recruitment of NK cells to lymph nodes provides IFN-gamma for T(H)1 priming. Nat Immunol 5:1260–1265
51. Assarsson E, Kambayashi T, Schatzle JD, Cramer SO, von Bonin A, Jensen PE, Ljunggren HG, Chambers BJ (2004) NK cells stimulate proliferation of T and NK cells through 2B4/CD48 interactions. J Immunol 173:174–180
52. Lucas M, Schachterle W, Oberle K, Aichele P, Diefenbach A (2007) Dendritic cells prime natural killer cells by trans-presenting interleukin 15. Immunity 26:503–517
53. Shah PD, Gilbertson SM, Rowley DA (1985) Dendritic cells that have interacted with antigen are targets for natural killer cells. J Exp Med 162:625–636
54. Degli-Esposti MA, Smyth MJ (2005) Close encounters of different kinds: dendritic cells and NK cells take centre stage. Nat Rev Immunol 5:112–124
55. Yoshida O, Akbar F, Miyake T, Abe M, Matsuura B, Hiasa Y, Onji M (2008) Impaired dendritic cell functions because of depletion of natural killer cells disrupt antigen-specific immune responses in mice: restoration of adaptive immunity in natural killer-depleted mice by antigen-pulsed dendritic cell. Clin Exp Immunol 152:174–181

56. Hanna J, Gonen-Gross T, Fitchett J, Rowe T, Daniels M, Arnon TI, Gazit R, Joseph A, Schjetne KW, Steinle A, Porgador A, Mevorach D, Goldman-Wohl D, Yagel S, LaBarre MJ, Buckner JH, Mandelboim O (2004) Novel APC-like properties of human NK cells directly regulate T cell activation. J Clin Invest 114:1612–1623
57. Blasius AL, Barchet W, Cella M, Colonna M (2007) Development and function of murine B220+CD11c+NK1.1+ cells identify them as a subset of NK cells. J Exp Med 204:2561–2568
58. Jennings P, Yuan D (2009) NK cell enhancement of antigen presentation by B lymphocytes. J Immunol 182:2879–2887
59. Kurt-Jones EA, Liano D, HayGlass KA, Benacerraf B, Sy MS, Abbas AK (1988) The role of antigen-presenting B cells in T cell priming in vivo. Studies of B cell-deficient mice. J Immunol 140:3773–3778
60. Lund FE, Hollifield M, Schuer K, Lines JL, Randall TD, Garvy BA (2006) B cells are required for generation of protective effector and memory CD4 cells in response to Pneumocystis lung infection. J Immunol 176:6147–6154
61. Crawford A, Macleod M, Schumacher T, Corlett L, Gray D (2006) Primary T cell expansion and differentiation in vivo requires antigen presentation by B cells. J Immunol 176:3498–3506
62. Galli G, Pittoni P, Tonti E, Malzone C, Uematsu Y, Tortoli M, Maione D, Volpini G, Finco O, Nuti S, Tavarini S, Dellabona P, Rappuoli R, Casorati G, Abrignani S (2007) Invariant NKT cells sustain specific B cell responses and memory. Proc Natl Acad Sci USA 104:3984–3989
63. Takahashi T, Strober S (2008) Natural killer T cells and innate immune B cells from lupus-prone NZB/W mice interact to generate IgM and IgG autoantibodies. Eur J Immunol 38:156–165
64. Linton PJ, Bautista B, Biederman E, Bradley ES, Harbertson J, Kondrack RM, Padrick RC, Bradley LM (2003) Costimulation via OX40L expressed by B cells is sufficient to determine the extent of primary CD4 cell expansion and Th2 cytokine secretion in vivo. J Exp Med 197:875–883
65. Niemeyer M, Darmoise A, Mollenkopf HJ, Hahnke K, Hurwitz R, Besra GS, Schaible UE, Kaufmann SH (2008) Natural killer T-cell characterization through gene expression profiling: an account of versatility bridging T helper type 1 (Th1), Th2 and Th17 immune responses. Immunology 123:45–56
66. Yuan D (2004) Interactions between NK cells and B lymphocytes. Adv Immunol 84:1–42
67. Gao N, Dang T, Dunnick WA, Collins JT, Blazar BR, Yuan D (2005) Receptors and counter-receptors involved in NK-B cell interactions. J Immunol 174:4113–4119
68. Shlomchik MJ (2008) Sites and stages of autoreactive B cell activation and regulation. Immunity 28:18–28
69. Kumar KR, Li L, Yan M, Bhaskarabhatla M, Mobley AB, Nguyen C, Mooney JM, Schatzle JD, Wakeland EK, Mohan C (2006) Regulation of B cell tolerance by the lupus susceptibility gene Ly108. Science 312:1665–1669
70. Jennings P, Chan A, Schwartzberg P, Wakeland EK, Yuan D (2008) Antigen-specific responses and ANA production in B6.Sle1b mice: a role for SAP. J Autoimmun 31:345–353
71. Gao N, Jennings P, Yuan D (2008) Requirements for the natural killer cell-mediated induction of IgG1 and IgG2a expression in B lymphocytes. Int Immunol 20:645–657
72. Gao N, Dang T, Yuan D (2001) IFN-gamma-dependent and -independent initiation of switch recombination by NK cells. J Immunol 167:2011–2018
73. Gramaglia I, Jember A, Pippig SD, Weinberg AD, Killeen N, Croft M (2000) The OX40 costimulatory receptor determines the development of CD4 memory by regulating primary clonal expansion. J Immunol 165:3043–3050
74. Jegerlehner A, Maurer P, Bessa J, Hinton HJ, Kopf M, Bachmann MF (2007) TLR9 signaling in B cells determines class switch recombination to IgG2a. J Immunol 178:2415–2420
75. Collins J, Dunnick W (1993) Germline transcripts of the murine immunoglobulin gamma 2a gene: structure and induction by IFN-gamma. Int Immunol 5:885–891
76. Shenoy-Scaria AM, Gauen LK, Kwong J, Shaw AS, Lublin DM (1993) Palmitylation of an amino-terminal cysteine motif of protein tyrosine kinases p56lck and p59fyn mediates interaction with glycosyl-phosphatidylinositol-anchored proteins. Mol Cell Biol 13:6385–6392

77. Solomon KR, Rudd CE, Finberg RW (1996) The association between glycosylphosphatidylinositol-anchored proteins and heterotrimeric G protein alpha subunits in lymphocytes. Proc Natl Acad Sci USA 93:6053–6058
78. Mantchev GT, Cortesao CS, Rebrovich M, Cascalho M, Bram RJ (2007) TACI is required for efficient plasma cell differentiation in response to T-independent type 2 antigens. J Immunol 179:2282–2288
79. Moore PA, Belvedere O, Orr A, Pieri K, LaFleur DW, Feng P, Soppet D, Charters M, Gentz R, Parmelee D, Li Y, Galperina O, Giri J, Roschke V, Nardelli B, Carrell J, Sosnovtseva S, Greenfield W, Ruben SM, Olsen HS, Fikes J, Hilbert DM (1999) BLyS: member of the tumor necrosis factor family and B lymphocyte stimulator. Science 285:260–263
80. Chu VT, Enghard P, Riemekasten G, Berek C (2007) In vitro and in vivo activation induces BAFF and APRIL expression in B cells. J Immunol 179:5947–5957
81. Litinskiy MB, Nardelli B, Hilbert DM, He B, Schaffer A, Casali P, Cerutti A (2002) DCs induce CD40-independent immunoglobulin class switching through BLyS and APRIL. Nat Immunol 3:822–829
82. Lund FE (2008) Cytokine-producing B lymphocytes-key regulators of immunity. Curr Opin Immunol 20:332–338
83. Salazar-Mather TP, Ishikawa R, Biron CA (1996) NK cell trafficking and cytokine expression in splenic compartments after IFN induction and viral infection. J Immunol 157:3054–3064
84. Wilder J, Yuan D (1995) Regulation of interferon mRNA production in natural killer cells. Int Immunol 7:575
85. Hoshino T, Winkler-Pickett RT, Mason AT, Ortaldo JR, Young HA (1999) IL-13 production by NK cells: IL-13-producing NK and T cells are present in vivo in the absence of IFN-gamma. J Immunol 162:51–59
86. Gao N, Schwartzberg P, Wilder JA, Blazar BR, Yuan D (2006) B cell induction of IL-13 expression in NK cells: role of CD244 and SLAM-associated protein. J Immunol 176:2758–2764
87. Hoshino T, Wiltrout RH, Young HA (1999) IL-18 is a potent coinducer of IL-13 in NK and T cells: a new potential role for IL-18 in modulating the immune response. J Immunol 162:5070–5077
88. Loza MJ, Peters SP, Zangrilli JG, Perussia B (2002) Distinction between IL-13+ and IFN-gamma+ natural killer cells and regulation of their pool size by IL-4. Eur J Immunol 32:413–423
89. Rosen DB, Cao W, Avery DT, Tangye SG, Liu YJ, Houchins JP, Lanier LL (2008) Functional consequences of interactions between human NKR-P1A and its ligand LLT1 expressed on activated dendritic cells and B cells. J Immunol 180:6508–6517
90. Velikovsky CA, Deng L, Chlewicki LK, Fernandez MM, Kumar V, Mariuzza RA (2007) Structure of natural killer receptor 2B4 bound to CD48 reveals basis for heterophilic recognition in signaling lymphocyte activation molecule family. Immunity 27:572–584
91. Garni-Wagner BA, Purohit A, Mathew PA, Bennett M, Kumar V (1993) A novel function-associated molecule related to non-MHC-restricted cytotoxicity mediated by activated natural killer cells and T cells. J Immunol 151:60–70
92. Schatzle JD, Sheu S, Stepp SE, Mathew PA, Bennett M, Kumar V (1999) Characterization of inhibitory and stimulatory forms of the murine natural killer cell receptor 2B4. Proc Natl Acad Sci USA 96:3870–3875
93. Boles KS, Stepp SE, Bennett M, Kumar V, Mathew PA (2001) 2B4 (CD244) and CS1: novel members of the CD2 subset of the immunoglobulin superfamily molecules expressed on natural killer cells and other leukocytes. Immunol Rev 181:234–249
94. Hershey GK (2003) IL-13 receptors and signaling pathways: an evolving web. J Allergy Clin Immunol 111:677–690; quiz 691
95. Young DA, Lowe LD, Booth SS, Whitters MJ, Nicholson L, Kuchroo VK, Collins M (2000) IL-4, IL-10, IL-13, and TGF-beta from an altered peptide ligand-specific Th2 cell clone down-regulate adoptive transfer of experimental autoimmune encephalomyelitis. J Immunol 164:3563–3572

The Regulatory Natural Killer Cells

Zhigang Tian and Cai Zhang

Abstract NK cells are effector lymphocytes of the innate immune system that control tumors and microbial infection by limiting their spread and subsequent tissue damage. Recently, more and more evidence has been obtained that NK cells also display a potent regulatory function by secreting various cytokines or cell-to-cell contact and thus regulate innate and adaptive immune responses and maintain immune homeostasis. In this review, we summarize the progress in studying the positive and negative regulatory effects of NK cells in immune responses as well as the NK subsets identified in humans and mice.

1 Introduction

The immune system evolves to protect the host against the attack of foreign pathogens by recognizing self-antigens and non-self-antigens, and to prevent the host from suffering autoimmune diseases via tolerance to self-antigen. Improper immune responses (e.g., immune response to self-antigens, immune tolerance or excessive immune response to foreign pathogens) may result in damage to the host [1, 2]. To sustain the immune homeostasis of the internal environment, a complex immune regulatory system (e.g., the cytokine network, the idiotype and antiidiotype network, the regulatory network among immune cells) has been developed during the evolutionary process of the immune system. Foxp3-expressing regulatory T cells (Treg), induced Treg (Tr1 cells), dendritic cells (DC) and natural killer T (NKT) cells are involved in maintaining immunological self-tolerance and immune homeostasis through several inhibitory mechanisms [3–6]. Natural killer (NK) cells

---

Z. Tian (✉)
School of Life Sciences, University of Science and Technology of China, Hefei City, Anhui 230027, China

C. Zhang
School of Pharmacy, Shandong University, Jinan City, Shandong 250012, China

are effector lymphocytes of the innate immune system that control several types of tumors and microbial infections by limiting their spread and subsequent tissue damage [7–9]. Although NK cells were originally defined by their capacity to lyse target cells and produce interferon gamma (IFNγ) without prior activation, more and more evidence has been obtained that NK cells also display a potent regulatory function by secreting various cytokines or cell-to-cell contact and maintain immune homeostasis. In this review, we summarize the progress in studying the positive and negative regulatory effects of NK cells in immune responses as well as the NK subsets identified in humans and mice.

2 The Positive Regulatory Effects of NK Cells

2.1 The Regulatory Effects of NK Cells on Dendritic Cells

NK cells discriminate target cells from other healthy "self" cells through a set of activating or inhibitory receptors, which recognize pathogen-encoded molecules ("non-self recognition"), self proteins whose expression is upregulated in transformed or infected cells ("stress induced-self recognition"), or self proteins that are expressed by normal cells but downregulated by infected or transformed cells ("missing-self recognition") [10–13]. NK cells and DC are two types of specialized cells of the innate immune system, the reciprocal interaction of which can potentially influence the maturation of each other and result in a potent, activating cross talk [14–16]. DC act during the priming phase of NK cell activation and shape the magnitude of innate immune responses by modulating the cytolytic effect of NK cells on tumors or virus-, bacteria- or parasite-infected cells. NK cells can provide signals that favor the generation of mature DC, promote DC maturation and cytokine production. In vivo, NK and DC interactions may occur in lymphoid organs as well as in sites of inflammation or in tumor tissues and at different stages of the innate and adaptive immune responses. By inducing DC activation, NK cell activation induced by tumor cells can indirectly promote antitumoral T cell responses.

NK cells could profoundly influence DC function by different mechanisms involving not only cell–cell contacts but also soluble factors [14–16]: (1) NK cell-mediated lysis of infected cells or tumor cells might release cell debris and microbial products that are taken up by DC, enabling cross-presentation of incoming antigens; (2) NK cell-mediated lysis of immature DC (iDC) might maximize the efficiency of antigen presentation by "cleaning-up" iDC that are unresponsive to particular infectious stimuli, thereby freeing up resources for reactive, maturing DC and promoting the efficiency of antigen presentation; (3) Activated NK cells boost the antigen processing and presentation of DC and the polarization of Th1 cells by producing IFNγ; (4) NK cells promote DC maturation and the production of IL-12, which in turn induces a more efficient cytotoxic T lymphocyte (CTL) response;

(5) NK cell-secreted granulocyte–macrophage-colony stimulating factor (GM-CSF) would promote DC survival and differentiation of monocytic precursors into DC.

The interaction of NK and DC plays a helper role in antitumor immune responses by promoting the induction of tumor-specific $CD4^+$ and $CD8^+$ T cell responses. Activated NK cells induce the maturation of DC into type-1 polarized DC (DC1), characterized by an up to 100-fold enhanced ability to produce IL-12. This process depends on IFNγ and TNFα produced by NK cells as well as NKG2D expressed on NK cells [17–21]. DC1 produce high levels of IL-12p70 even in the presence of immunosuppressive factors that abolish the IL-12p70 producing capacity of immature or mature DC [18]. In addition to tumor cell-derived signal (such as NKG2D–MICA interaction), the helper role of NK cells requires additional signals from type-I IFNs, products of virally infected cells, or from IL-2, produced by activated $CD4^+$ T helper (Th) cells. Other soluble factors, such as IL-18, exert synergistic effect with type-I IFNs or IL-2 during the process [17].

Efficient generation of cytotoxic T lymphocytes (CTL) from naïve $CD8^+$ T cells needs help from $CD4^+$ T cells, including secretion of cytokines and CD40/CD40L interactions, which lead to increased expression of costimulatory molecules on DC and induction of IL-12 [22, 23]. Interestingly, it was found that bone marrow-derived DC caused rejection of A20 lymphoma and induced tumor-specific long-term memory, even if they were not loaded with tumor-derived antigen. Surprisingly, experiments using CD40 knock-out mice and cell depletion indicated that this effect did not require the help of $CD4^+$ T cells [24]. Instead, the cross-talk between NK cells and DC plays key roles during this process. Activated NK cells stimulated by expression of NKG2D ligands on tumor cells produced IFNγ, which in turn induced the DC maturation. IL-12 produced by DC primed CTL responses. The primary rejection of A20 cells following immunization with unpulsed DC required NK cells and $CD8^+$ T cells, whereas long-term memories only need $CD8^+$ T cells [24, 25]. Therefore, a novel pathway linking innate and adaptive immunity was suggested in that the interplay between NK cells and DC could completely replace $CD4^+$ T cell help in the induction of $CD8^+$ CTL.

2.2 The Helper Role of NK Cells in Induction of Cytotoxic T Lymphocytes

NK cells fulfill essential accessory functions for the priming of antigen-specific CTL. Kos and his colleagues first postulated that the generation of human alloantigen-specific $CD8^+$ T cells required the participation of $CD3^-CD16^+CD56^+$ NK cells but did not need $CD4^+$ T helper cells [26, 27]. Depletion of NK cells from responders abolished the induction of alloantigen-specific CTL in vitro and in vivo. On the basis of a NKG2D-ligand-positive tumor model, results were obtained showing that NK-mediated regulatory as well as NK-mediated cytolytic activities played major roles in the initiation and persistence of CTL activity. $CD8^+$ T cell-dependent tumor rejection requires NK cell function in vivo, because tumors will progress both on depletion of NK cells or in the absence of optimal NK activity [28]. The absence of NK cells during subcutaneous tumor growth will abrogate generation of antitumor CTL responses. The accessory functions of NK cells in the initiation and persistence of CTL activity depend on the cross-talk of DC and NK cells. In addition, interaction of CD27 on NK cells and CD70 on tumor cells promoted the rejection of tumors by $CD8^+$ T cells, which is dependent on the production of IFNγ [29]. Wilcox and his colleagues found that IL-2 and IL-15 induced NK cells to express CD137 and ligation of CD137 stimulated NK cell proliferation and IFNγ secretion. CD137-stimulated NK cells promoted the expansion of activated $CD8^+$ T cells in vitro, demonstrating the immunoregulatory or helper activity for $CD8^+$ CTL [30].

$CD4^+$ T cells and $CD8^+$ T lymphocytes are important for elimination of intracellular pathogens. Limited information is available on the effect of NK cells on the $CD8^+$ CTL responses. Depletion of NK cells by antibodies (Ab) to asialo-GM1 or to NK1.1 suppressed the influenza virus-specific CTL responses, suggesting that NK cells are essential to initiation of CTL responses to infection [27, 31]. Depletion of NK cells from PBMC of healthy tuberculin reactors reduced the frequency of *Mycobacterium tuberculosis*-responsive $CD8^+$ $IFNγ^+$ T cells and decreased their ability to lyse *M. tuberculosis*-infected monocytes. Soluble factors secreted by activated NK cells can restore the frequency of $CD8^+$ $IFNγ^+$ T cells and this process depended on the presence of IFNγ, IL-15 and IL-18 [32]. Results showed that NK cells secreted IFNγ, which stimulated monocytes to produce IL-15 and IL-18, which in turn facilitated the expansion of $CD8^+$ T cells that produce IFNγ in response to *M. tuberculosis* infection. Meanwhile, the capacity of NK cells to prime CTL activity was also dependent on cell-to-cell contact between NK cells and infected monocytes because this process was inhibited by anti-CD40 and anti-CD40L [32]. NK cells link the innate and adaptive immune responses by promoting the capacity of $CD8^+$ T cells to produce IFNγ and to lyse infected cells, which is critical for protective immunity against *M. tuberculosis* and other intracellular pathogens.

2.3 NK Cells Promote the Differentiation of Th1 Cells by Secreting IFNγ

Previously, NK cells have been assumed to participate in adaptive immune responses by an indirect mechanism that involves their secretion of cytokines (such as IFNγ, GM-CSF and TNF-α) and chemokines (MIP-1α, MIP-1β, RANTES, etc) [7, 8, 33]. Recently, increasing evidence indicated that the direct cell-to-cell contact between NK and T cells also plays potential roles. Phillips and his colleagues first found that activated human NK cells expressed MHC class II molecules [34]. Later, Roncarolo and his colleagues also demonstrated that NK cells upregulated the expression of MHC class II molecules and presented antigen to $CD4^+$ T cells [35]. More recently, Zingoni et al. showed that human NK cells activated by innate cytokines, IL-12 and IL-15, expressed CD86 (a B7 family member and a ligand of the costimulatory receptor CD28). The ligation of activating NK receptors (e.g., CD16, NKG2D) induced the expression of OX40 ligand (OX40L) on activated NK cells, while OX40 is induced by TCR/CD3 signals and is mostly present on activated CD4$^+$ T cells. They further demonstrated that receptor-activated NK cells can costimulate TCR-induced proliferation and cytokine production of autologous CD4$^+$ T cells and that this process requires OX40-OX40L and CD28-B7 interaction [36, 37]. This interaction might happen in liver and secondary lymphoid organs (SLO). These findings suggest a novel link between the natural and adaptive immune responses, providing direct evidence for cross-talk between human CD4$^+$ T cells and NK receptor-activated NK cells.

By high-throughput proteomic analysis of NK cell membrane-enriched fractions and flow cytometry, Hanna and his colleagues showed that activated NK cells express significant levels of MHC class II molecules and ligands for TCR costimulatory molecules (e.g., B7-H3 and CD70). Incubation of activated NK cells resulted in a dose-dependent enhancement of CD4$^+$ T cell proliferation and secretion of IL-2 and IFNγ. NK cells possess multiple independent unique pathways for antigen uptake. The ligation of activating receptors (NKp30, NKp46, NKG2D and CD16) on NK cells and their ligands leads to lysis of target cells, and then NK cells become activated and acquire the ability to process pathogen-derived antigen and stimulate CD4$^+$ T cells. They also showed that NK cells isolated from inflamed tonsils express significant levels of HLA-DR, DP, DQ, CD86, CD70, and OX40L, which indicated that human NK cells acquire an APC-like phenotype in vivo in inflamed lymphoid organs [38]. These observations offer new insights into the direct interactions between NK and T cells and suggest novel APC-like properties of human NK cells.

Activated NK cells could boost the ongoing adaptive responses by producing IFNγ, which promotes the Th1 orientation of antigen-specific T cells. IL-12 plays an important role in promoting the production of Th1 type cytokines and inhibiting the production of Th2 type cytokines in antigen-specific adaptive immune responses. Activated NK cells produce IFNγ, which in turn activates macrophages to secrete IL-12. Both IFNγ and IL-12 can induce the differentiation of Th1 cells [15, 16]. In humans, tissue-resident NK cells from secondary lymphoid tissues supported Th1 polarization more efficiently than blood NK cells by their superior ability to produce IFNγ after stimulation with DC. Th1-polarized T cell responses have been found to be more effective in the immune control of tumors and viral infections in mice. Normal BALB/c mice are sensitive to *Leishmania major* infection and undergo fatal visceral dissemination when infected. During the early phase of infection, injection of IL-12 decreased the mortality rate and the mice acquired the capacity to resist rechallenging with *L. major*. However, depletion of NK cells abolished this effect of IL-12 and did not result in Th1 cytokine responses. Normal CH3 mice are resistant to *L. major* infection and sustain a type 1 cytokine status in vivo. Depletion of NK cells resulted in a type 2 cytokine status and dissemination of lethal infection in liver and lung [39]. Similar results were also obtained in mice with *Bordetella pertussis* infection [40]. Meanwhile, the type 2 cytokine status was observed in NK-deficient tumor-bearing mice, which affects the activity of CTL and suppresses subsequent memory CTL restimulation [28]. Increasing evidence indicated that the predominant effect of IL-12 on Th1 status depends on the presence of NK cells and the importance of NK cells in promoting the differentiation of Th1 cells and in the generation and persistence of adaptive immune responses. When the immune system tends to be type 2 oriented, NK cells can drive an efficient type 2 to type 1 switch in the population of antigen-presenting cells and Th cells to provide signaling for the generation of CTL.

2.4 NK Cells Regulate Immune Response via T Regulatory Cells

$CD4^+CD25^+$ Treg have been recognized as a major population of suppressing lymphocytes that maintain peripheral immune tolerance [3, 4]. Many studies have shown that Treg are involved in downregulating the immune response to organ transplants, tumors and intracellular pathogens, preventing the development of autoimmune diseases [41–44]. It is now accepted that Treg homeostasis depends in part on the peripheral conversion of naïve $CD4^+CD25^-$ T cells. This conversion implicates acquisition of the Treg-specific marker, forkhead winged helix protein 3 (Foxp3), after CD28 costimulation. Brillard and colleagues found that IL-2-activated NK cells decreased Treg conversion of adoptively transferred murine $CD4^+CD25^-$ T cells in vivo. Meanwhile, human activated NK cells decreased CD28-driven Foxp3 expression in $CD4^+CD25^-$ T lymphocytes [45]. These results suggest that activated NK cells interfere with CD28-mediated Foxp3 expression in $CD4^+CD25^-$ T lymphocytes. It has been recently found that NK cells lyse Treg that expand in response to the intracellular pathogen *M. tuberculosis*. Monokine (IL-12, IL-15, and IL-18) or TB lysate-stimulated monocytes-activated NK cells lyse TB lysate-expanded Treg, but not freshly isolated Treg. This effect is mediated through the NK cell-activating receptors, NKp46 and NKG2D. Meanwhile, the NKG2D ligand ULBP1 is upregulated in TB lysate-expanded Treg, and the lysis is determined to be mediated via interactions between NKG2D and ULBP1 [46]. The results suggest that NK cells may play an important role during the early phases of the immune response to *M. tuberculosis* and other intracellular pathogens by inhibiting Treg expansion.

3 The Negative Regulatory Effects of NK Cells

NK cells exert negative regulatory effects to maintain immune homeostasis in some physiological and pathological conditions. For example, they can kill over-stimulated macrophages, DC, and activated T and B cells. A small NK cell subset that expresses CD94/NKG2A, but not killer immunoglobulin-like receptors (KIR) in lymph nodes was capable of killing not only immature DC but also mature DC in order to prevent the over-activation of DC [47]. The CD94/NKG2A$^+$ KIR$^-$ NK cell phenotype is consistent with that of most CD56^{bright} NK cells. By limiting the supply and recruitment of iDC, activated NK cells exert the ability to control the subsequent innate and adaptive immune responses. The cross-talk between activated NK cells and iDC acts as a control/switch for the immune system. At low NK: DC ratio (1:5), the DC/NK interaction dramatically enhances DC cytokine production (IL-12, TNF-α) in a cell to cell contact-dependent manner and DC maturation which was dependent on endogenously produced TNF-α. While at high NK:DC ratio (5:1), inhibition of DC functions is the dominant feature due to the direct NK cell killing of immature DC [14–16, 18, 19]. NK cells may also play an important role in maintaining immune homeostasis by directly regulating clonal expansion of activated T cells. The role for NK cells in downregulation of T cell responses has been implicated in several studies. Trivedi and colleagues found that NK cells inhibit syngeneic T cell proliferation via upregulation of the cell cycle inhibitor, p21, resulting in a G0/G1 stage cell cycle arrest. The inhibition is cell–cell contact dependent, reversible, and antigen nonspecific [48]. In a skin transplant model, host NK cells can destroy graft-derived antigen presenting cells, inhibit priming of alloreactive T cells, and thus promote transplant tolerance [49, 50]. More recently, Deniz and his colleagues characterized an IL-10-secreting NK cell subset and found that the IL-10$^+$ NK cells significantly suppressed both allergen/antigen-induced CD4$^+$ T cell proliferation and secretion of IL-13 and IFN-γ, particularly due to secreted IL-10 as demonstrated by blocking of the IL-10 receptor [51]. A NKp46$^+$CD49b$^+$CD3$^+$ NK subset was identified and was found capable of being recruited into spleen and hepatic granulomas after *Leishmania donovani* infection and also inhibit host protective immunity in an IL-10-dependent manner [52]. These results suggest that the novel mechanism of T cell regulation by NK cells provides insight into NK cell-mediated regulation of adaptive immunity and provides a mechanistic link between NK cell function and suppression of T cell responses. The regulatory NK cells play major roles in autoimmunity, pregnancy tolerance and organ transplantation.

3.1 The Negative Regulatory Effects of NK Cells in Antoimmune Liver Injury

Hepatitis is a common worldwide disease with high mortality. Several hepatitis models have been reported by different groups to study hepatitis, mainly including Concanavalin A (Con A)-induced hepatitis [53–55] and LPS-induced liver injury [56]. The former was a well-described mouse model of T cell-dependent liver injury via intravenous injection with the T cell mitogen Con A, which can cause fulminant hepatitis [53–55]. The latter was considered to be associated with macrophages and was septic hepatitis [56, 57]. Activation of T cells or macrophages is one of the initial events in these two hepatitis models in which activated T cells and macrophages are directly cytotoxic as against hepatocytes or indirectly released proinflammatory cytokines, which mediate hepatocyte damage [58, 59]. Recently, it has been reported that natural killer T (NKT) cells, a population of T cells, play a key role in Con A-induced hepatitis via releasing a wide variety of cytokines and direct cytotoxicity against hepatocytes [60–62].

Recent research revealed that NK cells are involved in the pathogenesis of some autoimmune diseases (e.g., diabetes, multiple sclerosis) [63–65]. Although hepatitis is thought to be a kind of autoimmune disease, the critical role of NK cells in its pathogenesis is still undefined. It is found that the amount and proportion of NK cells in HBV- or HCV-infected liver increased significantly and that the liver injury is also related to NK cell activation [66–68]. Polyinosinic-polycytidylic acid (Poly I:C) is an artificial mimic of viral RNA and mimics the immune response during viral infection and can induce the activation of NK cells in vivo and in vitro [69]. We found that Poly I:C significantly induces NK cell accumulation and activation in the liver [70, 71]. Administration of Poly I:C induces a slight elevation of ALT/AST, mild inflammation and focal necrosis in the liver. Poly I:C-induced liver injury is much weaker than fulminant hepatitis caused by Con A injection. Depletion of NK cells by anti-AsGM1 Ab markedly attenuates Poly I:C-induced liver injury. We further demonstrated that Poly I:C-induced liver injury is dependent of NK cells, but independent of T, B and NKT cells, and independent of IFNγ, TNF-α and IL-6 by using gene knock-out mice [70]. The molecular mechanisms might be that NK cells expressing TNF-related apoptosis-inducing ligand (TRAIL) displayed strong cytotoxicity against primary hepatocytes in the liver injury [72]. These findings in Poly I:C-induced liver injury model suggest the involvement of NK cells in autoimmune hepatitis.

Interestingly, injection of mice with a subdose of Con A (inducing NKT activation but without liver injury) followed by Poly I:C leads to severe liver injury, which indicated that NKT cells upregulate the function of NK cells [73]. However, when the mice were injected with a low dose of Poly I:C followed by Con A, there was no injury to the liver, which indicates that Poly I:C pretreatment protects against T cell-mediated hepatitis, as evidenced by decreased mortality, hepatic necrosis, serum transaminase levels and inflammatory cytokines (IL-4, IFNγ). The protective effect of Poly I:C was diminished in NK-depleted mice, which could be partially restored by adoptive transfer of NK cells. Administration of Poly I:C caused NKT and T cell apoptosis via enhancing expression of Fas protein on these cells and expression of Fas ligand on NK cells [74]. These findings suggest that activation of NK cells by Poly I:C prevents Con A-induced T cell-hepatitis via downregulation of T cells or NKT cells and subsequent reduction of inflammatory cytokines. Our further investigation showed that injecting mice with a low dose of Con A led to no liver injury, but severe liver injury occurred when it was given to HBs antigen transgenic mice (NK cell deficient). The over-sensitiveness of HBs antigen transgenic mice to Poly I:C-induced liver injury was absolutely dependent on the presence of NK cells and IFNγ produced by intrahepatic NK cells, but independent of Kupffer cells and IL-12 [75, 76]. These results demonstrate that malfunction of NK cells in HBs antigen transgenic mice lost the negative effect on NKT cells. Further investigation demonstrated that NKG2D recognition of hepatocytes by NK cells plays a critical role in oversensitive liver injury during chronic HBV infection. In HBs antigen transgenic mice, low dose of Con A stimulation, with non hepatotoxic effect for wild-type mice, markedly increased expressions of NKG2D ligands (Rae-1 and Mult-1) in hepatocytes, which greatly activated hepatic NK cells via NKG2D/Rae-1 or Mult-1 recognition [77]. The Kupffer cell is an important innate immune cell for initiation of several hepatitis models. It is also found that TLR3 ligand Poly I:C pretreatment can alleviate subsequent liver injury induced by TLR4 ligand, LPS, in which Poly I:C treatment can downmodulate TLR4 expression on Kupffer cells [78]. In addition, NK cells are also found to prevent formation of liver fibrosis. During CCL4 induced-liver fibrosis, NK cells have an ability to kill activated stellate cells which have increased NKG2D ligand Rae-1, a stress-inducible molecule in addition, TRAIL is also responsible for the NK cell cytotoxicity [79].

3.2 The Role of NK Cells in Other Autoimmune Diseases

The role of NK cells in autoimmunity is attracting increased attention, although NK cells play seemingly opposite roles in autoimmune diseases and function as both regulators and inducers of autoimmune diseases [63–65, 80]. The role NK cells play depends on which cells become the targets for NK cell attack. If the targets are nontransformed or stressed organized tissues, NK cells might participate in their destruction in the initial stage of autoimmune diseases. If the targets are autoreactive immune cells, NK cells can act as regulators by killing the inflammatory cells mediating autoimmune diseases.

NK cells have been shown to regulate autoimmune responses by inhibiting the autoreactive T lymphocytes under some experimental conditions in animals. The mechanism of the NK cell's regulatory role in experimental autoimmune encephalomyelitis (EAE) was studied. Smeltz et al. showed that rat bone marrow-derived NK cells exhibited potent inhibitory effects on T cell proliferation to both Con A as well as the central nervous system antigen myelin basic protein, which indicates that NK cells may play an important role in regulating both normal and autoimmune T cell responses by exerting a direct effect on activated, autoantigen-specific T cells [81]. Furthermore, in vivo experiments showed that NK cell depletion by anti-NK1.1 monoclonal antibody treatment enhanced EAE in mice. To investigate the mechanism, proteolipid protein (PLP) (136–150) peptide-specific, encephalitogenic T cell lines, which were used as the NK cell target, were cultured. The results show that NK cells exert a direct cytotoxic effect on autoantigen-specific, encephalitogenic T cells. Furthermore, using enriched NK cells as effectors enhanced cytotoxicity to PLP-specific, encephalitogenic T cell lines [82]. The results indicate that NK cells play a regulatory role in EAE through killing of syngeneic T cells, which include myelin antigen-specific, encephalitogenic T cells, and thus ameliorate EAE. Takahashi and colleagues found that NK cells from $CD95^+NK^{high}$ multiple sclerosis (MS) patients could inhibit the antigen-driven secretion of

IFNγ by autologous myelin basic protein (MBP)-specific T cell clones in vitro, which indicate that NK cells may regulate activation of autoimmune memory T cells in an antigen nonspecific fashion to maintain clinical remission in $CD95^+NK^{high}$ MS patients [83]. Meanwhile, the type 2 cytokines, such as IL-5, secreted by NK cells (NK2 cells) also play a role in maintaining the remission of MS, while NK cells lose the NK2-like property when relapse of MS occurs, but regain it after recovery. It has been found that NK2 cells induced in vitro inhibit induction of Th1 cells, suggesting that the NK2-like cells in vivo may also prohibit autoimmune effector T cells [84]. In vivo blockade of the human IL-2R by mAb has been used for immunosuppression in transplantation, therapy for leukemia, and autoimmune diseases. During the course of IL-2Rα-targeted therapy (daclizumab) of uveitis and MS patients, it was found that regulatory $CD56^{bright}$ NK cells mediate immunomodulatory effects by inhibiting activated T cell survival [85, 86].

Type 1 diabetes is an autoimmune disease in which pancreatic islet β cells are destroyed by the cellular immune system, leading to β cell loss, insulin deficiency, and hyperglycemia [87, 88]. Much of the understanding of the pathogenesis of β cell destruction in type 1 diabetes has been obtained through the study of disease in the nonobese diabetic (NOD) mouse [89, 90]. Although some researchers found that depletion of NK cells prevented the loss of β cells following infection, and indicated that NK cells induced the progression of type 1 diabetes by stimulating autoreactive T and B cells [91, 92], some conflicting outcomes have been observed by other groups. Previous studies have shown that a single injection of CFA prevents diabetes in NOD mice, but the mechanisms of protection remain unknown. Lee and colleagues showed that this reduced incidence was associated with a decrease in the number of β cell-specific, autoreactive CTL. Their results demonstrated that CFA-induced NK-cell trafficking to the blood and spleen, induced IFNγ production by NK cells and decreased activation of β cell-specific T cells. Moreover, removal of NK cells abolished the protective effects of CFA while restoration of NK cells returned its protective effects, indicating that NK cells mediate protection from diseases [93]. It has been reported that NOD mice treated with Poly I:C have a markedly reduced incidence of diabetes [94], but the underlying mechanisms remain undefined. Recently, our investigation into autoimmune diabetes demonstrated that NK cells mediate the protective effect of Poly I:C through the promotion of Th2 bias of immune responses. We found that long-term Poly I:C-treated NK cells exhibit a NK3-like phenotype, and are involved in the induction of Th2 bias of spleen cells in response to islet autoantigens via TGF-β-dependent manner. These findings suggest that NK cells may participate in the regulation of autoimmune diabetes [95].

The number and activity of NK cells were found reduced in target organs or in peripheral blood within systemic lupus erythematosus (SLE), psoriasis, MS, rheumatoid arthritis (RA) and type 1 diabetes patients [96]. It has been hypothesized that NK cells act as a source of Th2 cytokines. When this type of NK cells is reduced, Th1 cytokines become abundant, which results in autoimmunity. It has been identified that there are associations between risk of autoimmune diseases and activating KIR or KIR/HLA genotypes accompanied by a lack of inhibition.

For example, risk of systemic sclerosis, rheumatoid vasculitis, and type 1 diabetes seemed to be associated with the presence of activating KIR2DS2 accompanied by diminished inhibitory interactions [96].

In conclusion, NK cells may either prevent autoimmune responses or may have a permissive role in autoimmunity. The discrepancy role might lie in the different functional NK subsets, the different phases of diseases, the localization of NK cells as well as the local cytokine environment.

3.3 The Regulatory Effects of Uterine NK Cells in Maternal Fetal Tolerance

NK cells constitute 50%–90% of lymphocytes in human uterine decidua in early pregnancy. These cells derive predominantly from a subset of peripheral blood NK cells, which gets recruited to the uterus under hormonal influence [97–100]. The uterine decidual NK cells are believed to promote placental and trophoblast growth and provide immunomodulation at the interface. The phenotypes and function of uterine NK (uNK) cells are different from those of peripheral blood NK cells. Human peripheral blood NK cells comprise two different subsets, the predominant $CD56^{dim}$ NK cell subset (90%–95%) and the much smaller $CD56^{bright}$ NK cell subset (5%–10%). However, the pattern of predominant human uterine decidual NK cells is $CD56^{bright}CD16^{-}$ with expression of inhibitory NK cell receptors CD94/NKG2A and KIR [97, 98, 101]. $CD56^{bright}$ NK cells are now thought to be an important NK cell subset for exerting immunoregulatory effect. The $CD56^{bright}CD16^{-}$ NK cells with expression of inhibitory NK cell receptors in uterus may play a major regulatory role in preventing the attack of NK cells on trophoblast cells and in promoting the implanting and growth of placenta and trophoblast. It is found that spontaneously aborting women usually have a greater percentage of $CD56^{dim}CD16^{+}$ cells and smaller percentage of $CD56^{bright}CD16^{-}$ NK cells in the secretory phase of endometrium [102]. These NK cells also have a limited repertoire of inhibitory receptors of the KIR family, and many of them lack KIR specificity for the fetal HLA-Cw antigens, which indicates that a limited maternal KIR repertoire and a lack of maternal KIR-fetal HLA-C epitope matching may be the cause of recurrent spontaneous abortion [103].

The cytokine profile of decidual $CD56^{bright}CD16^{-}$ NK cells is unique and different from $CD16^{+}$ NK cells and peripheral blood $CD56^{bright}CD16^{-}$ NK cells. Shigeru and colleagues found that the main populations of decidual $CD56^{bright}$ NK cells and $CD56^{dim}$ NK cells were TGF-β-producing, IFNγ, TNF-α, IL-4, IL-5 or IL-13-nonproducing NK3 cells, suggesting that TGF-β plays a very important role at the feto–maternal interface [102]. These cell populations increased in early pregnant decidua, but decreased accompanied with increased NK1 cell population in miscarriage cases. They also postulated that decidual NK cells might promote pregnancy tolerance by killing fetal DC and might downregulate the MHC class II expression on DC by TGF-β production and thus inhibit the CTL induction to fetal antigens [102]. Importantly, Blois et al. recently provided evidence that interaction between DC and NK cells promotes a tolerogenic microenvironment at the mouse maternal fetal interface by downregulation of the expression of activation markers on uNK cells and uterine DC (uDC) and dominance of Th2 cytokines. The secretion of IL-12p70 by uDC is dramatically abrogated in the presence of uNK cells [104]. It has been supposed that decidual NK cells might represent a distinct lineage of NK cells, possibly arising from a distinct hematopoietic precursor. Or, the decidual $CD56^{bright}CD16^-$ NK cells might mature from peripheral blood $CD56^{bright}CD16^-$ NK cells in the decidual microenvironment after recruitment into the uterus [98]. L-selectin and some chemokines, such as MIP-Iβ, IP-10 and SDF-1, are found involved in the adhesion and accumulation of peripheral blood $CD56^{bright}$ NK cells to endothelial cells [102, 105]. In addition, decidual $CD56^{bright}CD16^-$ NK3 cells were also found to play an important role in angiogenesis during pregnancy via producing a large amount of angiogenic growth factors, and insufficient NK3 function might induce poor angiogenesis in the uterus, resulting in preeclampsia [106, 107].

4 NK Cell Subsets and the "Regulatory" Function

Accumulating evidence shows that NK cells play positive or negative immune regulatory roles by secreting various cytokines or cell-to-cell contact and maintain immune homeostasis, which reveals that NK cells are not only "killers," but also "regulators." The malfunction or loss of NK cells is involved in the pathogenesis of many diseases (e.g., tumor, autoimmune diseases, infection, sterility, and recurrent spontaneous abortion). Increasing data have indicated the presence of different subsets of NK cells dependent on either the pattern of cytokines they produced or expression of cell surface markers and activating or inhibitory receptors. Therefore, we and others proposed that there may exist a "regulatory NK cell subset (NKreg subset)" [108, 109], although the exact phenotype and function are not clearly clarified.

It was originally proposed that human NK cells in peripheral blood can be divided into at least two subsets based on the expression of CD56 and CD16. A majority (approximately 90%) of human NK cells are $CD56^{dim}$ and express high levels of FcγRIII (CD16), whereas the minority (approximately 10%) are $CD56^{bright}$ and $CD16^{dim/neg}$ [110–113]. $CD56^{bright}CD16^-$ NK cells constitutively express the high- and intermediate-affinity IL-2 receptors and expand in vitro and in vivo in response to low doses of IL-2. In contrast, resting $CD56^{dim}CD16^+$ NK cells express only the intermediate affinity IL-2 receptor and proliferate weakly in response to high doses of IL-2 in vitro, even after induction of the high-affinity IL-2 receptor. $CD56^{bright}CD16^-$ NK cells produce significantly greater levels of IFNγ, TNF-β, GM-CSF, IL-10, and IL-13 protein in response to monokine stimulation than do $CD56^{dim}CD16^+$ NK cells, which produce negligible amounts of these cytokines. Resting $CD56^{dim}CD16^+$ NK cells are more cytotoxic against

NK-sensitive targets than CD56^{bright}CD16$^-$ NK cells. All resting CD56^{bright}CD16$^-$ NK cells express high levels of the inhibitory CD94/NKG2A complex recognizing HLA-E, but they do not express MHC class I allele-specific KIR that are, in contrast, expressed by CD56^{dim}CD16$^+$ NK cells. Both NK cell subsets in human peripheral blood express the activating receptors NKG2D and NKp30 as well as NKp46. CD56^{bright}CD16$^-$ NK cells express a functional CC-chemokine receptor 7 (CCR7), CD62L and have high-level expression of CXC-chemokine receptor 3 (CXCR3) and strong chemotactic responses to the ligands for these receptors, resulting in an enrichment of this subset in SLO and sites of inflammation. By contrast, CD56^{dim}CD16$^+$ NK cells lack expression of CCR7 but have high levels of expression of CXCR1 and CX3C-chemokine receptor 1 (CX3CR1), and show functional responses to IL-8 and fractalkine. It has been demonstrated that AIDS patients and some cancer patients have a lower percentage of CD56^{dim} NK-cell subset with decreased natural cytotoxicity [114, 115]. While in immune-tolerance organs, such as the human uterus and liver, NK cells were mostly composed of CD56^{bright} subset with relatively fewer CD56^{dim} cells [97, 116, 117]. These results indicate that CD56^{bright} NK cells have a immunoregulatory function and predominate in the lymph nodes and sites of inflammation, whereas CD56^{dim} NK cells play a key role in innate immunity against viruses and cancer.

The developmental relationship between CD56^{bright} and CD56^{dim} NK cells has not been clarified. It is commonly accepted that CD56^{bright} NK cells can be induced by IL-15 from NK cell precursors and may thereafter differentiate into CD56^{dim} NK cells [110, 111, 118]. However, the existence of a unique CD56^{dim} NK-cell precursor or an alternate signal that could induce the differentiation of CD56^{dim} cells from a common NK-cell precursor cannot be excluded. Recently, we found that both IL-2 and IL-15 support differentiation of NK cells from cord blood mononuclear cells. IL-15 improved the proliferation and activation of CD56^{dim} NK cells in long-term cord blood mononuclear cell culture, but IL-2 only maintained the survival of CD56^{bright} NK cells. IL-15 played a crucial role in sustaining the long-lasting functions of CD56^{dim} NK cells, while Bcl-xL is associated with the antiapoptotic effect [119]. The results revealed that IL-15, but not IL-2, maintained terminal differentiation and survival of CD56^{dim} NK cells from intermediate NK progenitor or precursor cells.

Similar to the Th1 and Th2 subsets of CD4$^+$ T cells, NK cells are also divided into NK1 and NK2 subpopulations according to the profile of cytokine secretion [120–122]. The NK1 subset showed a typical cytokine pattern with predominant expression of IFNγ, but almost no IL-4, IL-5 and IL-13. In contrast, the NK2 subset expresses IL-4, IL-5 and IL-13 but no IFNγ. Our and other results showed that the type 2 cytokines produced by NK2 cells are dominant in asthma and tumor microenvironment and are involved in pathogenesis of asthma and cancer [123–125]. The IL-10-secreting NK3 and TGF-β-secreting NKr1 cells also play major roles in immunoregulation and promote transplant and pregnancy tolerance [49, 102].

CD56 is not expressed in rodents, and there has been no similar clear evidence for functionally distinct mature NK cell subsets in mice. Recently, Smyth et al. showed in mice that the mature Mac-1^{high} NK cell pool can be further dissected into two functionally distinct $CD27^{high}$ and $CD27^{low}$ subsets [126, 127]. They demonstrated that $CD27^{high}$ and $CD27^{low}$ mouse NK cell subsets exhibit some similarities to but also some distinct differences from the human CD56 NK cell subsets in terms of their function and phenotype. $Mac-1^{high}CD27^{high}$ NK cells exert higher cytotoxicity against tumor target cells even in the presence of MHC class I expression and produce higher amount of IFNγ in response to IL-12 and IL-18 stimulation, which is demonstrating a predominant role in NK:DC cross talk. By contrast, $Mac-1^{high}CD27^{low}$ NK cells express higher level of inhibitory Ly-49C/I receptors that recognize self MHC class I molecules and display very low cytotoxicity and cytokine production. $Mac-1^{high}CD27^{high}$ NK cells were the main NK cells in LN, while $Mac-1^{high}CD27^{low}$ NK cells reside predominantly in lung and peripheral blood, which indicates the possible surveillance role outside of the lymphoid tissue environment. More importantly, $CD27^{high}$ and $CD27^{low}$ subsets also exist in humans [128, 129]. The majority of peripheral blood human NK cells were $CD27^{low}/CD56^{dim}$ NK cells, containing high levels of perforin and granzyme B, and were able to exert strong cytotoxic activity. In contrast, the minor $CD27^{high}$ NK cells were mostly $CD56^{bright}$, had significantly lower levels of perforin and granzyme B, and had a low cytolytic potential, whereas significantly higher frequencies of $CD27^+$ NK cells, most of them being $CD56^{bright}$, were found in spleen and in tonsils. Another murine NK subset with expression of CD127, the IL-7 receptor, and transcription factor GATA-3 was described in the thymus [130]. This subset exhibits similar properties like human $CD56^{bright}CD16^-$ NK cells in expression of CD127, low cytotoxicity, and higher cytokine production.

A $B220^+CD11c^+NK1.1^+$ cell population capable of producing higher levels of IFNγ has recently been found in secondary lymphoid tissues. This subset possesses both the lytic potential of NK cells and the antigen-presenting capacity of DC and thus has also been called interferon-producing killer dendritic cells (IKDC). Further investigation revealed that IKDC efficiently kill NK cell targets and proliferate extensively in response to IL-15 and IL-18, but induce little or no T cell proliferation relative to conventional DC when exposed to protein antigen or to MHC class II peptide. Existing data indicate that this population should be activated NK cells, but should not belong to DC [131–133]. More recently, a novel noncytolytic NK cell subset displaying a mature phenotype and remarkable immunoregulatory functions has been generated from peripheral blood $CD34^+$ progenitors [109]. These regulatory NK cells exhibit the phenotype of $CD56^+CD16^+NKp30^+NKp44^+NKp46^+CD94^+CD69^+CCR7^+HLA-G^+$ and secrete the immunoregulatory factors IL-10, IL-21, and HLA-G. They downregulate immune response by suppressing mature myeloid DC (mDC) into immature/tolerogenic DC, blocking cytolytic functions on conventional NK cells and inducing HLA-G membrane expression on peripheral blood-derived monocytes. The effect is dependent on reciprocal transpresentation of membrane-bound IL-15 forms expressed by human $CD34^+$ peripheral blood haematopoietic progenitors. Maroof and colleagues identified an $NKp46^+CD49b^+CD3^-$ NK subset, which can recruit to spleen and hepatic granulomas and inhibit host protective immunity in a IL-10-dependent manner [52].

In summary, more and more evidence has been obtained that NK cells are no longer simple "killers" against tumor and infected cells. More importantly, they fulfil essential immunoregulatory functions either by secreting various cytokines or through cell-to-cell contact with other immune cells, and thus participate in and influence adaptive immunity as well as maintain immune homeostasis. The diverse activities of NK cells might be decided by the diversity of NK cell subsets, the different tissue environment, and the cross-talk with other immune cells. Deeper investigation aimed at understanding the NK cell biology, the functional distinctions between NK cell subsets, and design of new tools to monitor NK cell activity, modulate NK cell trafficking patterns and dampen NK cell suppressors are needed to strengthen our ability to harness the power of NK cells for therapeutic aims.

References

1. Artavanis-Tsakonas K, Tongren JE, Riley EM (2003) The war between the malaria parasite and the immune system: immunity, immunoregulation and immunopathology. Clin Exp Immunol 133:145–152
2. von Herrath MG, Harrison LC (2003) Antigen-induced regulatory T cells in autoimmunity. Nat Rev Immunol 3:223–232
3. O'Garra A, Vieira P (2004) Regulatory T cells and mechanisms of immune system control. Nat Med 10:801–805
4. Jiang H, Chess L (2004) An integrated view of suppressor T cell subsets in immunoregulation. J Clin Invest 114:1198–1208
5. Reise Sousa C (2001) Dendritic cells as sensors of infection. Immunity 14:495–502
6. Godfrey DI, Kronenberg M (2004) Going both ways: immune regulation via CD1d-dependent NKT cells. J Clin Invest 114:1379–1388
7. Trinchieri G (1989) Biology of natural killer cells. Adv Immunol 47:187–376
8. French AR, Yokoyama WM (2003) Natural killer cells and viral infections. Curr Opin Immunol 15:45–51
9. Smyth MJ, Godfrey DI, Trapani JA (2001) A fresh look at tumor immunosurveillance and immunotherapy. Nat Immunol 2:293–299
10. Kärre K (2002) NK Cells, MHC class I molecules and the missing self. Scand J Immunol 55:221–228
11. Moretta L, Biassoni R, Bottino C, Mingari MC, Moretta A (2002) Natural killer cells: a mystery no more. Scand J Immunol 55:229–232
12. Wu P, Wei H, Zhang C, Zhang J, Tian Z (2005) Regulation of NK cell activation by stimulatory and inhibitory receptors in tumor escape from innate immunity. Front Biosci 10:3132–3142
13. Zhang C, Zhang J, Wei H, Tian Z (2005) Imbalance of NKG2D and its inhibitory counterparts: how does tumor escape from innate immunity? Int Immunopharmacol 5:1099–1111
14. Fernandez NC, Lozier A, Flament C, Ricciardi-Castagnoli P, Bellet D, Suter M, Perricaudet M, Tursz T, Maraskovsky E, Zitvogel L (1999) Dendritic cells directly trigger NK cell function: cross-talk relevant in innate anti-tumor immune responses in vivo. Nat Med 5:405–411
15. Cooper MA, Fehniger TA, Fuchs A, Colonna M, Caligiuri MA (2004) NK cell and DC interactions. Trends Immunol 25:47–52
16. Moretta A (2002) Natural killer cells and dendritic cells: rendezvous in abused tissues. Nat Rev Immunol 2:957–963

17. Raulet D (2004) Interplay of natural killer cells and their receptors with the adaptive immune responses. Nat Immunol 5:996–1002
18. Zitvogel L (2002) Dendritic and Natural killer cells cooperate in the control/switch of innate immunity. J Exp Med 195:F9–F14
19. Moretta L, Ferlazzo G, Mingari MC, Melioli G, Moretta A (2003) Human natural killer cell function and their interactions with dendritic cells. Vaccine 21:S38–S42
20. Kalinski P, Giermasz A, Nakamura Y, Basse P, Storkus WJ, Kirkwood JM, Mailliard RB (2005) Helper role of NK cells during the induction of anticancer responses by dendritic cells. Mol Immunol 42:535–539
21. Guan H, Moretto M, Bzik DJ, Gigley J, Khan IA (2007) NK cells enhance dendritic cell response against parasite antigens via NKG2D pathway. J Immunol 179:590–596
22. Vieira PL, de Jong EC, Wierenga EA, Kapsenberg ML, Kalinski P (2000) Development of Th1-inducing capacity in myeloid dendritic cells requires environmental instruction. J Immunol 164:4507–4512
23. Bennett SR, Carbone FR, Karamalis F, Flavell RA, Miller JF, Heath WR (1998) Help for cytotoxic-T-cell responses is mediated by CD40 signaling. Nature 393:478–480
24. Adam C, King S, Allgeier T, Braumuller H, Luking C, Mysliwietz J, Kriegeskorte A, Busch DH, Rocken M, Mocikat R (2005) DC-NK cell cross talk as a novel CD4+ T-cell-independent pathway for antitumor CTL induction. Blood 106:338–344
25. Westwood JA, Kelly JM, Tanner JE, Kershaw MH, Smyth MJ, Hayakawa Y (2004) Cutting Edge: Novel priming of tumor-specific immunity by NKG2D-triggered NK cell-mediated tumor rejection and Th1-independent CD4+ T cell pathway. J Immunol 172:757–761
26. Kos FJ, Engleman EG (1995) Requirement for natural killer cells in the induction of cytotoxic T cells. J Immunol 155:578–584
27. Kos FJ, Engleman EG (1996) Role of natural killer cells in the generation of influenza virus-specific cytotoxic T cells. Cell Immunol 173:1–6
28. Geldhof AB, Van Ginderachter JA, Liu Y, Noel W, Raes G, De Baetselier P (2002) Antagonistic effect of NK cells on alternatively activated monocytes: a contribution of NK cells to CTL generation. Blood 100:4049–4058
29. Kelly JM, Darcy PK, Markby JL, Godfrey DI, Takeda K, Yagita H, Smyth MJ (2002) Induction of tumor-specific T cell memory by NK cell-mediated tumor rejection. Nat Immunol 3:83–90
30. Wilcox RA, Tamada K, Strome SE, Chen L (2002) Signaling through NK cell-associated CD137 promotes both helper function for CD8+ cytolytic T cells and responsiveness to IL-2 but not cytolytic activity. J Immunol 169:4230–4236
31. Kos FJ, Engleman EG (1996) Immune regulation: a critical link between NK cells and CTLs. Immunol Today 17:174–176
32. Vankayalapati R, Klucar P, Wizel B, Weis SE, Samten B, Safi H, Shams H, Barnes PF (2004) NK cells regulate CD8+ T cell effector function in response to an intracellular pathogen. J Immunol 172:130–137
33. Biron CA, Nguyen KB, Pien GC, Cousens LP, Salazar-Mather TP (1999) Natural killer cells in antiviral defense: function and regulation by innate cytokines. Annu Rev Immunol 17:189–220
34. Phillips JH, Le AM, Lanier LL (1984) Natural killer cells activated in a human mixed lymphocyte response culture identified by expression of Leu-11 and class II histocompatibility antigens. J Exp Med 159:993–1008
35. Roncarolo MG, Bigler M, Haanen JB, Yssel H, Bacchetta R, de Vries JE, Spits H (1991) Natural killer cell clones can efficiently process and present protein antigens. J Immunol 147:781–787
36. Zingoni A, Sornasse T, Cocks BG, Tanaka Y, Santoni A, Lanier LL (2004) Cross-talk between activated human NK cells and CD4+ T cells via OX40-OX40L interaction. J Immunol 173:3716–3724
37. Zingoni A, Sornasse T, Cocks BG, Tanaka Y, Santoni A, Lanier LL (2005) NK cell regulation of T cell-mediated responses. Mol Immunol 42:451–454

38. Hanna J, Gonen-Gross T, Fitchett J, Daniels M, Heller M, Gonen-Gross T, Manaster E, Cho SY, LaBarre MJ, Mandelboim O (2004) Novel APC-like properties of human NK cells directly regulate T cell activation. J Clin Invest 114:1612–1623
39. Heinzel FP, Schoenhaut DS, Rerko RM, Rosser LE, Gately MK (1993) Recombinant interleukin 12 cures mice infected with Leishmania major. J Exp Med 177:1505–1509
40. Byrne P, McGuirk P, Todryk S, Mills KH (2004) Depletion of NK cells results in disseminating lethal infection with Bordetella pertussis associated with a reduction of antigen-specific Th1 and enhancement of Th2, but not Tr1 cells. Eur J Immunol 34:2579–2588
41. Sakaguchi S, Yamaguchi T, Nomura T, Ono M (2008) Regulatory T cells and immune tolerance. Cell 133:775–787
42. Kang SM, Tang Q, Bluestone JA (2007) CD4+CD25+ regulatory T cells in transplantation: progress, challenges and prospects. Am J Transplant 7:1457–1463
43. Wang HY, Wang RF (2007) Regulatory T cells and cancer. Curr Opin Immunol 19: 217–223
44. Costantino CM, Baecher-Allan CM, Hafler DA (2008) Human regulatory T cells and autoimmunity. Eur J Immunol 38:921–924
45. Brillard E, Pallandre JR, Chalmers D, Ryffel B, Radlovic A, Seilles E et al (2007) Natural killer cells prevent CD28-mediated Foxp3 transcription in CD4+CD25- T lymphocytes. Exp Hematol 35:416–425
46. Roy S, Barnes PF, Garg A, Wu S, Cosman D, Vankayalapati R (2008) NK cells lyse T regulatory cells that expand in response to an intracellular pathogen. J Immunol 180:1729–1736
47. Della Chiesa M, Vitale M, Carlomagno S, Ferlazzo G, Moretta L, Moretta A (2003) The natural killer cell-mediated killing of autologous dendritic cells is confined to a cell subset expressing CD94/NKG2A, but lacking inhibitory killer Ig-like receptors. Eur J Immunol 33:1657–1666
48. Trivedi PP, Roberts PC, Wolf NA, Swanborg RH (2005) NK cells inhibit T cell proliferation via p21-mediated cell cycle arrest. J Immunol 174:4590–4597
49. Yu G, Xu X, Vu MD, Kilpatrick ED, Li XC (2006) NK cells promote transplant tolerance by killing donor antigen-presenting cells. J Exp Med 203:1851–1858
50. Kroemer A, Edtinger K, Li XC (2008) The innate natural killer cells in transplant rejection and tolerance induction. Curr Opin Organ Transplant 13:339–343
51. Deniz G, Erten G, Kücüksezer UC, Kocacik D, Karagiannidis C, Aktas E et al (2008) Regulatory NK cells suppress antigen-specific T cell responses. J Immunol 180:850–857
52. Maroof A, Beattie L, Zubairi S, Svensson M, Stager S, Kayr PM (2008) Posttranscriptional regulation of il10 gene expression allows natural killer cells to express immunoregulatory function. Immunity 29:295–305
53. Tiegs G, Hentschel J, Wendel A (1992) A T cell-dependent experimental liver injury in mice inducible by concanavalin A. J Clin Invest 90:196–203
54. Dong Z, Zhang C, Wei H, Sun R, Tian Z (2005) Impaired NK cell cytotoxicity by high level of interferon-gamma in concanavalin A-induced hepatitis. Can J Physiol Pharmacol 83:1045–1053
55. Li B, Sun R, Wei H, Gao B, Tian Z (2006) Interleukin-15 prevents concanavalin A-induced liver injury in mice via NKT cell-dependent mechanism. Hepatology 43:1211–1219
56. Jirillo E, Caccavo D, Magrone T, Piccigallo E, Amati L, Lembo A, Kalis C, Gumenscheimer M (2002) The role of the liver in the response to LPS: experimental and clinical findings. J Endotoxin Res 8:319–327
57. Schumann J, Wolf D, Pahl A, Brune K, Papadopoulos T, van Rooijen N, Tiegs G (2000) Importance of Kupffer cells for T-cell-dependent liver injury in mice. Am J Pathol 157: 1671–1683
58. Hoebe KH, Witkamp RF, Fink-Gremmels J, Van Miert AS, Monshouwer M (2001) Direct cell-to-cell contact between Kupffer cells and hepatocytes augments endotoxin-induced hepatic injury. Am J Physiol Gastrointest Liver Physiol 280:G720–G728
59. Kusters S, Gantner F, Kunstle G, Tiegs G (1996) Interferon gamma plays a critical role in T cell-dependent liver injury in mice initiated by concanavalin A. Gastroenterology 111:462–741

60. Kaneko Y, Harada M, Kawano T, Yamashita M, Shibata Y, Gejyo F, Nakayama T, Taniguchi M (2000) Augmentation of Valpha14 NKT cell-mediated cytotoxicity by interleukin 4 in an autocrine mechanism resulting in the development of concanavalin A-induced hepatitis. J Exp Med 191:105–114
61. Takeda K, Hayakawa Y, Van Kaer L, Matsuda H, Yagita H, Okumura K (2000) Critical contribution of liver natural killer T cells to a murine model of hepatitis. Proc Natl Acad Sci USA 97:5498–5503
62. Dong Z, Wei H, Sun R, Tian Z (2007) The roles of innate immune cells in liver injury and regeneration. Cell Mol Immunol 4:241–252
63. Baxter AG, Smyth MJ (2002) The role of NK cells in autoimmune disease. Autoimmunity 35:1–14
64. Johansson S, Berg L, Hall H, Höglund P (2005) NK cells: elusive players in autoimmunity. Trends Immunol 26:613–618
65. Jie HB, Sarvetnick N (2004) The role of NK cells and NK cell receptors in autoimmune disease. Autoimmunity 37:147–153
66. Abe T, Kawamura H, Kawabe S, Watanabe H, Gejyo F, Abo T (2002) Liver injury due to sequential activation of natural killer cells and natural killer T cells by carrageenan. J Hepatol 36:614–623
67. Liu ZX, Govindarajan S, Okamoto S, Dennert G (2000) NK cells cause liver injury and facilitate the induction of T cell-mediated immunity to a viral liver infection. J Immunol 164:6480–6486
68. Chen Y, Wei H, Gao B, Hu Z, Zheng S, Tian Z (2005) Activation and function of hepatic NK cells in hepatitis B infection: an underinvestigated innate immune response. J Viral Hepat 12:38–45
69. Wiltrout RH, Salup RR, Twilley TA, Talmadge JE (1985) Immunomodulation of natural killer activity by polyribonucleotides. J Biol Response Mod 4:512–517
70. Dong Z, Wei H, Sun R, Hu Z, Gao B, Tian Z (2004) Involvement of natural killer cells in PolyI:C-induced liver injury. J Hepatol 41:966–973
71. Wang J, Xu J, Zhang W, Wei H, Tian Z (2005) TLR3 ligand-induced accumulation of activated splenic natural killer cells into liver. Cell Mol Immunol 2:449–453
72. Ochi M, Ohdan H, Mitsuta H, Onoe T, Tokita D, Hara H et al (2004) Liver NK cells expressing TRAIL are toxic against self hepatocytes in mice. Hepatology 39:1321–1331
73. Wang J, Sun R, Wei H, Dong Z, Tian Z (2006) Pre-activation of T lymphocytes by low dose of concanavalin A aggravates toll-like receptor-3 ligand-induced NK cell-mediated liver injury. Int Immunopharmacol 6:800–807
74. Wang J, Sun R, Wei H, Dong Z, Gao B, Tian Z (2006) Poly I:C prevents T cell-mediated hepatitis via an NK-dependent mechanism. J Hepatol 44:446–454
75. Chen Y, Wei H, Sun R, Tian Z (2005) Impaired function of hepatic natural killer cells from murine chronic HBsAg carriers. Int Immunopharmacol 5:1839–1852
76. Chen Y, Sun R, Jiang W, Wei H, Tian Z (2007) Liver-specific HBsAg transgenic mice are over-sensitive to Poly(I:C)-induced liver injury in NK cell- and IFN-gamma -dependent manner. J Hepatol 47:183–190
77. Chen Y, Wei H, Sun R, Dong Z, Zhang J, Tian Z (2007) Increased susceptibility to liver injury in hepatitis B virus transgenic mice involves NKG2D-ligand interaction and natural killer cells. Hepatology 46:706–715
78. Jiang W, Sun R, Wei H, Tian Z (2005) Toll-like receptor 3 ligand attenuates LPS-induced liver injury by down-regulation of toll-like receptor 4 expression on macrophages. Proc Natl Acad Sci USA 102:17077–17082
79. Radaeva S, Sun R, Jaruga B, Nguyen VT, Tian Z, Gao B (2006) Natural killer cells ameliorate liver fibrosis by killing activated stellate cells in NKG2D-dependent and tumor necrosis factor-related apoptosis-inducing ligand-dependent manners. Gastroenterology 130:435–452
80. Pazmany L (2005) Do NK cells regulate human autoimmunity? Cytokine 32:76–80

81. Smeltz RB, Wolf NA, Swanborg RH (1999) Inhibition of autoimmune T cell responses in the DA rat by bone marrow-derived NK cells in vitro: implications for autoimmunity. J Immunol 163:1390–1397
82. Xu W, Fazekas G, Hara H, Tabira T (2005) Mechanism of natural killer (NK) cell regulatory role in experimental autoimmune encephalomyelitis. J Neuroimmunol 163:24–30
83. Takahashi K, Aranami T, Endoh M, Miyake S, Yamamura T (2004) The regulatory role of natural killer cells in multiple sclerosis. Brain 127(Pt 9):1917–1927
84. Takahashi K, Miyake S, Kondo T, Terao K, Hatakenaka M, Hashimoto S, Yamamura T (2001) Natural killer type 2 bias in remission of multiple sclerosis. J Clin Invest 107:R23–R29
85. Li Z, Lim WK, Mahesh SP, Liu B, Nussenblatt RB (2005) Cutting edge: in vivo blockade of human IL-2 receptor induces expansion of CD56(bright) regulatory NK cells in patients with active uveitis. J Immunol 174:5187–5191
86. Bielekova B, Catalfamo M, Reichert-Scrivner S, Packer A, Cerna M, Waldmann TA et al (2006) Regulatory CD56(bright) natural killer cells mediate immunomodulatory effects of IL-2Ralpha-targeted therapy (daclizumab) in multiple sclerosis. Proc Natl Acad Sci USA 103:5941–5946
87. Tisch R, McDevitt H (1996) Insulin-dependent diabetes mellitus. Cell 85:291–297
88. Flodstrom M, Maday A, Balakrishna D, Cleary MM, Yoshimura A, Sarvetnick N (2002) Target cell defense prevents the development of diabetes after viral infection. Nat Immunol 3:373–382
89. Delovitch TL, Singh AB (1997) The nonobese diabetic mouse as a model of autoimmune diabetes: immune dysregulation gets the NOD. Immunity 7:727–738
90. Ogasawara K, Hamerman JA, Ehrlich LR, Bour-Jordan H, Santamaria P, Bluestone JA, Lanier LL (2004) NKG2D blockade prevents antoimmune diabetes in NOD mice. Immunity 20:757–767
91. Shi FD, Wang HB, Li H, Hong S, Taniguchi M, Link H, Van Kaer L, Ljunggren HG (2000) Natural killer cells determine the outcome of B cell-mediated autoimmunity. Nat Immunol 1:245–251
92. Poirot L, Benoist C, Mathis D (2004) Natural killer cells distinguish innocuous and destructive forms of pancreatic islet autoimmunity. Proc Natl Acad Sci USA 101:8102–8107
93. Lee I, Qin H, Trudeau J, Dutz J, Tan R (2004) Regulation of autoimmune diabetes by Complete Freund's Adjuvant is mediated by NK cells. J Immunol 172:937–942
94. Serreze DV, Hamaguchi K, Leiter EH (1989) Immunostimulation circumvents diabetes in NOD/Lt mice. J Autoimmun 2:759–776
95. Zhou R, Wei H, Tian Z (2007) NK3-like NK cells are involved in protective effect of polyinosinic-polycytidylic acid on type 1 diabetes in nonobese diabetic mice. J Immunol 178:2141–2147
96. Perricone R, Perricone C, Carlis CD, Shoenfeld Y (2008) NK cells in autoimmunity: a two-edg'd weapon of the immune system. Autoimmun Rev 7:384–390
97. Koopman LA, Kopcow HD, Rybalov B, Boyson JE, Orange JS, Schatz F, Masch R, Lockwood CJ, Schachter AD, Park PJ, Strominger JL (2003) Human decidual natural killer cells are a unique NK cell subset with immunomodulatory potential. J Exp Med 198:1201–1212
98. Zhang J, Croy BA, Tian Z (2005) Uterine natural killer cells: their choices, their missions. Cell Mol Immunol 2:123–129
99. Wu D, Zhang J, Sun R, Wei H, Tian Z (2007) Preferential distribution of NK cells into uteri of C57Bl/6J mice after adoptive transfer of lymphocytes. J Reprod Immunol 75:120–127
100. Zhang J, Tian Z (2007) UNK cells: their role in tissue re-modelling and preeclampsia. Semin Immunopathol 29:123–313
101. Eidukaite A, Siaurys A, Tamosiunas V (2004) Differential expression of KIR/NKAT2 and CD94 molecules on decidual and peripheral blood CD56bright and CD56dim natural killer cell subsets. Fertil Steril 81(Suppl 1):863–868
102. Shigeru S, Akitoshi N, Subaru MH, Shiozaki A (2008) The balance between cytotoxic NK cells and regulatory NK cells in human pregnancy. J Reprod Immunol 77:14–22

103. Varla-Leftherioti M (2005) The significance of the women's repertoire of natural killer cell receptors in the maintenance of pregnancy. Chem Immunol Allergy 89:84–95
104. Blois SM, Barrientos G, Garcia MG, Orsal AS, Tometten M, Cordo-Russo RI et al (2008) Interaction between dendritic cells and natural killer cells during pregnancy in mice. J Mol Med 86:837–852
105. Kitaya K, Nakayama T, Okubo T, Kuroboshi H, Fushiki S, Honjo H (2003) Expression of macrophage inflammatory protein-1β in human endometrium: its role in endometrial recruitment of natural killer cells. J Clin Endocrinol Metab 88:1809–1814
106. Hanna J, Goldman-Wohl D, Hamani Y, Avraham I, Greenfield C, Natanson-Yaron S et al (2006) Decidual NK cells regulate key developmental processes at the human fetal–maternal interface. Nat Med 12:1065–1074
107. Levine RJ, Lam C, Qian C, Yu KF, Maynard SE, Sachs BP et al (2006) Soluble endoglin and other circulating antiangiogenic factors in preeclampsia. N Engl J Med 355:992–1005
108. Zhang C, Zhang J, Tian Z (2006) The regulatory effect of natural killer cells: do "NK-reg cells" exist? Cell Mol Immunol 3:241–254
109. Giuliani M, Giron-Michel J, Negrini S, Vacca P, Durali D, Caignard A et al (2008) Generation of a novel regulatory NK cell subset from peripheral blood CD34+ progenitors promoted by membrane-bound IL-15. PLoS ONE 3:e2241
110. Cooper MA, Fehniger TA, Caligiuri MA (2001) The biology of human natural killer-cell subsets. Trends Immunol 22:633–640
111. Cooper MA, Fehniger TA, Turner SC, Chen KS, Ghaheri BA, Ghayur T, Carson WE, Caligiur MA (2001) Human natural killer cells: a unique innate immunoregulatory role for the CD56bright subset. Blood 97:3146–3151
112. Jacobs R, Hintzen G, Kemper A, Beul K, Kempf S, Behrens G, Sykora KW, Schmidt RE (2001) CD56bright cells differ in their KIR repertoire and cytotoxic features from CD56dim NK cells. Eur J Immunol 31:3121–3126
113. Fehniger TA, Cooper MA, Nuovo GJ, Cella M, Facchetti F, Colonna M, Caligiuri MA (2003) CD56 bright natural killer cells are present in human lymph nodes and are activated by T cell-derived IL-2: a potential new link between adaptive and innate immunity. Blood 101:3052–3057
114. Hu PF, Hultin LE, Hultin P, Hausner MA, Hirji K, Jewett A, Bonavida B, Detels R, Giorgi JV (1995) Natural killer cell immunodeficiency in HIV disease is manifest by profoundly decreased numbers of CD16+CD56+ cells and expansion of a population of CD16dimCD56- cells with low lytic activity. J Acquir Immune Defic Syndr Hum Retrovirol 10:331–340
115. Bauernhofer T, Kuss I, Henderson B, Baum AS, Whiteside TL (2003) Preferential apoptosis of CD56dim natural killer cell subset in patients with cancer. Eur J Immunol 33:119–124
116. Croy BA, Esadeg S, Chantakru S, van den Heuvel M, Paffaro VA, He H, Black GP, Ashkar AA, Kiso Y, Zhang J (2003) Update on pathways regulating the activation of uterine natural killer cells, their interactions with decidual spiral arteries and homing of their precursors to the uterus. J Reprod Immunol 59:175–191
117. Norris S, Collins C, Doherty DG, Smith F, Smith F, McEntee G, Traynor O, Nolan N, Hegarty J, O'Farrelly C (1998) Resident human hepatic lymphocytes are phenotypically different from circulating lymphocytes. J Hepatol 28:84–90
118. Chan A, Hong DL, Atzberger A, Kollnberger S, Filer AD, Buckley CD et al (2007) CD56^{bright} human NK cells differentiate into CD56^{dim} cells: role of contact with peripheral fibroblasts. J Immunol 179:89–94
119. Zheng X, Wang Y, Wei H, Ling B, Sun R, Tian Z (2008) Bcl-xL is associated with the anti-apoptotic effect of IL-15 on the survival of CD56(dim) natural killer cells. Mol Immunol 45:2559–2569
120. Peritt D, Robertson S, Gri G, Showe L, Aste-Amezaga M, Trinchieri G (1998) Cutting edge: differentiation of human NK cells into NK1 and NK2 subsets. J Immunol 161:5821–5824

121. Deniz G, Akdis M, Aktas E, Blaser K, Akdis CA (2002) Huamn NK1 and NK2 subsets determined by purification of IFN-γ-secreting and IFN-γ-nonsecreting NK cells. Eur J Immunol 32:879–884
122. Chakir H, Camilucci AA, Filion LG, Webb JR (2000) Differentiation of murine NK cells into distinct subsets based on variable expression of the IL-12Rβ2 subunit. J Immunol 165:4985–4993
123. Wei H, Zhang J, Xiao W, Feng J, Sun R, Tian Z (2005) Involvement of human natural killer cells in asthma pathogenesis: Natural killer cell 2 cells in type 2 cytokine predominance. J Allergy Clin Immunol 115:841–847
124. Zhang J, Sun R, Liu J, Wang L, Tian Z (2006) Reverse of NK cytolysis resistance of type II cytokine predominant-human tumor cells. Int Immunopharmacol 6:1176–1180
125. Wei H, Zheng X, Lou D, Zhang L, Zhang R, Sun R, Tian Z (2005) Tumor-induced suppression of interferon-gamma production and enhancement of interleukin-10 production by natural killer (NK) cells: paralleled to CD4+ T cells. Mol Immunol 42:1023–1031
126. Hayakawa Y, Smyth MJ (2006) CD27 dissects mature NK cells into two subsets with distinct responsiveness and migratory capacity. J Immunol 176:1517–1524
127. Hayakawa Y, Huntington ND, Nutt SL, Smyth MJ (2006) Functional subsets of mouse natural killer cells. Immunol Rev 214:47–55
128. Silva A, Andrews DM, Brooks AG, Smyth MJ, Hayakawa Y (2008) Application of CD27 as a marker for distinguishing human NK cell subsets. Int Immunol 20:625–630
129. Vossen MT, Matmati M, Hertoghs KM, Baars PA, Gent MR, Leclercq G et al (2008) CD27 defines phenotypically and functionally different human NK cell subsets. J Immunol 180:3739–3745
130. Vosshenrich CA, García-Ojeda ME, Samson-Villéger SI, Pasqualetto V, Enault L, Richard-Le Goff O et al (2006) A thymic pathway of mouse natural killer cell development characterized by expression of GATA-3 and CD127. Nat Immunol 7:1217–1224
131. Blasius AL, Barchet W, Cella M, Colonna M (2007) Development and function of murine B220+CD11c+NK1.1+ cells identify them as a subset of NK cells. J Exp Med 204:2561–2568
132. Vosshenrich CA, Lesjean-Pottier S, Hasan M, Richard-Le Goff O, Corcuff E, Mandelboim O, Di Santo JP (2007) CD11cloB220+ interferon-producing killer dendritic cells are activated natural killer cells. J Exp Med 204:2569–2578
133. Caminschi I, Ahmet F, Heger K, Brady J, Nutt SL, Vremec D et al (2007) Putative IKDCs are functionally and developmentally similar to natural killer cells, but not to dendritic cells. J Exp Med 204:2579–2590

NK Cells and Microarrays

Esther Wilk and Roland Jacobs

Abstract During the last decade, array technology has found its way into the laboratories and considerably changed the way of getting information on gene expression and metabolite production. Complex interactions can be observed and kinetics of cellular events can be simultaneously analysed. Due to the minimisation of the chips, thousands of biological reactions can be measured in one experiment, yielding information on protein production or gene expression [27]. In particular, gene arrays require elaborate bioinformatical processing of the vast amount of data obtained from a single experiment. Despite possible complex mathematical calculations, multiplex technology has revolutionised research of cells and living systems by enabling high-throughput analysis of gene regulation and protein production. Thus, array analysis has become a useful strategy for the investigation of cell dynamics and gene profiling in all cell types. This chapter will focus on gene and protein array technologies and their applications in NK cell research.

Abbreviations

GDNF	Glial cell-line-derived growth factor
EGF	Epidermal growth factor
NT-4	Neurotrophin-4
PIGF	Placenta growth factor
TGF	Transforming growth factor
TNF	Tumour necrosis factor
IFN	Interferon
IL	Interleukin

E. Wilk and R. Jacobs (✉)
Clinic for Immunology and Rheumatology, Hannover Medical School, 30625, Hannover, Germany
e-mail: jacobs.roland@mh-hannover.de

1 Introduction

Natural killer (NK) cells were initially described as a lymphocyte population with the innate ability to kill tumour and virus-infected cells [1]. As they neither expressed specific antigen receptors nor required sensitisation to perform killing they were named based on this prominent functional ability. They were considered to be a uniform lymphocyte subset and characterised as cells expressing a certain pattern of surface markers. However, molecules like NKp46 (CD335), which is exclusively restricted to NK cells, have only recently been described [2–6]. Therefore, human NK cells were previously defined as $CD56^+CD3^-$, $CD16^+CD56^+CD3^-$ or $CD16^+$ lymphocytes. Using these different marker combinations for identification of NK cells was not problematic as long as NK cells were thought to represent a homogeneous lymphocyte population. At least 95% of the NK cells could be identified by each of the earlier listed marker combinations. However, in the early 1990s the marker profiles of NK cells were analysed in much more detail. This marked the beginning of defining NK cell subpopulations, and differential expression patterns of surface markers were used to sort these subpopulations. Subsequent analyses of purified populations revealed diverse functional capacities [7–9]. For example, the majority (ca. 90%) of human blood NK cells present a $CD56^{dim}CD16^{bright}CD3^-$ phenotype, possess strong killing activity and produce moderate amounts of cytokines such as TNFα and IFNγ. In contrast, $CD56^{bright}$ $CD16^{dim/neg}CD3^-$ NK cells have very weak cytotoxic capacity but produce much more cytokines [10]. From these findings we have learned that the NK cell compartment comprises different NK cell subtypes in respect to surface marker expression as well as cellular function and tissue distribution [11].

Different to peripheral blood, $CD56^{bright}$ NK cells represent the main NK cell population in lymph nodes and inflamed tissues [12–14]. In lymph nodes, NK cells have recently been shown to interact with T cells and dendritic cells (DC), which are in close proximity, thereby modifying immune responses [13, 15, 16]. $CD56^{bright}$ NK cells accumulate at inflammatory sites, including exudative pleural fluid, peritoneal fluid from patients with peritonitis, and synovial fluid or tissue from patients with inflammatory arthritis [14]. In human decidua, $CD56^{bright}$ NK cells represent the main lymphocyte population. However, decidual $CD56^{bright}$ NK cells have been reported to express killer cell immunoglobulin-like receptors (KIR) and exhibit some other unique properties, suggesting that they constitute an NK cell population, which is remarkably different from either subset of peripheral blood NK (pNK) cells [17, 18]. Decidual $CD56^{bright}$ NK cells are thought to play an important role in decidualisation and implantation [19].

The more detailed the functions and phenotypes of NK cells were scrutinised, the more the heterogeneity of NK cells became evident. Although it is fairly easy to sort several NK cell populations by multicolour flow cytometry, the size of rare populations is often insufficient for subsequent physiological or functional analysis. However, modern array technology allows the investigation of the vast variability of cellular capabilities in spite of low cell numbers. There is a substantial variety of commercially available arrays for gene expression analysis, and these arrays are used to generate gene profiles. As genes are regulated according to immediate cellular needs, such gene profiles indicate functional and physiological states of the cells. Arrays can either cover the entire genome or be restricted to a certain set of genes of interest. For example, one might only be interested in certain cellular functions such as apoptosis, inflammation or proliferation. Although vast amounts of useful information are obtained from whole genome arrays, it must also be considered that handling and analysis of huge data sets is a challenging bioinformatical task. In addition, the selection of the most suitable array design is cost dependent and the more comprehensive an assay, the higher the price. Furthermore, in most cases the gene expression profiles obtained by array analysis must subsequently be confirmed by quantitative (real time) PCR.

In contrast to gene arrays, which only allow estimation of cellular functions based on gene activities, it is possible to directly measure cell activity by means of protein arrays. As the name implies, protein arrays can be used to assess the synthesis of multiple proteins from a single cell fraction, culture supernatant or serum sample. Depending on the design of the applied assay, the results will be primarily quantitative or qualitative. The number of simultaneously measurable proteins by arrays is usually restricted to less than 100 for qualitative (semi-quantitative) and 20 for quantitative assay kits.

In the following section, the technologies of gene and protein arrays will be described and examples from the practise will demonstrate their applicability in NK cell research. Array-based analysis of NK cells is applicable for various goals such as investigation of cellular changes in terms of gene expression and protein production under certain conditions or medical treatment. However, array analysis is also a very promising tool for the functional and phenotypical characterisation of NK cell subsets, and therefore the following examples are mostly restricted to this issue.

2 Gene Array Technology

Gene arrays or DNA microarrays are a multiplex technology applied in biology, allowing the concomitant analysis of a large number of genes. It consists of a solid slide of silicon, plastic or glass, only a few square centimetres in size, carrying up to several thousands of microscopic DNA oligonucleotide spots in known locations. These oligonucleotides are gene segments or other DNA sequences used as probes that will hybridise to corresponding elements (cDNA or cRNA) from the sample. The number of spots on a single chip depends on the aim of the gene analysis, and can be limited to a small number of selected genes or cover an entire genome (more than 30,000 genes in humans). Gene arrays are typically high-density arrays ($> 1,000$ elements/array). Due to the requirement of sophisticated devices, most

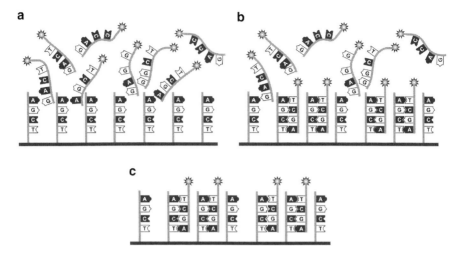

Fig. 1 Gene array technology. Fluorochrome-labelled segments of nucleic acids are incubated with the probes spotted on a chip (**a**). Corresponding strands hybridise and non-hybridised molecules can be removed (**b**). The signal intensity evolved by the dye of hybridised molecules correlates with the expression level of the respective gene in the sample

laboratories are unable to perform complete gene array analyses and prefer to take advantage of commercial or academic array facilities.

The principle of a gene array analysis is the hybridisation of sample oligonucleotides with the probe on the chip (Fig. 1). If genes, gene mutations or polymorphisms are analysed, the DNA of the target cells has to be transformed into single-strand DNA. For the analysis of gene activities, mRNA is isolated from the cells of interest and reverse transcribed into cDNA. The DNA oligonucleotides are then tagged with a fluorochrome enabling the fluorescence-based detection after hybridisation in a reader system. Some chip systems use two fluorochromes with different emission maxima allowing the analysis of two samples on the same chip. In general, each chip is equipped with a multiplicity of each gene probe (Fig. 2). Depending on the manufacturer, the probes of a particular gene can either be identical or represent different segments of the same gene. This design improves the specificity of the chip and enables quantification of the hybridisation events. Finally, positive and negative control nucleotide sequences are used to complete the chip system. The intensity of the fluorescence signal correlates with the number of hybridisation events for a certain probe and is thus used to determine the relative abundance of the oligonucleotide in the sample. The quantity of sample oligonucleotides must exceed a threshold in order to ensure saturation of the array capacity. In the case of insufficient quantity, the original oligonucleotide concentration of the sample can, in most cases, be increased by an amplification step. Of course, it is preferable to analyse samples without additional amplification in order to avoid the introduction of artefacts. Data obtained by array analysis are

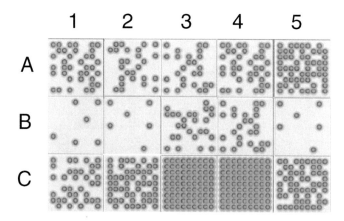

Fig. 2 Gene chip design. Multiple copies of the same or overlapping gene segments (1–5) of each gene are spotted at a known location on a solid surface. Depending on the manufacturer, a certain number of replicates are affixed for each gene (A–C). The differential signals obtained for each gene are finally computed by statistical calculations to determine the expression level of the respective gene

normally visualised as coloured maps representing the results as differential intensities of green to red shades according to the measured value (Fig. 3). Genes with similar expression kinetics under certain conditions are often subjected to cluster analyses which are displayed as dendrogram plots. If two genes are directly compared, the expression level of one gene is often given as x-folds of the second.

The array technique allows simultaneous monitoring of thousands of genes and thereby the determination of a gene profile of the analysed cells. Such a gene profile is very helpful to compare the same cells under different conditions, such as medical treatment, activation or maturation. However, it is also possible for example to compare different cell types in respect to functional capabilities or physiologic features. In NK cell research in particular, the comparative gene profiling of subpopulations was very helpful in uncovering functional differences among the various subsets.

3 Protein Array Technology

DNA array technique is a powerful tool, but is limited to detection on DNA and mRNA levels. Based on the amount of a specific mRNA, one can estimate gene activity and thereby the biosynthesis of the corresponding protein. However, protein synthesis is regulated not only at the transcriptional but also at the translational level. In addition, modification and degradation of mRNA can also affect protein expression [20, 21]. This indicates that protein expression levels can

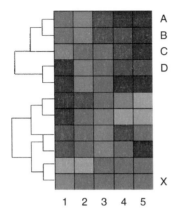 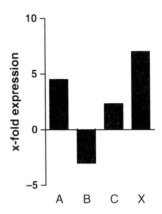

Fig. 3 Gene expression plot. Gene expression profiles are often depicted as heat charts, in which the expression level of each gene of a certain sample (1–5) corresponds to a specific ratio of *green* and *red* (*left panel*). After clustering the genes according to the similarity of the profiles, dendrogram plots can be used to illustrate relationships between the genes. Alternatively, values can be depicted as bar graphs where the expression levels of the genes (A–X) are given as *x*-fold expression of an appropriate reference gene (*right panel*)

accurately be measured only by direct protein analysis. For high-throughput approaches, protein arrays are the obvious method of choice. Protein arrays are excellent tools for the detection of a large number of proteins from sera, cell lysates or culture supernatants. Typically, these arrays have a low-density design (up to 100 elements/array) and can be performed also in non-specialised laboratories within a few hours. In contrast to gene arrays, this technology directly detects cellular products. In principle, this kind of technology can be adapted for any protein to study protein–protein interactions, although the most common means of identification of proteins are antibodies. Basically, the techniques applied are known from ELISA (Fig. 4). Capture antibodies, specific for each of the proteins of interest, are spotted on a certain location of a slide or membrane, which is then incubated with a specimen such as supernatant, serum or lysate containing the proteins that are to be determined. If the samples were labelled with detection molecules beforehand, the array can directly be analysed (Fig. 4a). Otherwise, the array matrix is washed after appropriate incubation, and a cocktail of enzyme-labelled detection antibodies is used for the next incubation step, resembling sandwich ELISA technique (Fig. 4b). This method is widely used for the detection of cytokine expression. Finally, the bound detection antibodies can be identified by colorimetric, fluorometric or chemiluminescent techniques, with the latter leaving visible blackened areas on X-ray films (Fig. 4c). Sizes and densities of the spots can be analysed by graphical and densiometric methods. Protein arrays used for fluorometric analyses are normally manufactured on glass chips. They exhibit lower background levels but require appropriate fluorescence reader systems. Background and maximal intensities of the assay are determined with the help of negative and positive control spots, respectively. Although the chip-based arrays can be adapted for quantitative

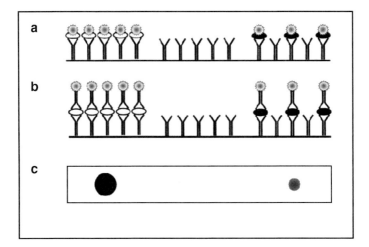

Fig. 4 Protein arrays analysis. Most arrays used for the detection of proteins are antibody-based systems. A capture antibody is spotted on a surface, such as a membrane, glass or plastic slide. The captured antigen is either labelled beforehand (**a**) or is recognised by a specific detection antibody in a sandwich like manner (**b**). Depending on the labelling system, the assays are analysed with fluorescence analysers or by exposure to X-ray films (**c**). The latter system leaves spots on the developed film, which correspond to the concentration of the particular protein in size and density. The spots can be transformed into values by graphical or densitometric methods

measurement, they tend to be used for semi-quantitative analyses. This means, due to dispensation with known standard concentrations, the data are not absolutely quantitative. However, normalised results from different arrays can be used for comparison of relative protein concentrations.

In addition to the detection of proteins, arrays can be used for analysing protein modifications. For example, this is an elegant method for studying the phosphorylation of signalling molecules. The chips are spotted with a capture antibody, which cannot distinguish between the modified and the unmodified protein, whereas the detection antibody can. Alternatively, pairs of antibodies, reactive with either the modified or the unmodified protein, are bound to the matrix and pan-specific antibodies are used for detection. Finally, many additional array designs can be constructed to study interactions of proteins with other molecules, such as DNA and RNA.

4 Quantitative Protein Arrays

The quantitative determination of produced proteins is possible by applying array systems optimised for this task. Most of them are bead-based array systems, which can be analysed by flow cytometric techniques. The beads are manufactured either differentially in size, colour or both, so that they can be distinguished during data acquisition. Furthermore, they serve as a solid phase, to which capture antibodies with the specificity of interest are bound. After incubation with the samples, the captured antigens can be identified by a fluorochrome-labelled detection antibody with the same specificity. For analysis of the experiments either a conventional flow cytometer (Cytometric Bead Assay, Becton Dickinson, Heidelberg, Germany; FlowCytomix Multiplex, Bender, Vienna, Austria) or a dedicated cytometric analyser (Luminex, Austin, TX) are employed, depending on the system. The arrays that can be used with a standard cytometer have a maximum capacity of about 30 distinct proteins/sample. Multiplex systems with specific detection systems can analyse up to 100 different proteins from a single sample. The main reason for the differences in the number of products that can be simultaneously measured is the resolution of the bead detection. The FACS-based systems use beads of different sizes, fluorescence intensities of one or two fluorochromes or a combination of all parameters (Fig. 5). This limits the number of beads that are clearly distinguishable and thus concomitantly analysable. The Luminex system uses beads, labelled with up to 100 different combinations of two dyes. The dedicated analyser is equipped with two lasers, of which one is used only for differentiation of the fluorescent shades of the beads. A second laser of a different wave length is used to measure the fluorescence intensities of the detection antibodies. The main advantage of the multiplex assays, in comparison to standard ELISA techniques,

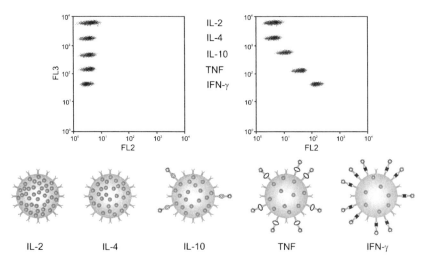

Fig. 5 Quantitative cytokine determination by flow cytometry. Beads are tagged with different dye concentrations enabling their identification according to mean fluorescence intensity (MFI) of beads by flow cytometry. Each of the bead populations is coated with capture antibodies for a particular cytokine. The captured cytokines are detected by antibodies labelled with a second fluorochrome. The samples are analysed in a dot plot graph showing the specificity of the beads on the y-axis and the quantity of the cytokines on the x-axis. The cytokine concentrations can be calculated from the MFIs of the samples and a standard curve that is generated in parallel to the assay

is due to the bead-based architecture. First, in a normal ELISA plate, the capture antibodies are coated to the plastic bottom and thus are limited to binding the antigen in a two-dimensional fashion. In contrast to solid matrices, beads are colloidally soluble in the sample fluid allowing a rapid reactivity while presenting a large surface area, due to the three dimensionality of the system. Second, the usage of bead-based technology is the basic prerequisite for high-speed analysis of the assay, allowing high throughput of samples while measuring a vast variety of parameters.

5 Array Technologies in NK Cell Research

NK cells are part of the innate immune system of many species. The following section will focus mainly on human NK cells. However, the described assay systems and experimental strategies are applicable in principle for all species, though commercial assay systems are mainly tailored for man and mouse.

NK cells were initially defined as large granular lymphocytes, which possess the inherent capacity to kill a wide range of tumour and virus-infected cells, and differ from T and B cells in that they lack specific antigen receptors. NK cell functions, such as cytotoxicity and cytokine production, can be modulated by several cytokines, including IL-2, IL-12, IL-15, IL-21. Dybkaer et al. used gene array technology in order to investigate the cellular events of NK cells after IL-2 activation. Therefore, they performed a genome-wide analysis of resting and IL-2-activated human NK cells [22]. This study revealed that resting NK cells express many genes associated with cellular quiescence. IL-2 stimulation rapidly induced the downregulation of these genes and upregulation of genes associated with proliferation and cell cycle progression. Concomitantly, many genes involved in immune function and responsiveness, such as activating receptors, chemokine/cytokine receptors, death receptor ligands and molecules participating in secretory pathways, were also upregulated. The expression profile indicates that NF-κB activation possibly plays a central role in pro-survival and pro-inflammatory function in IL-2-activated NK cells. This study shows that activation of NK cells is marked by a coordinated fine tuning of gene expression of those genes needed to exert specific functions. Additionally, the gene profile that is shaped in response to the stimulus gives detailed information on the regulation of molecules associated with cell signalling. Hence, gene data provide hints on the signal transduction pathways used by the cell in response to a particular stimulus. This is a good example of how a gene array approach can be very helpful in the selection of suitable functional experiments.

Although NK cells were initially thought to represent a homogenous lymphocyte fraction, it soon became obvious that the NK cell compartment comprises several functionally and phenotypically different subpopulations. Fascinating was the finding of the diverse functional capacities of $CD56^{bright}$ and $CD56^{dim}$ NK cells. $CD56^{dim}$ NK cells comprise the classical NK cell type exerting strong cytolytic activity. $CD56^{bright}$ cells, in contrast, are weak killers but are superior in terms of cytokine production. These functional differences were uncovered by applying standard experimental approaches to sorted subpopulations, for example, cytotoxicity assays and determination of cytokine production. Once gene array technology was fully established in biomedical research, this technique dominated for the detailed analysis of these two NK cell subpopulations. Several groups independently investigated these populations by using the gene array approach, though the experimental settings differed in some respect among the studies [17, 23–25]. Basically, the three studies revealed comparable results as it should be expected of course, since similar approaches were performed. Data were primarily assessed by gene array analyses, and the findings were confirmed for a selection of genes by quantitative PCR and/or on the protein level by flow cytometry (Fig. 6). Due to the differences of the study designs and the chosen foci, the work of three groups presents a more comprehensive analysis on NK cell populations than each of the single studies does. Koopman et al. compared the expression of approximately 10,000 genes in $CD56^{bright}$ uterine decidual NK (dNK) cells and $CD56^{bright}$ and also $CD56^{dim}$ pNK cells by microarray analysis using the HGU95Av2 chip (Affymetrix, Santa Clara, CA) [17]. They found that $CD56^{bright}$ dNK cells remarkably differ from either pNK cell subpopulation and they are more similar to $CD56^{dim}$ pNK cells.

Among the pNK cells, a relatively low number of differently expressed genes were detected with an equal distribution of overexpressed genes in both $CD56^{dim}$ and $CD56^{bright}$ NK cells. They propose two possible alternative explanations for the differential gene expression patterns between dNK and pNK cells. dNK cells may descend from a distinct haematopoietic precursor, or they might develop from pNK

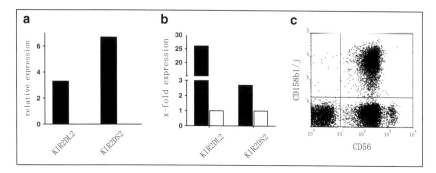

Fig. 6 Comparison of results obtained by different approaches. In order to estimate the reliability of the array analysis, the same genes were also measured by quantitative PCR. Furthermore, the expression was verified on the protein level by flow cytometry. Gene expression levels of KIR2DL2 (CD158b1) and KIR2DS2 (CD158j) as assessed by array analysis in $CD56^{dim}$ NK cells are depicted as x-folds (**a**). The levels obtained in $CD56^{bright}$ NK cells were below threshold and hence are classified as absent. Quantitative PCR analysis of the two genes revealed a much stronger expression in $CD56^{dim}$ NK cells (*black bars*). The values are given as x-folds of the values obtained in $CD56^{bright}$ NK cells (*open bars*) (**b**). Two-colour fluorescence confirms the differential expression of CD158b1/j (KIR2DL2 and KIR2DS2) on $CD56^{dim}$ and $CD56^{bright}$ NK cells. Cells were gated on $CD3^-$ lymphocytes. The inhibitory receptor (KIR2DL2) and the activating isoform (KIR2DS2) are identical in the extracellular portion of the molecule. Thus, the two functionally different receptors cannot be differentiated by the antibody

cells in the decidual environment under influence of stromal cells and hormones. The authors particularly emphasised the overexpression of CD9, NKG2C, NKG2E, Ly49L, KIR, galectin-1 and PP14 by dNK cells, as these molecules exhibit immunomodulatory functions during pregnancy. Thus, based on the array analysis, the study of Koopman et al. revealed that (a) $CD56^{bright}$ NK cells from uterus and peripheral blood differ considerably, and (b) uterine NK cells play an important role in pregnancy by establishing maternal–foetal tolerance.

Hanna et al. used CodeLink Human 20K I Bioarrays (Amersham Biosciences, Piscataway, NJ) to analyse ca. 20,000 genes of $CD16^-CD56^{bright}$, as well as resting and activated $CD16^+CD56^{dim}$ NK cells from peripheral blood [23]. Their data support the model that $CD56^{dim}$ and $CD56^{bright}$ NK cells represent functionally distinct subpopulations of mature NK cells. From the profiles of these subsets the authors concluded that $CD56^{bright}$ NK cells are predestined for immunoregulatory interactions with other immunocytes in lymphatic tissues such as lymph nodes. In contrast, $CD56^{dim}$ NK cells are armed with molecules needed for cytotoxic functions at inflammatory sites. In activated $CD16^+CD56^{dim}$ NK cells, they found upregulation of several molecules which have been reported by Koopman et al. on decidual $CD56^{bright}$ dNK cells. Hanna et al. suggest that this activation-like phenotype of dNK cells might be due to chronic exposure to semi-allogeneic extravillous trophoblasts that invade maternal decidua and the cytokine-enriched environment.

Wendt et al. analysed resting and activated $CD56^{dim}$ and $CD56^{bright}$ NK cells from peripheral blood in a whole genome approach (>39,000 transcripts) via U133A and U133B arrays (Affymetrix, Santa Clara, CA). Activation of the cells was induced by stimulation with PMA/ionomycin in order to enable similar activation of the two populations as Protein kinase (PK) C, the target enzyme of PMA, is equally expressed in all cells. In contrast, most receptors mediating stimulation under physiological conditions, such as CD2, CD16 and cytokine receptors, are differently expressed between $CD56^{dim}$ and $CD56^{bright}$ NK cells. The differential amount of such receptors would most likely lead to unequal activation of the two subpopulations. Also, this study confirmed that $CD56^{dim}$ NK cells express the molecules needed to perform cytotoxicity. In contrast, $CD56^{bright}$ lack this capability but do express receptors for migration into lymphatic tissues and molecules for interacting with other immune cells. It was frequently suggested that $CD56^{bright}$ NK cells might just represent an activated state of $CD56^{dim}$ NK cells. Wendt et al. investigated gene regulation in both NK cell populations in order to address the question whether or not $CD56^{bright}$ NK cells might be closely related to activated $CD56^{dim}$ NK cells following stimulation with PMA/ionomycin. They selected the 487 genes that were upregulated at least threefold in $CD56^{dim}$ NK cells upon activation. In parallel, expression signals of the same genes in resting $CD56^{bright}$ NK cells were compared with resting $CD56^{dim}$ NK cells. If $CD56^{bright}$ NK cells would simply represent an activated stage of $CD56^{dim}$ NK cells, the majority of the genes upregulated in activated $CD56^{dim}$ would also be expected to be increased in resting $CD56^{bright}$ as compared to resting $CD56^{dim}$ NK cells. However, expression patterns of these genes clearly differ between resting $CD56^{bright}$ and resting $CD56^{dim}$ NK cells. Although the expression of 117 genes was at least twofold higher in resting CD56^{bright} than in resting CD56^{dim} NK cells, 159 genes were concomitantly downregulated, arguing against a close relation between resting CD56^{bright} and activated CD56^{dim} NK cells.

As previously mentioned, all studies analysed some of the molecules on the protein level, in addition to the gene arrays, and in some cases found discrepancies between gene expression and protein abundance. This suggests that several factors, such as post-translational modification and differential mRNA stability, have considerable impacts on the protein composition of a cell. Thus, as extreme caution should be used when estimating protein expression based on mRNA levels alone, direct measuring of the proteins is preferable. Therefore, Wendt et al. used two different protein arrays [25]. In one approach, supernatants of activated CD56^{dim} and CD56^{bright} NK cells were subjected to a semi-quantitative array (Raybiotech, Norcross, GA) that detected a variety of approximately 80 cytokines and chemokines. This assay revealed distinct patterns of soluble factors that can be produced by the two populations. Some mediators (e.g. GDNF and EGF) were equally produced, while others were preferentially produced by CD56^{dim} (e.g. NT-4 and osteoprotegerin) or CD56^{bright} (e.g. PIGF and TGF-β2) NK cells. Furthermore, protein array experiments confirmed the preferential production of TNF and IFNγ by CD56^{bright} NK cells. However, the use of cytometric bead array (CBA) for quantification of secreted cytokines after 24-h and 48-h activation revealed that concentrations of IFNγ and TNF did not significantly differ between supernatants of activated CD56^{dim} versus CD56^{bright} NK cells. The lack of significance might be due to the power or duration of the stimulus, as kinetic experiments revealed considerably increased mRNA expression for both factors in CD56^{bright} NK cells as compared with CD56^{dim} NK cells within 4 h. After 24 h, both mRNAs were on a similar level, indicating a differential regulation of each cytokine. In addition, soluble mediators have short half-lives (on the order of minutes to hours) and are rapidly degraded [26]. Furthermore, it cannot be excluded that one or the other cytokine, which is induced in CD56^{dim} or CD56^{bright} NK cells by activation, can instantly be consumed and subsequently affect the production of other factors in a loop-back mechanism in the respective population. Other cytokines, which can be concomitantly detected by the CBA (IL-1β, IL-2, IL-4, IL-6, IL-10 and IL-12p70) remained below detection sensitivity levels (<20 pg/ml) in the same samples, indicating that the sorted NK cell populations were free of any contaminating T cells and monocytes.

The data from the three groups present a comprehensive overview on how NK cells can be analysed by modern array techniques. The gene expression profiles are stored in public data bases and can be used for the screening of new target molecules that have not yet been investigated in the context of NK cells. For researchers in the NK cell field, the detailed information on gene expression profiles in human NK cell populations will help to define NK cell populations in species other than humans. Neither murine nor rat NK cells express CD56; thus, other molecules selectively expressed by CD56^{dim} or CD56^{bright} can be used to find feasible markers for identifying distinct functional NK cell subpopulations in these animals. In addition, the number of distinguishable NK cell populations in humans is still increasing and these populations will surely be investigated in detail via applying array technology.

Acknowledgements We thank Rachel Thomas for thoroughly editing and improving the manuscript.

References

1. Trinchieri G (1989) Biology of natural killer cells. Adv Immunol 47:187–376
2. Walzer T, Blery M, Chaix J et al (2007) Identification, activation, and selective in vivo ablation of mouse NK cells via NKp46. Proc Natl Acad Sci U S A 104:3384–3389
3. Pessino A, Sivori S, Bottino C, Malaspina A, Morelli L, Moretta L, Biassoni R, Moretta A (1998) Molecular cloning of NKp46: a novel member of the immunoglobulin superfamily involved in triggering of natural cytotoxicity. J Exp Med 188:953–960
4. Biassoni R, Pessino A, Bottino C, Pende D, Moretta L, Moretta A (1999) The murine homologue of the human NKp46, a triggering receptor involved in the induction of natural cytotoxicity. Eur J Immunol 29:1014–1020
5. Sivori S, Pende D, Bottino C, Marcenaro E, Pessino A, Biassoni R, Moretta L, Moretta A (1999) NKp46 is the major triggering receptor involved in the natural cytotoxicity of fresh or cultured human NK cells. Correlation between surface density of NKp46 and natural cytotoxicity against autologous, allogeneic or xenogeneic target cells. Eur J Immunol 29:1656–1666
6. Falco M, Cantoni C, Bottino C, Moretta A, Biassoni R (1999) Identification of the rat homologue of the human NKp46 triggering receptor. Immunol Lett 68:411–414
7. Jacobs R, Stoll M, Stratmann G, Leo R, Link H, Schmidt RE (1992) CD16− CD56+ natural killer cells after bone marrow transplantation. Blood 79:3239–3244
8. Caligiuri MA, Zmuidzinas A, Manley TJ, Levine H, Smith KA, Ritz J (1990) Functional consequences of interleukin 2 receptor expression on resting human lymphocytes: identification of a novel natural killer cell subset with high affinity receptors. J Exp Med 171:1509–26
9. Cooper MA, Fehniger TA, Caligiuri MA (2001) The biology of human natural killer-cell subsets. Trends Immunol 22:633–640
10. Jacobs R, Hintzen G, Kemper A, Beul K, Kempf S, Behrens G, Sykora KW, Schmidt RE (2001) CD56^{bright} cells differ in their KIR repertoire and cytotoxic features from CD56^{dim} NK cells. Eur J Immunol 31:3121–3127
11. Wilk E, Kalippke K, Buyny S, Schmidt RE, Jacobs R (2008) New aspects of NK cell subset identification and inference of NK cells' regulatory capacity by assessing functional and genomic profiles. Immunobiology 213:271–283
12. Ferlazzo G, Pack M, Thomas D et al (2004) Distinct roles of IL-12 and IL-15 in human natural killer cell activation by dendritic cells from secondary lymphoid organs. Proc Natl Acad Sci USA 101:16606–16611
13. Fehniger TA, Cooper MA, Nuovo GJ, Cella M, Facchetti F, Colonna M, Caligiuri MA (2003) CD56^{bright} natural killer cells are present in human lymph nodes and are activated by T cell-derived IL-2: A potential new link between adaptive and innate immunity. Blood 101:3052–3057
14. Dalbeth N, Gundle R, Davies RJ, Lee YC, McMichael AJ, Callan MF (2004) CD56^{bright} NK cells are enriched at inflammatory sites and can engage with monocytes in a reciprocal program of activation. J Immunol 173:6418–6426
15. Hayakawa Y, Screpanti V, Yagita H, Grandien A, Ljunggren HG, Smyth MJ, Chambers BJ (2004) NK cell TRAIL eliminates immature dendritic cells in vivo and limits dendritic cell vaccination efficacy. J Immunol 172:123–129

16. Chiesa MD, Vitale M, Carlomagno S, Ferlazzo G, Moretta L, Moretta A (2003) The natural killer cell-mediated killing of autologous dendritic cells is confined to a cell subset expressing CD94/NKG2A, but lacking inhibitory killer ig-like receptors. Eur J Immunol 33:1657–1666
17. Koopman LA, Kopcow HD, Rybalov B et al (2003) Human decidual natural killer cells are a unique NK cell subset with immunomodulatory potential. J Exp Med 20(198):1201–1212
18. Lukassen HG, Joosten I, van Cranenbroek B, van Lierop MJ, Bulten J, Braat DD, van der Meer A (2004) Hormonal stimulation for IVF treatment positively affects the $CD56^{bright}$/$CD56^{dim}$ NK cell ratio of the endometrium during the window of implantation. Mol Hum Reprod 10:513–520
19. King A (2000) Uterine leukocytes and decidualization. Hum Reprod Update 6:28–36
20. Gygi SP, Rochon Y, Franza BR, Aebersold R (1999) Correlation between protein and mRNA abundance in yeast. Mol Cell Biol 19:1720–1730
21. Anderson L, Seilhamer J (1997) A comparison of selected mRNA and protein abundances in human liver. Electrophoresis 18:533–537
22. Dybkaer K, Iqbal J, Zhou G, Geng H, Xiao L, Schmitz A, d'Amore F, Chan WC (2007) Genome wide transcriptional analysis of resting and IL2 activated human natural killer cells: Gene expression signatures indicative of novel molecular signaling pathways. BMC Genomics 8:230
23. Hanna J, Bechtel P, Zhai Y, Youssef F, McLachlan K, Mandelboim O (2004) Novel insights on human NK cells' immunological modalities revealed by gene expression profiling. J Immunol 173:6547–6563
24. Wilk E, Wendt K, Schmidt RE, Jacobs R (2003) Different gene and protein expression in $CD56^{dim}$ and $CD56^{bright}$ natural killer cells reflecting effector and regulatory functions. Immunobiology 208:92
25. Wendt K, Wilk E, Buyny S, Buer J, Schmidt RE, Jacobs R (2006) Gene and protein characteristics reflect functional diversity of $CD56^{dim}$ and $CD56^{bright}$ NK cells. J Leukoc Biol 80:1529–1541
26. Friberg D, Bryant J, Shannon W, Whiteside TL (1994) In vitro cytokine production by normal human peripheral blood mononuclear cells as a measure of immunocompetence or the state of activation. Clin Diagn Lab Immunol 1:261–268
27. Lipshutz RJ, Morris D, Chee M et al (1995) Using oligonucleotide probe arrays to access genetic diversity. BioTechniques 19:442–447

Natural Killer Cells in the Treatment of Human Cancer

Karl-Johan Malmberg and Hans-Gustaf Ljunggren

Abstract NK cells may be exploited in the treatment of human cancer. One strategy aims for activation of endogenous NK cells in the cancer patient. Another takes advantage of the knowledge regarding "missing-self" recognition and KIR–HLA mismatches in settings of allogeneic stem cell transplantation (SCT) followed by, in some situations, the use of NK cell-based donor lymphocyte infusions. Other strategies employ direct adoptive transfer of NK cells. Here, we briefly discuss these and other prospects for the treatment of human cancer using NK cells either directly or indirectly.

1 Introduction

NK cells were originally identified on a functional basis, because of their ability to lyse certain tumor cells in vitro without the requirement for prior immune sensitization of the host [1]. Today, these cells are well characterized with respect to their origin, differentiation, receptor repertoire, and effector functions [2–5]. NK cells account for 5%–15% of peripheral blood lymphocytes. In addition to peripheral blood, they are found in the bone marrow, spleen, and lymph nodes, as well as in specific organs such as the liver and lungs. They are activated by cytokines and/or by interactions with specific molecules expressed on target cells [3, 5, 6]. Upon activation, NK cells produce cytokines and chemokines and can exert strong cytolytic effector functions [7, 8].

Human NK cells are broadly defined as $CD3^-CD56^+$ lymphocytes. They can be further subdivided into two main functional subsets on the basis of their surface expression of CD56. $CD56^{bright}$ NK cells have potent immunoregulatory properties,

---

K.-J. Malmberg and H.-G. Ljunggren (✉)
Center for Infectious Medicine, Department of Medicine, Karolinska Institutet, Karolinska University Hospital, S-141 86, Stockholm, Sweden
e-mail: kalle.malmberg@ki.se, hans-gustaf.ljunggren@ki.se

and CD56^{dim} NK cells have potent cytotoxic functions [9, 10]. These functions are, however, not strictly confined to each respective subset; a certain degree of functional overlap exists. The CD56^{dim} NK cells express high levels of FcγRIIIA (CD16) allowing them to mediate antibody dependent cellular cytotoxicity (ADCC). Although insight into NK cell development has been obtained, it is still not fully clear at which stage of differentiation CD56^{bright} and CD56^{dim} NK cells separate from each other [11, 12].

2 NK Cell Responses to Tumors

A large number of studies have demonstrated the capacity of NK cells to recognize many different types of murine and human tumor cell lines in vitro [13, 14]. Several experimental studies in mice have demonstrated a role for NK cells in the eradication of grafted murine tumor cell lines [15, 16]. NK cells have been shown to be involved in the rejection of experimentally induced and spontaneously developing tumors in mice [17]. Human NK cells adoptively transferred to mice can also participate in the rejection of grafted human tumors [18]. Direct evidence for NK cell targeting of human cancer has come from studies of NK cell interactions with primary tumor cells tested for susceptibility to NK cell-mediated lysis ex vivo. Such approaches have recently been taken for, e.g., neuroblastoma, ovarian carcinoma, and multiple myeloma [19–21]. Evidence for NK cell targeting of human tumors has also come from clinical studies in settings of SCT and adoptive transfer of NK cells to cancer patients [22–25]. Despite the large amount of studies that demonstrate the ability of NK cells to target tumor cells in vitro and in vivo in experimental models [13, 14], there is still only limited evidence for clinical efficacy of activated NK cells administered to patients with cancer.

All tumors are, however, not susceptible to NK cell-mediated lysis. A variety of reasons may underlie this resistance, including specific properties of some tumor cells themselves as well as effects imposed by certain tumors on NK cells. Data from both experimental models and from studies of human cancer have demonstrated tumor cell evasion from NK cells [26–28]. This includes the intriguing observations of, e.g., high MHC class I expression of some metastasizing human tumors and/or loss of ligands for NK cell activation receptors on other tumors [29–31]. This may argue for selection of tumor cell mutants or tumor cell modulation during an ongoing NK cell-mediated response. Downregulation of NK cell activating receptors can be mediated by, e.g., TGF-β which selectively downregulates the surface expression of some NK cell activating receptors [32]. Likewise, NK cell interactions with tumor target cells may specifically downregulate the expression of activating receptors, e.g., downregulation of DNAM-1 upon recognition of PVR-expressing ovarian carcinomas ([21]; our own unpublished results). Finally, tumor cells may also restrain NK cell effector function by promoting the expansion of CD4$^+$CD25$^+$ Treg cells. Recent evidence points to a critical role for Treg cells in dampening

NK cell immune responses by suppression of homeostatic proliferation, cytotoxicity, and IL-12-mediated IFN-γ production [33, 34].

The identification of NK cell germ-line-encoded activation and inhibitory receptors has in large part uncovered the molecular mechanisms used by NK cells to recognize tumor cells. This knowledge emerges from early observations that NK cell cytotoxicity is triggered by tumor cells lacking expression of certain (or all) self-MHC class I molecules [16, 35–37], a phenomenon referred to as "missing-self" recognition [38], which led in the early 1990s to the identification of NK cell inhibitory receptors that recognize MHC class I molecules [39, 40]. However, NK cells also need stimulation by target cell ligands to trigger activation via specific receptors. The identification of the latter remained elusive until some 10 years ago [3, 5]. Thus, we now know that NK cell recognition of tumors is tightly regulated by processes involving the integration of signals delivered from multiple activating and inhibitory receptors [3, 5, 6, 41, 42].

3 Molecular Interactions in NK Cell Responses to Tumors

One important group of human NK cell activation receptors is represented by the so-called natural cytotoxicity receptors (NCR); NKp46, NKp30, and NKp44 [43, 44]. Two of these, NKp46 and NKp30, are constitutively expressed on all peripheral blood NK cells. Despite considerable effort to characterize their ligands, their constituents on tumor cells remain poorly defined [45, 46]. However, the nuclear factor HLA-B-associated transcript 3 (BAT3) was shown to be released from tumor cells and to bind NKp30. BAT-3 triggered NKp30-mediated NK cell cytotoxicity and thus represents the first identified cellular ligand for any of the NCR [45, 46]. The NCR have a major role in NK cell-mediated lysis of various human tumor cell lines including melanomas, carcinomas, neuroblastomas, and myeloid or lymphoblastic leukemias, as well as EBV-transformed B cells [45]. Other well-characterized activation receptors are NKG2D and DNAM-1 [47, 48]. NKG2D recognizes the stress inducible molecules MICA and MICB as well as the UL16-binding proteins (ULBP). NKG2D ligands are expressed on a number of human epithelial tumor and leukemic cell lines and have a significant role in rendering these cells susceptible to NK cell-mediated lysis. DNAM-1 recognizes PVR (CD155) and Nectin-2 (CD122). These ligands are highly expressed in human carcinomas, melanomas, and neuroblastomas [49–51]. CD16 on NK cells binds the Fc-portion of IgG on opsonized cells, thus mediating ADCC. In addition, several other receptors, including 2B4 (CD244), NTBA, NKp80, CD2, CD11a/CD18, and CD59, have important coactivating or costimulatory functions in NK cell activation and tumor cell recognition [6].

Activation of NK cells is under control by inhibitory receptors [3, 5, 41]. Inhibitory receptors bind classical and/or nonclassical MHC class I molecules. These molecules are normally expressed on most healthy cells in the body, but are often lost upon transformation or during tumor evolution [52]. In humans, killer cell Ig-like inhibitory receptors (KIR) and CD94–NKG2A play major roles as HLA-class I-specific inhibitory NK cell receptors. KIR recognize groups of HLA-A, -B, and -C alleles [53–55], whereas CD94–NKG2A/B receptors recognize HLA-E molecules [56]. Individuals differ in the number and type of inherited KIR genes, and specific KIR gene products are expressed on distinct subsets of NK cells. Thus, many NK cells express only a few of many possible types of receptors. Most NK cells, however, express at least one inhibitory receptor that is specific for a self-MHC class I ligand. The pattern of KIR expressed creates a system allowing NK cells to detect cells lacking expression of single MHC class I alleles [54].

The relative importance of different NK cell activation receptors and their ligands in recognizing primary human tumors is only partially known. Interestingly, efficient natural cytotoxicity by resting (e.g., not preactivated by cytokines) NK cells usually requires coactivation by several types of receptors [6, 57]. In contrast, engagement of LFA-1 is sufficient to induce cytotoxicity by IL-2 activated NK cells [57]. Translated to the context of killing tumor cells, these findings indicate that tumor cells might elicit activation of NK cells by expressing an array of ligands for several activating receptors, any one of which alone would be incapable of triggering a response. For NK cells activated by IL-2, fewer qualitatively distinct receptor–ligand interactions might suffice for tumor cell recognition and killing provided MHC class I molecules do not confer inhibition. In some situations, the activation signals may override the inhibitory signals mediated by MHC class I molecules, as has been demonstrated, e.g., for NKG2D-mediated triggering of some MHC class I expressing tumor cell lines [58, 59]. The final degree of NK cell activation is also dependent on the relative density of ligands for the integration of signals from activating and inhibitory signals.

4 Effector Functions Involved in NK Cell Responses to Tumors

Like cytotoxic T cells, NK cells possess different effector mechanisms by which they mount antitumor responses [60]. NK cells use two major mechanisms to induce target cell apoptosis, the granule exocytosis and the death receptor pathway [61]. Granule exocytosis involves the release of perforin and granzymes [62]. The death receptor pathway is largely mediated by members of the TNF superfamily, in which FasL, TNF-α, and TRAIL are key apoptosis-inducing members [61, 63]. NK cells can also produce many different types of cytokines (e.g., IFN-γ, TNF-α, and GM-CSF) as well as chemokines, at least some of which have a direct effect on tumors. The best-studied cytokine in this respect is IFN-γ. IFN-γ can decrease tumor proliferation and metabolic activity and inhibit angiogenesis [64]. IFN-γ induces type-1 immunity and may by this means counteract tumor escape mechanisms actively promoted by cancer cells and regulatory T (Treg) cells through the secretion of type-2 cytokines [34].

5 Activation of NK Cells to Enhance Responses to Tumors

Several cytokines affect NK cell differentiation and activation such as IL-2, IL-12, IL-15, IL-18, IL-21, and type-1 IFN [65, 66]. Upon cytokine stimulation, NK cells proliferate, produce cytokines, and upregulate effector molecules such as adhesion molecules, NKp44, perforin, granzymes, FasL, and TRAIL. IL-2 stimulates NK cell progenitors and mature NK cells, and induces the production of NK cell effector molecules, enhancing NK cytolytic activity. In addition, IL-2 augments NK cell degranulation via a syntaxin 11-independent pathway [67], and may reduce the dependency of coactivation for some receptors implicated in tumor cell lysis [57]. Recent evidence suggests a unique role for IL-15 in the differentiation, proliferation, survival, and activation of NK cells [68, 69]. IL-15 may also protect lymphocytes from IL-2- induced activation-induced cell death (AICD) [69, 70]. This cytokine synergizes with Flt3-L and SCF in inducing human $CD56^{bright}$ NK cells [11, 12]. IL-12 and IL-18 act late during the NK cell differentiation, and synergistically enhance cytotoxicity against tumor targets and IFN-γ production by NK cells [71]. IL-21 is of particular interest in its ability to stimulate cytotoxic $CD56^{dim}$ NK cells and to enhance NK cell cytotoxicity [72, 73].

Cytokines have successfully been applied in the treatment of several human cancers and, in some instances, the mechanisms of action are through direct or indirect activation of NK cells [74, 75]. Several clinical trials have assessed the effects of IL-2 administration on activation of NK cells in patients with cancer [75, 76]. Unfortunately, high doses of IL-2 are associated with significant toxicity [77]. Irrespectively, IL-2 is frequently used in lower doses to promote NK cell activity in vivo. This cytokine has more recently also been used together with monoclonal antibodies that mediate ADCC to enhance NK cell activity [78]. In many respects, IL-15 may be a better cytokine. IL-15 is more efficient than IL-2 in expanding the NK cell compartment [79–81]. It promotes survival of NK cells and protects NK cells from AICD. However, high doses of IL-15 may be needed in vivo to obtain effective antitumor effects [81, 82]. New insight into the role of early hematopoetic growth factors, such as c-kit ligand and Flt-3 ligand, and their synergy with IL-15 in the development of human NK cells in the bone marrow, will likely permit studies of additional cytokine combinations for expansion of NK cells for clinical use. Nevertheless, better knowledge about how to best use cytokines for optimal activation of NK cells, either alone or in combination with other immune interventions [75], is clearly needed. It still remains to be studied whether the impressive results obtained in mice, using individual and combinations of cytokines to activate NK cells to kill tumor cells, can be translated to humans.

Apart from specific cytokines and/or growth factors, broad activators of immune function implicated in antitumor immunity, may also stimulate NK cells. For instance, in myeloma, NK cell activity has been shown to increase in response to thalidomide and its analog lenalidomide, explaining, in part, the mechanism of action of this drug [83]. Immunostimulatory DNA complexes have also been shown to enhance in vivo antitumor activity, mediated, at least in part, through the activation of NK cells [84]. Imatinib mesylate (Gleevec), a specific inhibitor of tyrosine kinase receptors may also lead to host antitumor effects mediated by the innate immune system and a new type of immune cell, referred to as IFN-producing killer dendritic cells (IKDC), that resembles natural killer cells [85–87]. Finally, NK cells may also contribute to the clinical efficacy of bacillus Calmette-Guerin (BCG) treatment of bladder cancer [88].

6 Modulating Receptor Signaling to Promote NK Cell Killing

Another possibility that has now reached the stage of clinical trials in humans is to block inhibitory KIR with monoclonal antibodies, thereby augmenting tumor cell recognition by NK cells [89]. Such reagents could ideally be used, e.g., in the treatment of hematopoetic cancers that are not amenable to SCT. Preclinical evidence in mouse models has demonstrated that this strategy may enhance antitumor activity in autologous [90] and allogeneic settings [91]. Along the same lines, the modulation of activating ligands on tumor cells may also improve the efficacy of NK cell recognition. NK cells may also be genetically engineered prior to adoptive transfer to the patients. One interesting possibility is to stably over-express chimeric receptors recognizing ligands expressed by tumors combined with signaling components that trigger NK cell function [92]. For example, NK cells engineered to express chimeric anti-CD19-CD3 signaling receptors became highly cytotoxic against autologous leukemia cells [93]. Genetically modified NK-92 cells, expressing a chimeric antigen receptor specific for the tumor-associated ErbB2 (HER2/neu) antigen, specifically lysed ErbB2-expressing tumor cells that were completely resistant to cytolysis by parental NK-92 cells [94].

7 NK Cell-Mediated Responses to Tumors in Settings of Hematopoetic Stem Cell Transplantation

Hematopoietic stem cell transplantation (SCT) is an established treatment strategy for several hematological malignancies. An intentional mismatch between donor KIR and recipient HLA ligands in hematopoetic SCT is predicted to allow for a graft-versus-tumor (GvT) effect by NK cells that develop in the recipient. In pioneering studies by Velardi and collaborators, adult patients with acute myeloid leukemia (AML) undergoing haploidentical SCT showed greatly improved disease-free survival time and low relapse rates when a KIR–ligand mismatch prevailed [25, 95]. This effect occurred in the absence of donor T cells, which had been removed prior to the transplant. Notably, hematopoetic grafts with NK cell alloreactivity in the graft-versus-host (GvH) direction also had increased rates of bone marrow engraftment and reduced rates of clinically significant GvH disease (GvHD). The latter effect may be caused by donor NK cell lysis of host antigen-presenting cells, impairing alloreactive donor T cell priming [25, 96]. Since the initial report by Velardi and collaborators [25], numerous studies have addressed the role of KIR–ligand mismatch in different settings of transplantation [97, 98]. Some, but not all, studies have demonstrated a beneficial role of NK cell alloreactivity. Discrepancies in outcome among the studies may depend on the transplantation protocol including differences such as the type of preconditioning, dose of stem cells, degree of T cell depletion, and posttransplantation immunosuppression. Further studies are warranted to better understand the conditions that steer NK cell maturation and receptor-acquisition following SCT, particularly relative to alloreactivity and GvT effects.

8 NK Cell-Mediated Responses to Tumors in Settings of Donor Lymphocyte Infusions Following Hematopoetic Stem Cell Transplantation

Many patients relapse after hematopoetic SCT. Donor lymphocyte infusions (DLI) can induce a direct and potent GvT effect in some of these patients [99–101]. The major risk of DLI is GvHD, which may be a severe, even lethal complication. To minimize the risk of GvHD, modified strategies have been developed such as partially T cell-depleted DLI. In haploidentical SCT, studies have been initiated where purified donor NK cells have been used as DLI with the aim to consolidate engraftment and induce GvT effects [102, 103]. In a related study, patients with solid tumors undergoing allogeneic SCT were infused with donor-derived ex vivo expanded NK cells (unpublished results). Although, no firm conclusions can be made regarding the clinical efficacy of NK cell-based DLI at this stage, available data indicate that NK cell infusions are safe and may generate antitumor responses. As normal tissues do not generally express ligands for activating NK cell receptors, alloreactive NK cells should not normally cause GvHD [45, 104]. The development of NK cell-based DLI represents new possibilities for treating relapses in patients undergoing haploidentical or cord blood SCT.

9 NK Cell-Mediated Responses in Prehematopoetic Stem Cell Transplantation Conditioning

Experimental data have demonstrated that NK cells can be administered directly after conditioning but before SCT. The potential benefits are threefold. First, NK cell-mediated GvL effects could enhance antitumor activity and reduce risk for relapse. Second, NK cell-mediated depletion of recipient DC could prevent the development of acute GvHD and perhaps allow higher numbers of alloreactive T cells in the graft, thus avoiding death from infections in the early posttransplant period [24, 25, 105]. Finally, NK cells may facilitate engraftment and promote a fast immune reconstitution, thereby reducing the need for myeloablative regimens and shortening the neutropenic period.

10 Adoptive Immunotherapy Using Short Term Ex Vivo Activated NK Cells

Because of the lack of significant clinical effects with past protocols based on autologous "LAK" cells or NK cells [106–108], and because of the promising effects observed in haploidentical T cell-depleted SCT, focus shifted towards the potential of using allogeneic (potentially alloreactive) NK cells in adoptive cell therapy. In recent studies, Miller and collaborators infused haploidentical NK cells together with IL-2 to 43 nontransplanted patients with advanced cancer. Low and high intensity preparative regimens were tested. The high-dose lymphodepleting regimen resulted in long-term survival and expansion of donor derived NK cells in vivo. The in vivo expansion was associated with increased levels of endogenous IL-15, possibly driving the proliferation of donor NK cells. Notably, one of these patients manifested a preferential expansion of the alloreactive NK cell subset. In general, these donor NK cell infusions were feasible and tolerated without unexpected toxicity. Moreover, 5/19 patients with AML achieved complete remission (CR) with this protocol [23].

11 Adoptive Immunotherapy Using NK Cell Lines and Ex Vivo Expanded NK Cells

In parallel, adoptive transfers are also being done with the NK-92 cell line [109]. This cell line can be grown continuously under GMP-conditions, expresses many NK cell activation receptors and low levels of KIR, and displays significant cytotoxicity towards many tumor targets. Following irradiation, more than 20 patients with advanced renal cell carcinoma and malignant melanomas have received NK-92 cells [110]. In general, these infusions are well tolerated and have yielded antitumor effects in some cases. Furthermore, this cell line is easily modified genetically which opens up interesting possibilities for future therapeutic trials. Whether adoptive immunotherapy with this cell line will produce more substantial clinical responses remains to be seen.

Several techniques have been developed for ex vivo expansion of NK cells [110]. A few of these protocols allow the expansion of NK cell-enriched cellular populations under GMP-conditions [111]. Using such protocols, it has been demonstrated that NK cells can be expanded ex vivo, also from tumor bearing patients [112, 113]. The latter opens up for expansion of autologous NK cells for adoptive immunotherapy, a strategy that may be developed further despite the earlier disappointments with autologous LAK cell therapy in the 1980s to patients with advanced solid tumors [107, 108]. Expansion protocols provide greater numbers of NK cells to be used for adoptive therapy that might be desired in some situations. For such expansions to be effective, it is important that the expansion of NK cells ex vivo is not associated with phenotypic changes, linage deviation, and/or selective expansion of specific subsets, such that their antitumor function will be affected. Another aspect to consider, apart from consequences of activation and proliferation, is that in vitro manipulation does not alter the NK cells' ability to mediate cell–cell interactions, trafficking, and homing to desired location. With respect to autologous NK cells, one may predict that they may be more effective in situations where tumors express low levels of MHC class I molecules.

12 Future Possibilities and Strategies for Adoptive NK Cell Immunotherapy

We have recently described critical questions that must be considered for the development of successful NK cell-based adoptive immunotherapy [22]. Below, we briefly discuss some issues with respect to the possible advantages, but also difficulties, of using allogeneic NK cells in future settings of adoptive NK cell-based immunotherapy.

As autologous NK cells are inhibited by self-MHC class I molecules, allogeneic NK cells may, in certain situations, represent a better cellular population for adoptive immunotherapy in vivo. The latter choice applies particularly to situations in which tumor targets express normal levels of MHC class I molecules in combination with low or only moderate expression of ligands for activating receptors. NK cell alloreactivity depends on "missing" KIR ligands (MHC class I) in the recipient. However, although NK cell alloreactivity is predicted by genetic algorithms based on KIR- and HLA-genotyping, the numbers of alloreactive NK cells in a given donor vary significantly, from 0% to 62% of the NK cells [114]. Predicting the effectiveness of therapy may thus be on the basis of assessment of the NK cell repertoire and selection of a donor with the largest alloreactive NK cell subset.

A prerequisite for survival of the infused cells is that they are not rejected by the recipient's immune system. If donor derived NK cells are infused at the time of transplantation they may engraft along with the stem cells because of the pretransplant conditioning. However, rejection of allogeneic NK cells represents a major challenge for specific NK cell therapy in the absence of myeloablative conditioning. It is likely that some type of conditioning will be required for effective transfer of allogeneic NK cells. Apart from preventing rejection, such regimens may also eradicate regulatory T cells that could otherwise interfere with the proliferation and function of the donor derived NK cells [115]. Moreover, there is reduced competition for growth factors during the homeostatic proliferation that follows lymphodepletion and the surge of cytokines, including IL-15, may promote proliferation, in vivo survival, and expansion of the infused NK cells. Indeed, in the studies by Miller and collaborators [23], NK cell expansion was dependent on an intense preparative regimen (high-dose cyclophosphamide/fludarabin). The latter regimen is similar to that used recently by Rosenberg and colleagues to induce homeostatic proliferation of adoptively transferred T cells [116]. As understanding of the conditions required for engraftment of NK cells improves, dosing of the preparative regimen will be more precise and the risks associated with high-dose myeloablative treatments will decrease.

13 Combination Therapies may Develop into Promising Treatment Options for Some Cancers

Finally, we predict that combination therapies including NK cells (directly or indirectly) will become ever more important in the future. Ligands for the activating receptor NKG2D are upregulated by genotoxic stress and stalled DNA replication, through activation of major DNA damage checkpoint pathways initiated by ATM or ATR protein kinases [117, 118]. Thus, the response to DNA damage alerts the immune system to the presence of potentially dangerous cells. As several of the currently used chemotherapeutic drugs, as well as ionizing irradiation, act via the DNA damage response pathway, a mild preconditioning using these drugs and/or local ionizing irradiation might sensitize tumor cells to immune recognition, leading to synergistic antitumor effects. Similarly, new generation cancer drugs such as the proteasome inhibitors and the histone deacetylase inhibitors can upregulate the death receptor DR5, sensitizing tumor cells to TRAIL-mediated killing by NK cells [119–121]. Histone deacetylase inhibitors induce MICA/B expression [122]. Imatinib mesylate, previously discussed as a potential stimulator of innate immunity to tumors, was also shown to influence the expression and shedding of the activating NK cell ligand MICA on Bcr/Abl positive targets [123]. Thus, although NKG2D expression on NK cells is restored upon Imatinib treatment, this may lead to decreased tumor targeting because of reduced MICA expression [123, 124].

As has been discussed, NK cells are major effectors in mediating ADCC. Rituximab (Mabthera), a chimeric mouse/human antibody that recognizes the CD20 antigen expressed on mature B lymphocytes [125], is currently given alone or combined with chemotherapy to patients with non-Hodgkin's lymphomas. One mechanism of this antibody's action is the induction of ADCC mediated by NK cells [126, 127]. Trials combining Rituximab with IL-2 to activate and expand the pool of NK cells available for ADCC are under way [128]. Several other antibodies are being evaluated in clinical practices and for many of them such as, e.g., Herceptin, at least part of their effector mechanism seems to be mediated by NK cells via ADCC [129, 130]. These and other findings suggest the possibility of using antibodies in conjunction with adoptive NK cell immunotherapy or NK cell stimulation-based protocols. A related therapeutic approach is the use of bispecific antibodies to promote NK cell targeting of tumors. Experimental and clinical data suggest that bispecific antibodies can be beneficial in tumor treatment. One approach is the use of antibodies specific for CD16 on NK cells and CD19 on B cell lymphomas or HER2/neu on breast cancers to target tumors expressing these, respective antigens [131]. Interestingly, clinical responses have been observed in patients with Hodgkin's lymphoma treated with bispecific antibodies against CD16 and CD30 [132].

14 Human Cancers that may be Subject to NK Cell Targeting

It is already evident from studies performed in vitro and even in some clinical studies that certain tumor types may be better suited than others for NK cell-based immunotherapy. The presence on human tumors of ligands for activating receptors provides an important prerequisite for NK cell activation, and thus for the potential of achieving good clinical results [51]. An illustration of this is the inefficient NK cell killing of lymphoid compared to myeloid leukemias that may be caused, at least in part, by the absence of LFA-1 ligand expression [95]. Likewise, low expression of MHC class I molecules, particularly in situations where KIR–ligand mismatching ("missing-self"-reactivity) does not prevail, is also important. Most immunotherapy trials have been performed in patients with significant tumor burdens, where conventional therapies were ineffective. The best clinical setting for most cellular therapies including NK cell-based immunotherapy is probably when the tumor burden is small, i.e., in minimal residual disease [133].

NK cell therapy against large solid tumors presents special problems including not only the size of the tumor per se but also the presumed necessity of NK cells to infiltrate the tumor. Despite the knowledge gained so far about the mechanisms that control trafficking of NK cells, we still know too little about the requirements for NK cell homing to and infiltration of tumors. It is known, however, that chemokines are required to attract NK cells to tumor sites. NK cells express a wide array of chemokine receptors on their cell surface. Different NK cell subsets can be identified on the basis of chemokine receptor expression and the pattern of expression is likely highly dependent upon the activation status of the NK cells [8].

15 Conclusion

As outlined above, we envisage many ways in which NK cells can be stimulated, manipulated, and used in settings of human cancer therapy. Strategies will not only be straight forward, and therapeutic results will not always be observed. Yet, we see a potential in the possibilities discussed. In particular, combination therapies involving NK cells, either directly or indirectly, may pave the way for new treatment strategies.

Acknowledgments We are supported by the Swedish Foundation for Strategic Research, the Swedish Research Council, the Swedish Cancer Society, the Royal Swedish Academy of Sciences, the Swedish Children's Cancer Society, the Cancer Society of Stockholm, the Karolinska Institutet, and the Karolinska University Hospital. This content of chapter is, in part, on the basis of a symposium paper published in Cancer Immunology Immunotherapy, Springer-Verlag, 2008 (Oct;57(10):1541-52. Epub 2008 Mar 4).

References

1. Kiessling R, Klein E, Wigzell H (1975) "Natural" killer cells in the mouse. I. Cytotoxic cells with specificity for mouse Moloney leukemia cells. Specificity and distribution according to genotype. Eur J Immunol 5:112
2. Huntington ND, Vosshenrich CA, Di Santo JP (2007) Developmental pathways that generate natural-killer-cell diversity in mice and humans. Nat Rev Immunol 7:703
3. Lanier LL (2005) NK cell recognition. Annu Rev Immunol 23:225
4. Moretta A, Bottino C, Mingari MC et al (2002) What is a natural killer cell? Nat Immunol 3:6
5. Moretta L, Moretta A (2004) Unravelling natural killer cell function: triggering and inhibitory human NK receptors. EMBO J 23:255
6. Bryceson YT, March ME, Ljunggren HG et al (2006) Activation, coactivation, and costimulation of resting human natural killer cells. Immunol Rev 214:73
7. Moretta A, Marcenaro E, Sivori S et al (2005) Early liaisons between cells of the innate immune system in inflamed peripheral tissues. Trends Immunol 26:668
8. Robertson MJ (2002) Role of chemokines in the biology of natural killer cells. J Leukoc Biol 71:173
9. Cooper MA, Fehniger TA, Caligiuri MA (2001) The biology of human natural killer-cell subsets. Trends Immunol 22:633
10. Farag SS, Fehniger TA, Ruggeri L et al (2002) Natural killer cell receptors: new biology and insights into the graft-versus-leukemia effect. Blood 100:1935
11. Di Santo JP (2006) Natural killer cell developmental pathways: a question of balance. Annu Rev Immunol 24:257
12. Farag SS, Caligiuri MA (2006) Human natural killer cell development and biology. Blood Rev 20:123
13. Smyth MJ, Hayakawa Y, Takeda K et al (2002) New aspects of natural-killer-cell surveillance and therapy of cancer. Nat Rev Cancer 2:850
14. Wu J, Lanier LL (2003) Natural killer cells and cancer. Adv Cancer Res 90:127
15. Algarra I, Ohlen C, Perez M et al (1989) NK sensitivity and lung clearance of MHC-class-I-deficient cells within a heterogeneous fibrosarcoma. Int J Cancer 44:675
16. Ljunggren HG, Karre K (1985) Host resistance directed selectively against H-2-deficient lymphoma variants. Analysis of the mechanism. J Exp Med 162:1745
17. Street SE, Hayakawa Y, Zhan Y et al (2004) Innate immune surveillance of spontaneous B cell lymphomas by natural killer cells and gammadelta T cells. J Exp Med 199:879
18. Guimaraes F, Guven H, Donati D et al (2006) Evaluation of ex vivo expanded human NK cells on antileukemia activity in SCID-beige mice. Leukemia 20:833
19. Carlsten M, Bjorkstrom NK, Norell H et al (2007) DNAX accessory molecule-1 mediated recognition of freshly isolated ovarian carcinoma by resting natural killer cells. Cancer Res 67:1317
20. Castriconi R, Dondero A, Corrias MV et al (2004) Natural killer cell-mediated killing of freshly isolated neuroblastoma cells: critical role of DNAX accessory molecule-1-poliovirus receptor interaction. Cancer Res 64:9180
21. El-Sherbiny YM, Meade JL, Holmes TD et al (2007) The requirement for DNAM-1, NKG2D, and NKp46 in the natural killer cell-mediated killing of myeloma cells. Cancer Res 67:8444

22. Ljunggren HG, Malmberg KJ (2007) Prospects for the use of NK cells in immunotherapy of human cancer. Nat Rev Immunol 7:329
23. Miller JS, Soignier Y, Panoskaltsis-Mortari A et al (2005) Successful adoptive transfer and in vivo expansion of human haploidentical NK cells in patients with cancer. Blood 105:3051
24. Ruggeri L, Aversa F, Martelli MF et al (2006) Allogeneic hematopoietic transplantation and natural killer cell recognition of missing self. Immunol Rev 214:202
25. Ruggeri L, Capanni M, Urbani E et al (2002) Effectiveness of donor natural killer cell alloreactivity in mismatched hematopoietic transplants. Science 295:2097
26. Hayakawa Y, Smyth MJ (2006) Innate immune recognition and suppression of tumors. Adv Cancer Res 95:293
27. Malmberg KJ (2004) Effective immunotherapy against cancer: a question of overcoming immune suppression and immune escape? Cancer Immunol Immunother 53:879
28. Malmberg KJ, Ljunggren HG (2006) Escape from immune- and nonimmune-mediated tumor surveillance. Semin Cancer Biol 16:16
29. Costello RT, Sivori S, Marcenaro E et al (2002) Defective expression and function of natural killer cell-triggering receptors in patients with acute myeloid leukemia. Blood 99:3661
30. Jager MJ, Hurks HM, Levitskaya J et al (2002) HLA expression in uveal melanoma: there is no rule without some exception. Hum Immunol 63:444
31. Salih HR, Rammensee HG, Steinle A (2002) Cutting edge: down-regulation of MICA on human tumors by proteolytic shedding. J Immunol 169:4098
32. Castriconi R, Cantoni C, Della Chiesa M et al (2003) Transforming growth factor beta 1 inhibits expression of NKp30 and NKG2D receptors: consequences for the NK-mediated killing of dendritic cells. Proc Natl Acad Sci U S A 100:4120
33. Ghiringhelli F, Menard C, Terme M et al (2005) CD4+CD25+ regulatory T cells inhibit natural killer cell functions in a transforming growth factor-beta-dependent manner. J Exp Med 202:1075
34. Smyth MJ, Teng MW, Swann J et al (2006) CD4+CD25+ T regulatory cells suppress NK cell-mediated immunotherapy of cancer. J Immunol 176:1582
35. Glas R, Sturmhofel K, Hammerling GJ et al (1992) Restoration of a tumorigenic phenotype by beta 2-microglobulin transfection to EL-4 mutant cells. J Exp Med 175:843
36. Hoglund P, Ljunggren HG, Ohlen C et al (1988) Natural resistance against lymphoma grafts conveyed by H-2Dd transgene to C57BL mice. J Exp Med 168:1469
37. Karre K, Ljunggren HG, Piontek G et al (1986) Selective rejection of H-2-deficient lymphoma variants suggests alternative immune defence strategy. Nature 319:675
38. Ljunggren HG, Karre K (1990) In search of the missing self: MHC molecules and NK cell recognition. Immunol Today 11:237
39. Karlhofer FM, Ribaudo RK, Yokoyama WM (1992) MHC class I alloantigen specificity of Ly-49+ IL-2-activated natural killer cells. Nature 358:66
40. Moretta A, Vitale M, Bottino C et al (1993) P58 molecules as putative receptors for major histocompatibility complex (MHC) class I molecules in human natural killer (NK) cells. Anti-p58 antibodies reconstitute lysis of MHC class I-protected cells in NK clones displaying different specificities. J Exp Med 178:597
41. Moretta L, Bottino C, Pende D et al (2004) Different checkpoints in human NK-cell activation. Trends Immunol 25:670
42. Zamai L, Ponti C, Mirandola P et al (2007) NK cells and cancer. J Immunol 178:4011
43. Pessino A, Sivori S, Bottino C et al (1998) Molecular cloning of NKp46: a novel member of the immunoglobulin superfamily involved in triggering of natural cytotoxicity. J Exp Med 188:953
44. Vitale M, Bottino C, Sivori S et al (1998) NKp44, a novel triggering surface molecule specifically expressed by activated natural killer cells, is involved in non-major histocompatibility complex-restricted tumor cell lysis. J Exp Med 187:2065
45. Bottino C, Castriconi R, Moretta L et al (2005) Cellular ligands of activating NK receptors. Trends Immunol 26:221

46. Pogge von Strandmann E, Simhadri VR, von Tresckow B et al (2007) Human Leukocyte Antigen-B-Associated Transcript 3 Is Released from Tumor Cells and Engages the NKp30 Receptor on Natural Killer Cells. Immunity 27:965
47. Bauer S, Groh V, Wu J et al (1999) Activation of NK cells and T cells by NKG2D, a receptor for stress-inducible MICA. Science 285:727
48. Bottino C, Castriconi R, Pende D et al (2003) Identification of PVR (CD155) and Nectin-2 (CD112) as cell surface ligands for the human DNAM-1 (CD226) activating molecule. J Exp Med 198:557
49. Chang CC, Ferrone S (2006) NK cell activating ligands on human malignant cells: molecular and functional defects and potential clinical relevance. Semin Cancer Biol 16:383
50. Costello RT, Fauriat C, Sivori S et al (2004) NK cells: innate immunity against hematological malignancies? Trends Immunol 25:328
51. Moretta L, Bottino C, Pende D et al (2006) Surface NK receptors and their ligands on tumor cells. Semin Immunol 18:151
52. Mendez R, Ruiz-Cabello F, Rodriguez T et al (2007) Identification of different tumor escape mechanisms in several metastases from a melanoma patient undergoing immunotherapy. Cancer Immunol Immunother 56:88
53. Moretta L, Moretta A (2004) Killer immunoglobulin-like receptors. Curr Opin Immunol 16:626
54. Parham P (2005) MHC class I molecules and KIRs in human history, health and survival. Nat Rev Immunol 5:201
55. Wagtmann N, Biassoni R, Cantoni C et al (1995) Molecular clones of the p58 NK cell receptor reveal immunoglobulin-related molecules with diversity in both the extra- and intracellular domains. Immunity 2:439
56. Braud VM, Allan DS, O'Callaghan CA et al (1998) HLA-E binds to natural killer cell receptors CD94/NKG2A, B and C. Nature 391:795
57. Bryceson YT, March ME, Ljunggren HG et al (2006) Synergy among receptors on resting NK cells for the activation of natural cytotoxicity and cytokine secretion. Blood 107:159
58. Cerwenka A, Baron JL, Lanier LL (2001) Ectopic expression of retinoic acid early inducible-1 gene (RAE-1) permits natural killer cell-mediated rejection of a MHC class I-bearing tumor in vivo. Proc Natl Acad Sci U S A 98:11521
59. Diefenbach A, Jensen ER, Jamieson AM et al (2001) Rae1 and H60 ligands of the NKG2D receptor stimulate tumour immunity. Nature 413:165
60. Wallace ME, Smyth MJ (2005) The role of natural killer cells in tumor control–effectors and regulators of adaptive immunity. Springer Semin Immunopathol 27:49
61. Smyth MJ, Cretney E, Kelly JM et al (2005) Activation of NK cell cytotoxicity. Mol Immunol 42:501
62. Trapani JA, Smyth MJ (2002) Functional significance of the perforin/granzyme cell death pathway. Nat Rev Immunol 2:735
63. Screpanti V, Wallin RP, Grandien A et al (2005) Impact of FASL-induced apoptosis in the elimination of tumor cells by NK cells. Mol Immunol 42:495
64. Smyth MJ, Crowe NY, Pellicci DG et al (2002) Sequential production of interferon-gamma by NK1.1(+) T cells and natural killer cells is essential for the antimetastatic effect of alpha-galactosylceramide. Blood 99:1259
65. Becknell B, Caligiuri MA (2005) Interleukin-2, interleukin-15, and their roles in human natural killer cells. Adv Immunol 86:209
66. Colucci F, Caligiuri MA, Di Santo JP (2003) What does it take to make a natural killer? Nat Rev Immunol 3:413
67. Bryceson YT, Rudd E, Zheng C et al (2007) Defective cytotoxic lymphocyte degranulation in syntaxin-11 deficient familial hemophagocytic lymphohistiocytosis 4 (FHL4) patients. Blood 110:1906
68. Mrozek E, Anderson P, Caligiuri MA (1996) Role of interleukin-15 in the development of human CD56+ natural killer cells from CD34+ hematopoietic progenitor cells. Blood 87:2632

69. Waldmann TA, Dubois S, Tagaya Y (2001) Contrasting roles of IL-2 and IL-15 in the life and death of lymphocytes: implications for immunotherapy. Immunity 14:105
70. Rodella L, Zamai L, Rezzani R et al (2001) Interleukin 2 and interleukin 15 differentially predispose natural killer cells to apoptosis mediated by endothelial and tumour cells. Br J Haematol 115:442
71. Lauwerys BR, Garot N, Renauld JC et al (2000) Cytokine production and killer activity of NK/T-NK cells derived with IL-2, IL-15, or the combination of IL-12 and IL-18. J Immunol 165:1847
72. Brady J, Hayakawa Y, Smyth MJ et al (2004) IL-21 induces the functional maturation of murine NK cells. J Immunol 172:2048
73. Parrish-Novak J, Dillon SR, Nelson A et al (2000) Interleukin 21 and its receptor are involved in NK cell expansion and regulation of lymphocyte function. Nature 408:57
74. Farag SS, Caligiuri MA (2004) Cytokine modulation of the innate immune system in the treatment of leukemia and lymphoma. Adv Pharmacol 51:295
75. Smyth MJ, Cretney E, Kershaw MH et al (2004) Cytokines in cancer immunity and immunotherapy. Immunol Rev 202:275
76. Rosenberg SA (2000) Interleukin-2 and the development of immunotherapy for the treatment of patients with cancer. Cancer J Sci Am 6(Suppl 1):S2
77. Fehniger TA, Cooper MA, Caligiuri MA (2002) Interleukin-2 and interleukin-15: immunotherapy for cancer. Cytokine Growth Factor Rev 13:169
78. Roda JM, Joshi T, Butchar JP et al (2007) The activation of natural killer cell effector functions by cetuximab-coated, epidermal growth factor receptor positive tumor cells is enhanced by cytokines. Clin Cancer Res 13:6419
79. Gosselin J, TomoIu A, Gallo RC et al (1999) Interleukin-15 as an activator of natural killer cell-mediated antiviral response. Blood 94:4210
80. Ozdemir O, Ravindranath, Y, Savasan S (2005) Mechanisms of superior anti-tumor cytotoxic response of interleukin 15-induced lymphokine-activated killer cells. J Immunother 28:44
81. Waldmann TA (2006) The biology of interleukin-2 and interleukin-15: implications for cancer therapy and vaccine design. Nat Rev Immunol 6:595
82. Kobayashi H, Dubois S, Sato N et al (2005) Role of trans-cellular IL-15 presentation in the activation of NK cell-mediated killing, which leads to enhanced tumor immunosurveillance. Blood 105:721
83. Hayashi T, Hideshima T, Akiyama M et al (2005) Molecular mechanisms whereby immunomodulatory drugs activate natural killer cells: clinical application. Br J Haematol 128:192
84. Fujii H, Trudeau JD, Teachey DT et al (2007) In vivo control of acute lymphoblastic leukemia by immunostimulatory CpG oligonucleotides. Blood 109:2008
85. Borg C, Terme M, Taieb J et al (2004) Novel mode of action of c-kit tyrosine kinase inhibitors leading to NK cell-dependent antitumor effects. J Clin Invest 114:379
86. Smyth MJ (2006) Imatinib mesylate–uncovering a fast track to adaptive immunity. N Engl J Med 354:2282
87. Ullrich E, Bonmort M, Mignot G et al (2007) Therapy-induced tumor immunosurveillance involves IFN-producing killer dendritic cells. Cancer Res 67:851
88. Brandau S, Riemensberger J, Jacobsen M et al (2001) NK cells are essential for effective BCG immunotherapy. Int J Cancer 92:697
89. Sheridan C (2006) First-in-class cancer therapeutic to stimulate natural killer cells. Nat Biotechnol 24:597
90. Koh CY, Blazar BR, George T et al (2001) Augmentation of antitumor effects by NK cell inhibitory receptor blockade in vitro and in vivo. Blood 97:3132
91. Koh CY, Ortaldo JR, Blazar BR et al (2003) NK-cell purging of leukemia: superior antitumor effects of NK cells H2 allogeneic to the tumor and augmentation with inhibitory receptor blockade. Blood 102:4067
92. Sentman CL, Barber MA, Barber A et al (2006) NK cell receptors as tools in cancer immunotherapy. Adv Cancer Res 95:249

93. Imai C, Iwamoto S, Campana D (2005) Genetic modification of primary natural killer cells overcomes inhibitory signals and induces specific killing of leukemic cells. Blood 106:376
94. Uherek C, Tonn T, Uherek B et al (2002) Retargeting of natural killer-cell cytolytic activity to ErbB2-expressing cancer cells results in efficient and selective tumor cell destruction. Blood 100:1265
95. Ruggeri L, Capanni M, Casucci M et al (1999) Role of natural killer cell alloreactivity in HLA-mismatched hematopoietic stem cell transplantation. Blood 94:333
96. Shlomchik WD, Couzens MS, Tang CB et al (1999) Prevention of graft versus host disease by inactivation of host antigen-presenting cells. Science 285:412
97. Farag SS, Bacigalupo A, Eapen M et al (2006) The effect of KIR ligand incompatibility on the outcome of unrelated donor transplantation: a report from the center for international blood and marrow transplant research, the European blood and marrow transplant registry, and the Dutch registry. Biol Blood Marrow Transplant 12:876
98. Ruggeri L, Mancusi A, Burchielli E et al (2007) Natural killer cell alloreactivity in allogeneic hematopoietic transplantation. Curr Opin Oncol 19:142
99. Kolb HJ, Mittermuller J, Clemm C et al (1990) Donor leukocyte transfusions for treatment of recurrent chronic myelogenous leukemia in marrow transplant patients. Blood 76:2462
100. Kolb HJ, Simoes B, Schmid C (2004) Cellular immunotherapy after allogeneic stem cell transplantation in hematologic malignancies. Curr Opin Oncol 16:167
101. Slavin S, Naparstek E, Nagler A et al (1996) Allogeneic cell therapy with donor peripheral blood cells and recombinant human interleukin-2 to treat leukemia relapse after allogeneic bone marrow transplantation. Blood 87:2195
102. Passweg JR, Stern M, Koehl U et al (2005) Use of natural killer cells in hematopoetic stem cell transplantation. Bone Marrow Transplant 35:637
103. Passweg JR, Tichelli A, Meyer-Monard S et al (2004) Purified donor NK-lymphocyte infusion to consolidate engraftment after haploidentical stem cell transplantation. Leukemia 18:1835
104. Gonzalez S, Groh V, Spies T (2006) Immunobiology of human NKG2D and its ligands. Curr Top Microbiol Immunol 298:121
105. Lundqvist A, McCoy JP, Samsel L, Childs R (2007) Reduction of GVHD and enhanced antitumor effects after adoptive infusion of alloreactive Ly49-mismatched NK cells from MHC-matched donors. Blood 109:3603
106. Burns LJ, Weisdorf DJ, DeFor TE et al (2003) IL-2-based immunotherapy after autologous transplantation for lymphoma and breast cancer induces immune activation and cytokine release: a phase I/II trial. Bone Marrow Transplant 32:177
107. Law TM, Motzer RJ, Mazumdar M et al (1995) Phase III randomized trial of interleukin-2 with or without lymphokine-activated killer cells in the treatment of patients with advanced renal cell carcinoma. Cancer 76:824
108. Rosenberg SA, Lotze MT, Muul LM et al (1985) Observations on the systemic administration of autologous lymphokine-activated killer cells and recombinant interleukin-2 to patients with metastatic cancer. N Engl J Med 313:1485
109. Tam YK, Martinson JA, Doligosa K et al (2003) Ex vivo expansion of the highly cytotoxic human natural killer-92 cell-line under current good manufacturing practice conditions for clinical adoptive cellular immunotherapy. Cytotherapy 5:259
110. Klingemann HG (2005) Natural killer cell-based immunotherapeutic strategies. Cytotherapy 7:16
111. Carlens S, Gilljam M, Chambers BJ et al (2001) A new method for in vitro expansion of cytotoxic human CD3-CD56+ natural killer cells. Hum Immunol 62:1092
112. Alici E, Sutlu, T, Bjorkstrand, B, et al (2008) Autologous anti-tumor activity by NK cells expanded from myeloma patients using GMP-compliant components. Blood 111:3155
113. Guven H, Gilljam M, Chambers BJ et al (2003) Expansion of natural killer (NK) and natural killer-like T (NKT)-cell populations derived from patients with B-chronic lymphocytic leukemia (B-CLL): a potential source for cellular immunotherapy. Leukemia 17:1973

114. Fauriat C, Andersson S, Bjorklund A et al (2008) Estimation of the size of the alloreactive NK cell repertoire: studies in individuals homozygous for the group A KIR haplotype. J Immunol 181:6010
115. Gattinoni L, Powell DJ Jr, Rosenberg SA et al (2006) Adoptive immunotherapy for cancer: building on success. Nat Rev Immunol 6:383
116. Dudley ME, Wunderlich JR, Robbins PF et al (2002) Cancer regression and autoimmunity in patients after clonal repopulation with antitumor lymphocytes. Science 298:850
117. Gasser S, Orsulic S, Brown EJ et al (2005) The DNA damage pathway regulates innate immune system ligands of the NKG2D receptor. Nature 436:1186
118. Gasser S, Raulet D (2006) The DNA damage response, immunity and cancer. Semin Cancer Biol 16:344
119. Lundqvist A, Abrams SI, Schrump DS et al (2006) Bortezomib and depsipeptide sensitize tumors to tumor necrosis factor-related apoptosis-inducing ligand: a novel method to potentiate natural killer cell tumor cytotoxicity. Cancer Res 66:7317
120. Sayers TJ, Brooks AD, Koh CY et al (2003) The proteasome inhibitor PS-341 sensitizes neoplastic cells to TRAIL-mediated apoptosis by reducing levels of c-FLIP. Blood 102:303
121. VanOosten RL, Moore JM, Karacay B et al (2005) Histone deacetylase inhibitors modulate renal cell carcinoma sensitivity to TRAIL/Apo-2L-induced apoptosis by enhancing TRAIL-R2 expression. Cancer Biol Ther 4:1104
122. Skov S, Pedersen MT, Andresen L et al (2005) Cancer cells become susceptible to natural killer cell killing after exposure to histone deacetylase inhibitors due to glycogen synthase kinase-3-dependent expression of MHC class I-related chain A and B. Cancer Res 65:11136
123. Cebo C, Da Rocha S, Wittnebel S et al (2006) The decreased susceptibility of Bcr/Abl targets to NK cell-mediated lysis in response to imatinib mesylate involves modulation of NKG2D ligands, GM1 expression, and synapse formation. J Immunol 176:864
124. Boissel N, Rea D, Tieng V et al (2006) BCR/ABL oncogene directly controls MHC class I chain-related molecule A expression in chronic myelogenous leukemia. J Immunol 176:5108
125. Reff ME, Carner K, Chambers KS et al (1994) Depletion of B cells in vivo by a chimeric mouse human monoclonal antibody to CD20. Blood 83:435
126. Dall'Ozzo S, Tartas S, Paintaud G et al (2004) Rituximab-dependent cytotoxicity by natural killer cells: influence of FCGR3A polymorphism on the concentration-effect relationship. Cancer Res 64:4664
127. Weng WK, Levy R (2003) Two immunoglobulin G fragment C receptor polymorphisms independently predict response to rituximab in patients with follicular lymphoma. J Clin Oncol 21:3940
128. Pitini V, Arrigo C, Naro C et al (2007) Interleukin-2 and lymphokine-activated killer cell therapy in patients with relapsed B-cell lymphoma treated with rituximab. Clin Cancer Res 13:5497
129. Adams GP, Weiner LM (2005) Monoclonal antibody therapy of cancer. Nat Biotechnol 23:1147
130. Carter PJ (2006) Potent antibody therapeutics by design. Nat Rev Immunol 6:343
131. Shahied LS, Tang Y, Alpaugh RK et al (2004) Bispecific minibodies targeting HER2/neu and CD16 exhibit improved tumor lysis when placed in a divalent tumor antigen binding format. J Biol Chem 279:53907
132. Hartmann F, Renner C, Jung W et al (2001) Anti-CD16/CD30 bispecific antibody treatment for Hodgkin's disease: role of infusion schedule and costimulation with cytokines. Clin Cancer Res 7:1873
133. Slavin S (2005) Allogeneic cell-mediated immunotherapy at the stage of minimal residual disease following high-dose chemotherapy supported by autologous stem cell transplantation. Acta Haematol 114:214

Index

A

Acceptance, 222, 224, 226, 228
Activation receptors, 406–408, 411–415
Adaptive immunity, 316, 321
Adoptive cell therapy, 412–414
Adoptive immunotherapy, 210, 212
Adoptive transfer, 209–212
Adoptive transfer of NK-92 cells, 212
Allergy, 191–196
Allogeneic bone marrow cells, 201, 203
Allogeneic HSCT, 200, 201, 203–205, 209, 210
Alopecia, 259–260
Amplification, 394
Antibody dependent cellular cytotoxicity (ADCC), 9, 17, 18, 346, 348, 349
Anti-CD20, 233
Anti-CD52, 233
Antimicrobial proteins/peptides
   cathelicidin(s), 159
   defensin(s), 158, 159, 161
   granulysin, 158, 159, 162
Antiphospholipid syndrome, 243, 245
Areata, 259–260
Asthma, 191–196
Atopy, 255–257
Autoantibodies, 355–357
Autoimmune diseases, 178–186, 241–251
Autoimmunity, 177–186, 375, 377–379
Autologous, 200, 206

B

2B4, 11–13, 15, 18, 30
Bacteria, 303–308
   Bordetella pertussis, 154
   Escherichia coli, 155, 158, 167, 168
   Franciscella tularensis, 158
   Helicobacter pylori, 170
   Klebsiella pneumoniae, 161
   Legionella pneumophilia, 154, 169
   Listeria innocua, 159
   Listeria monocytogenes, 155, 157, 158, 169, 170
   Mycobacterium avium, 155, 156, 158
   Mycobacterium bovis, 157, 161
   Mycobacterium tuberculoisis, 155, 158, 159, 165–167, 169, 170
   Pseudomonas aeruginosa, 155
   Salmonella enteritidis, 156
   Salmonella minesota, 167
   Salmonella typhimurium, 158, 169
   Staphylococcus aureus, 154, 155, 158, 169
   Streptococcus pneumoniae, 155
   Yersinia pestis, 170
B cell receptor, 346, 349
Bone marrow, 200–201, 203, 210, 212
Bone marrow transplantation, 201, 212
Bronchiolitis obliterans syndrome, 232
B6.Sle1b, 356

C

Cancer, 25–26, 31, 267, 268, 332, 334–336
Cardiac allograft vasculopathy (CAV), 225

CCL9, 226
CCL10, 226
CCL11, 227
CCR1, 227
CCR4, 226, 227
CCR7, 301, 302, 306
CD2, 7, 9, 11, 12, 17, 18
CD16, 5, 6, 8–11, 18–20, 23–24, 26–28
CD27, 27–29
CD48, 12–13, 30, 354, 356, 361–363
CD56, 317, 319
CD69, 352, 354, 357
CD86, 357, 358
CD117, 6–8, 29
CD154, 227
CD244, 354, 356, 361
CD56^{bright}, 5–8, 10, 16, 19, 28, 301, 302, 304, 307, 308
CD28 costimulation, 224
CD95L, 26
cDNA, 393, 394
CD94/NKG2C, 11, 20
CD94/NKG2 family, 128, 132, 134
CD4$^+$ T cells, 371–375, 381
CFSE, 354, 358, 359
Chemokine receptors, 226–227
Chemokines, 222, 226–228, 315–320
Cis interaction, 93–106
Combination therapy, 414–415
Conditioning, 200, 201, 203, 206, 207, 209, 211, 213
Cord blood, 200–201
Cord blood transplantation, 201
Costimulatory molecules, 320, 321
Coxackie virus, 182
cRNA, 393
CTL response, 370–372
Cutaneous lymphomas, 257
CXCL16, 229
CXCR3, 226
CXCR6, 229
Cyclosporin A, 227, 233
Cytokine circuit, 347, 348, 350–352, 356
Cytokines, 40–41, 44, 45, 52, 154, 156–166, 168–171, 178, 180, 182, 183, 186, 392, 396, 398–402, 405, 408, 409, 413
   complexes, 212
   IFN-α, 161, 164
   IFN-γ, 154, 157, 161, 164
   IL-2, 159, 163
   IL-12, 154, 161, 162, 163, 164, 169, 171
   IL-18, 162, 169
   IL-1β, 161, 163
   production, 180, 182, 183
   TGF-β, 159
   TNF-α, 157, 158, 161
Cytomegalovirus, 348
Cytotoxicity, 267
   Fas, 158, 162
   granzymes, 158
   perforin, 158, 159, 162
   TRAIL R, 158

D

DAP10, 11–12
DAP12, 11, 18, 20, 22, 28, 183
Dendritic cells (DC), 316–319, 321–323, 330–338, 369–375, 379–380, 382
   autoimmune diseases, 241–251
   lupus erythematosus, 249–250
Depletion, 181–183
Diabetes, 178, 181–182, 186, 376, 378, 379
DNAM-1, 21, 319, 323
Donor lymphocyte infusions (DLI), 411

E

EAT-2, 30
Effector mechanisms, 408, 414
Encephalomyelitis (EAE), 377
ERT, 30
Experimental autoimmune encephalomyelitis (EAE), 183

F

FceRI-γ, 11, 28
FcgRIII, 11
Foxp3, 369, 374

G

Gene
   activity, 393–395
   array, 393–396, 399, 400, 402
   expression, 393–396, 399–402

profile, 393, 395, 396, 399, 402
regulation, 393, 399, 401
Gene therapy, 211–212
Germline, 357–361
Germline transcripts, 357, 360, 361
GPI-anchored receptor, 361
gpUL18, 130–131
gpUL40, 134
Graft-*versus*-host disease (GVHD), 200, 201, 203, 205, 206, 208–210, 323
Granulocyte-macrophage colony-stimulating factor (GM-CSF), 12, 16

H
Hematopoietic stem cells (HSC), 200–201, 203, 204, 208
Hematopoietic stem cell transplantation (HSCT), 199, 208, 211, 212, 410–411
allogeneic, 200, 201, 203–205, 209, 210
conditioning, 201, 206, 207
Hepatitis, 375–377
Herpes simplex virus, 261–262
HIV, 22
HLA-G, 112, 113, 116–119
Homeostasis, 316, 321, 323

I
IFN-β, 20
IFN-γ, 2, 5–8, 10, 12–14, 16–17, 19, 22–26, 28, 30, 315–321, 347–353, 358, 360–362
IgG2a/c, 346, 347
Ig-like transcripts (ILT), 128
IL-4$^+$, 24
IL-5, 317, 319
IL-10, 24–25, 317, 323
IL-12, 14, 16–18, 21, 23–26, 303, 308, 309, 347, 350–353, 362
IL-13, 362–363
IL-15, 4–7, 12, 16, 19, 23, 29
IL-18, 16–18, 23, 25, 26, 353, 362
IL-22, 318
IL-15 cDNA, 211–212
Immune homeostasis, 369–370, 374–375, 380, 383
Immune regulation, 120
Immune tolerance, 369, 374, 381

Immunoreceptor tyrosine-based activation motif (ITAM), 10–12, 18, 22, 28
Immunoreceptor tyrosine-based inhibitory motifs (ITIM), 14, 15, 18, 22, 28, 126, 144
Immunosuppressants, 232
Immunotherapies, 210–213
Inflammatory diseases, 241–251
Inflammatory myopathy, 246
Influenza, 21
Inhibitory receptors, 126–128, 131, 134, 135, 140, 144, 407–408, 410
Islets, 181, 182

K
K5, 134, 140, 141
Killer cell immunoglobulin-like receptors (KIR), 5, 6, 8, 11, 14, 15, 19, 21, 22, 25, 26, 127, 128, 144, 182, 186, 221–234
genes, 276–278, 281, 282, 286, 289, 290
genotypes, autoimmune and inflammatory diseases, 244–245
haplotypes, 63–89, 276, 281, 286, 290
inhibitor, 276–278, 281, 283, 284, 286, 287
KIR A haplotype, 287
KIR B haplotype, 288, 289
KIR2DL3, 276–278, 284, 286, 288
KIR2DL4, 277, 286, 287
KIR3DS1, 278, 281, 283, 288
ligands, 277, 278, 284, 286, 288, 289, 290
polymorphism, 72, 79, 84, 86
population frequencies, 67–71, 79, 80, 83, 89
Killer immunoglobulin-like receptors (KIR), 374–375, 378–379, 381
Kinetics, 395, 402
KIR2DL1, 230, 231
KIR2DL4, 19
KIR3DL2, 230
KIR2DL2/2DL3, 230
KIR2DS4, 230
KIR-ligand mismatch, 410, 411, 415

L
Leukemia, 267, 269–273
LFA-1, 9, 17, 18

LFA-1/ICAM-1, 228
Licensing, 205
Listeria monocytogenes, 350
Liver injury, 375–377
Location, 180, 181, 185
LTb receptor (LTbR), 29
Lupus erythematosus, 242, 243, 249–250
Ly49, 127, 128, 134, 143, 144
Ly49A, 96–106
Ly49H, 127, 143, 144, 348
Lymph nodes, 2, 3, 5–9, 316–317
Lymphocyte function-associated antigen-1 (LFA-1), 227
Lymphoma, 267, 269–273
Lymphoproliferative diseases, 244
Ly49P, 127, 134, 144
Ly49 receptor, 27

M
m04, 127, 133, 134, 144
m06, 133, 134
m144, 132, 133
m152, 133, 134, 139
m157, 127, 143
Macrophages, 180, 182
Major histocompatibility complex (MHC)
   class I, 185
      group 1 HLA-C, 277–278, 284, 287–288
      HLA-B^{Bw4}, 277, 283–284
      HLA-G, 286
      homologues, 130–133
      receptor, autoimmune/inflammatory diseases, 245–246
MCP-1/CX3CL1, 226
Membrane transfers, 111–119
Methylprednisolone, 233
m138/fcr-1, 139
MICA and MICB, 140
Mice, 4, 11, 12, 14, 27, 29, 30
Microenvironment, 185
MicroRNA, 142–143
Missing self, 126, 129, 134, 135
Missing self hypothesis, 126, 135
Missing-self reactivity, 415
mRNA, 394, 395, 402
Multiple sclerosis (MS), 182–183

Myasthenia gravis, 180
Mycobacteria, 23

N
Natural cytotoxicity receptors (NCRs)
   NKp30, 164, 165
   NKp44, 164, 165
   NKp46, 164, 165
Natural killer (NK) cells, 93–106, 109–121, 177–186, 267–274
   activating and inhibitory receptors, 136
   alloreactive, 200, 205–207
   autoimmune and inflammatory diseases, 241–251
   definition, 405–406
   development, 39–54
   differentiation, 39–54
   identification, 405, 407, 415
   inhibitory receptor blockade, 211
   lupus erythematosus, 249–250
   receptor gene, 212
   responses, 180, 186
   Treg, 247–249
NCR receptors, 13
Negative factor (Nef), 134, 140
NK1, 192–196
NK2, 192–196
NK cell ligands
   AICL, 165, 167
   CD48, 166, 167
   LLT-1, 166
   MICA/B, 162, 167, 168
   ULBP(s), 159, 162, 167, 168
   vimentin, 15, 162, 167
NK cell subsets, 380–383
NKG2D, 128–129, 138–141, 143, 315, 318–323, 371, 373, 374, 376–377, 381
NKG2D ligands (NKG2DLs), 129, 138–141, 322, 323
NK$_p$30, 129, 137, 138, 319, 320, 323
NK$_p$46, 318, 320, 321, 373–375, 381, 382
NKR-P1 family, 129, 144
NOD mice, 181, 182, 184

O
Organ-specific, 178, 181–182, 185
OX40L, 318–320

P

Pancreas, 178, 181, 182, 184–186
Pathogen-associated molecular patterns (PAMPs), 13
 CpG DNA, 163
 D-γ-glutamyl-meso-DAP, 163–164
 flagellin, 162, 163
 KpOmpA, 161, 163
 LPS, 167, 168
 ManLAM, 170
 muramyl dipeptide (MDP), 162, 164
 poly (I:C), 167
 R848, 163
Pattern recognition receptors (PRRs)
 DC-SIGN, 170–171
 Nod-like receptors (NLRs), 160, 162
 Toll-like receptors (TLRs), 160, 162
Pemphigus vulgaris, 183, 243, 249
Perforin, 301, 302, 307, 308
Peripheral blood, 200–202, 205, 206
Plasmodium falciparum, 25
Poly(I:C), 347, 348, 350, 352, 362
Polyclonal antithymocyte globulin (rATG), 233
Polyclonal B cell response, 351
Polyoma virus, 349
pp65, 135–138
Preclinical models, 39
Pregnancy, 243, 245
Protein
 analysis, 396, 397
 array, 393, 395–399, 402
 expression, 395, 400, 402
 level, 395, 400, 402
Protozoan infections, 23–25
Psoriasis, 243, 261

R

Rapamycin, 226, 233
Receptors, 125–145, 405–415
Reconstitution, 200, 201, 207–211
Regulatory T (Treg) cells, 26, 318, 320–322, 329–338, 369, 374
Rejection, 200, 201, 203–206, 209, 221, 222, 224–232
Renal tubular epithelial cells, 225
Rheumatoid arthritis (RA), 178, 182, 243, 250

S

Secondary lymphoid organs, 301–302, 305
 lymph node, 299, 302, 304, 306–308
 tonsil, 302
Sentinels, 180
Severe combined immune deficient (SCID), 348
SH2D1A/DSHP/*SAP* gene, 356
Sjögren's syndrome, 184, 250
SLAM/CD2, 356
Spondylarthropathies, 243, 250
Subsets, 178, 182, 1848–186
Supramolecular activating clusters (SMAC), 27
Supramolecular inhibitory clusters (SMIC), 27
Switch recombination, 346, 358–361, 363
Systemic, 178, 182, 183, 185
Systemic Lupus Erythematosis (SLE), 178, 183
Systemic sclerosis, 243–245

T

T cells, 179–184, 302, 306, 315–324
 memory, 354–355
 polarization, 300, 308–309
 response, 300
T-dependent antigen, 353, 354
TGF-β, 317, 323
T helper 1 (Th1) cells, 316–318
T-independent (TI) antigen, 346–347, 349–352, 356, 360, 361
Tissues, 180, 185
TLR, 13–14, 18, 25
TLR2, 350, 352, 353
TLR3, 347, 348, 350
TLR4, 350, 353
TLR7, 349
TNF-α, 2, 10, 13–14, 16, 24, 26
Tolerance, 222, 227–230
Transcription factors, 39–41
Transforming growth factor (TGF-β), 14, 24–26
Transporter associated with antigen processing (TAP), 132
Trogocytosis, 108–121
Tumor cells, 405–410, 414

Tumor necrosis factor (TNF), 299, 301, 305, 307–309

U
UL142, 140
UL16-binding proteins (ULBP), 374
Ulcerative colitis, 243, 246
US2, 132–135
US6, 132–134
US11, 132–134
Uterine NK (uNK) cells, 379–380

V
Varicella zoster virus (VZV), 21
Viruses, 3, 12, 13, 16, 19–22, 26, 30

W
www.allelefrequencies.net, 68, 71, 79, 82